BRAIN EDEMA

BRAIN EDEMA

FROM MOLECULAR MECHANISMS TO CLINICAL PRACTICE

Edited by

Jérôme Badaut

CNRS-University of Bordeaux, Bordeaux, France;
Loma Linda University School of Medicine, Loma Linda, CA, United States

Nikolaus Plesnila

University of Munich Medical Center & Munich Cluster for Systems Neurology (Synergy), Munich, Germany

ELSEVIER

ACADEMIC PRESS
An imprint of Elsevier

Academic Press is an imprint of Elsevier
125 London Wall, London EC2Y 5AS, United Kingdom
525 B Street, Suite 1800, San Diego, CA 92101-4495, United States
50 Hampshire Street, 5th Floor, Cambridge, MA 02139, United States
The Boulevard, Langford Lane, Kidlington, Oxford OX5 1GB, United Kingdom

Notices

Knowledge and best practice in this field are constantly changing. As new research and experience broaden our understanding, changes in research methods, professional practices, or medical treatment may become necessary.

Practitioners and researchers must always rely on their own experience and knowledge in evaluating and using any information, methods, compounds, or experiments described herein. In using such information or methods they should be mindful of their own safety and the safety of others, including parties for whom they have a professional responsibility.

To the fullest extent of the law, neither the Publisher nor the authors, contributors, or editors, assume any liability for any injury and/or damage to persons or property as a matter of products liability, negligence or otherwise, or from any use or operation of any methods, products, instructions, or ideas contained in the material herein.

British Library Cataloguing-in-Publication Data
A catalogue record for this book is available from the British Library

Library of Congress Cataloging-in-Publication Data
A catalog record for this book is available from the Library of Congress

ISBN: 978-0-12-803196-4

Publisher:
Acquisition Editor: Natalie Farra
Editorial Project Manager: Kristi Anderson
Production Project Manager: Edward Taylor
Cover Designer: Alan Studholme

Typeset by MPS Limited, Chennai, India

Cover: Vectorized brain and cover design: © Antonia Weingart
Water background photo: © fotolia, srady

Contents

II

TECHNIQUES TO INVESTIGATE CEREBRAL BLOOD

III

MOLECULAR BASIS AND CONCEPTS IN BRAIN EDEMA FORMATION FOR NEW TREATMENT DEVELOPMENT

9. Role of Aquaporins in the Formation and Resolution of Brain Edema

BEATRIZ RODRIGUEZ-GRANDE, JAN-PIETER KONSMAN AND JÉRÔME BADAUT

10. Sur1–Trpm4-Promising Target for Brain Edema Treatment in Brain Edema: From Molecular Mechanisms to Clinical Practice

SEBASTIAN URDAY, KEVIN N. SHETH AND J. MARC SIMARD

11. Role of Matrix Metalloproteinases in Brain Edema

CHANGJUN YANG AND EDUARDO CANDELARIO-JALIL

IV
BRAIN EDEMA PROCESS IN PRECLINICAL MODELS

12. Edema and BBB Breakdown in Stroke

KATHLEEN E. SALMERON, DANIELLE N. EDWARDS, JUSTIN F. FRASER AND GREGORY J. BIX

17. Water Homeostasis Dysfunction in Epilepsy

DEVIN K. BINDER

18. Water Homeostasis Dysfunctions and Edema in Neuroinflammatory Disease

KLAUS G. PETRY AND ANDREAS BIKFALVI

19. Experimental Therapies for Brain Edema and Intracranial Hypertension

KATRIN RAUEN AND NIKOLAUS PLESNILA

V

CLINICAL FEATURES AND MANAGEMENT OF BRAIN EDEMA

20. Clinical Monitoring of Brain Edema

SANDRO M. KRIEG AND SEBASTIAN ILLE

Dedication

To our wives Isabelle and Carlotta, and to our children Océane, Victor, Linus, and Leonie for their support, love, and patience.

Jérôme Badaut and Nikolaus Plesnila

Acknowledgments

We would like to express our deep gratitude to the many people who helped us to carry out and finalize this book project.

First of all we would like to thank all authors for spending their valuable and often free time for sharing their results and ideas about brain edema with us and for contributing a chapter to this book. It was a long journey for all of us, but we hope the result was worth it.

Very special thanks to those who do not even know how much they contributed: to our wives and children and to our coworkers for who we had not enough time during the past months and years.

A big "Thank You" to Kristi Anderson and her team from Elsevier for their support, their patience, and their professionalism during the time it took to put this book together. Kristi, without you we would have never made it!

Last and not least: many thanks to all those who contributed with their research to the better understanding of the pathogenesis of brain edema. We humbly stand on the shoulders of giants.

Finally, we beg forgiveness to all those whose work we or the authors unintentionally have failed to mention.

Jérôme Badaut and Nikolaus Plesnila
Bordeaux and Munich, January 2017

List of Contributors

Abraham J. Al-Ahmad Texas Tech University Health Sciences Center, Amarillo, TX, United States

Jérôme Badaut CNRS-University of Bordeaux, Bordeaux, France; Loma Linda University School of Medicine, Loma Linda, CA, United States

Andreas Bikfalvi Bordeaux University INSERM U1029, Pessac, France

Devin K. Binder University of California, Riverside, CA, United States

Gregory J. Bix University of Kentucky, Lexington, KY, United States

Roger F. Butterworth University of Montreal, Montreal, QC, Canada

Eduardo Candelario-Jalil University of Florida, Gainesville, FL, United States

Yi-Je Chen University of California, Davis, CA, United States

Vibol Chhor Université Paris, Diderot, Sorbonne, Paris Cité, Paris, France; PremUP, Paris, France

Newton Cho University of Toronto Department of Surgery, Toronto, ON, Canada

Helle Damkier University of Copenhagen, Copenhagen, Denmark; Aarhus University, Aarhus, Denmark

Danielle N. Edwards University of Kentucky, Lexington, KY, United States

Petra Fallier-Becker Institute of Pathology and Neuropathology, University of Tübingen, Tübingen, Germany

Michael G. Fehlings University of Toronto Department of Surgery, Toronto, ON, Canada

Bobbi Fleiss Université Paris, Diderot, Sorbonne, Paris Cité, Paris, France; PremUP, Paris, France; St. Thomas' Hospital, London, United Kingdom

Justin F. Fraser University of Kentucky, Lexington, KY, United States

Jean-François Ghersi-Egea Lyon 1 University, Lyon, France

Pierre Gressens Université Paris, Diderot, Sorbonne, Paris Cité, Paris, France; PremUP, Paris, France; APHP, Robert Debré Hospital, Paris, France; St. Thomas' Hospital, London, United Kingdom

Laureen D. Hachem University of Toronto Department of Surgery, Toronto, ON, Canada

Sebastian Ille Klinikum rechts der Isar Technische Universität München, Munich, Germany

Oliver Kempski University of Mainz, Mainz, Germany

Jan-Pieter Konsman CNRS-University of Bordeaux, Bordeaux, France

Sandro M. Krieg Klinikum rechts der Isar Technische Universität München, Munich, Germany

Sighild Lemarchant CNRS-University of Bordeaux, Bordeaux, France

Ute Lindauer RWTH Aachen University, Aachen, Germany

Andreas F. Mack Institute of Clinical Anatomy and Cell Analysis, University of Tübingen, Tübingen, Germany

Raffaella Moretti Université Paris, Diderot, Sorbonne, Paris Cité, Paris, France; PremUP, Paris, France; Università degli studi di Udine, Udine, Italy; APHP, Robert Debré Hospital, Paris, France

Andre Obenaus Loma Linda University School of Medicine, Loma Linda, CA, United States; University of California, Irvine, CA, United States

Martha E. O'Donnell University of California, Davis, CA, United States

Stéphane Olindo CNRS-University of Bordeaux, Bordeaux, France

Klaus G. Petry Bordeaux University INSERM U1029, Pessac, France

Michelle E. Pizzo University of Wisconsin-Madison, Madison, WI, United States

Nikolaus Plesnila University of Munich Medical Center, Munich, Germany; Munich Cluster for Systems Neurology (Synergy), Munich, Germany

Katrin Rauen University Hospital Zurich, Zurich, Switzerland; University of Munich Medical Center, Munich, Germany

Beatriz Rodriguez-Grande CNRS-University of Bordeaux, Bordeaux, France

Kathleen E. Salmeron University of Kentucky, Lexington, KY, United States

Oliver Schnell Klinikum der Ludwig-Maximilians-Universität München, Campus Großhadern, Munich, Germany; University of Freiburg, Freiburg, Germany

Kevin N. Sheth Yale University School of Medicine, New Haven, CT, United States

Igor Sibon CNRS-University of Bordeaux, Bordeaux, France

J. Marc Simard University of Maryland School of Medicine, Baltimore, MD, United States

Robert G. Thorne University of Wisconsin-Madison, Madison, WI, United States

Luigi Titomanlio Université Paris, Diderot, Sorbonne, Paris Cité, Paris, France; PremUP, Paris, France; APHP, Robert Debré Hospital, Paris, France

Jörg-Christian Tonn Klinikum der Ludwig-Maximilians-Universität München, Campus Großhadern, Munich, Germany

Sebastian Urday Yale University School of Medicine, New Haven, CT, United States

Hartwig Wolburg Institute of Pathology and Neuropathology, University of Tübingen, Tübingen, Germany

Heike Wulff University of California, Davis, CA, United States

Changjun Yang University of Florida, Gainesville, FL, United States

Klaus Zweckberger University of Heidelberg, Heidelberg, Germany

Introduction: Brain Edema Formation—Significance for Patient Outcome

Jérôme Badaut[1] *and Nikolaus Plesnila*[2]

[1]CNRS-University of Bordeaux, Bordeaux, France [2]University of Munich Medical Center & Munich Cluster for Systems Neurology (Synergy), Munich, Germany

HISTORY OF BRAIN EDEMA

Edema is defined as a net increase of tissue water content leading to an increase in tissue volume. Usually tissue swelling has no major consequence and tissue function returns to normal as soon as swelling vanishes. The notable exception is when the swollen tissue cannot expand. Then tissue pressure increases, counteracts perfusion pressure, and results in ischemic tissue damage. In the periphery such a situation is known as compartment syndrome in the brain as intracranial hypertension.

The most frequent cause of increased intracranial pressure (ICP) is brain edema. Today we know that brain edema has a complex pathophysiology, as also evidenced by the content of this book, however, it took a long time and many partly vigorous and irreconcilable scientific discussions until the current concepts on the formation of brain edema were established.

Brain edema was first described by Robert Whytt, a Scottish physician and early neuroscientist in 1768 with the words "Altho there may be no obstruction in any part of the brain [hydrocephalus], a dropsy [accumulation of water] may be formed in it."[1] Whytt and later on in 1808 his countryman John Cheyne, better known as the first describer of the Cheyne–Stokes respiration, observed that edematous brain was soft and exuded fluid upon cutting. Both interpreted this observation as accumulation of water in the brain derived from abnormal vasculature.[1,2] This correct interpretation of vasogenic brain edema was rejected by John Abercrombie, another Scottish physician, in his at that time famous and influential textbook of neurology.[3] As a result the interest in brain edema ceased for almost 100 years.[4] Only in 1905 Martin Reichhardt, a German physician, described that the volume of the brain may also increase without exudation of water when cutting it and used the term "brain swelling" for this phenomenon in order to differentiate it from brain edema.[5] These and additional observations by Hugo Spatz, another German physician, in the 1920s[6] finally resulted in the classical classification of brain edema into vasogenic and cytotoxic, formerly termed "brain edema" and "brain swelling", respectively, by Igor Klatzo et al. 1958.[7] This and other technical and conceptual advances resulted in a plethora of novel findings on the mechanisms and the disease specific development of brain edema until the 1990s, which were also declared to "The decade of the brain." Thereafter, however, major advances in the field became scarce due to lack of technological developments and brain edema research became less popular.

RECENT ADVANCES IN THE FIELD OF BRAIN EDEMA RESEARCH

The interest in brain edema research resurrected with the development of novel technologies which allowed instigating edema development on the molecular, cellular, and tissue level in experimental animals and in patients. A major step forward was the availability of magnetic resonance imaging (MRI), which allowed studying the presence and movement of water dynamically, regionally, and non-invasively in tissues, including the brain. Different imaging modalities were even suggested to be able to differentiate between vasogenic and cytotoxic brain edema: T2 weighted imaging (T2WI) was related to vasogenic edema, while measuring the apparent diffusion coefficient (ADC) diffusion weighted imaging was believed to identify cytotoxic edema.[8] Consequently, neuroimaging became a very useful and nowadays routine clinical tool for the diagnosis and prognostic of patients suffering from brain edema and a valuable research tool for the investigation of the pathogenesis of brain edema in clinically relevant animal models of human disease (Chapter 3: Noninvasive Imaging Techniques for Brain Edema: From Basic Science to the Clinic). Another recently introduced very valuable research tool for the direct investigation of brain edema on the cellular is 2-photon microscopy, a microscopic technique using pulsed laser light to visualize processes occurring deep in the brain. Using this technology it is now possible to directly visualize opening of the blood–brain barrier and the evolving vasogenic brain edema and to measure cell swelling in the living brain (Chapter 4: Experimental Techniques to Investigate the Formation of Brain Edema In Vivo).[9,10]

Based on these advances in the methodology to study brain edema formation on the cellular level and under various pathological conditions, such as cerebral ischemia, traumatic brain injury (TBI), and cerebral malignancies, it was recognized that the process of brain edema formation is far more complex than previously anticipated and that the classical neuropathological definition of brain edema into cytotoxic and vasogenic may actually not be sufficient to explain the whole complexity of edema formation.[11] Cytotoxic and vasogenic edema may even represent two extremes of the same injury process and may be far more interconnected to the previously believed. Therefore it was time to bring together a group of experts in the field of brain edema research and gather the most contemporary views on the topic in one comprehensive publication. The book evolving from this joint effort is divided in five main sections: a general introduction on cerebral vessels and the blood–brain barrier, a second section on how to investigate edema, a third part on current molecular and cellular concepts of brain edema formation, a fourth part on disease specific edema formation, and a final fifth part on clinical features and management of brain edema. The following paragraphs will give the reader a short overview on which specific issues will be addressed in each of these sections.

THE PHYSIOLOGY OF CEREBRAL BLOOD VESSELS AND HOW TO INVESTIGATE THEM (PART 1 AND 2)

The importance of cerebral blood vessels has been proposed for several brain disorders. The brain has a high-energy demand and therefore needs a highly specialized vascular system which is actively regulated and fine-tuned in order to redirect blood flow to the areas of neuronal activity, a process termed neurovascular coupling (NVC). NVC is made possible by the neurovascular unit, a functional unit located around cerebral blood vessels composed by endothelium, smooth muscle cells, astrocytes, pericytes, and neurons.[12] The NVU, more specifically the endothelial tight junctions, also harbor the blood–brain barrier (BBB), which controls the traffic of solutes and cells between brain parenchyma and blood stream. The BBB protects the brain from deleterious compounds within blood, while providing the nutrients, hormones, and ions necessary for proper brain function.[13] Any morphological or functional disturbances of this well coordinated unit change CNS homeostasis and may result in an increase in brain water content, that is brain edema. In order to have a better understanding of these basic processes involved in the physiology of the BBB and the formation of brain edema, recent knowledge on the cerebral vasculature (cerebral blood flow and blood–brain interfaces) is reviewed in Chapters 1, 2, and 4, Physiology of Cerebral Blood Vessels, Blood–Brain Interfaces Organization in Relation to Inorganic Anion

Transport, CSF Secretion, and Circulation, and Experimental Techniques to Investigate the Formation of Brain Edema In Vivo.

As mentioned above, current brain edema research greatly benefited from the recent development of a variety of techniques. For example MRI has contributed to depict the temporal and spatial changes in edema formation in various disease states and to build the bridge between clinical and experimental observations of the edema process.[11] How changes in ADC and T2WI were attributed to cytotoxic and vasogenic edema, respectively, is described in Chapter 3: Noninvasive Imaging Techniques for Brain Edema: From Basic Science to the Clinic. The next step when using imaging technology to investigate brain edema is how the observed changes are associated with the molecular and cellular processes of edema formation and resolution. Therefore experimental models are now using new modalities of in vivo imaging associated with genetic and molecular tools to refine the understanding on edema process (Chapters 3 and 4, Noninvasive Imaging Techniques for Brain Edema: From Basic Science to the Clinic and Experimental Techniques to Investigate the Formation of Brain Edema In Vivo). Finally, in vitro models of the BBB are described, which are very helpful to understand the molecular mechanisms responsible for ion movements across the BBB (Chapters 5 and 7, In Vitro Models of the Blood–Brain Barrier to Better Understand the Pathophysiology of Brain Edema and Blood–Brain Barrier Mechanisms of Edema Formation: The Role of Ion Transporters and Channels).

MOLECULAR BASIS AND CONCEPTS IN BRAIN EDEMA FORMATION (PART 3)

As indicated previously, the proper function of the central nervous system is extraordinarily complex and therefore critically depends on a tightly controlled cellular and ionic environment. These functions are linked to short- and long-distance movements of fluid as present in the extracellular space and in cortico-spinal fluid (CSF)/perivascular space, respectively. Which and how movement of fluids in the brain is impaired in various neurological diseases and how this results in edema formation in described in Chapter 6: The Extracellular and Perivascular Spaces of the Brain.

The blood–brain interface represents a large surface area of exchange between the blood and CNS and it has been recognized to participate in the formation of the CSF and the brain interstitial fluid. As many barriers in the body, the BBB also carries a broad spectrum of ion transporters and channels. Until recently the contribution of these processes has not been studied in detail in the context of edema formation. How Na^+ transporters and K^+ channels drive transendothelial transport of ions and water from blood into brain thereby possibly contributing to edema formation during the early hours after injury is dealt with in Chapter 7: Blood Brain

Barrier Mechanisms of Edema Formation: The Role of Ion Transporters and Channels. Targeting these BBB ion transporters and channels appears to be a promising therapeutic avenue for reducing edema and brain damage in acute injuries.[14] In fact, several studies have recently highlighted the critical role of the sulfonylurea binding subunit of the inward-rectifier K^+ channels 1 and 2 ($K_{ir}6.1$ and $K_{ir}6.2$ or SUR1-TRPM4) in the pathogenesis of brain edema across various disease conditions.[15] Glibenclamide, a selective blocker of SUR1-TRPM4, results in improvement in edema, lesion volume, mortality, and neurological function in various experimental models and possibly also in patients (Chapter 10: Sur1-Trpm4-Promising Target for Brain Edema Treatment in Brain Edema: From Molecular Mechanisms to Clinical Practice).

The regulation of cell volume is critical in the brain, which is contained in a rigid outer shell. In fact, neuronal activation leads to constant changes of the ionic and metabolic composition of the brain's extracellular space. These changes are buffered by astrocytes on the expense of constant changes in cell volume. Under pathological conditions, the ability of astrocytes to maintain the homeostasis of the brain is overwhelmed and permanent cell swelling, known as cytotoxic edema, occurs. Since these processes are mediated by specific transporters within the cell membrane therapeutic interventions may be possible, but need to be implemented with care in view of the possible interactions with the neuroprotective activity of astrocytes (Chapter 8: Mechanisms of Cell-Volume Regulation in the Central Nervous System). To facilitate water movements across the cell membrane during cell volume regulation astrocytes express aquaporins, specific water channels, such as AQP4.[16] The role of AQP4 in the process of edema formation varies between injury models and may be beneficial and deleterious. Details are described in Chapter 9: Role of Aquaporins for the Formation and Resolution of Brain Edema.

In contrast to cytotoxic edema, vasogenic edema is associated with the dysfunction of the BBB. The increased permeability of the BBB is preceded by an increased proteolytic activity of matrix metalloproteinases (MMPs) with the direct consequence of degradation of tight junction proteins and basement membrane components, as well as interfering with cell–cell interactions in the neurovascular unit.[17] Understanding how the MMPs contribute to the pathogenesis of cerebral edema could help to identify novel targets to reduce its devastating consequences in many neurological diseases (Chapter 11, Role of Matrix Metalloproteinases in Brain Edema).

BRAIN EDEMA PROCESS IN PRE-CLINICAL MODELS (PART 4)

The present subdivision of brain edema in a cytotoxic and a vasogenic component is a simplified view of the edema process and only partly illustrates its complexity. This is very well demonstrated by the fact that various forms of edema are present in various diseases: edema is observed in ischemic stroke (Chapter 12, Edema and BBB Breakdown in Stroke), TBI (Chapter 13, Brain Edema Formation in Traumatic Brain Injury), spinal cord injury (SCI, Chapter 14, Spinal Cord Edema After Spinal Cord Injury: From Pathogenesis to Management), acute liver failure (ALF, Chapter 15, Brain Edema in Acute Liver Failure), brain tumors (Chapter 16, Blood–Brain Barrier and Edema in Brain Tumors), epilepsy (Chapter 17, Water Homeostasis Dysfunction in Epilepsy), and neuroinflammatory diseases (Chapter 18, Water Homeostasis Dysfunctions and Edema in Neuroinflammatory Disease).

Acute injuries (stroke, TBI, and SCI) are frequently exhibiting a core lesion (primary injury) which cannot be rescued because it has been already lethally damaged during the initial injury. However, there is a cascade of secondary events including inflammation, reactive oxygen species (ROS), lipid peroxidation, excitotoxicity, and edema formation which further exacerbates the primary damage (secondary injury). The cellular and molecular events are reviewed for each injury to provide some targets to new potential therapies for these diseases (Chapters 12, 13, and 14, Edema and BBB Breakdown in Stroke, Brain Edema Formation in Traumatic Brain Injury, and Spinal Cord Edema After Spinal Cord Injury: From Pathogenesis to Management).

Similar to acute brain injuries, malignant brain tumors such as glioblastomas are associated with edema, most of the time due to the modification of existing or the formation of novel blood vessel. In fact, the newly formed blood vessels show a blood-tumor barrier that is different from the blood–brain barrier in the healthy brain. Tight junctions are morphologically changed and do not express claudin-3, typical for brain endothelial cells. In the extracellular matrix agrin is downregulated and tenascin is upregulated (Chapter 16, Blood–Brain Barrier and Edema in Brain Tumors). It is interesting to highlight that in many of the brain disorders with edema formation, AQP4 expression is increased or decreased.[16] The changes of AQP4 are very likely contributing to water disturbance. Moreover, glial cells are involved in modulation of synaptic transmission, and AQP4 has been suggested to participate in modulation of excitability and therefore in epilepsy (Chapter 17, Water Homeostasis Dysfunction in Epilepsy). Epilepsy is frequently associated with brain tumors, stroke, and TBI. Similarly, neuroinflammation occurs in many pathological states of the CNS including chronic neuroinflammatory diseases (i.e., multiple sclerosis), with sequential processes responsible to alteration of CNS water homeostasis (Chapter 18, Water Homeostasis Dysfunctions and Edema in Neuroinflammatory Disease). Inflammation and infection components can be observed in ALF causing BBB dysfunction. In ALF, the brain edema formation remains the main cause of increase ICP and mortality (Chapter 15, Brain Edema in Acute Liver Failure). Brain edema and intracranial hypertension are significantly correlated with arterial ammonia concentrations and alterations in expression of genes coding for neuroglial proteins involved in cell volume regulation and in intercellular neural transport have been reported in ALF (Chapter 15, Brain Edema in Acute Liver Failure). Edema and consequently ICP increase are major causes of mortality and long-term disability. All this knowledge on the molecular and cellular mechanisms facilitated the development of promising experimental therapies for brain edema formation (Chapter 19, Experimental Therapies for Brain Edema and Intracranial Hypertension).

CLINICAL FEATURES AND MANAGEMENT OF BRAIN EDEMA (PART 5)

The ultimate goal of brain edema research is to translate findings from the laboratory into clinical management thereby improving patient care. Therefore the last but not least part of the current book (Part 5) deals with the contemporary state-of-the-art clinical movement of brain edema written by academic clinicians with hands-on experience on the topic. Due to the lack of specificity of all currently available treatment modalities against brain edema, clinical brain edema management is mostly independent from the original pathology and focuses on diagnosis and management of ICP, maintenance of cerebral perfusion pressure, and optimal cerebral oxygenation levels. Accordingly, in most cases the first clinical goal is to reduce the space-occupying effects of brain edema by draining CSF using osmotic therapies, remove necrotic tissue, or decompress the brain by craniotomy. The clinical application of all these diagnostic and therapeutic strategies for monitoring of brain edema and the treatment of brain edema in children and in adults suffering from cerebral ischemia, TBI, or brain tumors is described in Chapters 21, 22, 23, 24, and 25: Brain Edema in Developing Brain Diseases, Cerebral Edema in Cerebrovascular Diseases, Traumatic Brain Injury and Edema Treatment, Treatment of Edema Formation in Oncology, and Perspectives on Future Translational Research on Brain Edema.

PERSPECTIVES IN FUTURE TRANSLATIONAL RESEARCH ON BRAIN EDEMA

Finally, the book will end with a personal, but nevertheless as objective as possible, view of the editors and one of the authors on what needs to be done in the future in order to promote translational brain edema research, to develop compounds for the treatment of brain edema, and to move these compounds from the bench to the bedside. In view of recent methodological developments and the knowledge acquired in the field of brain edema research we are confident that such progress can be achieved within the foreseeing future.

References

1. Whytt R. *Observations on the dropsy in the brain*. Edinburgh: Belfour; 1768.
2. Chene J. *Hydrocephalus acutus, dropsy in the brain*. Edinburgh, London: Mundell, Doig, Stevenson, and Murray; 1808.
3. Abercrombie J. *Pathological and practical researches on diseases of the brain and the spinal cord*. Edinburgh: Waugh and Innes; 1828.
4. Torack RM. Historical aspects of normal and abnormal brain fluids. III. Cerebral edema. *Arch Neurol* 1982;**39**:355–7.
5. Reichhardt M. Zur Entstehung des Hirndrucks bei Hirngeschwülsten und anderen Hirnkrankheiten und über eine bei diesen zu beobachtende besondere Art der Hirnschwellung. Dtsch Z Nervenheilkd 1905;28:306–55.
6. Spatz H. Die Bedeutung der "symptomatischen Hirnschwellung" für die Hirntumoren und für andere raumbeengende Prozesse in der Schädelgrube. *Arch Psychiatr* 1929;**88**:790–4.
7. Klatzo I, Piraux A, Laskowski EJ. The relationship between edema, blood-brain-barrier and tissue elements in a local brain injury. *J Neuropathol Exp Neurol* 1958;**17**:548–64.
8. Obenaus A, Ashwal S. Magnetic resonance imaging in cerebral ischemia: focus on neonates. *Neuropharmacology* 2008;**55**:271–80.
9. Schwarzmaier SM, Gallozzi M, Plesnila N. Identification of the vascular source of vasogenic brain edema following traumatic brain injury using in vivo 2-photon microscopy in mice. *J Neurotrauma* 2015;**32**:990–1000.
10. Risher WC, Andrew RD, Kirov SA. Real-time passive volume responses of astrocytes to acute osmotic and ischemic stress in cortical slices and in vivo revealed by two-photon microscopy. *Glia* 2009;**57**:207–21.
11. Simard JM, Kent TA, Chen M, Tarasov KV, Gerzanich V. Brain oedema in focal ischaemia: molecular pathophysiology and theoretical implications. *Lancet Neurol* 2007;**6**:258–68.
12. Zhang JH, Badaut J, Tang J, Obenaus A, Hartman R, Pearce WJ. The vascular neural network—a new paradigm in stroke pathophysiology. *Nat Rev Neurol* 2012;**8**:711–6.
13. Abbott NJ, Ronnback L, Hansson E. Astrocyte-endothelial interactions at the blood–brain barrier. *Nat Rev Neurosci* 2006;**7**:41–53.
14. O'Donnell ME. Blood–brain barrier Na transporters in ischemic stroke. *Adv Pharmacol* 2014;**71**:113–46. http://dx.doi.org/10.1016/bs.apha.2014.06.011. Epub;%2014 Aug 23: 113–46.
15. Caffes N, Kurland DB, Gerzanich V, Simard JM. Glibenclamide for the treatment of ischemic and hemorrhagic stroke. *Int J Mol Sci* 2015;**16**:4973–84.
16. Badaut J, Fukuda AM, Jullienne A, Petry KG. Aquaporin and brain diseases. *Biochim Biophys Acta* 2014;**1840**:1554–65.
17. Seo JH, Guo S, Lok J, et al. Neurovascular matrix metalloproteinases and the blood–brain barrier. *Curr Pharm Des* 2012;**18**:3645–8.

SECTION I

GENERAL INTRODUCTION

CHAPTER

1

Physiology of Cerebral Blood Vessels

Ute Lindauer

RWTH Aachen University, Aachen, Germany

OUTLINE

Brain Edema.

DOI: http://dx.doi.org/10.1016/B978-0-12-803196-4.00001-1

INTRODUCTION

Approximately 20% of the body's oxygen and 25% of the body's glucose supply are consumed by the brain. Thus, in the brain a disproportionately high percentage of oxygen and glucose is consumed compared to its small fraction of approximately 2% of total body weight. Only very low amounts of glucose are stored as glycogen in astrocytes and most of the glucose is metabolized via oxidative phosphorylation, making the uninterrupted delivery of substrates for energy consumption an important prerequisite for proper brain function. The supply of the main energy substrates oxygen and glucose to the brain depends on the vasculature, and the brain receives approximately 20% of the cardiac output.[1,2] The vasculature in the brain is not a stiff line system comparable to the plumbing system in a residential building. Rather, it is an actively regulated organ which, on the one hand, is influenced by global changes of systemic parameters (systemic blood pressure, blood gases, and blood pH) and, on the other hand, is capable of locally directing the blood to the regions of demand with high spatial and temporal resolution.[3] Together with pericytes and astrocytes, as part of the neurovascular unit, cerebral vascular endothelium forms the tight blood-brain barrier and has an important immuno-regulatory function.[4,5]

CEREBROVASCULAR ARCHITECTURE AND PRINCIPLE OF BLOOD PERFUSION

Four main vessels are feeding the brain, the internal carotid arteries originating from the common carotid arteries, and the vertebral arteries. They form the Circle of Willis at the base of the brain, from which the main cerebral arteries are advancing and branching at the surface of the brain.[6] Here they form the pial arteries, which send penetrating arterioles into the brain parenchyma with further ramification in precapillary arterioles and the capillary bed. Penetrating arterioles are responsible for approximately 40% of total cerebral vascular resistance;[7] however, recent findings of large-scale microscopy data analysis combined with hemodynamic simulations suggest that the capillary bed is the largest contributor to vascular resistance.[8] The venous blood is collected in small postcapillary venules followed by larger venules which reach the brain surface. The main draining systems are the deep and superficial cerebral veins which merge in dural venous sinuses and later on form the jugular vein, finally draining the blood into the superior vena cava.

Oxygen diffusion into the parenchyma occurs not only within the capillary bed but also to a significant amount at the level of penetrating arterioles.[9] The oxygen content of the blood is lowest in high-branching-order capillary segments and postcapillary venules (pO_2 ~ 30–35 mmHg) and slightly increases while passing the venules. It has been shown that under resting conditions, most of the oxygen is released to the tissue at the level of precapillary arterioles and low-branching-order capillaries, whereas high-branching-order capillaries only show a very low oxygen saturation gradient along their course toward the postcapillary venules. This phenomenon may be explained by a reuptake of oxygen from the tissue back into high-branching-order capillaries[9] and/or into venules.[10] Glucose is transported into the tissue mainly within the capillary bed by specific transporters.[2] Compared to other vascular systems, the capillaries in the brain are not fenestrated but the endothelial cells are connected by tight junctions, forming the blood-brain barrier together with adjacent pericytes and astrocytic foot processes (for more detail, see Section "Blood-Brain Barrier Function"). Capillaries, by definition, lack smooth muscle cells (SMCs). However, at distinct locations they are in close contact with pericytes which may take over the role of smooth muscle cells for adjusting blood flow to neuronal activity or metabolic needs at the very regional level.[11–13]

Microvascular oxygen delivery strongly depends on the erythrocyte flux within the vessels. In contrast to the vasculature of skeletal muscles, where total closing and reopening of capillaries occurs ("morphological capillary recruitment"), in the brain all capillaries are perfused at least by blood plasma all the time.[14,15] Perfusion by red blood cells is characterized by a significant heterogeneity under resting conditions, meaning that many capillaries are not maximally perfused. Within the capillary network, rapid fluctuations, spatial heterogeneity of velocity, and density of erythrocytes can be observed. During global as well as local stimulation of cerebral blood flow (CBF), the mean blood cell flux and velocity increases, capillary perfusion with red blood cells becomes more homogenous, and the number of poorly cell-perfused capillaries decreases.[14,16,17] Hence, within the brain microcirculation a dynamic adjustment of red blood cell perfusion in continuously plasma-perfused capillaries takes place.

CELLULAR COMPONENTS OF CEREBRAL VASCULATURE AND THE NEUROVASCULAR UNIT

The blood vessels of the cerebral macrocirculation (pial arteries and arterioles, larger branches of the penetrating arterioles, and larger venules and veins) are separated from the parenchyma by the Virchow–Robin space and are thus not in direct contact to changes of the milieu within the brain tissue. Within the microcirculation (smaller branches of penetrating arterioles, precapillary arterioles, capillaries, and postcapillary venules), the vessels come in close contact with parenchymal cells, forming the well-known entity of the neurovascular unit, a structure where neuronal-activity dependent parenchymal changes also have impact on mural cells of the vessels.[18,19]

The innermost cell barrier toward the blood is formed by endothelial cells. Separated from the endothelial cells by a basement membrane, the middle layer is composed of contractile cells. At the level of arteries and arterioles, diameter changes are induced by contraction or relaxation of smooth muscle cells which form a continuous layer within the vessel wall. Further down the vascular tree at the level of capillaries, smooth muscle cells disappear and are replaced by pericytes which are spatially isolated cells with contractile properties.[20,21] Larger arterioles and arteries within the pial vasculature are surrounded by the adventitial layer bearing a dense network of perivascular nerve endings of extracerebral origin, whereas intraparenchymal vessels are in contact with intrinsic nerves originating from subcortical areas or with local interneurons.[19] In the following a more detailed description of the different cell types of the neurovascular unit will be provided.

Cerebrovascular Endothelial Cells

The endothelial cells form the innermost cell layer of the barrier separating the blood from the parenchyma. On the luminal (apical) side, the cells are coated by the glycocalyx, a negatively charged dense network of membrane bound molecules (proteoglycans, glycoproteins, and glycolipids), and supplemented by absorbed plasma proteins and glycosaminoglycanes. This coating has a thickness of 0.4–0.5 μm,[22] and is known as the endothelial surface layer. It has important physiological roles in regulating cerebral blood flow by significantly contributing to hemodynamic resistance[23] and by mechanotransduction of blood flow-induced forces (shear stress).[24] It impacts the permeability of solutes and water,[25–27] as well as the adhesion and emigration of leukocytes[28] and coagulation[29,30] (for more detailed information, see Refs. 31,32).

Blood-Brain Barrier Function

The endothelial cells of cerebral vessels are tightly connected by tight junctions and adherence junctions, continuously blocking permeability via the paracellular route and by this morphologically building the blood-brain barrier. As summarized by three very profound reviews (see Refs. 33–35), the main transmembrane components of the tight junctions are claudins, junctional adhesion molecules, and occludin, which are anchored within the cytoplasm by zonula occludens proteins and by this connected to the actin cytoskeleton. Adherence junctions are composed of the cytoplasmic components ß-catenin and afadin and the transmembrane proteins nectins and cadherins. The tightness of the blood-brain barrier is not accomplished by the endothelial cells alone. At their abluminal side the endothelial cells are covered by the basal lamina (basal membrane), with extracellular matrix proteins being involved in proper tight junction protein expression and by this in barrier function.[36] The dense coverage of capillary endothelial cells with astrocytic end-feet[37] plays an important role in building up transendothelial resistance. In addition, pericytes are essential for blood-brain barrier function, and it has been shown that pericyte deficiency in the brain leads to increased endothelial transcytosis and increased permeability to immunoglobulins and other plasma proteins.[38,39] Both astrocytes and pericytes interact with endothelial cells, and by growth factor signaling and transcriptional regulation of tight and adherence junction proteins, the tightness of the blood-brain barrier is dynamically regulated.[35] Beside tight junctions, endothelial cells are connected with each other by gap junctions formed by connexins, responsible for the direct passage of ions and small molecules between adjacent cells.[40] Gap junction proteins are very often colocalized with tight junction proteins and a dynamic interaction of connexins with other junction proteins has been described.[41,42] A cross-talk of the regulation of these two intercellular junctions has been suggested, with gap junction proteins probably also playing a role in blood barrier function.

Vasoactive Factors Released by Endothelial Cells: Endothelium-Dependent Cerebrovascular Reactivity

For several vasoactive humoral or parenchymal stimuli an intact cerebrovascular endothelium is a precondition to induce diameter changes of cerebral vessel. Under resting conditions as well as upon receptor or ion channel activation under physiological or pathophysiological conditions the endothelium is capable of releasing vasoactive factors acting on the underlying smooth muscle cells inducing vasodilation or vasoconstriction.

Probably the most important and best known molecule released by endothelial cells is the vasodilator nitric oxide (NO), identified as the molecule behind the "endothelium-derived relaxing factor," released by endothelial cells upon acetylcholine (ACh) receptor activation and acting via cGMP production in vascular smooth muscle cells.[43–46] Along with ACh, other molecules like bradykinin, substance P, or arginine vasopressin (only in selected vascular beds)[47–49] also induce NO-dependent vasodilation. Beside this receptor activation-dependent augmentation of NO production, NO is also continuously produced and perivascularly released by endothelial cells and perivascular neurons (see below) by the constitutively expressed NO synthases (NOS) (isoforms I and III), inducing a basal vasodilative tonus of the vessels. Endothelial-derived NO also provides important antiinflammatory properties preventing coagulation and leukocyte aggregation and adhesion to the vessel wall, for example, by reducing endothelial expression of adhesion molecules and pro-inflammatory cytokines[50–52] (for review, see Refs. 53,54).

In addition to NO, arachidonic acid metabolites are important endothelial mediators of vasodilation. Besides the enzyme responsible for NO production, endothelial cells also constitutively express cyclooxygenase 1 (COX-1, also called prostaglandin H synthase[55,56]), converting arachidonic acid, liberated from membrane phospholipids by phospholipases, to prostacyclin (prostaglandin I2), inducing relaxation of the underlying vascular smooth muscle cells by activation of adenylate-cyclase and production of cAMP.[57,58] As with NO synthase within endothelial cells, COX-1 is continuously activated by blood flow-induced shear stress or can be further activated by receptor-dependent activation.[59] Under pathophysiological conditions, arachidonic acid can also be metabolized by the COX-pathway to prostaglandin PGH_2/thromboxaneA_2 (TXA_2) with vasoconstricting properties.[60]

Another pathway of arachidonic acid metabolism represents the formation of epoxyeicosatrienoic acids (EETs) by cytochrome P450 epoxygenase located within endothelial cells. According to a recent study EETs relax smooth muscle cells by activation of a novel Ca^{2+} signaling complex in cerebral vascular smooth muscle cells, consisting of the calcium-permeable member of the transient receptor potential vanilloid cation channels TRPV4, of ryanodine receptors, and of large-conductance Ca^{2+}-activated potassium channels (BK_{Ca}) located on smooth muscle cells.[61,62] However, suppression of EETs production by NO within endothelial cells has been suggested.[63]

A third important endothelium-dependent vasodilating mechanism is "endothelium-derived hyperpolarization" (EDH), also significantly contributing to resting diameters of cerebral vessels and thus to resting CBF. Following a long search for identifying a factor mediating the hyperpolarization of smooth muscle cells, the current knowledge argues for a mechanism rather than a real mediator molecule. Activation of small- and intermediate-conductance Ca^{2+}-activated potassium channels (SK_{Ca}, $K_{Ca}2.3$, and IK_{Ca}, $K_{Ca}3.1$) within the membrane of endothelial cells (e.g., by stimulation of endothelial P2 purinoceptors[64]) induces endothelial cell hyperpolarization. This hyperpolarization propagates towards the vascular smooth muscle probably through myo-endothelial gap junctions, inducing relaxation by closure of voltage gated Ca^{2+} channels at the smooth muscle cells (see Section "Cerebrovascular Smooth Muscle Cells").[59,65–68] In addition to their role for EDH, IK_{Ca} channels may also play an important role in endothelial barrier dysfunction and edema formation in brain pathologies.[69]

Two other gaseous mediators or modulators of vascular tone, carbon monoxide (CO), and hydrogen sulfide (H_2S), may also be produced by endothelial cells via the constitutively expressed enzymes heme oxygenase-2[70,71] and cystathionine-γ-lyase/cystathionine β-synthase/3-mercaptopyruvate sulfurtransferase,[72] respectively. Acute elevations of CO in newborn pigs dilate cerebral vessels by directly acting on BK_{Ca} channels on smooth muscle cells by elevating the channel-apparent Ca^{2+} sensitivity[73] and by elevating BK_{Ca} channel activity by increasing the coupling of Ca^{2+} sparks to BK_{Ca} channels.[74,75] In addition, a permissive role of endothelial NO has been reported for CO-induced dilation in piglet pial arterioles.[76] In contrast to the acute effects of CO in piglets, prolonged CO production in the adult rat cerebral microcirculation has been shown to induce vasoconstriction, apparently by inhibiting NOS.[77] Species and age related differences in the anatomical localization of the NO- and CO-gas-producing enzymes and their reception systems may be responsible for these discrepant observations. For H_2S it has been convincingly shown within the mesenteric circulation and the aorta that H_2S is a major EDHF, causing vascular endothelial and smooth muscle cell hyperpolarization and relaxation by activating ATP-sensitive (K_{ATP}) as well as IK_{Ca} and SK_{Ca} potassium channels through cysteine S-sulfhydration.[78,79] Within the cerebral circulation H_2S has been shown to dilate cerebral arterioles mainly by activating K_{ATP} channels on smooth muscle cells,[80] which may become particularly important during NO deficiency.[81]

Endothelial cells are also capable of releasing vasoconstricting molecules. Beside the arachidonic acid metabolites PGH_2/TXA_2 (see above), one of the most potent endothelium-derived constrictors is endothelin-1, which is produced as pre-pro-ET-1, cleaved to big ET-1 by an endopeptidase and then converted to ET-1 by endothelin-converting enzyme.[82] Vasoconstriction is induced by activation of endothelin-A and endothelin-B receptors located on vascular smooth muscle cells. Beside this strong predominant vasoconstrictive action, endothelin-1 at very low concentrations can also induce endothelium-dependent and NO-mediated vasodilation by activation of endothelin-B receptors located at endothelial cell membranes.[83,84]

Mechanotransduction of Shear Stress

Endothelial cells are exposed to mechanical frictional forces by the circulating blood. This shear stress induces vasodilation and is thus also called flow-induced vasodilation. Membrane molecules and cellular microdomains (e.g., the glycocalyx, adhesion proteins, G proteins of G protein-coupled receptors, and the cytoskeleton) contribute to sensing of shear stress.[85] Ca^{2+} influx via activation of transient receptor potential (TRP) channels[24] leads to eNOS activation and NO production and thus vasodilation. In addition, a autocrine/paracrine action of endogenous ACh may be involved, where ACh is produced and released by endothelial cells following increased shear stress, again resulting in enhanced NO production.[86]

Cerebrovascular Smooth Muscle Cells

Vascular smooth muscle cells are the main effectors responsible for the regulation of global blood flow at the level of larger pial arteries as well as for the spatially defined regulation at the level of penetrating and precapillary arterioles. Besides the intrinsic tone of the arterial circulation induced by intraluminal pressure and sheer stress at the endothelial cell layer, the balance between vasoconstrictive and vasodilative molecules released by vascular endothelial or brain parenchymal cells and acting on smooth muscle cells determines the diameters and thus the blood flow within the vasculature. Dependent on the activity of myosin light chain kinase (MLCK) and myosin light chain phosphatase (MLCP), the balance of phosphorylation and de-phosphorylation of myosin light chain determines the contractility of the smooth muscle cell.[87]

Principle Mechanisms of SMC Constriction or Relaxation

Constriction or relaxation of SMCs occurs mainly by two mechanisms: (1) via changes of the intracellular Ca^{2+} concentration due to exchange from the extracellular compartment or due to release and reuptake from and to internal stores, determining the activity of the MLCK,[88] and (2) via alteration of the Ca^{2+} sensitivity of the contractile apparatus without changes in Ca^{2+} concentration (so-called "Ca^{2+} sensitization"), determined by RhoA/Rho-associated protein kinase (ROCK)-dependent phosphorylation status of MLCP with ROCK promoting contraction by phosphorylating and thus inhibiting MLCP.[89,90] Beside induction of Ca^{2+}-sensitization, it has recently been shown that the RhoA/ROCK pathway can also induce membrane depolarization via activation of transient receptor potential cation channels (subfamily M member 4, TRPM4) and thus opening of voltage-dependent Ca^{2+} channels (VDCC).[91]

Intracellular Ca^{2+} increases via flux through receptor operated ion channels (ROCs) or by activation of VDCC following depolarization of the smooth muscle cell membrane. ROCs are nonselective cation channels, mainly members of the transient receptor potential canonical (TRPC) channels family. They simultaneously induce the entry of Na^+ and Ca^{2+} and thus triggering cell membrane depolarization. Whether the increased Ca^{2+} flux through the channels directly activate the contractile process is not known so far.[92]

Membrane potential changes are responsible for the degree of Ca^{2+} influx through VDCC. In cerebral vasculature, L-type ($Ca_V1.2$) and T-type ($Ca_V3.1$ and $Ca_V3.2$) voltage-dependent Ca^{2+} channels are expressed in smooth muscle cells.[93]

Potassium Channels on VSMCs

In addition to the activity of ROCs, membrane potential changes are mainly induced by changes of the conductance of potassium channels within the cell membrane. Upon activation of potassium channels, cell membrane hyperpolarization occurs due to potassium outward flow, inducing closure of VDCCs followed by a decrease in intracellular Ca^{2+} concentration and vascular relaxation (for detailed review on potassium channels of VSMCs, see Ref. 53). The main types of classical potassium channels within smooth muscle cells are ATP-sensitive potassium channels (K_{ATP}), large-conductance Ca^{2+}-activated potassium channels (BK_{Ca}), voltage-dependent potassium channels (K_V), and inward rectifier potassium channels (K_{IR}). Whereas K_{IR} channels are mainly activated by low concentrations of potassium (up to 20 mM) and K_V channels by membrane depolarization, K_{ATP} and K_{Ca} channels can be activated by agonists following receptor activation as well as second messengers (cGMP and cAMP). K_{ATP} channels are also sensitive to hypoxia and energy failure induced ATP reduction. BK_{Ca} channels are also activated by local Ca^{2+}-release events (Ca^{2+} sparks) through ryanodine-sensitive channels in the sarco/endoplasmic reticulum[94] probably gated by Ca^{2+} released through L-type as well as T-type VDCCs residing near ryanodine receptor,[95] resulting in spontaneous transient outward currents (STOCs).[96–98] Following membrane depolarization and rise in cytosolic Ca^{2+} concentration, K_V and BKCa-mediated K^+ conductances drive a negative feedback mechanisms to prevent excessive myogenic constriction.[96,98]

In addition to the "classical" K^+ channels with a single-pore region mentioned above, the expression of "nontypical," two-pore domain K^+ (K_{2P}) channels has recently been confirmed in cerebrovascular SMCs, which can be directly activated by unmetabolized arachidonic acid inducing K^+ outward flux, membrane hyperpolarization, and vasodilation.[99] The overall significance of these channels is not known so far and requires further investigation.

Intracellular Ca^{2+} concentration can also be changed by activation of recently identified acid sensing ion channels expressed within the SMC membrane of cerebral arteries.[100] It has been shown that they are involved in regulation of smooth muscle cell migration,[101] and they may play a role in pressure-induced constriction and basal myogenic tone.[102]

Role of Internal Ca^{2+} Stores

Beside Ca^{2+} entry from extracellular space through ion channels located within the cell membrane, Ca^{2+} can also be released from internal stores via inositol 1,4,5-trisphosphate (IP_3) receptor channels located in the membrane of the sarco/endoplasmic reticulum. Activation of smooth muscle cell membrane bound receptors followed by phospholipase C activation leads to hydrolysis of phosphatidylinosol 4,5-bisphosphate (PIP_2) to IP_3. IP_3 receptors located on the endoplasmic reticulum respond to this by releasing Ca^{2+} and thus inducing contraction.[103,104] (Re-)loading of the sarco/endoplasmic reticulum with Ca^{2+} occurs via activity of the Ca^{2+} pump sarco/endoplasmic reticulum Ca^{2+}-ATPase (SERCA).[105] Depletion of sarco/endoplasmic reticulum Ca^{2+} stores, on the other hand, activates store operated Ca^{2+} channels increasing cytosolic Ca^{2+} concentration specifically within "noncontractile" compartments without vasoconstriction.[106]

Mechanisms of NO-Induced Vasodilation

As mentioned above, NO released by endothelial cells and perivascular neurons provides a basic vasodilative tonus within the vasculature in the brain. NO stimulates soluble guanylate cyclase within SMCs converting GTP to cGMP followed by activation of cGMP-dependent protein kinase and target protein phosphorylation, resulting in vasodilation by various mechanisms (BK_{Ca} channel activation, inhibition of Ca^{2+} release from endoplasmic reticulum, and inhibition of Rho kinase/ROCK pathway by phosphorylation of Rho).[107] In addition, cGMP-independent effects exist: Besides inducing nitrosylation of target proteins (e.g., causing activation of potassium channels), NO is known to inhibit cytochrome P450 (CYP450, CYP) ω-hydroxylase activity (CYP 4A and 4F families) that builds 20-hydroxyeicosatetraenoic acid (20-HETE) from arachidonic acid in cerebral arteries.[108] When perivascular NO is reduced or absent, 20-HETE accumulates and potently constricts VSMCs by inhibiting BK_{Ca} channels; by activating L-type VDCCs; and by stimulating protein kinase C, Rho kinase, and the mitogen-activated protein kinase pathway (for detailed review, see Ref. 109). In addition, NO has been shown to suppress the Ca^{2+} currents of T-type and L-type VDCCs and by this reducing Ca^{2+} concentration in SMCs.[110,111] Furthermore, NO may also induce vasodilation by increasing the uptake of Ca^{2+} by SERCA.[112]

Concluding Remarks

The above mentioned mechanisms leading to VSMC contraction or relaxation and by this vasoconstriction or vasodilation are not uniformly active in all vascular beds within the cerebral vasculature. They may significantly differ in their importance and specificity between larger pial arteries and intraparenchymal arterioles. For a comprehensive overview of these differences with respect to the control of smooth muscle membrane potential see the recent review by Longden et al.[113]

Pericytes

Pericytes are contractile cells located within leaves of the basement membrane. They form bulging cell bodies with distinct circular and longitudinal processes and sheet the capillaries with variable density and morphology along the capillary bed. The coverage is most compact at the transition from precapillary arterioles to first-order capillaries, forming a mesh-like structure, whereas further on they become more delicate with thin circular and small longitudinal strands. Toward the postcapillary venules they regain their mesh-like structure, however, with lower degree of coverage compared to the first-order level.[20,21,114] Pericytes are in close contact with endothelial cells, connected by cap junctions and adherence junctions,[115,116] and reside in close proximity to astrocytic end-feet and perivascular neuronal processes, and thus, together with VSMCs, being an integral part of the neurovascular unit.

Pericytes are characterized by expressing contractile and cytoskeletal proteins (amongst others: α-smooth muscle actin and myosin), and are commonly identified by their expression of the cell surface antigens transmembrane chondroitin sulfate proteoglycan NG2, platelet-derived growth factor receptor-β, and aminopeptidase A (for overview, see Ref. 117).

They play an important role in tightening of the blood-brain barrier by supporting and controlling endothelial cells expressing tight and adherence junction proteins and by inducing polarization of astrocytic end-feet.[38] In addition, they seem to be involved in the regulation of signaling for leukocyte adhesion and transmigration at endothelial cells and communication of inflammatory signals within the neurovascular unit.[118,119] Furthermore, pericytes react to many vasodilative and vasoconstrictive molecules and mechanisms comparable to VSMCs.[120,121] Based on this, in recent years, profound experimental evidence has been described for a prominent role in regulation of cerebral blood flow at the level of capillaries,[11,13,122] with pericytes probably being responsible for the very early increase of capillary blood flow to neuronal activation, even before precapillary and penetrating arterioles dilate.[12] Discrepant results[21] have opened a discussion on the precise definition of capillaries at the transition zone from precapillary arterioles to capillaries and by this of pericytes and VSMCs within the microvasculature.[20] It is worth to notice in this context that pericytes can constrict and by this can narrow the lumen of capillaries to an extent leading to the complete blockade of red blood cell passage under pathophysiological conditions.[11,12]

Many signaling pathways and molecules (e.g., the platelet-derived growth factor (PDGF)-BB—PDGF receptor β pathway, the transforming growth factor-β (TGF-β)—TGF-β receptor-2 pathway, the notch pathway, the vascular endothelial growth factor (VEGF)-A-VEGF receptor-2 pathway, the arachidonic acid pathway, NO, and potassium) are involved in the above mentioned functions of pericytes mediating the communication to and receiving information from endothelial cells, astrocytes, and perivascular neurons and thus within the neurovascular unit. A detailed description of these pathways is beyond the scope of this chapter and is excellently provided in a recent review by Sweeney et al.,[117] which can be recommended for further reading.

Astrocytes

Astrocytes are ideally located within the neurovascular unit with having contact to neurons at the synapses as well as to the vasculature by astrocytic end-feet. At the level of larger penetrating arteries they form the glia limitans, with their end-feet layer bordering the Virchow–Robin space which separates the arteries from the parenchyma, contains interstitial fluid, and builds the so-called "cerebral glymphatic system."[35,123] Further down the vascular tree, the Virchow–Robin space becomes smaller and disappears, and astrocytes come in close contact to the SMCs at the vessel wall of arterioles and venules and to the endothelial cells and pericytes at the level of capillaries, respectively.

Astrocytes have an important role in supporting neuronal activity by uptake and recycling of neurotransmitters and, via gap junction coupling, in redistribution of K^+ ions.[124,125] In addition, they substantially participate in glucose metabolism, the main fuel of neurons.[126]

For vascular function, the signaling from and to their end-feet is essential. Besides being involved in blood-brain barrier tightening, astrocytes are the key contributors to water exchange from and to the brain parenchyma passing through the blood-brain barrier. Aquaporin (AQP) channels permit passive bidirectional water flux along the osmotic diffusion gradient, with AQP4 being highly expressed in astrocytic perivascular end-feet.[127]

Finally, astrocytes are important players in blood flow regulation.[4,128–133] The key event of vasoactive signaling of astrocytes is the increase of Ca^{2+} within the cells. Activation of metabotropic glutamate receptors located at the membrane of the cell bodies induce phospholipase C activation, IP_3 production and release of Ca^{2+} from endoplasmic reticulum stores resulting in increased Ca^{2+} concentration in somata and end-feet.[133,134] Recently, the proposed key role of IP_3-dependent Ca^{2+} signaling for vascular regulation has been questioned,[135] and besides IP_3 receptor activation-dependent Ca^{2+} signaling, transmembrane Ca^{2+} fluxes may be involved in wave and microdomain Ca^{2+} fluctuations in astrocytic processes.[136] Other neurotransmitters like ATP released by neurons may also induce astrocytic Ca^{2+} elevation by purinergic receptor stimulation via IP_3-dependent (purinergic metabotropic P2Y receptors[137]) or IP_3-independent pathways (purinergic ionotropic P2X receptors[138,139]). Several mechanisms have been proposed to induce vasodilation due to Ca^{2+} increases in astrocyte somata and end-feet processes. Activation of phospholipase A or phospholipase D[139] liberates arachidonic acid from membrane phospholipid pools. Metabolism of arachidonic acid by cytochrome P450 epoxygenase results in astrocytic EETs production,[63] which may dilate the vessel by activation of BK_{Ca} channels directly on SMCs[140] or on astrocytic end-feet membranes, followed by K^+ release with subsequent activation of K_{IR} channels on SMCs,[141] both inducing membrane hyperpolarization and thus relaxation of the cells. The role of astrocytic EETs production in blood flow regulation is not fully proven so far, as vasodilation due to local Ca^{2+} uncaging in astrocytic end-feet was not prevented by CYP450 epoxygenase inhibition.[132] Arachidonic acid can also be metabolized by COX forming prostaglandin E2. However, it is not resolved whether astrocytes uniformly express COX-1 or cyclooxygenase 2 (COX-2) and whether prostaglandin E2 (PGE2) dilates or rather constricts parenchymal arterioles (for review, see Ref. 113). In addition to arachidonic acid metabolism induced vasodilation, Ca^{2+} waves reaching the end-feet processes may also induce K^+-dependent vascular dilation by activation of astrocytic BK_{Ca} channels followed by K^+ release with subsequent activation of K_{IR} channels on SMCs. In this context, it has recently been proposed that Ca^{2+}-sensitive TRPV4 channels, which are expressed in the astrocytic end-feet membranes, may also be activated by Ca^{2+} waves, stimulating further Ca^{2+} influx and by this locally boosting the effect on BK_{Ca} channels and amplifying the relaxation.[142]

Perivascular Neurons—Extrinsic and Intrinsic

Depending on the cerebrovascular segment, vessels are supplied by perivascular nerve fibers originating from extracerebral ganglia (extrinsic innervation) or by neurons located within the brain (intrinsic innervation). In the following a summary will be provided according to a recent main review on this topic[19]: The extrinsic innervation supplies the pial surface arteries with sympathetic (from the superior cervical ganglion, transmitters: norepinephrine and neuropeptide Y (NPY)), parasympathetic [from sphenopalatine and otic ganglia, transmitters: vasoactive intestinal peptide (VIP), ACh, and NO from NOS], and sensory nerve fibers (from the trigeminal ganglion, transmitters: calcitonin gene-related peptide, substance P, neurokinin A, and pituitary adenylate-cyclase activating polypeptide). The role of the extrinsic innervation lies in adjusting the diameters of the main feeding vessels and by this the global blood flow to the brain according to systemic needs (sympathetic innervation: adjustment of autoregulatory limits) or to counteract pathological constricting mechanisms (sensory innervation) and by this protect the brain from hypoperfusion. Further down the vascular tree, while the vessels penetrating into the parenchyma, the Virchow–Robin space disappears, and the arterioles lose their extrinsic innervation. Within the parenchyma the arteries get in contact to neurons located in the brain. This intrinsic innervation is provided by nerve fibers originating from subcortical areas (mainly from the nucleus basalis, the locus coeruleus, or the raphe nucleus, transmitters: ACh, norepinephrine, and 5-Hydroxytryptamin = Serotonin (5-HT)), or by local GABA(gamma-aminobutyric acid)-ergic interneurons, with subtype specific release of vasodilating (NO, ACh, VIP, GABA, and PGE2) or vasoconstricting (NPY, somatostatin (SOM), 5-HT, and 20-HETE) molecules. The intrinsic nerve fibers from subcortical areas only occasionally directly contact the basal lamina of the vasculature but most often connect with interneurons or astrocytes. The contribution of the respective component of the intrinsic perivascular innervation is highly dependent on the specific network activated by neuronal activity, and changes in network activity due to pathophysiological events may also have impact on the signaling at the vessel wall and thus on local blood flow in the brain.

AUTOREGULATION AND MYOGENIC TONE

Resistance vessels respond to changes of intravascular pressure with diameter changes: increased intravascular pressure elicits vasoconstriction and vasodilation occurs in response to a decrease in pressure. This phenomenon, first described by Bayliss, is known as the myogenic response mediated by SMCs and is particularly important for cerebral autoregulation.[7,143,144] Along the physiological range of systemic blood pressure, cerebral blood flow is kept constant, protecting the brain from blood pressure induced fluctuations. Below the lower limit of autoregulation the arterial diameter passively follows the perfusion pressure and the brain tissue is at risk to become hypo-perfused, whereas above the upper limit of autoregulation, the capacity of the SMC to generate force is exhausted, leading to uncontrolled augmentation of perfusion followed by damage of capillary structure and loss of blood-brain barrier integrity. The mechanisms of myogenic tone induced autoregulation are for the most part known and can be summarized as follows.[113,145] Increased wall tension is sensed by mechanosensors, with TRP channels most probably involved beside intergrins and G protein-coupled receptors. As described above, opening of TRP channels leads to entry of Na^+ and Ca^{2+} and thus cell membrane depolarization with increasing VDCC activity, Ca^{2+} influx and constriction. Several negative feedback mechanisms provide protection against overwhelming contraction, with membrane depolarization induced stimulation of K_V channels and Ca^{2+} sparks induced STOCs via BK_{Ca} channel activation both resulting in counteracting hyperpolarization currents. Beside these Ca^{2+} concentration-dependent mechanisms, Ca^{2+} sensitization via MLCP inhibition by the ROCK pathway and actin polymerization via PKC is also involved in the development of the myogenic response. Modulation of these intrinsic myogenic mechanisms may occur by extrinsic factors from the endothelium, from astrocytes and from perivascular nerves.

Along the vascular tree each resistance vessel segment has its own operating range for autoregulation with pial arteries developing less myogenic tone compared with parenchymal arterioles.

FURTHER MECHANISMS OF GLOBAL BLOOD FLOW REGULATION: REACTION TO PCO_2/PH AND PO_2 CHANGES

During ventilation insufficiency, inhalation of CO_2 or ischemia/hypoxia-induced tissue lactic acidosis, acidification of brain tissue occurs. Over a wide range of pCO_2/pH changes, cerebral vessels linearly react with diameter changes, with acidosis inducing vasodilation, and with alkalosis inducing vasoconstriction. The basal perivascular NO level has been shown to be a significant prerequisite for the physiological pH reactivity of cerebral vessels.[146,147] NO from perivascular nerve fibers seems to be the major source of basal NO responsible for pH reactivity, which occurs independent of vascular endothelium.[146] In larger pial arteries, combined activation of K_{ATP} and K_{Ca} channels seems to be responsible for dilation,[148] whereas in penetrating arterioles the specific role of K_{ATP} and K_{Ca} channels in the vasodilator response to reduced pH is controversly discussed. A recent study has established a role for ryanodine receptor-mediated Ca^{2+} spark activation of BK_{Ca} channels in penetrating arterioles, confirming the contribution of K_{Ca} channels.[149]

One of the central aims of cerebral blood perfusion is to guarantee the delivery of oxygen according to metabolic needs. Within the normal range of physiological arterial pO_2 there is only little change in cerebral blood flow, because arterial content of oxygen rather than paO_2 determines the blood flow.[150] However, when pO_2 decreases below a critical level (around <60 mmHg),[151] cerebral vessels dilate, mostly via activation of K_{ATP} channels at VSMCs,[152] to increase cerebral blood flow and by this maintain the delivery of oxygen to the tissue. Hyperoxia with pO_2 levels above the physiological range, on the other hand, induces a slight vasoconstriction with reduction of CBF. Several mechanisms may be involved in hyperoxia-induced vasoconstriction, for example, NOS inhibition[153] or inhibition of endothelial prostaglandin synthesis.[154] The purpose of the hyperoxia-induced vasoconstriction may be to protect the brain from excess oxygen to limit the production of oxygen free radicals.

MECHANISMS OF LOCAL BLOOD FLOW REGULATION—NEUROVASCULAR COUPLING

Within the cerebral microcirculation blood flow is locally directed to the regions of neuronal metabolic demand with high spatial and temporal resolution. This process is called neurovascular coupling with mainly feedforward mechanisms responsible for diameter changes of the vessels.[4] Whereas for a long time penetrating arterioles have been thought to solely react to the mediators released during increased neuronal activity and by this being mainly responsible for the initiation of the blood flow response, only recently a role for capillaries with active relaxation of pericytes has been established for the very early response.[12] Dilation of upstream arterioles occurs later after the initial capillary dilation, probably via identical signaling pathways or via conducted vasodilation within pericytes and/or endothelial cells.[155]

For most of the above mentioned signaling molecules a role in neurovascular coupling has been described, and the most important pathways will be shortly summarized. The basal perivascular NO level, provided by perivascular neurons (NOS containing interneurons or extrinsic parasympathetic nerve fibers) is an important permissive factor allowing other signaling pathways to act properly. However, in contrast to its mediator function in the cerebellar cortex,[156,157] NO is not thought to be a real mediator molecule in neurovascular coupling in the cerebral cortex.[158] Within the main mechanism involving postsynaptic pyramidal cells, glutamate activates N-methyl-D-aspartate (NMDA) receptors followed by Ca^{2+} increase, phospholipase A2 activation and generation of arachidonic acid, which is mainly metabolized to PGE2 by neuronal COX-2 and prostaglandin E2 synthase, inducing vasodilation at the level of arterioles.[139,159,160] Comparable to NO, COX-2-derived PGE2 may also at least partially act as a modulator/permissive factor of cerebrovascular function.[161] Several pathways involving astrocytes, as already described above, have been proposed. Astrocytic EETs production and subsequent activation of BK_{Ca} channels at astrocytes or VSMCs,[162] COX-1-dependent prostaglandin production,[133] as well as K^+ related mechanisms, via opening of BK_{Ca} channels in astrocyte end-feet, release of K^+ and activation of K_{IR} on VSMCs[163,164] may be the main astrocytic pathways involved in neurovascular coupling. Recent results postulate that astrocytic signaling mainly mediates vasodilation at the level of capillaries targeting pericytes rather than at the level of arterioles.[139] Further mediators like adenosine which increases as a neurotransmitter or as a metabolic product during energy metabolism and ATP consumption and dilates vessels via adenosine receptor activation at VSMCs or via adenosine induced enhancement of EETs production in astrocytes, as well as lactate released during increased glycolysis within astrocytes, have been shown to be involved in neurovascular coupling.[165–167]

SUMMARY

The physiologic functioning of the cerebral circulation is not only important for satisfying the metabolic needs of the brain tissue by providing the supply of nutrients and removal of waste products. With its

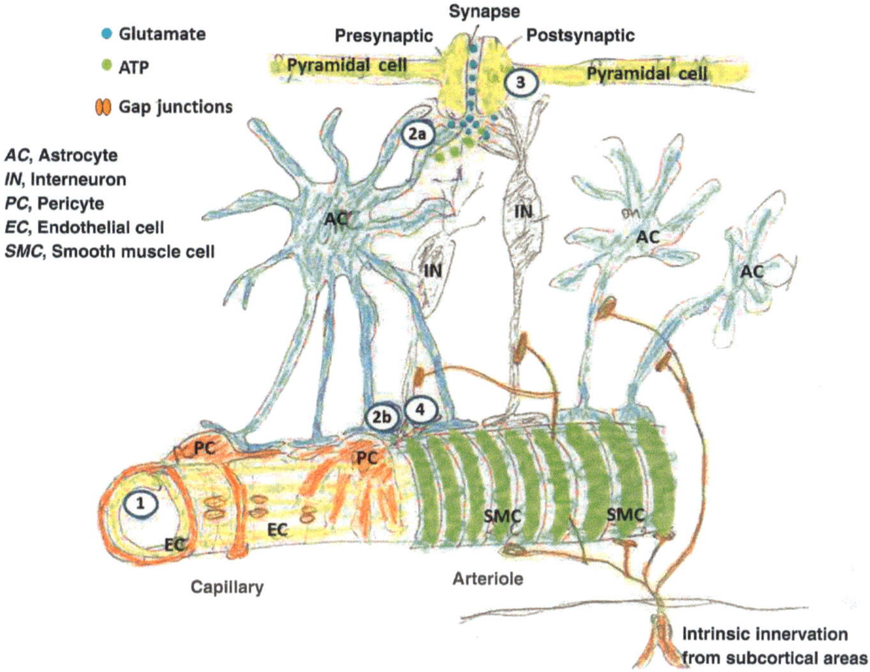

FIGURE 1.1 **The neurovascular unit: (1) At the capillary level, endothelial cells are in close contact to pericytes and astrocytic end-feet, forming the blood-brain barrier.** Local regulation of blood flow involves astrocytic as well as neuronal mechanisms. (2a) During neuronal activation, glutamate and ATP are released at the synapse, activating astrocytes in close proximity to the synapse (details see Section "Astrocytes"), leading to (2b) release of vasoactive factors at the end-feet (details see Section "Astrocytes"). (3) Postsynaptic NMDA receptor activation at pyramidal cells induces the release of prostaglandins with subsequent relaxation of VSMCs (details see Section "Mechanisms of Local Blood Flow Regulation—Neurovascular Coupling"). (4) Local interneurons as well as neurons from subcortical areas projecting to cortical microvessels also contribute to local blood flow regulation (details see Section "Perivascular Neurons—Extrinsic and Intrinsic").

highly complex structure of the blood-brain barrier, the cerebral vasculature also prevents the brain tissue from entering of inflammatory cells and pathogens and controls the passage of macro- and micro-molecules from the blood. The cooperation of the different cell types at the neurovascular unit is highly sophisticated (Fig. 1.1), and destabilization of this interplay by dysfunction of one or more parts finally leads to disturbance of neuronal networks and by this alters brain function under pathophysiological conditions.

References

1. Hoyer S. The young-adult and normally aged brain. Its blood flow and oxidative metabolism. A review—Part I. *Arch Gerontol Geriatr* 1982;1(2):101–16.
2. Mergenthaler P, Lindauer U, Dienel GA, Meisel A. Sugar for the brain: the role of glucose in physiological and pathological brain function. *Trends Neurosci* 2013;36(10):587–97.
3. Kuschinsky W, Wahl M. Local chemical and neurogenic regulation of cerebral vascular resistance. *Physiol Rev* 1978;58(3):656–89.
4. Attwell D, Buchan AM, Charpak S, Lauritzen M, Macvicar BA, Newman EA. Glial and neuronal control of brain blood flow. *Nature* 2010;468(7321):232–43.
5. Faraci FM. Protecting against vascular disease in brain. *Am J Physiol Heart Circ Physiol* 2011;300(5):H1566–1582.
6. Lee RM. Morphology of cerebral arteries. *Pharmacol Ther* 1995;66(1):149–73.
7. Faraci FM, Heistad DD. Regulation of large cerebral arteries and cerebral microvascular pressure. *Circ Res* 1990;66(1):8–17.
8. Gould IG, Tsai P, Kleinfeld D, Linninger A. The capillary bed offers the largest hemodynamic resistance to the cortical blood supply. *J Cereb Blood Flow Metab* 2017;37(1):52–68.
9. Sakadzic S, Mandeville ET, Gagnon L, et al. Large arteriolar component of oxygen delivery implies a safe margin of oxygen supply to cerebral tissue. *Nat Commun* 2014;5:5734.
10. Lecoq J, Parpaleix A, Roussakis E, et al. Simultaneous two-photon imaging of oxygen and blood flow in deep cerebral vessels. *Nat Med* 2011;17(7):893–8.
11. Fernandez-Klett F, Offenhauser N, Dirnagl U, Priller J, Lindauer U. Pericytes in capillaries are contractile in vivo, but arterioles mediate functional hyperemia in the mouse brain. *Proc Natl Acad Sci USA* 2010;107(51):22290–5.
12. Hall CN, Reynell C, Gesslein B, et al. Capillary pericytes regulate cerebral blood flow in health and disease. *Nature* 2014;508(7494):55–60.
13. Peppiatt CM, Howarth C, Mobbs P, Attwell D. Bidirectional control of CNS capillary diameter by pericytes. *Nature* 2006;443(7112):700–4.
14. Villringer A, Them A, Lindauer U, Einhaupl K, Dirnagl U. Capillary perfusion of the rat brain cortex. An in vivo confocal microscopy study. *Circ Res* 1994;75(1):55–62.
15. Kuschinsky W, Paulson OB. Capillary circulation in the brain. *Cerebrovasc Brain Metab Rev* 1992;4(3):261–86.
16. Gutierrez-Jimenez E, Cai C, Mikkelsen IK, et al. Effect of electrical forepaw stimulation on capillary transit-time heterogeneity (CTH). *J Cereb Blood Flow Metab* 2016;36(12):2072–86.
17. Angleys H, Ostergaard L, Jespersen SN. The effects of capillary transit time heterogeneity (CTH) on brain oxygenation. *J Cereb Blood Flow Metab* 2015;35(5):806–17.
18. Jessen NA, Munk AS, Lundgaard I, Nedergaard M. The glymphatic system: a beginner's guide. *Neurochem Res* 2015;40(12):2583–99.
19. Hamel E. Perivascular nerves and the regulation of cerebrovascular tone. *J Appl Physiol (1985)* 2006;100(3):1059–64.
20. Attwell D, Mishra A, Hall CN, O'Farrell FM, Dalkara T. What is a pericyte? *J Cereb Blood Flow Metab* 2016;36(2):451–5.
21. Hill RA, Tong L, Yuan P, Murikinati S, Gupta S, Grutzendler J. Regional blood flow in the normal and ischemic brain is controlled by arteriolar smooth muscle cell contractility and not by capillary pericytes. *Neuron* 2015;87(1):95–110.
22. Vink H, Duling BR. Identification of distinct luminal domains for macromolecules, erythrocytes, and leukocytes within mammalian capillaries. *Circ Res* 1996;79(3):581–9.
23. Vogel J, Sperandio M, Pries AR, Linderkamp O, Gaehtgens P, Kuschinsky W. Influence of the endothelial glycocalyx on cerebral blood flow in mice. *J Cereb Blood Flow Metab* 2000;20(11):1571–8.

24. Dragovich MA, Chester D, Fu BM, et al. Mechanotransduction of the endothelial glycocalyx mediates nitric oxide production through activation of TRP channels. *Am J Physiol Cell Physiol* 2016;311(6):C846–53.
25. Adamson RH, Clough G. Plasma proteins modify the endothelial cell glycocalyx of frog mesenteric microvessels. *J Physiol* 1992;445:473–86.
26. Michel CC, Phillips ME, Turner MR. The effects of native and modified bovine serum albumin on the permeability of frog mesenteric capillaries. *J Physiol* 1985;360:333–46.
27. Turner MR, Clough G, Michel CC. The effects of cationised ferritin and native ferritin upon the filtration coefficient of single frog capillaries. evidence that proteins in the endothelial cell coat influence permeability. *Microvasc Res* 1983;25(2):205–22.
28. Ley K, Gaehtgens P, Fennie C, Singer MS, Lasky LA, Rosen SD. Lectin-like cell adhesion molecule 1 mediates leukocyte rolling in mesenteric venules in vivo. *Blood* 1991;77(12):2553–5.
29. Benedict CR, Pakala R, Willerson JT. Endothelial-dependent procoagulant and anticoagulant mechanisms. Recent advances in understanding. *Tex Heart Inst J* 1994;21(1):86–90.
30. Lijnen HR, Collen D. Endothelium in hemostasis and thrombosis. *Prog Cardiovasc Dis* 1997;39(4):343–50.
31. Pries AR, Kuebler WM. Normal endothelium. *Handb Exp Pharmacol* 2006;176 Pt 1:1–40.
32. Pries AR, Secomb TW, Gaehtgens P. The endothelial surface layer. *Pflugers Arch* 2000;440(5):653–66.
33. Erdo F, Denes L, de Lange E. Age-associated physiological and pathological changes at the blood-brain barrier: A review. *J Cereb Blood Flow Metab* 2016.
34. Hawkins BT, Davis TP. The blood–brain barrier/neurovascular unit in health and disease. *Pharmacol Rev* 2005;57(2):173–85.
35. Zhao Z, Nelson AR, Betsholtz C, Zlokovic BV. Establishment and dysfunction of the blood–brain barrier. *Cell* 2015;163(5):1064–78.
36. Tilling T, Korte D, Hoheisel D, Galla HJ. Basement membrane proteins influence brain capillary endothelial barrier function in vitro. *J Neurochem* 1998;71(3):1151–7.
37. Kacem K, Lacombe P, Seylaz J, Bonvento G. Structural organization of the perivascular astrocyte endfeet and their relationship with the endothelial glucose transporter: a confocal microscopy study. *Glia* 1998;23(1):1–10.
38. Armulik A, Genove G, Mae M, et al. Pericytes regulate the blood–brain barrier. *Nature* 2010;468(7323):557–61.
39. Ghosh M, Balbi M, Hellal F, Dichgans M, Lindauer U, Plesnila N. Pericytes are involved in the pathogenesis of cerebral autosomal dominant arteriopathy with subcortical infarcts and leukoencephalopathy. *Ann Neurol* 2015;78(6):887–900.
40. Soon AS, Chua JW, Becker DL. Connexins in endothelial barrier function-novel therapeutic targets countering vascular hyperpermeability. *Thromb Haemost* 2016;116(5):852–67.
41. Derangeon M, Spray DC, Bourmeyster N, Sarrouilhe D, Herve JC. Reciprocal influence of connexins and apical junction proteins on their expressions and functions. *Biochim Biophys Acta* 2009;1788(4):768–78.
42. Laird DW. Life cycle of connexins in health and disease. *Biochem J* 2006;394(Pt 3):527–43.
43. Furchgott RF, Zawadzki JV. The obligatory role of endothelial cells in the relaxation of arterial smooth muscle by acetylcholine. *Nature* 1980;288(5789):373–6.
44. Arnold WP, Mittal CK, Katsuki S, Murad F. Nitric oxide activates guanylate cyclase and increases guanosine 3′:5′-cyclic monophosphate levels in various tissue preparations. *Proc Natl Acad Sci USA* 1977;74(8):3203–7.
45. Ignarro LJ, Buga GM, Wood KS, Byrns RE, Chaudhuri G. Endothelium-derived relaxing factor produced and released from artery and vein is nitric oxide. *Proc Natl Acad Sci USA* 1987;84(24):9265–9.
46. Palmer RM, Ferrige AG, Moncada S. Nitric oxide release accounts for the biological activity of endothelium-derived relaxing factor. *Nature* 1987;327(6122):524–6.

47. Katusic ZS. Endothelial L-arginine pathway and regional cerebral arterial reactivity to vasopressin. *Am J Physiol* 1992;262(5 Pt 2):H1557–1562.
48. Onoue H, Kaito N, Tomii M, Tokudome S, Nakajima M, Abe T. Human basilar and middle cerebral arteries exhibit endothelium-dependent responses to peptides. *Am J Physiol* 1994;267(3 Pt 2):H880–886.
49. Rosenblum WI, Shimizu T, Nelson GH. Endothelium-dependent effects of substance P and calcitonin gene-related peptide on mouse pial arterioles. *Stroke* 1993;24(7):1043–7 [discussion 1047-1048].
50. De Caterina R, Libby P, Peng HB, et al. Nitric oxide decreases cytokine-induced endothelial activation. Nitric oxide selectively reduces endothelial expression of adhesion molecules and proinflammatory cytokines. *J Clin Invest* 1995;96(1):60–8.
51. Kubes P, Suzuki M, Granger DN. Nitric oxide: an endogenous modulator of leukocyte adhesion. *Proc Natl Acad Sci USA* 1991;88(11):4651–5.
52. Rosenblum WI, Nishimura H, Ellis EF, Nelson GH. The endothelium-dependent effects of thimerosal on mouse pial arterioles in vivo: evidence for control of microvascular events by EDRF as well as prostaglandins. *J Cereb Blood Flow Metab* 1992;12(4): 703–6.
53. Faraci FM, Heistad DD. Regulation of the cerebral circulation: role of endothelium and potassium channels. *Physiol Rev* 1998;78(1):53–97.
54. Faraci FM. Endothelium-derived vasoactive factors and regulation of the cerebral circulation. *Neurosurgery* 1993;33(4):648–58 [discussion 658-649].
55. Smith WL, Garavito RM, DeWitt DL. Prostaglandin endoperoxide H synthases (cyclooxygenases)-1 and -2. *J Biol Chem* 1996;271(52):33157–60.
56. Davidge ST. Prostaglandin H synthase and vascular function. *Circ Res* 2001;89(8):650–60.
57. Ellis EF, Wei EP, Kontos HA. Vasodilation of cat cerebral arterioles by prostaglandins D2, E2, G2, and I2. *Am J Physiol* 1979;237(3):H381–385.
58. Armstead WM. Role of nitric oxide and cAMP in prostaglandin-induced pial arterial vasodilation. *Am J Physiol* 1995;268(4 Pt 2):H1436–1440.
59. Bryan Jr. RM, You J, Golding EM, Marrelli SP. Endothelium-derived hyperpolarizing factor: a cousin to nitric oxide and prostacyclin. *Anesthesiology* 2005;102(6):1261–77.
60. Katusic ZS, Shepherd JT. Endothelium-derived vasoactive factors: II. Endothelium-dependent contraction. *Hypertension* 1991;18(5 Suppl):III86–92.
61. Earley S, Heppner TJ, Nelson MT, Brayden JE. TRPV4 forms a novel Ca2+ signaling complex with ryanodine receptors and BKCa channels. *Circ Res* 2005;97(12):1270–9.
62. Iliff JJ, Jia J, Nelson J, Goyagi T, Klaus J, Alkayed NJ. Epoxyeicosanoid signaling in CNS function and disease. *Prostaglandins Other Lipid Mediat* 2010;91(3-4):68–84.
63. Medhora M, Narayanan J, Harder D. Dual regulation of the cerebral microvasculature by epoxyeicosatrienoic acids. *Trends Cardiovasc Med* 2001;11(1):38–42.
64. You J, Johnson TD, Marrelli SP, Bryan Jr. RM. Functional heterogeneity of endothelial P2 purinoceptors in the cerebrovascular tree of the rat. *Am J Physiol* 1999;277(3 Pt 2):H893–900.
65. Hannah RM, Dunn KM, Bonev AD, Nelson MT. Endothelial SK(Ca) and IK(Ca) channels regulate brain parenchymal arteriolar diameter and cortical cerebral blood flow. *J Cereb Blood Flow Metab* 2011;31(5):1175–86.
66. Bryan Jr. RM, Marrelli SP. Endothelium-dependent hyperpolarization: out of the dish and into the brain. *J Cereb Blood Flow Metab* 2011;31(5):1173–4.
67. Kohler R, Olivan-Viguera A, Wulff H. Endothelial small- and intermediate-conductance K channels and endothelium-dependent hyperpolarization as drug targets in cardiovascular disease. *Adv Pharmacol* 2016;77:65–104.
68. Cipolla MJ, Smith J, Kohlmeyer MM, Godfrey JA. SKCa and IKCa channels, myogenic tone, and vasodilator responses in middle cerebral arteries and parenchymal arterioles: effect of ischemia and reperfusion. *Stroke* 2009;40(4):1451–7.

69. Chen YJ, Wallace BK, Yuen N, Jenkins DP, Wulff H, O'Donnell ME. Blood–brain barrier KCa3.1 channels: evidence for a role in brain Na uptake and edema in ischemic stroke. *Stroke* 2015;46(1):237–44.
70. Maines MD. Heme oxygenase: function, multiplicity, regulatory mechanisms, and clinical applications. *FASEB J* 1988;2(10):2557–68.
71. Trakshel GM, Kutty RK, Maines MD. Resolution of the rat brain heme oxygenase activity: absence of a detectable amount of the inducible form (HO-1). *Arch Biochem Biophys* 1988;260(2):732–9.
72. Kimura H. Hydrogen sulfide: its production, release and functions. *Amino Acids* 2011;41(1):113–21.
73. Xi Q, Tcheranova D, Parfenova H, Horowitz B, Leffler CW, Jaggar JH. Carbon monoxide activates KCa channels in newborn arteriole smooth muscle cells by increasing apparent Ca2+ sensitivity of alpha-subunits. *Am J Physiol Heart Circ Physiol* 2004;286(2):H610–618.
74. Leffler CW, Parfenova H, Jaggar JH. Carbon monoxide as an endogenous vascular modulator. *Am J Physiol Heart Circ Physiol* 2011;301(1):H1–H11.
75. Leffler CW, Nasjletti A, Yu C, Johnson RA, Fedinec AL, Walker N. Carbon monoxide and cerebral microvascular tone in newborn pigs. *Am J Physiol* 1999;276(5 Pt 2):H1641–1646.
76. Barkoudah E, Jaggar JH, Leffler CW. The permissive role of endothelial NO in CO-induced cerebrovascular dilation. *Am J Physiol Heart Circ Physiol* 2004;287(4): H1459–1465.
77. Ishikawa M, Kajimura M, Adachi T, et al. Carbon monoxide from heme oxygenase-2 Is a tonic regulator against NO-dependent vasodilatation in the adult rat cerebral microcirculation. *Circ Res* 2005;97(12):e104–114.
78. Mustafa AK, Sikka G, Gazi SK, et al. Hydrogen sulfide as endothelium-derived hyperpolarizing factor sulfhydrates potassium channels. *Circ Res* 2011;109(11):1259–68.
79. Beltowski J, Jamroz-Wisniewska A. Hydrogen sulfide and endothelium-dependent vasorelaxation. *Molecules* 2014;19(12):21183–99.
80. Liang GH, Adebiyi A, Leo MD, McNally EM, Leffler CW, Jaggar JH. Hydrogen sulfide dilates cerebral arterioles by activating smooth muscle cell plasma membrane KATP channels. *Am J Physiol Heart Circ Physiol* 2011;300(6):H2088–2095.
81. Han J, Chen ZW, He GW. Acetylcholine- and sodium hydrosulfide-induced endothelium-dependent relaxation and hyperpolarization in cerebral vessels of global cerebral ischemia-reperfusion rat. *J Pharmacol Sci* 2013;121(4):318–26.
82. Rubanyi GM, Polokoff MA. Endothelins: molecular biology, biochemistry, pharmacology, physiology, and pathophysiology. *Pharmacol Rev* 1994;46(3):325–415.
83. Faraci FM. Effects of endothelin and vasopressin on cerebral blood vessels. *Am J Physiol* 1989;257(3 Pt 2):H799–803.
84. Gray GA, Webb DJ. The endothelin system and its potential as a therapeutic target in cardiovascular disease. *Pharmacol Ther* 1996;72(2):109–48.
85. Ando J, Yamamoto K. Vascular mechanobiology: endothelial cell responses to fluid shear stress. *Circ J.* 2009;73(11):1983–92.
86. Wilson C, Lee MD, McCarron JG. Acetylcholine released by endothelial cells facilitates flow-mediated dilatation. *J Physiol* 2016;594(24):7267–307.
87. Kim HR, Appel S, Vetterkind S, Gangopadhyay SS, Morgan KG. Smooth muscle signalling pathways in health and disease. *J Cell Mol Med* 2008;12(6A):2165–80.
88. Takashima S. Phosphorylation of myosin regulatory light chain by myosin light chain kinase, and muscle contraction. *Circ J* 2009;73(2):208–13.
89. Chrissobolis S, Sobey CG. Evidence that Rho-kinase activity contributes to cerebral vascular tone in vivo and is enhanced during chronic hypertension: comparison with protein kinase C. *Circ Res* 2001;88(8):774–9.

90. Somlyo AP, Somlyo AV. Ca2+ sensitivity of smooth muscle and nonmuscle myosin II: modulated by G proteins, kinases, and myosin phosphatase. *Physiol Rev* 2003;83(4):1325–58.
91. Li Y, Brayden JE. Rho kinase activity governs arteriolar myogenic depolarization. *J Cereb Blood Flow Metab* 2017;37:140–52.
92. Earley S, Brayden JE. Transient receptor potential channels in the vasculature. *Physiol Rev* 2015;95(2):645–90.
93. Abd El-Rahman RR, Harraz OF, Brett SE, et al. Identification of L- and T-type Ca2+ channels in rat cerebral arteries: role in myogenic tone development. *Am J Physiol Heart Circ Physiol* 2013;304(1):H58–71.
94. Nelson MT, Cheng H, Rubart M, et al. Relaxation of arterial smooth muscle by calcium sparks. *Science* 1995;270(5236):633–7.
95. Hashad AM, Mazumdar N, Romero M, et al. Interplay among distinct Ca2+ conductances drives Ca2+ sparks/spontaneous transient outward currents in rat cerebral arteries. *J Physiol* 2017;595(4):1111–1126.
96. Jaggar JH, Wellman GC, Heppner TJ, et al. Ca2+ channels, ryanodine receptors and Ca(2+)-activated K+ channels: a functional unit for regulating arterial tone. *Acta Physiol Scand* 1998;164(4):577–87.
97. Perez GJ, Bonev AD, Nelson MT. Micromolar Ca(2+) from sparks activates Ca(2+)-sensitive K(+) channels in rat cerebral artery smooth muscle. *Am J Physiol Cell Physiol* 2001;281(6):C1769–1775.
98. Perez GJ, Bonev AD, Patlak JB, Nelson MT. Functional coupling of ryanodine receptors to KCa channels in smooth muscle cells from rat cerebral arteries. *J Gen Physiol* 1999;113(2):229–38.
99. Bryan Jr. RM, You J, Phillips SC, et al. Evidence for two-pore domain potassium channels in rat cerebral arteries. *Am J Physiol Heart Circ Physiol* 2006;291(2):H770–780.
100. Lin LH, Jin J, Nashelsky MB, Talman WT. Acid-sensing ion channel 1 and nitric oxide synthase are in adjacent layers in the wall of rat and human cerebral arteries. *J Chem Neuroanat* 2014;61-62:161–8.
101. Grifoni SC, Jernigan NL, Hamilton G, Drummond HA. ASIC proteins regulate smooth muscle cell migration. *Microvasc Res* 2008;75(2):202–10.
102. Gannon KP, Vanlandingham LG, Jernigan NL, Grifoni SC, Hamilton G, Drummond HA. Impaired pressure-induced constriction in mouse middle cerebral arteries of ASIC2 knockout mice. *Am J Physiol Heart Circ Physiol* 2008;294(4):H1793–1803.
103. Berridge MJ. Inositol trisphosphate and calcium signalling mechanisms. *Biochim Biophys Acta* 2009;1793(6):933–40.
104. Zhao G, Adebiyi A, Blaskova E, Xi Q, Jaggar JH. Type 1 inositol 1,4,5-trisphosphate receptors mediate UTP-induced cation currents, Ca2+ signals, and vasoconstriction in cerebral arteries. *Am J Physiol Cell Physiol* 2008;295(5):C1376–1384.
105. Levitsky DO, Clergue M, Lambert F, et al. Sarcoplasmic reticulum calcium transport and Ca(2+)-ATPase gene expression in thoracic and abdominal aortas of normotensive and spontaneously hypertensive rats. *J Biol Chem* 1993;268(11):8325–31.
106. Flemming R, Cheong A, Dedman AM, Beech DJ. Discrete store-operated calcium influx into an intracellular compartment in rabbit arteriolar smooth muscle. *J Physiol* 2002;543(Pt 2):455–64.
107. Munzel T, Feil R, Mulsch A, Lohmann SM, Hofmann F, Walter U. Physiology and pathophysiology of vascular signaling controlled by guanosine 3′,5′-cyclic monophosphate-dependent protein kinase [corrected]. *Circulation* 2003;108(18):2172–83.
108. Sun CW, Alonso-Galicia M, Taheri MR, Falck JR, Harder DR, Roman RJ. Nitric oxide-20-hydroxyeicosatetraenoic acid interaction in the regulation of K+ channel activity and vascular tone in renal arterioles. *Circ Res* 1998;83(11):1069–79.
109. Miyata N, Roman RJ. Role of 20-hydroxyeicosatetraenoic acid (20-HETE) in vascular system. *J Smooth Muscle Res* 2005;41(4):175–93.

110. Sharma N, Bhattarai JP, Hwang PH, Han SK. Nitric oxide suppresses L-type calcium currents in basilar artery smooth muscle cells in rabbits. *Neurol Res* 2013;35(4):424–8.
111. Harraz OF, Brett SE, Welsh DG. Nitric oxide suppresses vascular voltage-gated T-type Ca2+ channels through cGMP/PKG signaling. *Am J Physiol Heart Circ Physiol* 2014;306(2):H279–285.
112. Cohen RA, Weisbrod RM, Gericke M, Yaghoubi M, Bierl C, Bolotina VM. Mechanism of nitric oxide-induced vasodilatation: refilling of intracellular stores by sarcoplasmic reticulum Ca2+ ATPase and inhibition of store-operated Ca2+ influx. *Circ Res* 1999;84(2):210–9.
113. Longden TA, Hill-Eubanks DC, Nelson MT. Ion channel networks in the control of cerebral blood flow. *J Cereb Blood Flow Metab* 2016;36(3):492–512.
114. Hartmann DA, Underly RG, Grant RI, Watson AN, Lindner V, Shih AY. Pericyte structure and distribution in the cerebral cortex revealed by high-resolution imaging of transgenic mice. *Neurophotonics* 2015;2(4):041402.
115. Fujimoto K. Pericyte-endothelial gap junctions in developing rat cerebral capillaries: a fine structural study. *Anat Rec* 1995;242(4):562–5.
116. Diaz-Flores L, Gutierrez R, Varela H, Rancel N, Valladares F. Microvascular pericytes: a review of their morphological and functional characteristics. *Histol Histopathol.* 1991;6(2):269–86.
117. Sweeney MD, Ayyadurai S, Zlokovic BV. Pericytes of the neurovascular unit: key functions and signaling pathways. *Nat Neurosci* 2016;19(6):771–83.
118. Jansson D, Rustenhoven J, Feng S, et al. A role for human brain pericytes in neuroinflammation. *J Neuroinflammation* 2014;11:104.
119. Kovac A, Erickson MA, Banks WA. Brain microvascular pericytes are immunoactive in culture: cytokine, chemokine, nitric oxide, and LRP-1 expression in response to lipopolysaccharide. *J Neuroinflammation* 2011;8:139.
120. Fernandez-Klett F, Priller J. Diverse functions of pericytes in cerebral blood flow regulation and ischemia. *J Cereb Blood Flow Metab* 2015;35(6):883–7.
121. Rucker HK, Wynder HJ, Thomas WE. Cellular mechanisms of CNS pericytes. *Brain Res Bull* 2000;51(5):363–9.
122. Chaigneau E, Oheim M, Audinat E, Charpak S. Two-photon imaging of capillary blood flow in olfactory bulb glomeruli. *Proc Natl Acad Sci USA* 2003;100(22):13081–6.
123. Iliff JJ, Wang M, Liao Y, et al. A paravascular pathway facilitates CSF flow through the brain parenchyma and the clearance of interstitial solutes, including amyloid beta. *Sci Transl Med* 2012;4(147):147ra111.
124. Steinhauser C, Grunnet M, Carmignoto G. Crucial role of astrocytes in temporal lobe epilepsy. *Neuroscience* 2016;323:157–69.
125. Allen NJ. Astrocyte regulation of synaptic behavior. *Annu Rev Cell Dev Biol* 2014;30:439–63.
126. Gandhi GK, Cruz NF, Ball KK, Dienel GA. Astrocytes are poised for lactate trafficking and release from activated brain and for supply of glucose to neurons. *J Neurochem* 2009;111(2):522–36.
127. Stokum JA, Kurland DB, Gerzanich V, Simard JM. Mechanisms of astrocyte-mediated cerebral edema. *Neurochem Res* 2015;40(2):317–28.
128. Filosa JA, Morrison HW, Iddings JA, Du W, Kim KJ. Beyond neurovascular coupling, role of astrocytes in the regulation of vascular tone. *Neuroscience* 2016;323:96–109.
129. Howarth C. The contribution of astrocytes to the regulation of cerebral blood flow. *Front Neurosci* 2014;8:103.
130. Mulligan SJ, MacVicar BA. Calcium transients in astrocyte endfeet cause cerebrovascular constrictions. *Nature* 2004;431(7005):195–9.
131. Petzold GC, Murthy VN. Role of astrocytes in neurovascular coupling. *Neuron* 2011;71(5):782–97.

132. Takano T, Tian GF, Peng W, et al. Astrocyte-mediated control of cerebral blood flow. *Nat Neurosci* 2006;9(2):260–7.
133. Zonta M, Angulo MC, Gobbo S, et al. Neuron-to-astrocyte signaling is central to the dynamic control of brain microcirculation. *Nat Neurosci* 2003;6(1):43–50.
134. Wang X, Lou N, Xu Q, et al. Astrocytic Ca2+ signaling evoked by sensory stimulation in vivo. *Nat Neurosci* 2006;9(6):816–23.
135. Nizar K, Uhlirova H, Tian P, et al. In vivo stimulus-induced vasodilation occurs without IP3 receptor activation and may precede astrocytic calcium increase. *J Neurosci* 2013;33(19):8411–22.
136. Srinivasan R, Huang BS, Venugopal S, et al. Ca(2+) signaling in astrocytes from Ip3r2(-/-) mice in brain slices and during startle responses in vivo. *Nat Neurosci* 2015;18(5):708–17.
137. Simard M, Arcuino G, Takano T, Liu QS, Nedergaard M. Signaling at the gliovascular interface. *J Neurosci* 2003;23(27):9254–62.
138. Carrasquero LM, Delicado EG, Bustillo D, Gutierrez-Martin Y, Artalejo AR, Miras-Portugal MT. P2X7 and P2Y13 purinergic receptors mediate intracellular calcium responses to BzATP in rat cerebellar astrocytes. *J Neurochem* 2009;110(3):879–89.
139. Mishra A, Reynolds JP, Chen Y, Gourine AV, Rusakov DA, Attwell D. Astrocytes mediate neurovascular signaling to capillary pericytes but not to arterioles. *Nat Neurosci* 2016;19(12):1619–27.
140. Campbell WB, Gebremedhin D, Pratt PF, Harder DR. Identification of epoxyeicosatrienoic acids as endothelium-derived hyperpolarizing factors. *Circ Res* 1996;78(3):415–23.
141. Higashimori H, Blanco VM, Tuniki VR, Falck JR, Filosa JA. Role of epoxyeicosatrienoic acids as autocrine metabolites in glutamate-mediated K+ signaling in perivascular astrocytes. *Am J Physiol Cell Physiol* 2010;299(5):C1068–1078.
142. Dunn KM, Hill-Eubanks DC, Liedtke WB, Nelson MT. TRPV4 channels stimulate Ca2+ -induced Ca2+ release in astrocytic endfeet and amplify neurovascular coupling responses. *Proc Natl Acad Sci USA* 2013;110(15):6157–62.
143. Bayliss WM. On the local reactions of the arterial wall to changes of internal pressure. *J Physiol* 1902;28(3):220–31.
144. Davis MJ, Hill MA. Signaling mechanisms underlying the vascular myogenic response. *Physiol Rev* 1999;79(2):387–423.
145. Walsh MP, Cole WC. The role of actin filament dynamics in the myogenic response of cerebral resistance arteries. *J Cereb Blood Flow Metab* 2013;33(1):1–12.
146. Lindauer U, Kunz A, Schuh-Hofer S, Vogt J, Dreier JP, Dirnagl U. Nitric oxide from perivascular nerves modulates cerebral arterial pH reactivity. *Am J Physiol Heart Circ Physiol* 2001;281(3):H1353–1363.
147. Iadecola C, Zhang F. Permissive and obligatory roles of NO in cerebrovascular responses to hypercapnia and acetylcholine. *Am J Physiol* 1996;271(4 Pt 2):R990–1001.
148. Lindauer U, Vogt J, Schuh-Hofer S, Dreier JP, Dirnagl U. Cerebrovascular vasodilation to extraluminal acidosis occurs via combined activation of ATP-sensitive and Ca2+ -activated potassium channels. *J Cereb Blood Flow Metab* 2003;23(10):1227–38.
149. Dabertrand F, Nelson MT, Brayden JE. Acidosis dilates brain parenchymal arterioles by conversion of calcium waves to sparks to activate BK channels. *Circ Res* 2012;110(2):285–94.
150. Johnston AJ, Steiner LA, Gupta AK, Menon DK. Cerebral oxygen vasoreactivity and cerebral tissue oxygen reactivity. *Br J Anaesth* 2003;90(6):774–86.
151. Gupta AK, Menon DK, Czosnyka M, Smielewski P, Jones JG. Thresholds for hypoxic cerebral vasodilation in volunteers. *Anesth Analg* 1997;85(4):817–20.
152. Tomiyama Y, Brian Jr. JE, Todd MM. Cerebral blood flow during hemodilution and hypoxia in rats: role of ATP-sensitive potassium channels. *Stroke* 1999;30(9):1942–7 [discussion 1947–1948].

153. Schmetterer L, Findl O, Strenn K, et al. Role of NO in the O2 and CO2 responsiveness of cerebral and ocular circulation in humans. *Am J Physiol* 1997;273(6 Pt 2):R2005–2012.
154. Messina EJ, Sun D, Koller A, Wolin MS, Kaley G. Increases in oxygen tension evoke arteriolar constriction by inhibiting endothelial prostaglandin synthesis. *Microvasc Res* 1994;48(2):151–60.
155. Dietrich HH, Kajita Y, Dacey Jr. RG. Local and conducted vasomotor responses in isolated rat cerebral arterioles. *Am J Physiol* 1996;271(3 Pt 2):H1109–1116.
156. Akgoren N, Dalgaard P, Lauritzen M. Cerebral blood flow increases evoked by electrical stimulation of rat cerebellar cortex: relation to excitatory synaptic activity and nitric oxide synthesis. *Brain Res* 1996;710(1–2):204–14.
157. Yang G, Zhang Y, Ross ME, Iadecola C. Attenuation of activity-induced increases in cerebellar blood flow in mice lacking neuronal nitric oxide synthase. *Am J Physiol Heart Circ Physiol* 2003;285(1):H298–304.
158. Lindauer U, Megow D, Matsuda H, Dirnagl U. Nitric oxide: a modulator, but not a mediator, of neurovascular coupling in rat somatosensory cortex. *Am J Physiol* 1999;277(2 Pt 2):H799–811.
159. Lacroix A, Toussay X, Anenberg E, et al. COX-2-derived prostaglandin E2 produced by pyramidal neurons contributes to neurovascular coupling in the rodent cerebral cortex. *J Neurosci* 2015;35(34):11791–810.
160. Niwa K, Araki E, Morham SG, Ross ME, Iadecola C. Cyclooxygenase-2 contributes to functional hyperemia in whisker-barrel cortex. *J Neurosci* 2000;20(2):763–70.
161. Stefanovic B, Bosetti F, Silva AC. Modulatory role of cyclooxygenase-2 in cerebrovascular coupling. *Neuroimage* 2006;32(1):23–32.
162. Peng X, Carhuapoma JR, Bhardwaj A, et al. Suppression of cortical functional hyperemia to vibrissal stimulation in the rat by epoxygenase inhibitors. *Am J Physiol Heart Circ Physiol* 2002;283(5):H2029–2037.
163. Leithner C, Royl G, Offenhauser N, et al. Pharmacological uncoupling of activation induced increases in CBF and CMRO2. *J Cereb Blood Flow Metab* 2010;30(2):311–22.
164. Filosa JA, Bonev AD, Straub SV, et al. Local potassium signaling couples neuronal activity to vasodilation in the brain. *Nat Neurosci* 2006;9(11):1397–403.
165. Gordon GR, Choi HB, Rungta RL, Ellis-Davies GC, MacVicar BA. Brain metabolism dictates the polarity of astrocyte control over arterioles. *Nature* 2008;456(7223):745–9.
166. Shi Y, Liu X, Gebremedhin D, Falck JR, Harder DR, Koehler RC. Interaction of mechanisms involving epoxyeicosatrienoic acids, adenosine receptors, and metabotropic glutamate receptors in neurovascular coupling in rat whisker barrel cortex. *J Cereb Blood Flow Metab* 2008;28(1):111–25.
167. Dirnagl U, Niwa K, Lindauer U, Villringer A. Coupling of cerebral blood flow to neuronal activation: role of adenosine and nitric oxide. *Am J Physiol* 1994;267 (1 Pt 2):H296–301.

CHAPTER

2

Blood–Brain Interfaces Organization in Relation to Inorganic Ion Transport, CSF Secretion, and Circulation

Jean-François Ghersi-Egea[1] *and Helle Damkier*[2,3]

[1]Lyon 1 University, Lyon, France [2]University of Copenhagen, Copenhagen, Denmark [3]Aarhus University, Aarhus, Denmark

OUTLINE

Brain Edema.
DOI: http://dx.doi.org/10.1016/B978-0-12-803196-4.00002-3

INTRODUCTION

Exchanges between the blood and the brain are regulated by several cellular interfaces. These interfaces maintain the cerebral inorganic ion homeostasis necessary for neuronal activity. They also protect the brain from potentially harmful compounds, while ensuring continuous supply of nutrients, micronutrients, and hormones necessary for normal brain development and functions. Blood–brain interface investigations are often restricted to the blood–brain barrier (BBB) located at the endothelium of the cerebral and spinal microvessels. The brain, however, is an anatomically complex organ characterized by interconnected but individualized structures, and by the existence of an internal (ventricular) and external (subarachnoidal and cisternal) fluid circulation. As a correlate, cellular interfaces other than the BBB participate in the regulation of the molecular and cellular exchanges between the blood and the central nervous system. These include in particular the choroid plexuses which form the barrier between the blood and the ventricular cerebrospinal fluid (CSF) that they secrete (the blood-cerebrospinal fluid barrier, BCSFB), the endothelium of meningeal and brain penetrating arteries and veins which separates the blood from the subarachnoid and perivascular CSF, the arachnoid membrane which separates the subarachnoid CSF from the venous blood, and the circumventricular organs that specialize in sensing blood signals for central regulation of physiological functions. This review focuses on the BBB and BCSFB which are the main gates of entry into the brain, and play a major role in the regulation of cerebral fluid flow and ion homeostasis.

PART I: BLOOD–BRAIN BARRIERS AND FLUID FLOW IN THE BRAIN

Blood–Brain Barrier and Blood–Cerebrospinal Fluid Barrier Functions and Specificities

The histological organization and barrier properties of the blood–brain interfaces have been described extensively in several recent reviews.[1,2] As for the BBB, pericytes are embedded in the basal membrane of the capillary endothelium that forms the structural barrier, and glial endfeet abut most of the microvessels, the complete structure being referred to as the functional neurovascular unit (Fig. 2.1). The pericyte clearly plays a role in barrier differentiation during early brain development,[4] while its functions as contractile cell regulating blood flow remains a matter of debate, especially at the level of the cerebral capillary. The astrocytic covering has also been associated with the maintenance of the barrier phenotype. As astrocytes develop during the postnatal period, while the

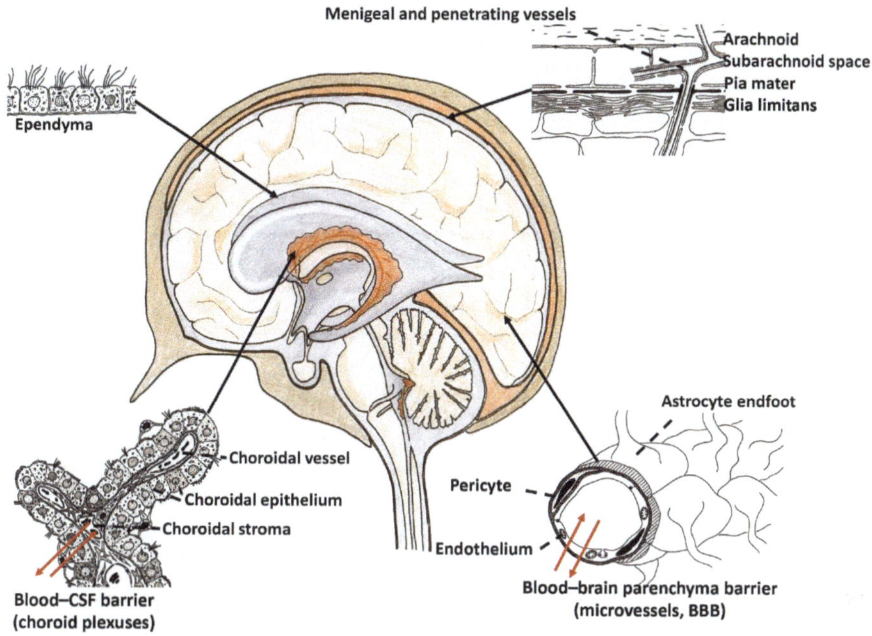

FIGURE 2.1 **Schematic representation of the human brain showing the CSF circulation and the different interfaces between the blood, the brain parenchyma, and the CSF.** The different CSF compartments are represented in *blue. Source: Modified from Strazielle N, Ghersi-Egea JF. Factors affecting delivery of antiviral drugs to the brain. Rev Med Virol 2005;***15***: 105–33 (Ref. 3).*

BBB is tight much earlier during development, the exact function of these cells remains to be ascertained with precision.

The four choroid plexuses, located in the two lateral ventricles, the third ventricle and the fourth ventricle, display a different gross morphology, but share the same histological organization. They are formed by numerous villi organized as a tight epithelium originating from the ependyma and covering a highly vascularized stroma that also harbors resident dentritic cells and macrophages. The end-loops of the choroidal vessels in close contact with the epithelium are fenestrated and leaky, and the barrier function is shifted to the epithelium (Fig. 2.1). The choroid plexuses appear very early during development. The factors inducing choroid plexus growth and the appearance of the tight phenotype of the epithelium are not clearly defined yet. They may involve signaling molecules such as bone morphogenic proteins and sonic hedgehog, as well as different hormones acting by an autocrine mechanism (reviewed in Strazielle and Ghersi-Egea[2]).

The BBB develops a very large surface area of exchange, estimated to be around 100 cm^2/g in mammals.[5] The BCSFB also develops a large surface area of exchange, owing to the basolateral membrane interdigitations and to the apical brush border that characterize the epithelium. The apical surface area has been estimated to be 75 cm^2 in a one month old rat.[6] In human, this surface may represent 10% of the BBB surface. While the exchanges between blood and brain can occur across large surface-areas, they are highly restricted for compounds that are not transported by specific mechanisms. The basis for this "barrier" function is the presence of tight junctions that seal the barrier cells together. The main proteins forming the intercellular core of the tight junctions include Jam proteins, tricellulin (in epithelia), occludin, and various claudins. The molecular organization of the tight junctions, however, differs between the endothelial and the epithelial barriers, and this translates into functional differences. The transendothelial electric resistance is between 1000 and 5000 $\Omega \times cm^2$ at the BBB, while the transepithelial electric resistance is in the range of 100–200 $\Omega \times cm^2$ at the BCSFB. The paracellular diffusion is virtually null at the BBB, while at the BCSFB restricted but significant paracellular diffusion may occur, at least for some inorganic anions and possibly small polar compounds. These differences are in part accounted for by the differential expression of claudin proteins between the two barriers. Claudin 5, a barrier-tightening protein is the main claudin expressed at the BBB. Claudin 3, 4, 12, 16, and possibly 1 (specifically in human) are also present. At the choroid plexuses, the main claudins expressed are Claudin 1, 2, and 3, both in rodent and human. Claudin 9, 19, and possibly 22 and 11 are also part of the choroidal junctions.[7–9] The properties conferred to an intercellular junction by claudin proteins depend on the identity of all claudins expressed at the junction, and of their ability to form cis and trans homo- and heterodimeric interactions. Claudin 2 is of special relevance to the difference in barrier properties observed between brain interfaces. It is a pore-forming protein, responsible for the low transelectric resistance of tight epithelia in which it is expressed, and allows the paracellular passage of selected inorganic cations and water.[10,11] Within the brain, its expression is highly specific of the choroid plexuses, and claudin 2 may allow the paracellular flux of K^+, and as such be connected to the mechanisms of CSF secretion (see infra).

At both barriers, the cerebral entry of polar organic compounds is strongly impeded by the tight junctions, unless they are transported by specific influx transporters. This forms the prime basis of the barrier phenotype associated with these cell layers. In addition, many liposoluble molecules, both endogenous compounds and potentially toxic xenobiotics including drugs and environmental pollutants, are prevented to diffuse into the brain across the barrier cell membranes by multispecific efflux transporters, or are metabolized by some detoxifying enzymes. These different mechanisms (Fig. 2.2) all contribute to brain protection. The

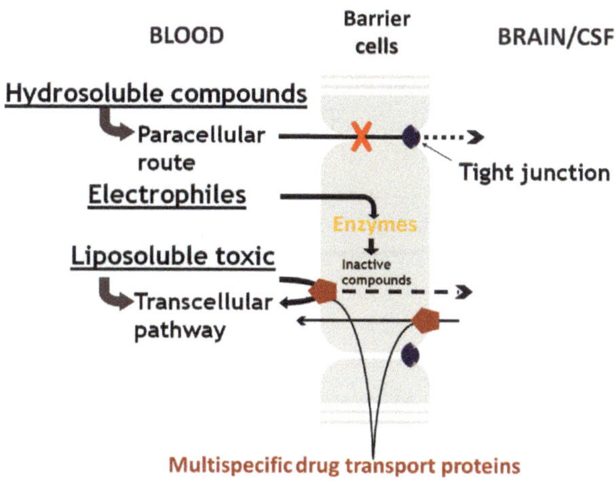

FIGURE 2.2 Mechanisms of brain protection at blood–brain interfaces.

differential expression of the transporters and enzymes between the BBB and BCSFB explains the complementary role played by these barriers in protecting the brain and its fluid environment (see infra).

The barriers also differ from one another with regard to the other functions they fulfill. The BBB is largely involved in supplying the brain with oxygen and nutrients such as glucose, amino acids, lipids, and in regulating mono and multivalent inorganic ion homeostasis (see infra). The BCSFB secretes most of the CSF (see infra), and is also a site for supplying or regulating the levels of hormone and various neuroactive polypeptides in the brain.[12] For instance the choroid plexuses are the main cerebral source of transthyretin, a thyroid hormone carrier, which regulates the intracerebral concentration and activity of these hormones.[13] The choroid plexuses also transport retinoic acid from blood to CSF,[14] and, especially during development, are a source of morphogenic and guidance molecules involved in brain formation (reviewed in Strazielle and Ghersi-Egea[15]). The implication of the blood–CSF barrier in supplying nutrients to the brain is probably restricted to the developmental period.

Finally, the BBB and BCSFB are differently involved in neuro-immune interactions. The BBB is virtually not permissive to immune cells in physiological conditions. This property of the BBB led to the concept of an immune privilege of the brain.[16] Only in pathological contexts such as neuroinfection or neuroinflammation, when adhesion molecules are up-regulated at the endothelium, lymphocytes or/and monocytes traffic across the barrier and participate in the pathophysiology of related diseases, including multiple sclerosis.[17] A physiological immune surveillance of the brain occurs nonetheless. Besides a role of resident microglial cells, it seems to involve memory T cells circulating within the CSF after trafficking possibly through meningeal venules, and through the choroid plexuses[18] via a mechanism which is not fully elucidated but could involve the choroidal endothelial P-selectin.[19] In addition to this role in physiological blood–CNS traffic of immune cells, the choroid plexus-CSF system can relay immune signals generated during both peripheral and central inflammation. This translates into the rapid secretion of different chemokines and cytokines by the choroid plexus that can reinforce the immune surveillance of the brain when the organism is challenged. For instance, following brain trauma in areas distant from the ventricles, a strong increase in the expression levels of Il-1b, and then in the secretion of chemokines such as CXCL1 and CCL2 is observed in the choroid plexus, followed by an accumulation of neutrophils and monocytes in this tissue.[20,21] When the organism is exposed to a peripheral pro-inflammatory challenge, the number of myeloid cells increases in the cisternal spaces of the brain.[22] During experimental autoimmune encephalitis, when animals are exposed to a brain antigen, the CSF circulation seems to orchestrate the spreading of immune cells within the brain.[22] These examples illustrate that cell trafficking via the choroid plexus-CSF system may be amplified in different brain diseases involving an inflammatory process before the BBB itself becomes permissive to immune cells. This choroid plexus-CSF system may also have a role in later stages of diseases, as a trafficking site for macrophages involved in repair processes.[23]

Cerebrospinal Fluid Circulation, CSF Brain Exchanges, and Perivascular Flow

More than a passive drainage system for the brain, the CSF circulation is now recognized as participating in the volume transmission of various polypeptides and other biologically active molecules throughout the brain, and may also be important for the clearance of brain metabolites, due to its connection to the perivascular spaces. The majority of CSF is produced by the choroid plexuses located in each of the four brain ventricles. CSF flows from each of the two lateral ventricles to the third ventricle through the foramina of Monroe. From the third ventricle, CSF flows through the narrow cerebral aqueduct (aqueduct of Sylvius) to the fourth ventricle. The CSF then leaves the ventricular system and enters the central canal and the subarachnoid space (SAS) though the foramina of Luschka and Magendie. The CSF in the SAS covers the entire brain and spinal cord. At the base of the brain, it flows in a caudo-rostral direction along the basal cisterns of the brain up to the cistern of the laminae terminalis and olfactory bulb SAS. The CSF also feeds a number of less known ventricular and extraventricular compartments. For instance, the optic, infundibular, and mammillary recesses are connected to the ventral part of the third ventricle within the hypothalamus. Extraventricular fluid spaces include the "internal" ambient and quadrigeminal cisterns, as well as the velum interpositum and anterior medullary velum which are connected to the SAS. Illustrations of these CSF flow pathways are found in Ghersi-Egea et al.[24] and Schmitt et al.[25] The classical pathway for CSF elimination is through the arachnoid granulations into the superior sagittal sinus. For many years it has also been known that CSF is connected to perivascular spaces (the Virchow–Robin spaces) and that a substantial portion of CSF also drains via perineural spaces of the olfactory and possibly other cranial nerves to lymphatic vessels, in particular those in the nasal mucosa underneath the cribriform plate draining toward the cervical lymph nodes.[26]

The flow of CSF is brisk along the ventricular system. It slows down when reaching the subarachnoid and cisternal spaces,[24] but is still observable by MRI in the live rat as demonstrated by Iliff et al., who refer to these spaces as a "paravascular route".[27] Exchanges between the CSF and the brain parenchyma occur at different levels. First, paracellular diffusion across the permissive ependyma occurs along most of the ventricular system, except across the tanicytes that border circumventricular organs such as the median eminence, as these tanicytes are sealed by tight junctions.[24,28] Because diffusion is a slow process, delivery in the CSF-to-tissue direction at distance from the ventricles is mainly significant for molecules continuously secreted (or chronically administered) into the CSF. Cellular retrograde transport of CSF-borne molecules can also occur. A typical example is manganese, entering the CSF through the choroid plexus and accumulating in deeper structures through this pathway.[25,29] At the subarachnoid/cisternal level, exchanges between the CSF and underlying parenchyma are complex. The glia limitans, at places where it forms a thick multilayered structure, seems to act both as a barrier restricting diffusion processes, and as a reservoir from which accumulated compounds can be released gradually.[24] Finally, CSF mixes with perivascular fluids surrounding both penetrating arteries and veins. The width of Virchow–Robin space is of several microns, and the perivascular route represents a low resistance fluid flow pathway within the brain tissue. CSF flow from the SAS to the brain tissue was described by Rennels et al.[30] and confirmed by in vivo two-photon microscopy.[31] However, the direction of CSF movement along arteries and veins is still debated, as experimental design can introduce bias in fluid movement within the brain (reviewed in Badaut and Ghersi-Egea[32]). Deeper into the tissue, the perivascular spaces narrow along the microvascular bed. At this level, exchanges between the perivascular fluid and the interstitial fluid may be influenced by aquaporin 4 which is enriched in the astrocytic endfeet surrounding the microvessels, through a mechanism which is not clearly defined yet.[31,33] Aquaporin 4 is also highly enriched in the ependyma bordering the ventricles and in the glia limitans underneath the pial membrane separating the SAS from the brain.[33,34] It may also play a role in exchange processes between interstitial fluid and CSF at these interfaces.

PART II: TRANSPORT AND SECRETION PROCESSES AT THE BRAIN BARRIERS

Influx of Organic Compounds, Polypeptides, and Proteins Into the Brain

It has been known for decades that nutrients that are too hydrophilic to permeate through tight barriers get access to the brain via specific transporters. Most of them belong to the solute carrier (SLC) superfamily, and are located at the blood-facing membrane, or at both the blood-facing and brain-facing membranes of barrier cells (reviewed in Redzic[1] and Strazielle and Ghersi-Egea[2]). Typical examples include the glucose carrier SLC2A1 (GLUT-1) or carriers for neutral and cationic amino acids respectively SLC7A5 (LAT1) and SLC7A1 (Na^+-dependent system y+). Other carriers can function both as influx and efflux transporters. For instance SLC16A1 allows the use of ketone bodies as a source of energy for the brain during development and in adult nonphysiological situations such as in the context of ketogenic diet or hypoxia, and also carries lactate produced by astrocytes out of the adult brain.[1,2] Different hormones and micronutrients also require transporters to regulate their brain levels. These include the Na^+-dependent vitamin C transporter-2 (SLC23A1) located at the blood–CSF barrier, or SLC16A2 (MCT8) and SLC21A14 (OATP14) which are both involved in the transport of thyroid hormones across the two barriers. Of note, the cerebral homeostasis of thyroid hormones also involves their binding to transtyretin, a thyroid hormone-binding protein secreted by the choroid plexus into the CSF and circulating throughout the brain. Finally, a number of larger polypeptides that are necessary for CNS function, either as a hormone (insulin), inorganic ion carrier (transferrin), or nutrient carrier (lipoproteins), seem to get access to the brain by a regulated receptor-mediated transcytosis mechanism. The insulin, transferrin, and low density lipoprotein receptors, as well as the low density lipoprotein receptor-related protein-1 (LRP1, which transports molecules such as the insulin-like growth factor I) are involved in these transcellular pathways.[35,36] A large number of studies have been set to explore the potential usefulness of these pathways to deliver drugs into the brain across the BBB. Yet, the precise intracellular trafficking pathways through the endothelial cells, the mechanisms of abluminal release, and the reversibility of transport are only partially known. In particular, the actual efficiency (transcytotic transport rate) is often unreported. With regard to the blood–CSF barrier, at which electron microscopy highlights a large intracellular vesicular transport network,[2] very little has been documented about the efficacy of transcytotic vs endocytic pathways that follow receptor-mediated endocytosis, and the directionality of transport.

Efflux of Organic Compounds, Polypeptides, and Proteins Out of the Brain

A number of other transport processes are associated with the efflux out of the brain of various potentially deleterious molecules. Some are highly specific for selected endogenous metabolites produced by the brain in physiological or pathological situations. For instance SLC1A1-3 carriers, located at the brain-facing membrane of cerebral endothelial cells, actively transport excitotoxic amino acids such as glutamate out of the brain.[37] As mentioned before, SLC16A1 can efflux lactate from the brain. Di- and tripeptides are eliminated from the CNS across the choroid plexus by SLC15A2 (PEPT2), a carrier specifically located at the apical membrane of the choroidal epithelium.[38] With regard to larger polypeptides, amyloid beta peptides cleaved from the amyloid precursor protein, and involved in the pathophysiology of Alzheimer's disease, are transported across the BBB into the blood. The mechanism described involves the LRP1 receptor.[39] This questions the directionality and reversibility of LRP1-mediated transcytosis, as the same receptor may participate in the supply of some growth factors to the brain (see supra). Antibody transport also occurs at blood–brain interfaces. Functional evidence shows that the barrier cells display a strong polarized capacity to carry antibodies in the brain/CSF-to-blood direction. This efflux transport likely involves the neonatal receptor for the Fc fragment FcRn which is expressed at high level at both the blood–brain and blood–CSF barriers, and contributes to the low level of antibodies present in the brain.[2,40]

Besides these rather specific efflux mechanisms, the blood–brain interfaces harbor a transport machinery which protects the brain in a more general way against many potentially deleterious xenobiotics and endobiotics. This machinery uses multispecific efflux transporters that belong to either the ATP binding cassette (ABC) superfamily of transporters (mainly to ABCB, ABCC, and ABCG families), or to the SLC superfamily (mainly organic anion and cation transporters of the SLC21/SLCO and SLC22 families). ABC transporters are primary energy-dependent carriers located on the blood-facing membrane of the barrier cells. They prevent the entry into the brain of numerous lipophilic or amphiphilic compounds that could otherwise cross the brain barriers by diffusion across cellular membranes. Substrates include many environmental pollutants as well as pharmacologically active drugs, including antitumor, antiinfectious, immunosuppressant, antiepileptic, or else psychotropic drugs.[41]

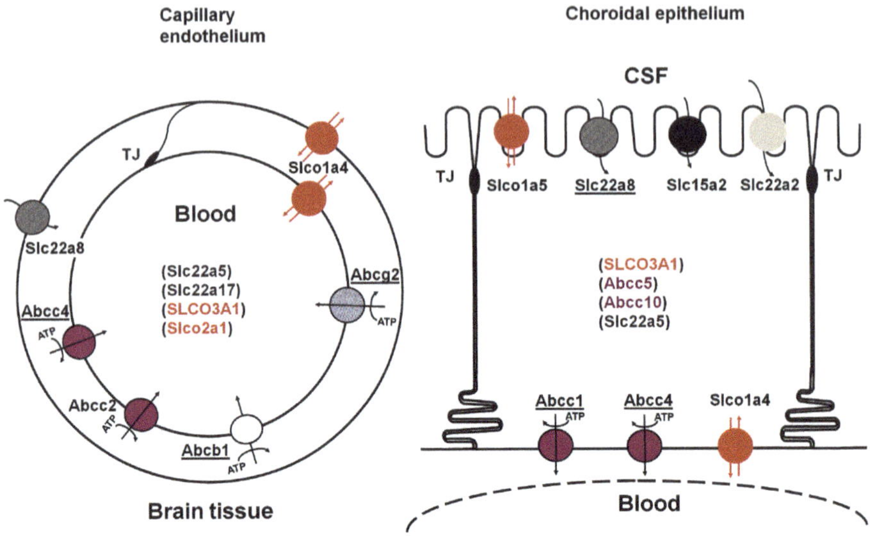

FIGURE 2.3 **Schematic representation of the main multispecific transporters involved in neuroprotective efflux at the blood–brain (left) and blood–CSF (right) barriers.** Underlined names represent transporters for which evidence has also been reported in human. The ATP-dependency is shown only for primary ATP-dependent transporters. Transporters in parenthesis are carriers for which evidence of their involvement in CNS–blood efflux processes is still limited. SLC15A2 is shown here as it also transports peptidomimetic drugs in addition to natural di- and tri-peptide substrates. See Ref. 44 for references. *TJ*, Tight junction. *Source: Reprinted from Strazielle N, Ghersi-Egea JF. Efflux transporters in blood–brain interfaces of the developing brain. Front Neurosci 2015;9:21.*

This explains in part why blood–brain interfaces are a major issue in the development of therapeutic drugs targeting the CNS. SLC multispecific efflux transporters are usually secondarily energy-dependent carriers, i.e., coupled to Na^+ transport and thus to Na^+K^+ATPase, and function mainly to remove their substrates from the brain fluid environment. They also accept numerous compounds as substrates, including nonsteroidal antiinflammatory drugs, nucleoside antiviral agents, antibiotics to name a few.[42,43] These different multispecific ABC and SLC transporters have different but overlapping and broad substrate specificities, explaining their efficacy to protect the brain against numerous structurally unrelated compounds. They are differently localized between the two barriers, adding to the complementary role of the BBB and BCSFB in protecting the brain. A nonexhaustive description of the main transporters involved in brain protection of the barrier cells is shown in Fig. 2.3. Some species-to species variations in transporter expression occur at brain interfaces, but ABCB1 (P-glycoprotein) and ABCG2 (breast cancer resistance protein) are landmark of the BBB, while ABCC1 (multidrug-related resistance protein

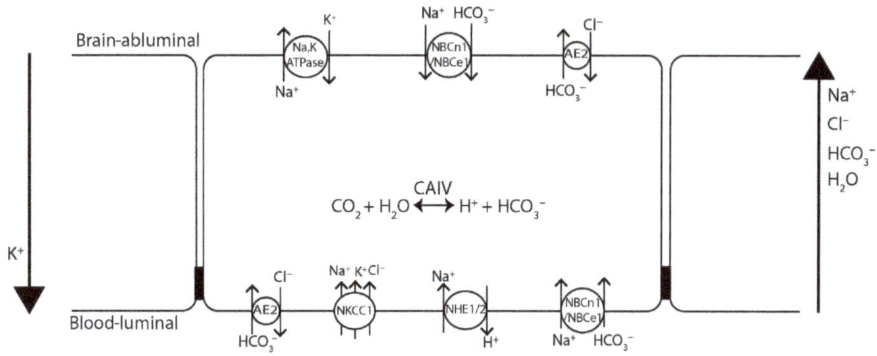

FIGURE 2.4 **Transporters and channels in the blood–brain barrier.** Brain endothelial cells with a brain-facing abluminal membrane and a blood-facing luminal membrane are shown. The gradients for net ion movement are indicated on each side of the cells. See text for details and protein abbreviations.

1, MRP1) and SLC15A2 (PEPT2), as well as Slco1a5 in rodent, are mainly expressed at the BCSFB. Other transporters, such as SLC21A8 (OAT3) and ABCC4 (MRP4) are present at both barriers (reviewed in Refs. 2,44).

Of note the expression of efflux transporters at the BBB tends to be lower in the developing brain than in the adult brain, especially ABCB1, while at the BCSFB the expression of most of them is already high during perinatal development, suggesting a specific relevance of the BCSFB to brain protection in the neonate and young child.[44]

Transporters and Channels for Inorganic Ions at the Blood–Brain Barrier

The tight BBB which restricts the paracellular passage of solutes between blood and brain is characterized by a luminal side facing the blood and an abluminal side facing the brain. Transporters located on these membranes (Fig. 2.4) maintain a net movement of inorganic ions into the brain, which contribute to maintaining interstitial fluid in the healthy brain. The water permeability of the BBB is much lower than that of the choroid plexus. The BBB does not express water channels and the movement of water is maintained by simple diffusion. A number of transporters and channels responsible for the movement of mono- and divalent ions from blood to brain have been described. Their exact subcellular localization and thereby transport direction is far from resolved; however, functional and visualization techniques have indicated the presence and, to some extent, physiological role of these transporters. The fluid secretion of interstitial fluid across the BBB entails a net movement of Na^+, Cl^-, and most likely also HCO_3^- and water from blood to brain. Na^+ is primarily thought to enter the endothelial cell from the blood through the luminal membrane presumably by the electroneutral Na^+,K^+,Cl^- cotransporter, *NKCC1* (SLC12A2) that transports 1 Na^+ and 1 K^+ along with 2 Cl^-. At the opposite side of the endothelial cell, the Na^+ gradient is maintained by the abluminally located *Na^+/K^+ ATPase* that moves 3 Na^+ out of the cell in exchange for 2 K^+. Other Na^+ transporters have been located in the BBB. The Na^+/H^+ exchangers *NHE1* (SLC9A1) and *NHE2* (SLC9A2) are both expressed in the BBB. These transporters import Na^+ in exchange for H^+. Both transporters have been located primarily in the luminal membrane.[45] Two other Na^+-dependent acid extruders have been located in the BBB; the electrogenic $Na^+:HCO_3^-$ cotransporter, *NBCe1* (SLC4A4) and the electroneutral $Na^+:HCO_3^-$ cotransporter, *NBCn1* (SLC4A7). NBCe1 has been shown to exhibit different $Na^+:HCO_3^-$ stoichiometry depending on cell type transporting either 1:3 into cells (proximal tubule of the kidney) or 1:2 out of cells (pancreas). NBCn1 transports 1 Na^+ along with 1 HCO_3^- into the cell. In the BBB the subcellular localization, stoichiometry and net direction of NBCe1 and NBCn1 are yet to be determined.[46] HCO_3^- can enter the cell from the blood either directly through one of the cotransporters or indirectly by generation of HCO_3^- from CO_2. The indirect transport of HCO_3^- is mediated by diffusion of CO_2 across the membrane. Inside the cell CO_2 is hydrated to carbonic acid by carbonic anhydrase IV and dissociates into H^+ and HCO_3^-. H^+ is removed by the Na^+/H^+ exchangers and HCO_3^- is transported across the endothelium by bicarbonate extruders such as the anion exchanger, AE2. The exact role of AE2 is not known and neither is its subcellular location; however, the transporter is proposed to participate in the net secretion of HCO_3^- from blood to brain. In addition to extruding HCO_3^-, AE2 imports Cl^-. The net direction of Cl^- is similar to Na^+ from blood to brain. If AE2 is placed in the abluminal membrane it could participate in the movement of HCO_3^- from blood to the cell but would transport Cl^- in the opposite direction. If AE2 is placed in the luminal membrane it could participate in the uptake of Cl^- along with NKCC1 but in this case the movement of HCO_3^- across the BBB would be back to the blood. Net HCO_3^- movement into the brain is necessary to buffer the drops in pH mediated by blood-borne CO_2 that freely diffuses across the membrane. Studies have shown that the brain is better at protecting against acidosis rather than against alkalosis, most likely because of the net movement of HCO_3^- into the brain. A number of other Cl^- channels such as ClC2, 3 and 5 have been detected in BBB cells; however, the subcellular localization is again unknown.

K^+ is believed to be moved from brain to blood, based on the fact that K^+ concentration in the interstitial fluid is lower than in blood; however, the mechanisms behind this assumption are not fully elucidated. It is possible that K^+ is removed by other pathways such as outflow of interstitial fluid. The Na^+/K^+ ATPase imports large volumes of K^+ across the abluminal membrane. At the luminal membrane NKCC1 also imports a volume of K^+. Several K^+ channels have been localized in the BBB that could account for the net removal of K^+ such as the voltage-dependent K^+ channel Kv1, Ca^{2+}-acitivated K^+ channels, and the inward rectifying K^+ channel Kir2 that both export K^+. There could possibly also be K^+, Cl^- cotransporters transporting K^+ along with Cl^- out of the cell. The K^+ permeability of the BBB is 10-fold higher than the Na^+ and Cl^- permeability which could indicate passive transport routes for K^+ in the BBB.

The net permeability of water molecules through the BBB is 10-fold lower than through peripheral blood vessels. The endothelial cells of the BBB do not express water channels but water is transported by passive diffusion mediated by the osmotic gradient created by solute transport. The water channel AQP4 is expressed in the astrocytes especially in the membrane domains facing the vessels called the perivascular endfeet. AQP4 is proposed to be implicated in the buffering of K^+ from the astrocytes[47] and the exchange of water between the CSF and the interstitial fluid (see supra). There are no specific inhibitors of AQP4, which has complicated the research area. The research is based mainly on the study of genetically modified mouse models which is difficult to interpret as the inborn genetic modification can lead to an array of compensatory mechanisms that complicate the interpretation of data. In addition different mouse models have shown different results, further complicating the area.

Calcium concentration is tightly regulated in the brain and changes in serum calcium do not affect brain levels. Calcium is primarily thought to enter the brain through the CSF and thus across the blood–CSF barrier. The plasma membrane Ca^{2+}-ATPase, PMCA, is strongly expressed in the choroid plexus; however, there is also evidence that the BBB expresses the transporter especially during hypocalcemic states.[48] In the hypocalcemic state the abundance of PMCA is increased in the BBB. The subcellular location is not fully established but suggested to be in the abluminal membrane.

Transporters and Channels for Inorganic Ions and Water at the Blood–Cerebrospinal Fluid Barrier in Relation to CSF Secretion

The secretion of CSF by the choroid plexus epithelial (CPE) cells is mediated by a number of transporters (reviewed in Damkier et al.[49]). The epithelial tissue is very active amounting to approximately 500 ml of CSF per day in humans. The main ions transported from blood to CSF are Na^+, Cl^-, and HCO_3^- (Fig. 2.5). K^+ is transported in the opposite direction. CSF secretion is driven by the luminally located *Na^+, K^+ ATPase*[50] actively pumping Na^+ from the cell into the CSF in exchange for K^+. This creates the Na^+ gradient necessary to catalyze the secondary active transporters

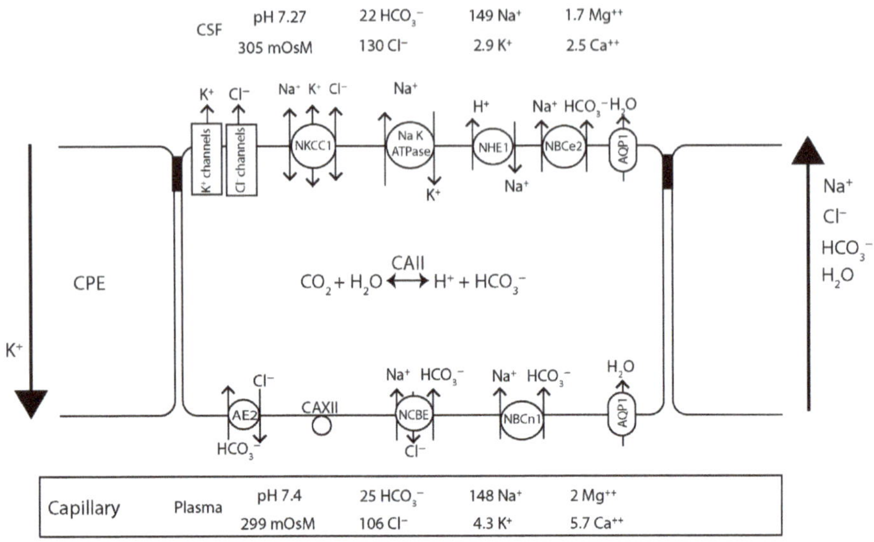

FIGURE 2.5 **Transporters and channels in the CPE cell.** CPE cells with the luminal membrane facing the CSF and a basolateral membrane facing the capillaries are shown. The gradients for net ion movement in CSF secretion are indicated on each side of the epithelial cells. See text for details and protein abbreviations.

important for secretion. In the basolateral membrane Na^+ is imported by the electrogenic Na^+-dependent Cl^-/HCO_3^- exchanger, *NCBE/NBCn2* (SLC4A10).[51] The exact stoichiometry of this transporter is not fully known. However, it is certain that an import of Na^+ and HCO_3^- from blood to the CPE cell occurs, that is driven by the Na^+ gradient into the cell. Another sodium importing transporter in the basolateral membrane is the electroneutral Na^+:HCO_3^- cotransporter *NBCn1* (SLC4A7).[51] The localization of this transporter is, however, not consistently basolateral as some mouse strains for example express the protein in the apical membrane. The role of NBCn1 in secretion, if any, has not been established. In addition to the Na^+/K^+ ATPase a number of other Na^+ exchangers as well as Na^+-dependent cotransporters are found in the luminal membrane of the CPE. Two transporters move acid/base equivalents along with Na^+. The Na^+/H^+ exchanger *NHE1* (SLC9A1),[52] imports Na^+ from the CSF in exchange for H^+. The electrogenic Na^+:HCO_3^- cotransporter, *NBCe2* (SLC4A5),[53,54] moves 1 Na^+ along with 3 HCO_3^- out of the cell and these transporters are therefore most likely involved in pH regulation in the CSF. CSF pH is highly dependent on the transport of the CO_2/HCO_3^- buffers since it lacks proteins to buffer pH. The CP expresses carbonic anhydrases II and XII. The Na^+, K^+, $2Cl^-$ cotransporter, *NKCC1* (SLC12A2),[55] is also located in the luminal membrane, but the direction of transport by this protein is not fully known. Both transport directions have been described and the specific role of this transporter in CSF secretion is still unclear. Other Na^+ transporters such as the epithelial Na^+ channel ENaC[56] and the Na^+ Borate transporter *NaBC1* (SLC4A11)[57] have similarly been localized to the CPE. However, the functional significance of these carriers is unknown.

The concentration of K^+ in the CSF is relatively constant despite fluctuations in the blood. The majority of K^+ imported by the CP via the Na^+/K^+ ATPase is recycled across the luminal membrane by K^+ channels belonging to the group of inward rectifying conductance (Kir) and outward rectifying conductance (Kv) channels.[58] In addition to these luminal K channels, the K^+ Cl^- cotransporter, KCC4,[59] is also present in this membrane, exporting both K^+ and Cl^-. Nevertheless, there is an overall net K^+ transport from CSF to blood. In addition to the import mediated by the Na^+/K^+ ATPase K^+ is imported via NKCC1. This transporter increases absorption when K^+ is increased in the CSF. At the basolateral membrane K^+ is transported to the blood by another K^+ Cl^- cotransporter, KCC3.[60]

Cl^- is the major anion in the CSF and is transported across the cell from blood to CSF via Cl^- channels in the luminal membrane, especially inward rectifying anion channels (Clir) that are constitutively active.[61] The molecular identities of these, however, are not known. Other Cl^- channels belong to the volume-regulated anion channel family but the contribution to net Cl^- efflux is minute. Cl^- is imported from the basolateral side primarily by the Cl^-/HCO_3^- exchanger, AE2.[62] HCO_3^- is the other major anion in the CSF. The two previously mentioned Na^+-dependent HCO_3^- cotransporters, NCBE and NBCn1, are located at the basolateral membrane.[51] The previously mentioned electrogenic Na^+:HCO_3^- cotransporter, NBCe2, is located at the luminal membrane. The Clir channels in the luminal membrane are also highly permeable to HCO_3^-; however, the significance of this to overall HCO_3^- export is most likely minute compared to NBCe2.

Ca^{2+} is vital for many brain functions and a large amount of Ca^{2+} is lost from the brain through CSF drainage. The main route for Ca^{2+} to enter the CNS is through the CP and CSF by a carrier mediated pathway,[63] which at least in part is shared by Mn^{2+}.[25] The molecular mechanism is not fully established. TRPV4 has been located in the luminal membrane[64] where it is involved in reabsorption of Ca^{2+} and thus not export of Ca^{2+}. Current–voltage experiments suggest the presence of TRPM3 in the basolateral membrane;[65] however, the localization of this transporter has not been established. The Ca^{2+} ATPases PMCA3 and PMCA4 have been localized in the luminal membrane but the role of this transporter is not known. Additionally *SLC25A5* encoding the $Na^+/Ca^{2+}–K^+$ exchanger as well as the Ca^{2+} binding protein calbindin have been located in the choroid plexus;[66] however, the exact mechanism these undertake is unknown. In addition to these transporters, the antiaging protein α-Klotho[67] which is involved in balancing systemic Ca^{2+} is highly expressed in the CP and could play an important role in the regulation of Ca^{2+} in CNS though the CP. The mechanism of Mg^{2+} transport in the CP is equally unknown. Mg^{2+} shares a common pathway with Ca^{2+} but the molecular basis for this transport is far from resolved.

The osmotic gradient created by the choroid plexus is small and transport of water across this epithelium is close to isosmotic. Water-transport through water channels generally requires an osmotic gradient. It is not fully established how water moves across the CP epithelium. The water channel aquaporin 1 is found in the luminal and to some extent in the basolateral membranes.[68] Knockout of AQP1 in mice reduces water permeability by 80% but decreases CSF secretion only by 35%.[69] This suggests a movement of water either through the paracellular pathway or mediated by cotransporters such as KCC found in the luminal membrane. The former may involve claudin 2 which is a pore-forming tight junction protein highly specific for the choroid plexus.[7] The latter has been shown to be able to move a considerable amount of water molecules along with the solutes.[70]

CONCLUSIONS

Underlying mechanisms by which blood–brain interfaces play a crucial role in protecting the brain, controlling its homeostasis, and promoting intracerebral fluid flow are only partially deciphered. Investigating the role of specific or multispecific transporters in the BBB and CP are complicated by the lack of specific inhibitors and the compensatory mechanisms in the inborn knockout animals. Growing in vitro cells of the BBB and the BCSFB is also not trivial. Taking the cells out of the brain environment, and/or immortalizing them, leads to a rearrangement of their unique polarization, loss of barrier properties, and changes in the expression of transporters. A better understanding of the transport functions of these interfaces is necessary to devise new strategies aiming at treating diseases that involve a dysregulation of neuroprotection or of fluid dynamics, such as cerebral edema.

In recent years new visualization techniques have rekindled the investigation of fluid movement in the brain giving rise to new discoveries of the important function of the brain flow in clearing waste from the brain during sleep, or in participating to volume transmission of biologically active compounds. It is of growing interest to target the CSF secretion by pharmacological means to either increase or decrease CSF production. It is known that the CSF secretion rate decreases with age. This has been implicated in the development of Alzheimer's disease and other dementias. Early targeting of the CP in this respect could potentially postpone the progression of disease. At the other end of the scale, increased CSF production is a relatively rare occurrence but has been described in for instance adenomas in the CP and possibly also in the disorder idiopathic intracranial hypertension (pseudotumor cerebri). In the latter disorder, acetazolamide is the drug of choice for treatment. In addition to being a diuretic with function in the kidney, this drug also inhibits CSF production. It is, however, not a well-tolerated drug and therefore identifying specific targets for regulating CSF production at the CP remains a challenge to be met.

Acknowledgments

This work was supported by ANR-10-IBHU-0003 Cesame grant (to J.F.G.E.) and the Faculty of Health and Medical Sciences, University of Copenhagen (to H.D.).

References

1. Redzic Z. Molecular biology of the blood–brain and the blood–cerebrospinal fluid barriers: similarities and differences. *Fluids Barriers CNS* 2011;**8**:3.
2. Strazielle N, Ghersi-Egea JF. Physiology of blood–brain interfaces in relation to brain disposition of small compounds and macromolecules. *Mol Pharm* 2013;**10**:1473–91.
3. Strazielle N, Ghersi-Egea JF. Factors affecting delivery of antiviral drugs to the brain. *Rev Med Virol* 2005;**15**:105–33.
4. Armulik A, Genove G, Betsholtz C. Pericytes: developmental, physiological, and pathological perspectives, problems, and promises. *Dev Cell* 2011;**21**:193–215.
5. Fenstermacher JD. The pharmacology of the blood–brain barrier. In: Neuwelt E, editor. *Implications of the blood–brain barrier and its manipulation*. New York: Plenum; 1989. p. 137–55.
6. Keep RF, Jones HC. A morphometric study on the development of the lateral ventricle choroid plexus, choroid plexus capillaries and ventricular ependyma in the rat. *Brain Res Dev Brain Res* 1990;**56**:47–53.
7. Kratzer I, Vasiljevic A, Rey C, et al. Complexity and developmental changes in the expression pattern of claudins at the blood–CSF barrier. *Histochem Cell Biol* 2012;**138**:861–79.
8. Goncalves A, Ambrosio AF, Fernandes R. Regulation of claudins in blood–tissue barriers under physiological and pathological states. *Tissue Barriers* 2013;**1**:e24782.
9. Tietz S, Engelhardt B. Brain barriers: crosstalk between complex tight junctions and adherens junctions. *J Cell Biol* 2015;**209**:493–506.
10. Amasheh S, Meiri N, Gitter AH, et al. Claudin-2 expression induces cation-selective channels in tight junctions of epithelial cells. *J Cell Sci* 2002;**115**:4969–76.
11. Rosenthal R, Milatz S, Krug SM, et al. Claudin-2, a component of the tight junction, forms a paracellular water channel. *J Cell Sci* 2010;**123**:1913–21.
12. Chodobski A, Szmydynger-Chodobska J. Choroid plexus: target for polypeptides and site of their synthesis. *Microsc Res Tech* 2001;**52**:65–82.
13. Richardson SJ. Evolutionary changes to transthyretin: evolution of transthyretin biosynthesis. *FEBS J* 2009;**276**:5342–56.

14. Angelow S, Haselbach M, Galla HJ. Functional characterisation of the active ascorbic acid transport into cerebrospinal fluid using primary cultured choroid plexus cells. *Brain Res* 2003;**988**:105–13.
15. Strazielle N, Ghersi-Egea JF. Choroid plexus in the central nervous system: biology and physiopathology. *J Neuropathol Exp Neurol* 2000;**59**:561–74.
16. Piccio L, Rossi B, Scarpini E, et al. Molecular mechanisms involved in lymphocyte recruitment in inflamed brain microvessels: critical roles for P-selectin glycoprotein ligand-1 and heterotrimeric G(i)-linked receptors. *J Immunol* 2002;**168**:1940–9.
17. Ransohoff RM, Kivisakk P, Kidd G. Three or more routes for leukocyte migration into the central nervous system. *Nat Rev Immunol* 2003;**3**:569–81.
18. Kivisakk P, Mahad DJ, Callahan MK, et al. Human cerebrospinal fluid central memory CD4+ T cells: evidence for trafficking through choroid plexus and meninges via P-selectin. *Proc Natl Acad Sci USA* 2003;**100**:8389–94.
19. Carrithers MD, Visintin I, Kang SJ, Janeway Jr. CA. Differential adhesion molecule requirements for immune surveillance and inflammatory recruitment. *Brain* 2000;**123**(Pt 6):1092–101.
20. Szmydynger-Chodobska J, Strazielle N, Gandy JR, et al. Posttraumatic invasion of monocytes across the blood–cerebrospinal fluid barrier. *J Cereb Blood Flow Metab* 2012;**32**:93–104.
21. Szmydynger-Chodobska J, Strazielle N, Zink BJ, Ghersi-Egea JF, Chodobski A. The role of the choroid plexus in neutrophil invasion after traumatic brain injury. *J Cereb Blood Flow Metab* 2009;**29**:1503–16.
22. Schmitt C, Strazielle N, Ghersi-Egea JF. Brain leukocyte infiltration initiated by peripheral inflammation or experimental autoimmune encephalomyelitis occurs through pathways connected to the CSF-filled compartments of the forebrain and midbrain. *J Neuroinflamm* 2012;**9**:187.
23. Shechter R, Miller O, Yovel G, et al. Recruitment of beneficial M2 macrophages to injured spinal cord is orchestrated by remote brain choroid plexus. *Immunity* 2013;**38**:555–69.
24. Ghersi-Egea JF, Finnegan W, Chen JL, Fenstermacher JD. Rapid distribution of intraventricularly administered sucrose into cerebrospinal fluid cisterns via subarachnoid velae in rat. *Neuroscience* 1996;**75**:1271–88.
25. Schmitt C, Strazielle N, Richaud P, Bouron A, Ghersi-Egea JF. Active transport at the blood–CSF barrier contributes to manganese influx into the brain. *J Neurochem* 2011;**117**:747–56.
26. Krisch B, Leonhardt H, Oksche A. Compartments and perivascular arrangement of the meninges covering the cerebral cortex of the rat. *Cell Tissue Res* 1984;**238**:459–74.
27. Iliff JJ, Lee H, Yu M, et al. Brain-wide pathway for waste clearance captured by contrast-enhanced MRI. *J Clin Invest* 2013;**123**:1299–309.
28. Langlet F, Mullier A, Bouret SG, Prevot V, Dehouck B. Tanycyte-like cells form a blood–cerebrospinal fluid barrier in the circumventricular organs of the mouse brain. *J Comp Neurol* 2013;**521**:3389–405.
29. Bock NA, Paiva FF, Nascimento GC, Newman JD, Silva AC. Cerebrospinal fluid to brain transport of manganese in a non-human primate revealed by MRI. *Brain Res.* 2008;**1198**:160–70 [Epub 2008 Jan 2004].
30. Rennels ML, Gregory TF, Blaumanis OR, Fujimoto K, Grady PA. Evidence for a "paravascular" fluid circulation in the mammalian central nervous system, provided by the rapid distribution of tracer protein throughout the brain from the subarachnoid space. *Brain Res* 1985;**326**:47–63.
31. Iliff JJ, Wang M, Liao Y, et al. A paravascular pathway facilitates CSF flow through the brain parenchyma and the clearance of interstitial solutes, including amyloid beta. *Sci Transl Med* 2012;**4**:147ra111.

32. Badaut J, Ghersi-Egea JF. The choroid plexus and cerebrospinal fluid system: roles in neurodegenerative diseases. In: Neman J, Chen TC, editors. *The choroid plexus and cerebrospinal fluid. Emerging roles in CNS development, maintenance, and disease progression.* Amsterdam: Elsevier, Academic Press; 2015. p. 129–54.
33. Badaut J, Fukuda AM, Jullienne A, Petry KG. Aquaporin and brain diseases. *Biochim Biophys Acta* 2014;**1840**:1554–65.
34. Jimenez AJ, Dominguez-Pinos MD, Guerra MM, Fernandez-Llebrez P, Perez-Figares JM. Structure and function of the ependymal barrier and diseases associated with ependyma disruption. *Tissue Barriers* 2014;**2**:e28426.
35. Nishijima T, Piriz J, Duflot S, et al. Neuronal activity drives localized blood-brain-barrier transport of serum insulin-like growth factor-I into the CNS. *Neuron* 2010;**67**:834–46.
36. Smith MW, Gumbleton M. Endocytosis at the blood–brain barrier: from basic understanding to drug delivery strategies. *J Drug Target* 2006;**14**:191–214.
37. O'Kane RL, Martinez-Lopez I, DeJoseph MR, Vina JR, Hawkins RA. Na(+)-dependent glutamate transporters (EAAT1, EAAT2, and EAAT3) of the blood–brain barrier. A mechanism for glutamate removal. *J Biol Chem* 1999;**274**:31891–5.
38. Smith DE, Hu Y, Shen H, Nagaraja TN, Fenstermacher JD, Keep RF. Distribution of glycylsarcosine and cefadroxil among cerebrospinal fluid, choroid plexus, and brain parenchyma after intracerebroventricular injection is markedly different between wild-type and Pept2 null mice. *J Cereb Blood Flow Metab* 2011;**31**:250–61.
39. Bell RD, Sagare AP, Friedman AE, et al. Transport pathways for clearance of human Alzheimer's amyloid beta-peptide and apolipoproteins E and J in the mouse central nervous system. *J Cereb Blood Flow Metab* 2007;**27**:909–18.
40. Schlachetzki F, Zhu C, Pardridge WM. Expression of the neonatal Fc receptor (FcRn) at the blood–brain barrier. *J Neurochem* 2002;**81**:203–6.
41. Schinkel AH, Jonker JW. Mammalian drug efflux transporters of the ATP binding cassette (ABC) family: an overview. *Adv Drug Deliv Rev* 2003;**55**:3–29.
42. Hagenbuch B, Stieger B. The SLCO (former SLC21) superfamily of transporters. *Mol Aspects Med* 2013;**34**:396–412.
43. Koepsell H. The SLC22 family with transporters of organic cations, anions and zwitterions. *Mol Aspects Med* 2013;**34**:413–35.
44. Strazielle N, Ghersi-Egea JF. Efflux transporters in blood–brain interfaces of the developing brain. *Front Neurosci* 2015;**9**:21.
45. O'Donnell ME. Blood–brain barrier Na transporters in ischemic stroke. *Adv Pharmacol* 2014;**71**:113–46.
46. Mokgokong R, Wang S, Taylor CJ, Barrand MA, Hladky SB. Ion transporters in brain endothelial cells that contribute to formation of brain interstitial fluid. *Pflugers Arch* 2014;**466**:887–901.
47. Amiry-Moghaddam M, Ottersen OP. The molecular basis of water transport in the brain. *Nat Rev Neurosci* 2003;**4**:991–1001.
48. Keep RF, Ulanski 2nd LJ, Xiang J, Ennis SR, Lorris Betz A. Blood–brain barrier mechanisms involved in brain calcium and potassium homeostasis. *Brain Res* 1999;**815**:200–5.
49. Damkier HH, Brown PD, Praetorius J. Cerebrospinal fluid secretion by the choroid plexus. *Physiol Rev* 2013;**93**:1847–92.
50. Ernst SA, Palacios 2nd JR, Siegel GJ. Immunocytochemical localization of Na^+,K^+-ATPase catalytic polypeptide in mouse choroid plexus. *J Histochem Cytochem* 1986;**34**:189–95.
51. Praetorius J, Nejsum LN, Nielsen S. A SCL4A10 gene product maps selectively to the basolateral plasma membrane of choroid plexus epithelial cells. *Am J Physiol Cell Physiol* 2004;**286**:C601–610.

52. Damkier HH, Prasad V, Hubner CA, Praetorius J. Nhe1 is a luminal Na^+/H^+ exchanger in mouse choroid plexus and is targeted to the basolateral membrane in Ncbe/Nbcn2-null mice. *Am J Physiol Cell Physiol* 2009;**296**:C1291–1300.
53. Bouzinova EV, Praetorius J, Virkki LV, Nielsen S, Boron WF, Aalkjaer C. Na^+-dependent HCO_3^- uptake into the rat choroid plexus epithelium is partially DIDS sensitive. *Am J Physiol Cell Physiol* 2005;**289**:C1448–1456.
54. Millar ID, Brown PD. NBCe2 exhibits a 3 HCO3(−):1 Na+ stoichiometry in mouse choroid plexus epithelial cells. *Biochem Biophys Res Commun* 2008;**373**:550–4.
55. Plotkin MD, Kaplan MR, Peterson LN, Gullans SR, Hebert SC, Delpire E. Expression of the Na(+)-K(+)-2Cl-cotransporter BSC2 in the nervous system. *Am J Physiol* 1997;**272**:C173–183.
56. Leenen FH. The central role of the brain aldosterone-"ouabain" pathway in salt-sensitive hypertension. *Biochim Biophys Acta* 2010;**1802**:1132–9.
57. Damkier HH, Nielsen S, Praetorius J. Molecular expression of SLC4-derived Na^+-dependent anion transporters in selected human tissues. *Am J Physiol Regul Integr Comp Physiol* 2007;**293**:R2136–2146.
58. Millar ID, Bruce J, Brown PD. Ion channel diversity, channel expression and function in the choroid plexuses. *Cerebrospinal Fluid Res* 2007;**4**:8.
59. Karadsheh MF, Byun N, Mount DB, Delpire E. Localization of the KCC4 potassium-chloride cotransporter in the nervous system. *Neuroscience* 2004;**123**:381–91.
60. Pearson MM, Lu J, Mount DB, Delpire E. Localization of the K(+)-Cl(−) cotransporter, KCC3, in the central and peripheral nervous systems: expression in the choroid plexus, large neurons and white matter tracts. *Neuroscience* 2001;**103**:481–91.
61. Kajita H, Whitwell C, Brown PD. Properties of the inward-rectifying Cl-channel in rat choroid plexus: regulation by intracellular messengers and inhibition by divalent cations. *Pflugers Arch* 2000;**440**:933–40.
62. Lindsey AE, Schneider K, Simmons DM, Baron R, Lee BS, Kopito RR. Functional expression and subcellular localization of an anion exchanger cloned from choroid plexus. *Proc Natl Acad Sci USA* 1990;**87**:5278–82.
63. Murphy VA, Smith QR, Rapoport SI. Homeostasis of brain and cerebrospinal fluid calcium concentrations during chronic hypo- and hypercalcemia. *J Neurochem* 1986;**47**:1735–41.
64. Liedtke W, Friedman JM. Abnormal osmotic regulation in trpv4−/− mice. *Proc Natl Acad Sci USA* 2003;**100**:13698–703.
65. Oberwinkler J, Lis A, Giehl KM, Flockerzi V, Philipp SE. Alternative splicing switches the divalent cation selectivity of TRPM3 channels. *J Biol Chem* 2005;**280**:22540–8.
66. Ho HT, Dahlin A, Wang J. Expression profiling of solute carrier gene families at the blood–CSF barrier. *Front Pharmacol* 2012;**3**:154.
67. Imura A, Tsuji Y, Murata M, et al. Alpha-Klotho as a regulator of calcium homeostasis. *Science* 2007;**316**:1615–8.
68. Nielsen S, Smith BL, Christensen EI, Agre P. Distribution of the aquaporin CHIP in secretory and resorptive epithelia and capillary endothelia. *Proc Natl Acad Sci USA* 1993;**90**:7275–9.
69. Oshio K, Watanabe H, Song Y, Verkman AS, Manley GT. Reduced cerebrospinal fluid production and intracranial pressure in mice lacking choroid plexus water channel Aquaporin-1. *FASEB J* 2005;**19**:76–8.
70. Zeuthen T. Secondary active transport of water across ventricular cell membrane of choroid plexus epithelium of *Necturus maculosus*. *J Physiol* 1991;**444**:153–73.

SECTION II

TECHNIQUES TO INVESTIGATE CEREBRAL BLOOD

CHAPTER

3

Noninvasive Imaging Techniques for Brain Edema: From Basic Science to the Clinic

Andre Obenaus[1,2] *and Jérôme Badaut*[1,3]

[1]Loma Linda University School of Medicine, Loma Linda, CA, United States [2]University of California, Irvine, CA, United States [3]CNRS-University of Bordeaux, Bordeaux, France

OUTLINE

Brain Edema.
DOI: http://dx.doi.org/10.1016/B978-0-12-803196-4.00003-5

INTRODUCTION

Brain edema is a common hallmark observed in various brain disorders (acute traumatic brain injury to neurodegenerative diseases) and can potentially impact the long-term outcome of patients. Noninvasive imaging, such as magnetic resonance imaging (MRI) and computed tomography (CT) of the brain, provide the unique ability to evaluate edema evolution after disease/injury onset, resolution, or after therapeutic intervention(s). In parallel to the development and establishment of the noninvasive neuroimaging in clinical setting, research using preclinical rodent models provides the ability to link the alterations observed on neuroimaging to molecular and cellular mechanisms. Detailed descriptions of the mechanisms underlying edema will be expanded in various chapters and how they relate to different brain disorders. This chapter will provide an overview of the current state of the neuroimaging techniques to visualize edema and its temporal resolution. The focus will be primarily brain ischemia due to the expansive literature available.

COMPARTMENTS THAT CONTRIBUTE TO WATER MOVEMENT IN THE BRAIN

In brain tissues the extracellular space is composed of multiple cell types (neurons, glia, etc.), compartments (brain vasculature, perivascular space, ventricles, etc.), connective tissues, and maintains an ionic milieu that is required for normal brain function. Several comprehensive reviews on the biophysics of the extracellular space are available and the reader is referred to these for more in-depth discussion[1–3] and Chapter 6, The Extracellular and Perivascular Space of Brain. Critical to water movement and water homeostasis in the brain are a number of ionic and nonionic channels, such as aquaporins (see below).[4]

The role of glial cells in modulating extracellular space constituents, including water movement, is becoming well described and accepted by the neuroscience community. Glial cells (i.e., astrocytes) function as an integral part in maintaining the properties of the blood-brain barrier (BBB), controlling extracellular ionic components, and are known to provide trophic support for neurons. Modulation of the extracellular space (shrinkage or expansion) can have significant effects on neuronal physiological processes by affecting the diffusion of the extracellular molecules such as neurotransmitters or ions (e.g., glutamate, K^+). For example, chemically shrinking or expanding the extracellular space can result in induction or prevention of epileptiform activity.[5,6] The various cellular membranes and molecules in the extracellular space all contribute to providing a barrier to diffusion of water, ions, and other molecules.[7,8] These barriers in addition to various channel proteins facilitate communication between cells and the extracellular space and are known to have a significant role in the formation and evolution of edema.[9]

One critical compartment in modulating edema formation is the BBB of which endothelial cells are the primary cellular component. Endothelial cells exhibit tight junctions (with specific protein expression such as occludins, claudins, and cadherins) between cells to create a physical and molecular barrier using specific transporters and limited transcytosis (see Ref. 10). Cerebral endothelial cells are a component of the physiological unit collectively termed the neurovascular unit (NVU).[11,12] Other constituents of the NVU are pericytes, smooth muscle cells, astrocytes, and neurons. The interaction between these cells comprising the NVU, confers the BBB its properties, and prevents movement of harmful components from the blood into the brain parenchyma. Another key function of the BBB is the transport of metabolites (nutrients, energy) to brain tissue constituents. During neurological injury, such as in stroke, the tight junctions between endothelial cells can become dysregulated and are thought to play a significant role in the development of edema.[13,14] It is important to note, however, that the concept of tight junction leakage and edema formation after stroke may not be as important as previously reported, but rather endothelial cell viability may have a far larger role that heretofore expected.[15]

In summary the extracellular space, the NVU of the BBB, and its constituent channels and junctions all function together to limit and regulate water movement within the brain parenchyma. Dysregulation of any one component of these cellular compartments can lead to the formation of edema. While clinically relevant neuroimaging modalities, such as MRI and CT, cannot directly assess the molecular mechanisms underlying edema formation, they do provide a clear measure of the presence and evolution of edema within brain tissues. We now provide an overview of these imaging modalities in use for clinical and translational basic science in the evaluation of edema.

FORMS OF EDEMA RELEVANT TO NEUROIMAGING

Klatzo and collaborators originally defined two major classes of cerebral edema: cytotoxic and vasogenic (for reference see Unterberg et al.[16]). These broad categories are still used on an everyday basis. However, recent work related to the molecular processes involving edema has required a revision of this classification scheme with three proposed classes: cytotoxic, ionic, and vasogenic edema (Fig. 3.1). We will review the different edema classes in regard to the changes that are observed on MRI neuroimaging.

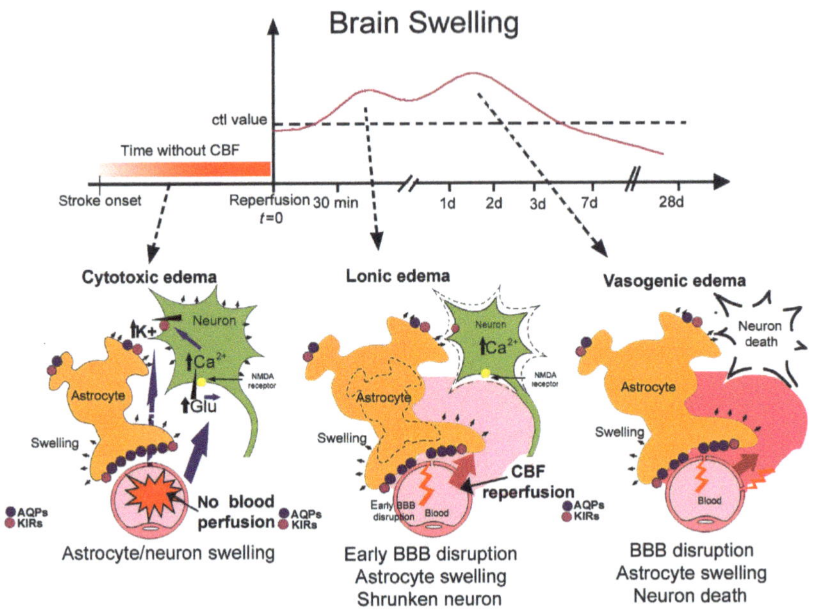

FIGURE 3.1 **Classes of edema.** Schematic drawing of the time-course of edema formation (top of the figure) highlighting three different edema phases: cytotoxic, ionic, and vasogenic edema. During occlusion the first few minutes are characterized of cytotoxic edema with loss of ionic gradients. The shear stress induced by the reperfusion is followed by two peaks of brain swelling one around 30 min after occlusion coinciding with ionic edema and a second at 48 h reflecting vasogenic edema. The initial cytotoxic edema is characterized as a swelling of the astrocytes and the neuronal dendrites caused by a disruption of the cellular ionic gradients and the entry of ions followed by water leading to cellular swelling. In ionic edema, astrocytes become swollen and neuronal death starts to occur resulting in shrinkage of the neurons and shear stress on the nonperfused vascular tree that results in early transient leakage of the BBB. Vasogenic edema is a result of disruption of the tight junctions, leading to increased permeability of the cerebral blood-vessels to albumin and other plasma proteins, further contributing to swelling of astrocytes and subsequent neuronal cell death. *Source: Modified from Badaut et al. (2011).*

Cytotoxic Edema

Cytotoxic edema is defined by intracellular water accumulation without BBB disruption while vasogenic edema appears after BBB disruption, leading to a diffusion of proteins from the blood to the tissue followed by water accumulation in the extracellular space.[16] Cytotoxic edema occurs within minutes after cessation of cerebral blood flow in brain ischemia; and it is historically characterized as initial swelling of astrocytes and neuronal dendrites.[17] The early cellular swelling within the first 10 min is a result of oxygen and glucose deprivation followed by a slow rise in extracellular K^+.[18] In brain ischemia the disruption of the cellular ionic gradients due to the absence of energy leads to entry of ions into astrocytes which try to buffer the extracellular space. This uptake of ions and neurotransmitters are accompanied by water entry into cells which results in cellular swelling (for details see Chapter 8: Mechanisms of Cell Volume Regulation in the Central Nervous System). Cellular water channels including potassium channels (i.e., Kir4.1) may contribute to water entry via astrocytic endfeet (for mechanisms, see Chapter 9: Role of Aquaporins for the Formation and Resolution of Brain Edema, and Chapter 10: Sur-1-Trpm4 A Promising Target for Brain Edema Treatment). Cytotoxic edema can also be present during the initial stage when the lesion expands as part of the edema process. Cytotoxic edema can be observed on diffusion-weighted imaging (DWI) and its quantitative parameter the apparent diffusion coefficient (ADC). Both parameters are thought to reflect water mobility within brain tissues; changes in these values could be, in part, the consequence of the cytotoxic edema process as well as the level of expression of AQP4 (see Chapter 9: Role of Aquaporins for the Formation and Resolution of Brain Edema).

Vasogenic Edema

The disruption of cerebrovascular endothelial tight junctions leads to increased permeability to albumin and other plasma proteins that are associated with a net water entrance in the brain tissue resulting in vasogenic edema.[16,19–23] The net entry of water into the brain parenchymal tissues can be visualized with various neuroimaging modalities as described below. In conjunction with DWI, T2-weighted imaging (T2WI) can also be used to estimate water content within brain tissues. However, it is important to note that there is possibly a temporal continuum between cytotoxic and vasogenic edema without clear delineation of each of these events. Before the final rupture of the endothelial tight junctions during vasogenic edema, there is transcapillary water and electrolyte transport into the brain parenchyma that can be called ionic edema (for details see Chapter 10: Sur-1-Trpm4 A Promising Target for Brain Edema Treatment). Ionic edema was viewed as part of cytotoxic edema, but is now seen as an independent component with a distinct physiological process (see additional details in Chapter 9: Role of Aquaporins for the Formation and Resolution of Brain Edema, and Chapter 10: Sur-1-Trpm4 A Promising Target for Brain Edema Treatment).

Interstitial or Ionic Edema

Recent advances describing the molecular mechanisms of underlying edema formation and its effects on BBB properties has led to a third edema subtype. Simard et al. have proposed ionic as a third subtype of edematous processes that is most often observed in cerebrovascular diseases,[24] described as ionic edema. Ionic edema is defined by a transcapillary sodium flux in endothelial cells occurs just prior to the disruption of their tight junctions.[24] Cytotoxic edema in endothelial cells which may evolve to become ionic edema due to the absence of oxygen and nutrients that further alters the endothelial cell's ionic gradients, including transcapillary flux of Na^+ and transcytosis.[24,25] In stroke models, early transient leakage of the BBB has been observed,[26,27] possibly due to an increase of the transcytosis mechanisms accompanied by an increase of caveolae.[28] This results in further entry of water via endothelial cells resulting in brain swelling within 30 min that further increases BBB permeability after reperfusion.[26,27,29] Astrocytes remain swollen and neuronal death starts to occur resulting in neuronal shrinkage.[17] The potential molecular mechanisms involved in this cascade involve aquaporin's, Kir4.1 and other water channels are described in more detail in other chapters. Neuroimaging of ionic edema is not straightforward due to the temporal evolution of edema and its associated complexity. However, it is possible to detect the ionic edema phase using a combination of T2WI and DWI, particularly in stroke pathophysiology (see below). It is interesting to note that the magnitude of these MRI observations can be correlated to patient outcomes demonstrating that ADC/T2 values are valuable biomarkers of outcomes and treatments.[30] The cellular and molecular events underlying these multimodal imaging changes remain to be fully explored.[9]

NEUROIMAGING MODALITIES FOR EDEMA

Advances over the last 40 years in noninvasive imaging have dramatically altered our ability to detect, diagnose, and treat human disease (see Chapter 23: Traumatic Brain Injury and Edema Treatment). Clinically at the forefront of these advances has been the development of MRI, CT and nuclear medicine techniques such as positron emission tomography (PET). Each imaging approach has strengths and weaknesses for imaging the evolution and resolution of brain edema and each is described in detail below.

Magnetic Resonance Imaging

As early as 1982 several publications reported using MRI (at that time MRI was denoted as nuclear magnetic resonance) to evaluate the evolution of brain edema[31] (Fig. 3.2). The authors described the clinical attributes of different sequences used in MRI at that time which could provide novel and informative contrast in normal brain tissues. The authors also reported the first data on the MRI characteristics observed in a variety of

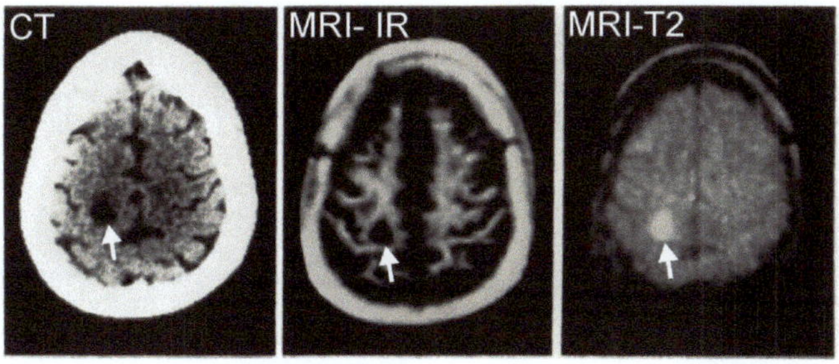

FIGURE 3.2 Early work demonstrated the superior ability of T2 weighted (T2WI) MRI to detect edema in a patient with an infarct that was not visible on other imaging modalities, including CT. The presence of hemorrhage was visible on CT. *Source: Modified from Bydder et al. (1982).*

neurological injuries including acute infractions. They were further able to demonstrate that MRI was more sensitive than CT for evaluation of edema and cysts from stroke, tumors, and other clinical syndromes.

In addition to these first clinical reports, basic science studies in animal models of injury also appeared that clearly demonstrated the ability of MRI to visualize edema.[32] Naruse et al. described that T2 was most effective in visualizing edema and also reported "water-rich" and "protein-rich" edema subtypes. Since these early descriptions using MRI to image edema, there has been a steady rise in the number of publications using MRI to monitor edema. Thus MRI has radically transformed our ability to noninvasively image edema formation and resolution in response to brain injury.

One of the advantages of MRI is that there are a range of contrast types (T2-weighted, T1-weighted, diffusion-weighted, etc.) that can be selected to image edema. Indeed the classical imaging modalities for evaluation of edema have been T2WI and DWI (Fig. 3.3).

In hyperacute clinical settings, DWI, and its quantitative values, the ADC map are the imaging modalities that best reflect the early appearance of edema[33] (Figs. 3.3 and 3.4). DWI is thought to reflect the ability of water molecules to diffuse through tissues and can report alterations such as cell swelling (restricted diffusion) or death (increased diffusion) (see Ref. 17,34). Alterations in cell swelling or death can be visualized following acute cerebrovascular infarction.[35,36] Increases or decreases in diffusion can often be seen within the same tissues, but is time dependent. This is true in stroke/ischemia where early (<3 days) after injury there is a decrease in the ADC but at later time points (>3 days) a prolonged increase exists (Figs. 3.3 and 3.4). Questions of the source of these signal changes (intracellular vs. extracellular) are still largely unresolved[37–39] and discussed.[17] However, at present it is thought that the early DWI increases

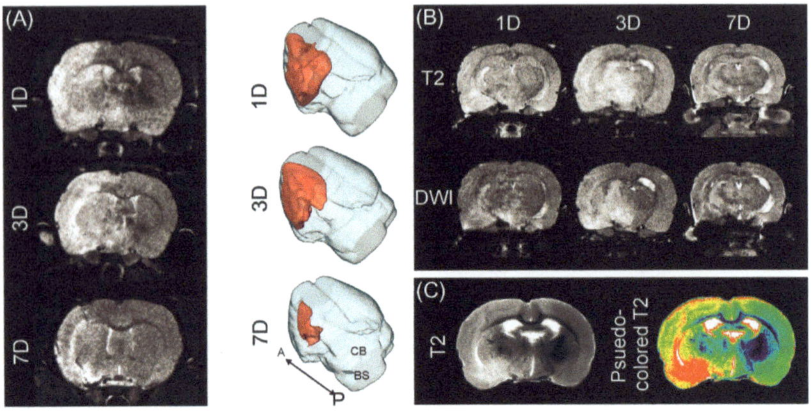

FIGURE 3.3 **Appearance of edema on stroke neuroimaging.** (A) Temporal evolution of T2WI in an MCAO rat model of stroke. There is a progressive increase in T2 signal over time in the cortex and striatum consistent with temporal evolution and subsequent resolution of edema. The 3D reconstructions of the T2hyperintensities (red) reveal a progressive decrease in lesion (edema) volume over 7 days. Also note that edema induces hemispheric asymmetry that is often prominent at 1 and 3 days post stroke. (B) DWI is the primary clinical imaging tool to map stroke related injury at early time points. In this example it is clear that DWI hyperintensities (decreased ADC values) peak at 3 days post stroke. C) Other methods for visualization of the edema on neuroimaging is the use of mapping absolute values or pseudo-coloring MRI images to assist in identification of edematous regions. *Source: (A) and (B) data collected at 4.7T in vivo, while C was ex vivo at 11.7T; Data courtesy of Drs. Belayev and Bazan, Louisiana State University and Obenaus, Loma Linda University.*

(ADC decreases) are thought to reflect cytotoxic edema wherein there is a rapid shift of water molecules from the extracellular to the intracellular compartment that is driven osmotically by sodium and calcium fluxes into the cell. After cessation of blood flow, for example in stroke, this process can occur very rapidly (within mins).[40,41]

At the more acute stages of edema formation, between 6h and 24h in brain ischemia, T2WI images begin to exhibit MR image hyperintensities as a result of a potential mix between cytotoxic and transition to vasogenic edema (Fig. 3.3) likely due to ionic edema processes (see below). At subacute time points after injury (>24h) T2WI hyperintensities are still highly prominent and edema continues to progress for the next 48–72h depending on the severity of injury. However, it should be noted that this evolution can be heterogeneous in the clinical population and even in well controlled studies with animal models of injury.[30,42] At later time points (>7 days) there is transition as dying tissue becomes frank necrosis that ultimately results in cystic lesions. At these later times, edema is generally considered to be resolved and now cystic lesions can be observed on T2WI images and also on DWI ("T2 shine-through").[43]

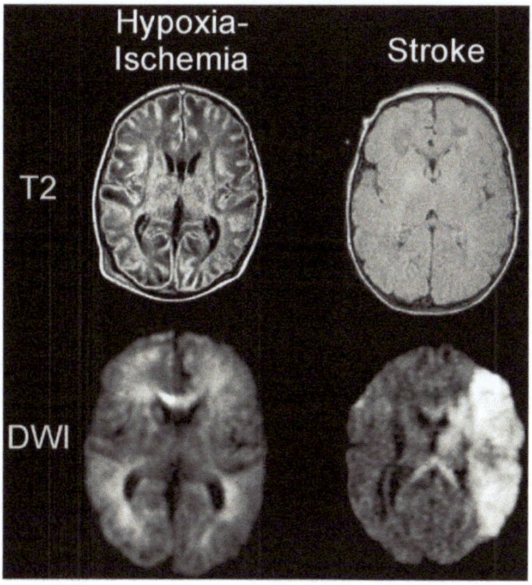

FIGURE 3.4 Neuroimaging of edema in hypoxia-ischemic injury and arterial ischemic stroke in the human neonate. DWI in the neonate clearly can identify the ischemic/stroke regions better than T2 early after injury. This is due in part to the increased water content within the neonatal brain. *Source: All data acquired at 1.5T; data courtesy of Drs. Ashwal and Obenaus, Loma Linda University.*

Other MR imaging modalities can also be utilized to assess edema and its effects.[44] Fluid attenuated inversion recovery (FLAIR) T2WI imaging is often used to suppress the cerebrospinal fluid (CSF) signal to image early and evolving edema.[45] FLAIR has the ability, particularly at hyperacute and acute time points to delineate smaller infarcts, uniquely those that are in close approximation to ventricular spaces. At more chronic imaging times the cystic lesion will have the appearance of CSF-like characteristics and will not be directly observed, although regions of ongoing gliosis within tissues may have FLAIR hyperintensities.[46]

T1WI has also been used to monitor edema evolution, but is predominately used in the presence of contrast-enhanced studies, such as monitoring edema as a consequence of tumorigenesis.[47,48] In stroke studies, T1WI is less frequently utilized but is occasionally used with contrast studies to evaluate BBB leak. Edema is typically seen as hypointense on T1WI in stroke patients.[44]

The existence of edema and its resolution, principally in the context of therapeutic treatments, has gained significance. Of interest is whether therapies can rescue tissues at risk like the penumbra in brain ischemia which is potentially salvageable in contrast to the nonrecoverable core. MRI has been successfully applied to resolve some of these questions.

Recently, combined DWI and perfusion-weighted imaging (PWI) has been applied to evaluation of ischemic core and penumbra.[49,50] A mismatch between DWI and PWI of the infarcted area provides some measure of the volume of tissue impacted by the core/penumbra. The core is considered to be the result of cytotoxic edema and its subsequent loss of parenchymal cellular constituents that is not rescuable by pharmacological or other therapies. In contrast the penumbra is typically not considered purely edematous tissue, but rather tissue that is at risk for imminent infarction if adequate reperfusion is not established. Thus the core is likely not rescuable, whereas penumbral tissue can be rescued if treatments to enhance perfusion to penumbral tissues are initiated quickly.[51] Indeed, if penumbral tissues are not rescued, tissues will become edematous and ultimately die and become cystic. Much effort is being spent in determining the optimal imaging strategies for visualize rescue of the penumbra.[52] Concordance between clinical and animal studies have been reported with future emphasis on automated computer vision methods for determination of core/penumbra tissues.[50,53,54]

Other MR imaging methods to study stroke and its evolution, such as magnetization transfer and susceptibility imaging, have not been as successful as standardized T2 and DWI imaging protocols in monitoring edema formation and evolution.[55,56] A relatively new extension of DWI, diffusion tensor imaging (DTI, a directionally encoding variant of DWI) has not been used solely for edema identification. A recent study found that DTI and diffusion kurtosis imaging (DKI) could differentiate brain tumor types based on DKI peritumoral edema signatures.[57] Research using DTI to identify edema into tumors and traumatic brain injury have been partially successful but typically the primary goal of these studies have not been to identify edema per se.[58,59] Future and ongoing studies will validate the significance of these emerging imaging techniques for edema evaluation.

Computed Tomography

CT has been used extensively in the acute setting for many acquired brain injuries, largely in part, due to its rapid acquisition techniques, its availability, and by its proximity to the emergency room. In addition, the lack of concerns about issues related to high-field magnetism from MRI make it an ideal imaging modality for patients on life support, those with implants or those with pace makers. Thus for acute neurological injuries CT is often the first imaging study, particularly for evaluation of hemorrhage.[60]

Imaging of edema using CT is not optimal compared to DWI/T2 MRI. However, in acute stroke during the evolution of edema, CT can be helpful.[61] It is important to note that CT is thought not to be able to visualize cytotoxic edema where there is no net increase of water, but only a shift of water from extracellular to intracellular compartments, i.e., ionic edema. No net change in water content during this phase does not affect X-ray attenuation in injured tissues and results in little or no change in the final image. However, during the early injury phase CT can evaluate for the presence of hemorrhage and may predict regions of developing edema.[60]

As edema moves from the cytotoxic to the vasogenic phase with accumulation of extracellular water, CT imaging can prove to be beneficial in monitoring acute stroke,[62] tumors,[63,64] intracerebral hemorrhage,[65] and traumatic brain injury.[66] Vasogenic edema typically appears as a region of hypodensity and often exaggerates the gray-white matter interface in clinical studies. Depending upon the neurological condition being assessed, occasionally the addition of imaging studies utilizing CT-contrast agents are performed, wherein increased enhancement can be seen either as a solid region intensity change or exhibiting a rim or patchy area of enhancement.[67] Another contraindication for use of CT is that if the abnormality of interest is close to surrounding bones, beam-hardening artifacts may be present and obscure subtle regions of edema or brain injury.

While CT may be the initial imaging modality of choice particularly for emergent neurological disorders, MRI (either T2 or DWI) is often ordered to confirm and extend the findings of the CT results. Thus MRI has the improved ability to localize and stratify location and type of edema with better resolution and temporal efficiency than CT.

Ultrasound

Use of ultrasound (US) for examining neonatal brain abnormalities and lesions is an established imaging modality.[68] However, given the anatomical resolution of MRI and CT in localizing edematous regions, the use of US has been limited despite its relative ease of use and low cost. There are some reports in the neonate of using US for edema localization particularly in comparative studies.[68,69] Head ultrasonography may provide an edge in the in acute postnatal period to assess neonatal hypoxia-ischemia for discriminating edema, hemorrhage, and regions of brain tissue abnormalities.[70] More traditionally, US is often used to measure cerebrovascular function instead of edema. However, more work is needed within the neonatal clinical care setting to determine clinical utility.

Nuclear Medicine Techniques

Neuroimaging using nuclear medicine modalities to visualize and identify edema after neurological injury have not used in routine clinical practice. However, nuclear medicine imaging typically requires the use of an exogenous agent to achieve contrast and depending on the radioligand of choice, imaging may be limited to specialized clinical centers. PET and single photon emission tomography (SPECT) imaging have been primarily utilized as confirmation of injury related processes. For example SPECT imaging can be used to assess blood flow to injured tissues which might be a surrogate marker for tissues at risk for or undergoing edema.[71]

PET imaging despite having poor anatomical resolution has excellent specificity and sensitivity that is dependent upon the radioligand of choice. Further considerations being explored are combined PET/CT and PET/MRI imaging modalities that can enhance diagnosis and elucidate the appearance of edema.[72,73] This is particularly true for neuroimaging of inflammation using radioligands such as translocater protein (TSPO) markers (i.e., ^{11}C-PBR28, etc.). Fujita et al. reported increased TSPO binding in the perilesional region that exhibited edema consistent with ongoing inflammation.[74] Other studies have posited that use of ^{11}C-methionine in tumor studies can discriminate between tumor infiltrative and vasogenic edema.[75] In an intracerebral hemorrhage model, metabolic imaging using 18F-fluorodeoxyglucose was also be used to identify regions of vasogenic edema.[76]

However, PET imaging is an emerging research area where some case reports have reported interesting concurrence of PET ligand binding and regions of edema that were confirmed either by CT or MRI.[77,78] However, given the relative lack of larger scale studies, it awaits to be seen if PET imaging can play a significant role in imaging edema formation and evolution. Certainly, in its current form, it appears that more direct methods such as those afforded by MRI appear to have a diagnostic and research edge.

Advanced and Emerging Imaging Approaches

Scientific and neuroimaging advances for the detection of edema are slowly evolving with the development of new imaging techniques, contrast agents and others. However, one area of rapid advancement is the use of computer techniques to rapidly and accurately to identify not only the regional localization of edema but also its volume- and tissue-level characteristics. Such approaches have been used in identification of edema related to macular degeneration,[79] radiogenomics for tumors,[80] edema segmentation,[81] and stroke.[82]

The goal of these computer vision approaches is to localize and extract various tissue parameters in an unbiased manner. There are numerous methods by which this can be performed computationally. For example, in tumors where there is a clear boundary edema segmentation has been performed using a hybrid level set approach.[83] In 10 patients when compared to a manual gold-standard these authors reported an 85% and 79%, percent matching and correspondence respectively, demonstrating excellent concordance for edema location and volumes. A more recent study in tumors and stroke utilized a hidden Markov random field model and associated expectation–maximization algorithms embedded within software as part of an automated segmentation approach.[84] Using appropriate thresholds they reported ~18% volume differences and ~26% volumetric overlap errors in tumor patients. Alternate computational algorithms have also been applied to brain tumors, such as optimization of cost functions.[81]

In computer vision techniques applied to stroke or ischemic injury, many studies simply attempt to define the region of the ischemic lesion without specificity of edema only regions. However, much of the "lesion" in stroke or ischemia is in fact considered to be edematous brain tissues. Our own recent series of publications demonstrated that in both animal models and human neonatal hypoxic ischemic injury we were able to demarcate regions of injury from normal appearing brain tissues using quantitative tissue values.[53,85,86] We elected to use hierarchical region splitting (HRS) in which unsupervised recursive evaluations are performed based on modeling of the whole brain histogram from MR images (Fig. 3.5). Using HRS at 1 and 28 days after hypoxic injury, we found that we could reliably identify (compared to manual gold standards) the edema containing regions with a high degree of correlation (r^2>0.95) from T2WI.[86]

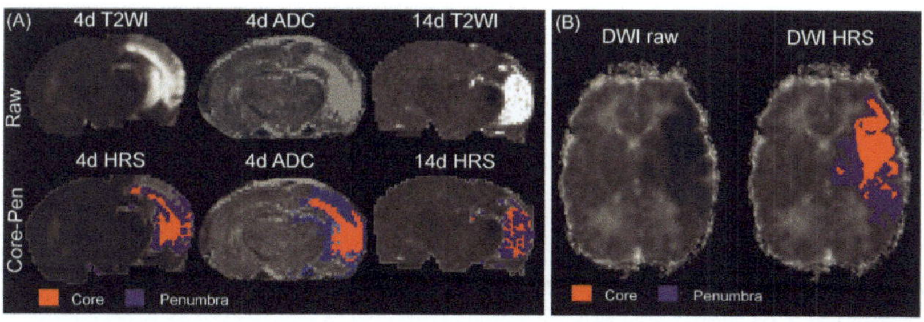

FIGURE 3.5 Computational assessment of edema. (A) In a rodent neonatal model of hypoxic-ischemic injury Hierarchical Region Splitting (HRS) was used on both T2 maps and DWI quantitative ADC maps to determine core/penumbra regions. In this example one can observe that the ADC has a much larger apparent core along with a larger penumbra than T2WI. At 14d when the lesion starts to become cystic the core/penumbra has become smaller. (B) HRS can also be used to determine core/penumbra regions in human stroke as in this example of neonatal hypoxic ischemic injury. HRS has been validated in both animal models and humans against standard DWI/perfusion weighted imaging (PWI) mismatch. *Source: Data courtesy of Drs. Ashwal, Ghosh and Obenaus, Loma Linda University.*

The initial success of this approach was then extended to determination of the core and penumbra tissues, where the penumbra is known to be highly edematous (vasogenic).[85] We successfully identified core/penumbra across a range of imaging time points (1–31 days postinjury) using a variety of MRI field strengths (1.5, 4.7, and 11.7T) in both human and rodent neonates (Fig. 3.5). We also confirmed our computational approach to the clinical standard of MRI DWI/PWI mismatch for core/penumbra determination. Our HRS method was adept at localizing core and penumbra volumes from DWI and PWI imaging and allowed monitoring of the evolution of the lesion and its edematous constituents.[53] Finally, we were also able to demonstrate that T2WI only was sufficient for computational core/penumbra determinations with excellent concordance to DWI/PWI mismatch. Thus computer methods can accelerate identification and quantitative assessments of edema in brain tissues. One important caveat is that no single method is optimal and in support of this we recently reported that combined methods may actually improve sensitivity and specificity in defining ischemia lesions and edema from MR images.[85] Considerable more research is warranted into computational approaches and can have the potential to impact delivery of clinical care. Interestingly, despite the existence of three edema phases as discussed in the beginning of this chapter, the imaging community is still proposing interpretation of the neuroimaging alterations based on a two model concept of edema. This could be due to the absence of new technology, development of new MRI sequences or adequate resolution in which to able to visualize the ionic component of the edema process. The future development of noninvasive should focus on measuring ionic edema part.

Optical coherence tomography (OCT) is an established form of imaging for edema in the eye.[87] OCT is excellent for imaging edema when optical transmission is not hindered by bone structures but there virtually no studies using OCT for evaluation of brain edema. OCT has been used in animal models to monitor vascular and hemodynamic changes and has the potential to monitor edema evolution.[88] Exploration of OCT for edema visualization should be explored.

Future directions in imaging of edema also point to combined imaging modalities. For example the combination of PET-CT and PET-MRI will further our understanding and ability to monitor the presence and evolution of edema.[73] In addition, use of computational tools to accelerate our knowledge about edema will be critical as use of therapeutics and imaging techniques are translated to the clinic.

SUMMARY

Neuroimaging of cytotoxic, ionic, and vasogenic edema is important for clinical diagnosis and therapy and for understanding translational scientific and mechanistic studies. Imaging modalities such as MRI and to a less extent CT have contributed greatly to our ability to visualize and quantify onset, development, and resolution of edema. MRI of DWI and T2 describes the early and late edematous responses after brain injury. Quantitative analysis of DWI/T2 measures provides additional evidence on the time course of edema formation and resolution and are particularly useful in the context of therapies that may reduce edema. Other imaging modalities, such as US, OCT, or PET, are not routinely being used for edema imaging but future developments may enhance their capabilities to provide additive and novel information as it relates to edema. Finally, computational methods for visualization of tissue dynamics (core vs. penumbra) will continue to contribute to our understanding of the subtleties of edema in diseases such as stroke. Thus neuroimaging provides an avenue to visualize the emergence of edema and its resolution, predominantly as directed therapies against edema are tested.

Acknowledgments

The authors wish to acknowledge the following funding: Department of Defense DCMRP #DR080470 (AO), NIH P01 NS082184-01 (AO Co-I) and NIH 1RO1HD061946-01 (JB), Fondation des gueules cassées (JB), ANR-TRAIL (JB) and eraNET neuron CNSaFlame (JB).

References

1. Nicholson C. Diffusion and related transport properties in brain tissue. *Rep Prog Phys* 2001;**64**:815–84.
2. Sykova E, Nicholson C. Diffusion in brain extracellular space. *Physiol Rev* 2008;**88**(4):1277–340.
3. Nicholson C, Sykova E. Extracellular space structure revealed by diffusion analysis. *Trends Neurosci* 1998;**21**(5):207–15.
4. Amiry-Moghaddam M, Ottersen OP. The molecular basis of water transport in the brain. *Nat Rev Neurosci* 2003;**4**(12):991–1001.
5. Obenaus A. Neuroimaging biomarkers for epilepsy: advances and relevance to glial cells. *Neurochem Int* 2013;**63**(7):712–8.
6. Dudek FE, Obenaus A, Tasker JG. Osmolality-induced changes in extracellular volume alter epileptiform bursts independent of chemical synapses in the rat: importance of non-synaptic mechanisms in hippocampal epileptogenesis. *Neurosci Lett* 1990;**120**(2):267–70.
7. Basser PJ. Inferring microstructural features and the physiological state of tissues from diffusion-weighted images. *NMR Biomed* 1995;**8**(7-8):333–44.
8. Le Bihan D, Basser PJ. Molecular diffusion and nuclear magnetic resonance. In: LeBihan D, editor. Diffusion and Perfusion Magnetic Resonance Imaging. Raven Press; New York: 1995. pp. 5–17.
9. Badaut J, Fukuda AM, Jullienne A, Petry KG. Aquaporin and brain diseases. *Biochim Biophys Acta* 2014;**1840**(5):1554–65.
10. Abbott NJ, Ronnback L, Hansson E. Astrocyte-endothelial interactions at the blood-brain barrier. *Nat Rev Neurosci* 2006;**7**(1):41–53.
11. Zhang JH, Badaut J, Tang J, Obenaus A, Hartman R, Pearce WJ. The vascular neural network—a new paradigm in stroke pathophysiology. *Nat Rev Neurol* 2012;**8**(12):711–6.

12. Zlokovic BV. The blood-brain barrier in health and chronic neurodegenerative disorders. *Neuron* 2008;**57**(2):178–201.
13. Tan F, Fu W, Cheng N, Meng DI, Gu Y. Ligustrazine reduces blood–brain barrier permeability in a rat model of focal cerebral ischemia and reperfusion. *Exp Ther Med* 2015;**9**(5):1757–62.
14. Jiao H, Wang Z, Liu Y, Wang P, Xue Y. Specific role of tight junction proteins claudin-5, occludin, and ZO-1 of the blood-brain barrier in a focal cerebral ischemic insult. *J Mol Neurosci* 2011;**44**(2):130–9.
15. Krueger M, Bechmann I, Immig K, Reichenbach A, Hartig W, Michalski D. Blood–brain barrier breakdown involves four distinct stages of vascular damage in various models of experimental focal cerebral ischemia. *J Cereb Blood Flow Metabol* 2015;**35**(2):292–303.
16. Unterberg AW, Stover J, Kress B, Kiening KL. Edema and brain trauma. *Neuroscience* 2004;**129**(4):1021–9.
17. Badaut J, Ashwal S, Obenaus A. Aquaporins in cerebrovascular disease: a target for treatment of brain edema? *Cerebrovasc Dis* 2011;**31**(6):521–31.
18. Risher WC, Andrew RD, Kirov SA. Real-time passive volume responses of astrocytes to acute osmotic and ischemic stress in cortical slices and in vivo revealed by two-photon microscopy. *Glia* 2009;**57**(2):207–21.
19. Unterberg AW, Stover J, Kress B, Kiening KL. Edema and brain trauma. *Neuroscience* 2004;**129**(4):1019–27.
20. Baethmann A, Van Harreveld A. Water and electrolyte distribution in gray matter rendered edematous with a metabolic inhibitor. *J Neuropathol Exp Neurol* 1973;**32**(3): 408–23.
21. Klatzo I, Piraux A, Laskowski EJ. The relationship between edema, blood–brain-barrier and tissue elements in a local brain injury. *J Neuropathol Exp Neurol* 1958;**17**(4):548–64.
22. Norris JW, Pappius HM. Cerebral water and electrolytes. Effect of asphyxia, hypoxia, and hypercapnia. *Arch Neurol* 1970;**23**(3):248–58.
23. Van Harreveld A, Collewijn H, Malhotra SK. Water, electrolytes, and extracellular space in hydrated and dehydrated brains. *Am J Physiol* 1966;**210**(2):251–6.
24. Simard JM, Kent TA, Chen M, Tarasov KV, Gerzanich V. Brain oedema in focal ischaemia: molecular pathophysiology and theoretical implications. *Lancet Neurol* 2007;**6**(3):258–68.
25. O'Donnell ME, Lam TI, Tran L, Anderson SE. The role of the blood-brain barrier Na-K-2Cl cotransporter in stroke. *Adv Exp Med Biol* 2004;**559**:67–75.
26. de Castro Ribeiro M, Badaut J, Price M, et al. Thrombin in ischemic neuronal death. *Exp Neurol* 2006;**198**(1):199–203.
27. Hirt L, Ternon B, Price M, Mastour N, Brunet JF, Badaut J. Protective role of early aquaporin 4 induction against postischemic edema formation. *J Cereb Blood Flow Metabol* 2009;**29**(2):423–33.
28. Knowland D, Arac A, Sekiguchi KJ, et al. Stepwise recruitment of transcellular and paracellular pathways underlies blood-brain barrier breakdown in stroke. *Neuron* 2014;**82**(3):603–17.
29. Strbian D, Durukan A, Pitkonen M, et al. The blood–brain barrier is continuously open for several weeks following transient focal cerebral ischemia. *Neuroscience* 2008;**153**(1):175–81.
30. Obenaus A, Ashwal S. Magnetic resonance imaging in cerebral ischemia: focus on neonates. *Neuropharmacology* 2008;**55**(3):271–80.
31. Bydder GM, Steiner RE, Young IR, et al. Clinical NMR imaging of the brain: 140 cases. *AJR Am J Roentgenol* 1982;**139**(2):215–36.
32. Naruse S, Horikawa Y, Tanaka C, Hirakawa K, Nishikawa H, Yoshizaki K. Proton nuclear magnetic resonance studies on brain edema. *J Neurosurg* 1982;**56**(6):747–52.
33. Warach S, Mosley M, Sorensen AG, Koroshetz W. Time course of diffusion imaging abnormalities in human stroke. *Stroke* 1996;**27**(7):1254–6.

34. Obenaus A, Ashwal S. Neuroimaging of stroke and ischemia in animal models. *Trans Stroke Res* 2012;**3**(1):4–7.
35. van der Toorn A, Sykova E, Dijkhuizen RM, et al. Dynamic changes in water ADC, energy metabolism, extracellular space volume, and tortuosity in neonatal rat brain during global ischemia. *Magn Reson Med* 1996;**36**(1):52–60.
36. Huisman TA. Diffusion-weighted and diffusion tensor imaging of the brain, made easy. *Cancer Imaging* 2011(10 Spec no A):S163–171.
37. Duong TQ, Fisher M. Applications of diffusion/perfusion magnetic resonance imaging in experimental and clinical aspects of stroke. *Curr Atheroscler Rep* 2004;**6**(4):267–73.
38. Ackerman JJ, Neil JJ. The use of MR-detectable reporter molecules and ions to evaluate diffusion in normal and ischemic brain. *NMR Biomed.* 2010;**23**(7):725–33.
39. Sotak CH. Nuclear magnetic resonance (NMR) measurement of the apparent diffusion coefficient (ADC) of tissue water and its relationship to cell volume changes in pathological states. *Neurochem Int* 2004;**45**(4):569–82.
40. Silva MD, Omae T, Helmer KG, Li F, Fisher M, Sotak CH. Separating changes in the intra- and extracellular water apparent diffusion coefficient following focal cerebral ischemia in the rat brain. *Magn Reson Med* 2002;**48**(5):826–37.
41. Moseley ME, Cohen Y, Mintorovitch J, et al. Early detection of regional cerebral ischemia in cats: comparison of diffusion- and T2-weighted MRI and spectroscopy. *Magn Reson Med* 1990;**14**(2):330–46.
42. Coats JS, Freeberg A, Pajela EG, Obenaus A, Ashwal S. Meta-analysis of apparent diffusion coefficients in the newborn brain. *Pediatr Neurol* 2009;**41**(4):263–74.
43. Silvera S, Oppenheim C, Touze E, et al. Spontaneous intracerebral hematoma on diffusion-weighted images: influence of T2-shine-through and T2-blackout effects. *AJNR Am J Neuroradiol* 2005;**26**(2):236–41.
44. Krieger DA, Dehkharghani S. Magnetic resonance imaging in ischemic stroke and cerebral venous thrombosis. *Top Mag Resonan Imaging* 2015;**24**(6):331–52.
45. Giraud M, Cho TH, Nighoghossian N, et al. Early blood brain barrier changes in acute ischemic stroke: a sequential MRI study. *J Neuroimaging* 2015;**25**(6):959–63.
46. Rollins N, Winick N, Bash R, Booth T. Acute methotrexate neurotoxicity: findings on diffusion-weighted imaging and correlation with clinical outcome. *Am J Neuroradiol* 2004;**25**(10):1688–95.
47. Artzi M, Bokstein F, Blumenthal DT, et al. Differentiation between vasogenic-edema versus tumor-infiltrative area in patients with glioblastoma during bevacizumab therapy: a longitudinal MRI study. *Eur J Radiol* 2014;**83**(7):1250–6.
48. Piper RJ, Mikhael S, Wardlaw JM, Laidlaw DH, Whittle IR, Bastin ME. Imaging signatures of meningioma and low-grade glioma: a diffusion tensor, magnetization transfer and quantitative longitudinal relaxation time MRI study. *Magn Reson Imaging* 2015.
49. Fisher M, Albers GW. Advanced imaging to extend the therapeutic time window of acute ischemic stroke. *Ann Neurol* 2013;**73**(1):4–9.
50. Scalzo F, Nour M, Liebeskind DS. Data science of stroke imaging and enlightenment of the penumbra. *Front Neurol* 2015;**6**:8.
51. Jung S, Gilgen M, Slotboom J, et al. Factors that determine penumbral tissue loss in acute ischaemic stroke. *Brain* 2013;**136**(Pt 12):3554–60.
52. Hirano T. Searching for salvageable brain: the detection of ischemic penumbra using various imaging modalities? *J Stroke Cerebrovasc Dis* 2014;**23**(5):795–8.
53. Ghosh N, Yuan X, Turenius CI, et al. Automated core-penumbra quantification in neonatal ischemic brain injury. *J Cereb Blood Flow Metabol* 2012;**32**(12):2161–70.
54. Werner P, Saur D, Zeisig V, et al. Simultaneous PET/MRI in stroke: a case series. *J Cereb Blood Flow Metabol* 2015;**35**(9):1421–5.
55. Sibon I, Tourdias T, Felix S, et al. Magnetisation transfer parameters and stroke outcome. *J Clin Neurosci* 2015;**22**(6):1012–7.

56. Polan RM, Poretti A, Huisman TA, Bosemani T. Susceptibility-weighted imaging in pediatric arterial ischemic stroke: a valuable alternative for the noninvasive evaluation of altered cerebral hemodynamics. *Am J Neuroradiol* 2015;**36**(4):783–8.
57. Tan Y, Wang XC, Zhang H, et al. Differentiation of high-grade-astrocytomas from solitary-brain-metastases: comparing diffusion kurtosis imaging and diffusion tensor imaging. *Eur J Radiol* 2015;**84**(12):2618–24.
58. Chiang CW, Wang Y, Sun P, et al. Quantifying white matter tract diffusion parameters in the presence of increased extra-fiber cellularity and vasogenic edema. *NeuroImage* 2014;**101**:310–9.
59. Hoefnagels FW, De Witt Hamer P, Sanz-Arigita E, et al. Differentiation of edema and glioma infiltration: proposal of a DTI-based probability map. *J Neurooncol* 2014;**120**(1):187–98.
60. Fink KR, Benjert JL. Imaging of nontraumatic neuroradiology emergencies. *Radiol Clin North Am* 2015;**53**(4):871–90.
61. von Kummer R, Dzialowski I, Gerber J. Therapeutic efficacy of brain imaging in acute ischemic stroke patients. *JNeuroradiol* 2015;**42**(1):47–54.
62. Copen WA. Multimodal imaging in acute ischemic stroke. *Curr Treat Opt Cardiovasc Med* 2015;**17**(3):368.
63. Urday S, Kimberly WT, Beslow LA, et al. Targeting secondary injury in intracerebral haemorrhage–perihaematomal oedema. *Nat Rev Neurol* 2015;**11**(2):111–22.
64. Dhermain F. Radiotherapy of high-grade gliomas: current standards and new concepts, innovations in imaging and radiotherapy, and new therapeutic approaches. *Chin J Cancer* 2014;**33**(1):16–24.
65. Naidech AM. Intracranial hemorrhage. *Am J Respir Crit Care Med* 2011;**184**(9):998–1006.
66. Kubal WS. Updated imaging of traumatic brain injury. *Radiol Clin North Am* 2012;**50**(1):15–41.
67. Gutierrez LG, Rovira A, Portela LA, Leite Cda C, Lucato LT. CT and MR in non-neonatal hypoxic-ischemic encephalopathy: radiological findings with pathophysiological correlations. *Neuroradiology* 2010;**52**(11):949–76.
68. Yoon HJ, Kim JH, Jeon TY, Yoo SY, Eo H. Devastating metabolic brain disorders of newborns and young infants. *Radiographics* 2014;**34**(5):1257–72.
69. Shen W, Pan JH, Chen WD. Comparison of transcranial ultrasound and cranial MRI in evaluations of brain injuries from neonatal asphyxia. *Int J Clin Exp Med* 2015;**8**(10):18319–26.
70. Pinto PS, Tekes A, Singhi S, Northington FJ, Parkinson C, Huisman TA. White-gray matter echogenicity ratio and resistive index: sonographic bedside markers of cerebral hypoxic-ischemic injury/edema? *J Perinatol* 2012;**32**(6):448–53.
71. Kanemaru K, Kinouchi H, Yoshioka H, et al. Cerebral hemodynamic disturbance in dural arteriovenous fistula with retrograde leptomeningeal venous drainage: a prospective study using (123)I-iodoamphetamine single photon emission computed tomography. *J Neurosurg* 2015;**123**(1):110–7.
72. Salmon E, Bernard Ir C, Hustinx R. Pitfalls and limitations of PET/CT in brain imaging. *Semin Nucl Med* 2015;**45**(6):541–51.
73. Fink JR, Muzi M, Peck M, Krohn KA. Multimodality brain tumor imaging: MR imaging, PET, and PET/MR imaging. *J Nucl Med* 2015;**56**(10):1554–61.
74. Fujita M, Mahanty S, Zoghbi SS, et al. PET reveals inflammation around calcified Taenia solium granulomas with perilesional edema. *PLoS One*. 2013;**8**(9):e74052.
75. Kinoshita M, Goto T, Arita H, 'et al. Imaging (1)(8)F-fluorodeoxy glucose/(1)(1) C-methionine uptake decoupling for identification of tumor cell infiltration in peritumoral brain edema. *J Neurooncol* 2012;**106**(2):417–25.
76. Lin X, Tang Y, Sun B, et al. Cerebral glucose metabolism: Influence on perihematomal edema formation after intracerebral hemorrhage in cat models. *Acta Radiol.* 2010;**51**(5):549–54.

77. Kim YJ, Kwon SU. Recurrent steroid-responsive cerebral vasogenic edema in status migrainosus and persistent aura. *Cephalalgia* 2015;**35**(8):728–34.
78. Susheela SP, Revannasiddaiah S, Madhusudhan N, Bijjawara M. The demonstration of extension of high-grade glioma beyond magnetic resonance imaging defined edema by the use of (11) C-methionine positron emission tomography. *J Cancer Res Therap* 2013;**9**(4):715–7.
79. Sundaresan V, Ram K, Joshi N, Sivaprakasam M, Gandhi R. Computer-assisted grading of diabetic macular edema on retinal color fundus images. Conference proceedings: annual international conference of the IEEE engineering in medicine and biology society. 2015;2015:4330–3.
80. Ellingson BM. Radiogenomics and imaging phenotypes in glioblastoma: novel observations and correlation with molecular characteristics. *Curr Neurol Neurosci Rep* 2015;**15**(1):506.
81. Njeh I, Sallemi L, Ayed IB, et al. 3D multimodal MRI brain glioma tumor and edema segmentation: a graph cut distribution matching approach. *Comput Med Imaging Graphics* 2015;**40**:108–19.
82. Mohamed FB, Vinitski S, Gonzalez CF, et al. Increased differentiation of intracranial white matter lesions by multispectral 3D-tissue segmentation: preliminary results. *Magn Reson Imaging* 2001;**19**(2):207–18.
83. Xie K, Yang J, Zhang ZG, Zhu YM. Semi-automated brain tumor and edema segmentation using MRI. *Eur J Radiol* 2005;**56**(1):12–19.
84. Artzi M, Aizenstein O, Jonas-Kimchi T, Myers V, Hallevi H, Ben Bashat D. FLAIR lesion segmentation: application in patients with brain tumors and acute ischemic stroke. *Eur J Radiol* 2013;**82**(9):1512–8.
85. Ghosh N, Sun Y, Bhanu B, Ashwal S, Obenaus A. Automated detection of brain abnormalities in neonatal hypoxia ischemic injury from MR images. *Med Image Anal* 2014;**18**(7):1059–69.
86. Ghosh N, Recker R, Shah A, Bhanu B, Ashwal S, Obenaus A. Automated ischemic lesion detection in a neonatal model of hypoxic ischemic injury. *J Magn Reson Imaging* 2011;**33**(4):772–81.
87. Nolan RC, Narayana K, Galetta SL, Balcer LJ. Optical coherence tomography for the neurologist. *Semin Neurol* 2015;**35**(5):564–77.
88. Liu G, Chen Z. Advances in Doppler OCT. *Chin Opt Lett* 2013;**11**(1):11702.

CHAPTER

4

Experimental Techniques to Investigate the Formation of Brain Edema In Vivo

Nikolaus Plesnila

University of Munich Medical Center & Munich Cluster for Systems Neurology (Synergy), Munich, Germany

OUTLINE

KEY PHRASES

- Brain water content (BWC) is the key parameter for the detection of brain edema.
- The assessment of the volume of the brain's extracellular space (ECS) is an accepted measure for cytotoxic brain edema.
- The ECS can be quantified by electrical impedance or by diffusion measurement using ion-selective microelectrodes (TAM^+ method).

Brain Edema.
DOI: http://dx.doi.org/10.1016/B978-0-12-803196-4.00004-7

- Vasogenic brain edema occurs if the blood–brain barrier (BBB) becomes leaky and BWC increases.
- Vasogenic edema can be identified by an increase in BWC together with the presence of plasma proteins or tracers in the brain parenchyma.

INTRODUCTION

Brain edema formation is still one of the most life threatening complications following brain injury caused by ischemic stroke, cerebral hemorrhage, cardiac arrest, or traumatic brain injury.[1–4] Brain edema results in an overall increase in brain volume or, simply, in brain swelling which due to the rigid cranial vault may result in intracranial hypertension, reduced cerebral perfusion pressure, global cerebral ischemia, and finally in brain death. In parallel, excessive brain swelling may also result in acute herniation of the brain stem and death due to respiratory arrest. Accordingly, already decades ago brain edema was identified as one of the most important targets for the prevention of secondary brain damage following acute brain injury.[5–9] Despite considerable research efforts, however, even today no established medical therapy is available for the treatment of brain edema. Therefore, more experimental research with improved methodology and technology are at urgent need in order to generate novel therapeutic concepts and innovative approaches toward the establishment of new therapies for the prevention and reduction of brain swelling.

IDENTIFICATION OF BRAIN EDEMA

Brain edema is defined as an increase in brain water content (BWC).[10] BWC may increase through various mechanisms; however, there are only two sources of water within the cranial cavity which may result in an increase of water content in the brain parenchyma: One is the vascular system and the other one is the glymphatic system, a recently described circulation system within the brain which drives cerebro-spinal fluid (CSF) along arteriolar perivascular routes through the brain parenchyma and drains it into the venular perivascular space[11,12] (for details see Chapter 7: Blood–Brain Barrier Mechanisms of Edema Formation: The Role of Ion Transporters and Channels). Hence, an increase of BWC can only occur though an increased flow of water from the intravascular space into the brain parenchyma by an increase in glymphatic influx or by a reduced glymphatic efflux. Although reduced glymphatic efflux was recently demonstrated to increase BWC chronically,[13] it has to be determined whether the glymphatic system is also involved in the formation of acute brain edema formation and to what extent. Therefore, in this chapter we will mainly assume that the main source of water entering the brain during acute pathological conditions derives from the intravascular space. Independent of the source of water leading to the formation of brain edema, the primary parameter to determine whether brain edema is present or not is in any case the increase in BWC.

Measurement of Brain Water Content

The classical method to determine BWC experimentally is the wet–dry method, i.e., removing the water from the brain and determining the BWC by subtracting the dry weight of the brain from its wet weight. The easiest and therefore the most widely used technique to remove water from brain tissue is to dry the brain by heating to 90–110°C for 24–48 hours.[14–16] Results are then expressed as the share of water to the total weight of the brain using the following formula:

$$\%\,Brain\,water\,content\,(BWC) = 100 \times (wet\,weight - dry\,weight)/wet\,weight$$

Using this method the typical BWC of rodent brain is determined to be around 78.5% with a variability between laboratories of about ±0.5%.[14–16] The variability within the same laboratory as assessed by the standard error (SD) is usually less than 0.25%, i.e., the SD is less than 0.3% of the mean. This is a very good value for the determination of a biological variable.

The method of heat drying has two important limitations: (1) the sample is burned during the drying procedure and cannot be used for further analysis and (2) the variability increases with decreasing sample weight, i.e., the method already reaches its limits when measuring the BWC of whole mouse brain. In order to overcome these limitations, at least partially, it was recently suggested to use vacuum drying for the determination of BWC.[17] This technique seems to be able to determine BWC in smaller brain samples as comparted to the heat drying method (30 vs 60 mg). In addition, vacuum drying does not destroy the sample during the drying process and further analysis can be performed without the necessity to perform additional experiments.

Under pathological conditions such as cerebral ischemia or traumatic brain injury the BWC of the injured hemisphere typically increases by 3%–4% from 78.5% in control animals to 82% in injured animals.[14–16] BWC in samples mainly containing injured brain may even increase by 8%.[17] These values may seem small at first, however, when considering that tissue swelling increases exponentially with BWC and therefore a 4% increase in BWC results in a 25% increase in brain volume, it becomes clear that even small therapeutic reductions in BWC may have highly significant effects on edema formation, intracranial pressure, and midline shift and, hence, on overall outcome.[16]

In addition to the determination of BWC in experimental animals, a relatively new development is the ability to measure BWC in absolute values and in real time in patients using thermal conductivity (TC). TC is a parameter which is acquired as a control parameter when measuring cerebral blood flow in patients using a thermodilution probe.[18] Using this technique Ko and colleagues were able to measure BWC in the white matter of comatose brain injured patients. BWC was around 70% in normal white matter, a value corresponding quite well to that obtained experimentally in cats,[19] and increased to 79% in brain regions appearing edematous on computed tomography. Osmotherapy with mannitol or hypertonic saline decreased this value by more than 1%.[18] The method is not widely used so far; however, it may have significant clinical potential since the TC device is commercially available and already present in many neuro-intensive care units and the measurements seem to be reliable and are able to monitor therapeutic effects.

The disadvantage of the various wet–dry methods is that relatively large samples are required to obtain accurate and reproducible BWC values and rapid assessment of tissue water is not possible since the time necessary to dry brain tissue usually requires at least 24 hours. These disadvantages of the determination of BWC can be overcome by determining the specific gravity (SG) of a tissue sample using gravimetric techniques. First used by Laboite and Weber in 1965 and later improved by Nelson and colleagues and Marmarou and colleagues in the 1970s and 1980s,[19–22] in this method, the specimen is submerged in a layered mixture of kerosene and bromobenzene of known density. The sample will sink in this mixture until its SG equals that of the surrounding liquid. BWC of the sample can then be calculated with the following formula[22]:

$$BWC\,(g\,H_2O\,per\,g\,tissue) = 1 - \{(SG_{wet\,tissue} - 1)/(1 - 1/SG_{dry\,tissue}) \times SG_{wet\,tissue})$$

Nelson and colleagues validated this formula for the determination of BWC in mouse brain injected with water. Since the SG of dry tissue was not affected by the injection of water, this value was determined experimentally and inserted in the formula as a constant. Hence, the determination of BWC required only the SG of the wet tissue, a value which can be determined within a few minutes. Later on Marmarou and colleagues realized that when the formula is used for the determination of BWC in edematous brain tissue, Nelson's assumption that the SG of dry tissue remains constant is not fully valid since edema fluid consists also of a considerable amount of proteins. In a series of elegant experiments injecting serum instead of water into the brain parenchyma they corrected Nelson's formula for the change of the SG of dry tissue during edema formation and came up with the following simplified formula[21]:

$$BWC\ (g\ H_2O\ per\ g\ tissue) = (1597.3/SG_{wet\ tissue}) - 1464.9$$

Using this formula BWC of injured brain tissue can be reliably measured between 68% and 85%.[21]

The disadvantage of SG measurements is that they require a very precise preparation of the density columns,[19] a skill which seems to be lost since this technique is not used very often nowadays.

Taken together, in the laboratory BWC can be measured with two techniques. The wet–dry method is relatively easy to perform and does not need specialized laboratory equipment, but needs relatively large samples, takes time (>24 hours), and usually destroys the sample; the SG method needs only small pieces of tissue for reliable measurements and preserves the sample for further analysis, but requires the very precise preparation of the gradient columns.

DETECTION OF CYTOTOXIC BRAIN EDEMA

Inhibition of brain metabolism or shortage of cerebral energy supply, e.g., by uncoupling of mitochondrial respiration using 2,4-dinitrophenol or during cardiac arrest, stroke, and brain trauma results mainly in swelling of astrocytes and some neuronal processes and subsequent shrinkage of the brain's extracellular space (ECS).[5,23] Based on these considerations swelling of astrocytes and shrinkage of the ECS of the brain are the appropriate parameters to determine cytotoxic edema.

The most direct way to determine the volume of astrocytes and the brain's ECS is tissue fixation and visualization of these parameters by electron microscopy (EM). Unfortunately, most if not all fixation procedures cause some degree of tissue ischemia and may therefore cause artefactual changes of astrocytic and ECS volume.[24] The only reasonable technical approach seems to be freeze-substitution, a technique which allows us to study the upper 10–15 µm of the cerebral cortex without significant artifacts.[25] However, even when using this technique only the determination of the ECS may be possible by EM, since even massive brain swelling is believed to change the diameter of astrocytes by less than 6%, a value which cannot be quantified by EM.[26] By using the freeze-substitution technique it was at least possible to estimate that the ECS of the brain is about 20% of the total brain volume, that the main cellular components affected by swelling are astrocytes, and that the ECS shrinks significantly under conditions of asphyxia or global cerebral ischemia.[25,26]

Since EM allows us to study only very small areas of the brain and to perform only end-point measurements, it is understandable that researchers were looking for alternative techniques to study such a dynamic process as cytotoxic edema. The first dynamic method to estimate the volume of the ECS in the brain used electrical impedance.[27,28] Impedance, the reciprocal of electrical conductivity, is measured by placing electrodes within the brain parenchyma, passing constant current pulses through the tissue, and measuring the evolving conductivity. At first a variety of set-ups were used;[27–30] however, later on mostly four- or multielectrode systems were utilized which avoided the problem of electrode polarization.[31–33] In most cases these systems consisted of two pairs of one current injecting electrode and one voltage-sensing electrode placed within the gray or the white matter at a distance of approximately 4–10 mm. The lower the conductivity and the higher the impedance, the smaller the ECS and the more pronounced cytotoxic edema. With these set-ups it was shown that the ECS of the brain shrinks almost immediately by approximately 50% following global and focal cerebral ischemia and recovers within minutes upon normalization of cerebral blood flow.[31,34,35]

Since the above described electrode set-ups used relatively large metal electrodes they were only able to measure cytotoxic edema in large tissue volumes, i.e., measurements on the cellular level were not possible. This limitation was overcome in the late-1990s by using multibarreled glass microelectrodes for impedance measurements.[36] The additional advantage of this technique was that the impedance could be measured simultaneously to the local injection of pharmacologically active substances through an open channel of the glass pipette. Thereby local swelling of cells could be investigated, e.g., under conditions of high extracellular glutamate or potassium.

The brain's ECS can not only be measured by "injection" of electricity, but also by injection of tracers into the brain parenchyma and recording of their spatial distribution. This technique was pioneered by Fenstermacher and Patlak in the early 1970s using radioactive tracers.[37,38] Due to long incubation times and the inability to measure dynamic changes of the ECS this technique was replaced in the early 1980s by the real-time iontophoretic method using teramethylammonium (TMA^+) as a tracer.[39] TMA^+ is applied to the ECS of the brain parenchyma with a micropipette using iontophoresis (tip diameter 2–10 μm) and then recorded at a distance of 100–200 μm with an ion-selective microelectrode. The diffusion of TMA^+ in brain tissue is then used to calculate the volume fraction α which is a direct measure of ECS volume. Additionally, also the diffusion of TMA^+ within the brain can be quantified by calculating the apparent diffusion coefficient (ADC or D*). Using the TMA^+ method the ECS was determined to be around 20% of the total tissue volume in mammalian cerebral cortex, cerebellum, and spinal cord.[40] After hypoxia, terminal hypoxia, or global cerebral ischemia the volume fraction shrank from 0.2 to 0.04, i.e., the ECS decreased from 20% to 4%.[41–43] Based on the accuracy and the validity of the TMA^+ method for the determination of the ECS during physiological and pathological conditions, it is quite surprising that this technique has not been used more frequently to study the degree and dynamics of cytotoxic brain edema and to evaluate therapeutic options.

More recently, also optical methods were used to determine the diffusion of substances within the ECS in vivo and in real time by using fluorescent tracers. One approach, termed integrative optical imaging (IOI), pressure injected high molecular weight markers such as FITC dextrans into the ECS of the brain and measured the spatial diffusion profile of the tracers with a CCD camera.[44,45] A more recent approach used a noninvasive transdural dye loading approach to load the whole cortical ECS with fluorescein isothiocyanate (FITC) dextrans of various molecular weights, bleached the fluorescence in a portion of the cortex with a laser, and monitored the return of the fluorescent signal using a photomultiplier, a technique also known as fluorescence recovery after photobleaching (FRAP).[46] Using this elegant, noninvasive technique the authors could nicely demonstrate hindered diffusion after application of glutamate and epileptic seizures suggesting shrinkage of the ECS during these conditions.[46]

Taken together, the current methods to measure cytotoxic edema in vivo do not measure cell swelling directly because volume changes of individual cells are too small to be reliably detected. Instead all methods determine the volume of the ECS as an indirect parameter of cytotoxic edema formation since all available data generated within the last four decades indicate that changes of the ECS under pathological conditions reflect cytotoxic cell swelling. A variety of electrophysiological and optical methods are available to measure the ECS either directly (electrical impedance and the TMA^+ method) or indirectly (IOI and FRAP) and await application for the multiple open question in field of cytotoxic brain edema research.

DETECTION OF VASOGENIC BRAIN EDEMA

Vasogenic brain edema results from dysfunction of the blood–brain barrier (BBB), a process often termed "opening of the BBB" despite the fact that the BBB is a highly selective barrier (for details see Chapter 3: Noninvasive Imaging Techniques for Brain Edema: From Basic Science to the Clinic) and its dysfunction may therefore cause very selective permeability changes which are not compatible with the conceptual idea that the BBB opens or closes like a simple gate. This dysfunction results in a general or selective increase of the permeability of the BBB toward soluble molecules located in the intravascular space and is followed by an increase in BWC. Accordingly, the measurement of vasogenic brain edema is performed by detecting an increase in BWC together with the presence of molecules within the brain parenchyma which are usually only present in blood. Various techniques have been developed for this purpose over the past decades.

The first method to investigate the BBB also resulted in its discovery. In 1885 Paul Ehrlich, the latter Nobel prize laureate and discoverer of the immune system, injected hydrophilic dyes into the blood stream of experimental animals in order to investigate the affinity of these substances to various tissues. He observed that all tissues except the brain took up the dyes. Ehrlich's interpretation that the brain lacks affinity for these dyes was later on refuted by one of his coworkers, Edwin Goldman, who injected the same dyes into the CSF and observed selective staining of the parenchyma of brain and spinal cord. In 1900 these and other experiments resulted in the conclusion that cerebral blood vessels were responsible for keeping dyes out of the brain[47] and, finally, resulted in the concept that brain vessels form a "barrier" by Stern and Gautier in 1918.[48] Refer to Saunders et al. for a detailed review on this topic.[49]

In the past century various dyes and tracers were developed and used for the assessment of BBB permeability.[50,51] Trypan blue, synthetized by Paul Ehrlich, was among the first ones used for this purpose, while Evans Blue (EB) and horseradish peroxidase (HRP) became the most popular ones. EB is a blue azo dye which is also fluorescent and binds with particularly high affinity to several serum proteins including albumin making it essentially a marker for serum proteins. EB can be applied intravenously or intraperitoneally and can be detected visually or by fluorescence microscopy in brain slices. Using colorimetric or fluorescence detection it can also be quantified in brain homogenates. Despite its frequent use, EB is far away from being an ideal plasma marker and therefore also an ideal indicator substance for BBB permeability.[51] The main reasons are that EB is unstable in saline solutions, that it binds to brain tissue, that after injection into the blood stream it exists in bound and free form, and that it detects mainly the permeability of the BBB to large molecules. In view of better suited molecules (see below) some authors discourage the further use of EB for studying BBB permeability.[51]

The second "historic" molecule for the study of the BBB is HRP, a protein with a molecular weight of 40 kDa. HRP can be injected intravenously or intracerebroventricularly and its localization can be visualized by incubation of brain slices with one of its substrates, e.g., diaminobenzidine (DAB). This results in the production of a visible or electron dense product which can then be investigated by light or electron microscopy. By using this technique Brightman and Reese established that intercellular tight junctions between cerebrovascular endothelial cells formed the ultrastructural basis for the BBB.[52] Despite its clear value for this landmark finding, the use of HRP has also some technical problems. The most serious one is that HRP causes release of histamine and serotonin in rodents from degranulating mast cells thereby causing increased vascular permeability in various organs.[51] Since HRP should only be used at low doses and with coadministration of histamine and serotonin receptor antagonists and better suited markers are available (see below), also the use of HRP for the assessment of vasogenic brain edema is not recommended any more.

An alternative technique for detecting the extravasation of plasma into the brain parenchyma is the use of radio-labeled plasma markers or plasma proteins, e.g., inulin, sucrose, or albumin. The main advantage of this approach is that it allows the quantitative determination of the permeability of the BBB with high accuracy as shown by Woodbury, Davson, and Oldendorf in the 1960s.[53] Unfortunately the accuracy of the technique makes it quite vulnerable for artifacts caused by contamination of brain tissue samples with blood. This is particularly the case when animal models of disease are used which involve vascular stasis or bleedings as often observed following cerebral ischemia or brain trauma. Since the other important disadvantage of this technique is that radiolabels cannot be visualized in tissue, radio-labeled plasma markers or plasma proteins are not often used in experimental studies assessing vasogenic brain edema.

An easy to perform technique for the detection of plasma extravasation into the brain parenchyma is the detection of endogenous proteins, e.g., albumin, immunoglobulins, or fibrinogen, by immunohistochemistry. Detection can be performed with readily available laboratory equipment, with high spatial resolution, and in 3-D when using thick sections and confocal microscopy.[54] The concentration of tracers in blood is stable, no toxicity is present, and animals do not need to be handled during the experiment. Since the tracers are present in blood during the whole experiment such an approach, however, does not allow us to detect the precise time point of BBB dysfunction. Accordingly, endogenous tracers are specifically suitable for long-term experiments where chronic dysfunction of the BBB rather than acute vasogenic edema is expected, e.g., studies on slowly occurring neuronal or vascular degeneration.[54–57]

Most, if not all, disadvantages of the plasma tracers described so far can be overcome by the use of exogenously applied fluorescence or biotin-labeled plasma markers as first demonstrated with sodium-fluorescein already in the 1960.[58] Such markers, specifically since the introduction of dextrans, are available in a whole variety of molecular weights and colors and are therefore suited to investigate the permeability of the BBB for small and large molecules and at various time points. At high plasma concentrations dextrans may have the same adverse effects as HRP (see above), but improvements in labeling intensities and detection techniques reduced the concentrations necessary for reliable detection of the tracers ex vivo and in vivo to such low level that no short- or long-term adverse effects are observed nowadays.[51]

The main advantage of exogenously applied plasma markers is that they can be visualized by fluorescence or electron microscopy, i.e., from the mesoscopic tissue level down to capillaries, single cells, and subcellular structures. This was previously demonstrated for the whole brain, brain slices, and brain homogenates.[50] Most importantly, however, the low toxicity of the compounds, their acceptable plasma half-life, and their bright fluorescence makes these markers ideally suitable for in vivo applications. Such an approach has the important advantage that extravasation of plasma tracers can be visualized in individual animals over time, i.e., in a dynamic manner. Visualization can be performed with little to no invasiveness on the whole brain by using noninvasive fluorescence imaging (NFI) or intravital microscopy.[59–62] NFI and other epifluorescence techniques, e.g., epifluorescence in vivo microscopy,[61,62] integrate fluorescence signals coming from different depth of the brain and are therefore not able to quantify the amount of extravagated marker molecules due to signal overlay. This problem, however, was overcome when 2-photon microscopy became available.[60] 2-PM microscopy is able to penetrate brain tissue up to 600 μm, i.e., fluorescence signals are acquired from different depth of the brain and can therefore be reconstructed three-dimensionally. Thereby the exact localization of extravasation and the exact amount of extravagated fluorophores can be quantified in a dynamic manner over time.[59,60] Using this technique we could recently show that vasogenic edema fluid following traumatic brain injury does not only extravasate from arterioles into the brain parenchyma—as previously believed—but to the almost same amount also from venules.[60]

Taken together, a plethora of well-established and valid experimental techniques are available for the detection of brain edema and for the differentiation of the various forms of brain edema, i.e., cytotoxic and vasogenic. In view of this solid technical basis and the increased interest in BBB function and dysfunction, instructive advances in the field of brain edema research which are urgently needed for the development of novel therapeutic options for patients suffering from brain swelling should be expected in the near future.

References

1. Zheng H, Chen C, Zhang J, Hu Z. Mechanism and therapy of brain edema after intracerebral hemorrhage. *Cerebrovasc Dis* 2016;**42**:155–69.
2. Brogan ME, Manno EM. Treatment of malignant brain edema and increased intracranial pressure after stroke. *Curr Treat Options Neurol* 2015;**17**:327.
3. Michinaga S, Koyama Y. Pathogenesis of brain edema and investigation into anti-edema drugs. *Int J Mol Sci* 2015;**16**:9949–75.

4. Kuroiwa T. International brain edema symposia 1967–2011. *Acta Neurochir Suppl* 2013;**118**:1–9. http://dx.doi.org/10.1007/978-3-7091-1434-6_1.
5. Klatzo I. Presidental address. Neuropathological aspects of brain edema. *J Neuropathol Exp Neurol* 1967;**26**:1–14.
6. Katzman R, Clasen R, Klatzo I, Meyer JS, Pappius HM, Waltz AG. Report of Joint Committee for Stroke Resources. IV. Brain edema in stroke. *Stroke* 1977;**8**:512–40.
7. Go KG. Pathophysiological aspects of brain edema. *Clin Neurol Neurosurg* 1984;**86**:77–80.
8. Baethmann A, Maier-Hauff K, Kempski O, Unterberg A, Wahl M, Schurer L. Mediators of brain edema and secondary brain damage. *Crit Care Med* 1988;**16**:972–8.
9. Hossmann KA. The pathophysiology of experimental brain edema. *Neurosurg Rev* 1989;**12**:263–80.
10. Pappius HM. The distribution of water in brain tissues swollen in vitro and in vivo. *Prog Brain Res* 1965;**15**:135–54.
11. Iliff JJ, Lee H, Yu M, et al. Brain-wide pathway for waste clearance captured by contrast-enhanced MRI. *J Clin Invest* 2013;**123**:1299–309.
12. Jessen NA, Munk AS, Lundgaard I, Nedergaard M. The glymphatic system: a beginner's guide. *Neurochem Res* 2015;**40**:2583–99.
13. Vindedal GF, Thoren AE, Jensen V, et al. Removal of aquaporin-4 from glial and ependymal membranes causes brain water accumulation. *Mol Cell Neurosci* 2016;**77**:47–52. http://dx.doi.org/10.1016/j.mcn.2016.10.004.
14. Groger M, Lebesgue D, Pruneau D, et al. Release of bradykinin and expression of kinin B2 receptors in the brain: role for cell death and brain edema formation after focal cerebral ischemia in mice. *J Cereb Blood Flow Metab* 2005;**25**:978–89.
15. Vakili A, Kataoka H, Plesnila N. Role of arginine vasopressin V1 and V2 receptors for brain damage after transient focal cerebral ischemia. *J Cereb Blood Flow Metab* 2005;**25**:1012–19.
16. Keep RF, Hua Y, Xi G. Brain water content. A misunderstood measurement? *Transl Stroke Res* 2012;**3**:263–5.
17. Sebastiani A, Hirnet T, Jahn-Eimermacher A, Thal SC. Comparison of speed-vacuum method and heat-drying method to measure brain water content of small brain samples. *J Neurosci Methods* 2016:10.
18. Ko SB, Choi HA, Parikh G, et al. Real time estimation of brain water content in comatose patients. *Ann Neurol* 2012;**72**:344–50.
19. Marmarou A, Poll W, Shulman K, Bhagavan H. A simple gravimetric technique for measurement of cerebral edema. *J Neurosurg* 1978;**49**:530–7.
20. Labotite H, Weber B. Physiolopathology of cerebral edema. Experimental study of the metabolic mechanism, its appearance and its treatment. *Agressologie* 1965;**6**:743–59.
21. Marmarou A, Tanaka K, Shulman K. An improved gravimetric measure of cerebral edema. *J Neurosurg* 1982;**56**:246–53.
22. Nelson SR, Mantz ML, Maxwell JA. Use of specific gravity in the measurement of cerebral edema. *J Appl Physiol* 1971;**30**:268–71.
23. Baethmann A, Van HA. Water and electrolyte distribution in gray matter rendered edematous with a metabolic inhibitor. *J Neuropathol Exp Neurol* 1973;**32**:408–23.
24. Rosenblum WI. Cytotoxic edema: monitoring its magnitude and contribution to brain swelling. *J Neuropathol Exp Neurol* 2007;**66**:771–8.
25. VanHarreveld A, Malhotra SK. Extracellular space in the cerebral cortex of the mouse. *J Anat* 1967;**101**:197–207.
26. VanHarreveld A, Collewijn H, Malhotra SK. Water, electrolytes, and extracellular space in hydrated and dehydrated brains. *Am J Physiol* 1966;**210**:251–6.
27. VanHarreveld A, Ochs S. Cerebral impedance changes after circulatory arrest. *Am J Physiol* 1956;**187**:180–92.
28. VanHarreveld A, Hooper NK, Cusick JT. Brain electrodes and cortical impedance. *Am J Physiol* 1961;**201**:139–43.

29. Herrmann HD, Hort W, Loew F, Palleske H. The determination of the specific tissue impedance—a simple method of measuring the experimental cerebral edema. *Cesk Neurol* 1968;**31**:105–11.
30. van der Veen PH, Zuiderveen GG, Buiter D, van der Meer J. Electrical impedance of cat brain with cold-induced edema. *Exp Neurol* 1973;**40**:675–82.
31. Hossmann KA. Cortical steady potential, impedance and excitability changes during and after total ischemia of cat brain. *Exp Neurol* 1971;**32**:163–75.
32. Gazendam J, Go KG, Van der Meer JJ, Zuiderveen F. Changes of electrical impedance in edematous cat brain during hypoxia and after intracerebral ouabain injection. *Exp Neurol* 1979;**66**:78–87.
33. Fujita S, Ueda T, Yagi M. Detection of experimental and clinical brain edema using an electrical impedance method. *J Neurosurg* 1972;**37**:156–63.
34. Schuier FJ, Hossmann KA. Experimental brain infarcts in cats. II. Ischemic brain edema. *Stroke* 1980;**11**:593–601.
35. Matsuoka Y, Hossmann KA. Cortical impedance and extracellular volume changes following middle cerebral artery occlusion in cats. *J Cereb Blood Flow Metab* 1982;**2**:466–74.
36. Demirci M, Ayata C, Dalkara T, Erdemli G, Onur R. Monitoring cellular edema at single-neuron level by electrical resistance measurements. *J Neurosci Methods* 1997;**72**:175–81.
37. Levin VA, Fenstermacher JD, Patlak CS. Sucrose and inulin space measurements of cerebral cortex in four mammalian species. *Am J Physiol* 1970;**219**:1528–33.
38. Fenstermacher JD, Li CL, Levin VA. Extracellular space of the cerebral cortex of normothermic and hypothermic cats. *Exp Neurol* 1970;**27**:101–14.
39. Nicholson C, Phillips JM. Ion diffusion modified by tortuosity and volume fraction in the extracellular microenvironment of the rat cerebellum. *J Physiol* 1981;**321**:225–57.
40. Nicholson C, Sykova E. Extracellular space structure revealed by diffusion analysis. *Trends Neurosci* 1998;**21**:207–15.
41. Lundbaek JA, Hansen AJ. Brain interstitial volume fraction and tortuosity in anoxia. Evaluation of the ion-selective micro-electrode method. *Acta Physiol Scand* 1992;**146**:473–84.
42. Kumura E, Dohmen C, Graf R, Yoshimine T, Heiss WD. Significant shrinkage of extracellular space during global cerebral ischemia: differences in gray and white matter ischemia. *Acta Neurochir Suppl* 2003;**86**:67–70.
43. Vorisek I, Sykova E. Ischemia-induced changes in the extracellular space diffusion parameters, K+, and pH in the developing rat cortex and corpus callosum. *J Cereb Blood Flow Metab* 1997;**17**:191–203.
44. Nicholson C, Tao L. Hindered diffusion of high molecular weight compounds in brain extracellular microenvironment measured with integrative optical imaging. *Biophys J* 1993;**65**:2277–90.
45. Nicholson C, Tao L. Diffusion properties of brain tissue measured with electrode methods and prospects for optical analysis. *Adv Exp Med Biol* 1993;**333**:213–23.
46. Binder DK, Papadopoulos MC, Haggie PM, Verkman AS. In vivo measurement of brain extracellular space diffusion by cortical surface photobleaching. *J Neurosci* 2004;**24**:8049–56.
47. Lewandowsky M. Zur Lehre von der Cerebrospinal Fluessigkeit. *Zentralbl Klin Med* 1900;**40**:480–94.
48. Stern L, Gautier R. Le passage dans le liquide céphalo-rachidien de substances introduites dans la circulation et leur action sur le système nerveux central chez les différentes espèces animales. *R C R de la Soc Phys Hist Nat Geneve* 1918;**35**:91–4.
49. Saunders NR, Dreifuss JJ, Dziegielewska KM, et al. The rights and wrongs of blood-brain barrier permeability studies: a walk through 100 years of history. *Front Neurosci* 2014;**8**:404. http://dx.doi.org/10.3389/fnins.2014.00404.

50. Wunder A, Schoknecht K, Stanimirovic DB, Prager O, Chassidim Y. Imaging blood-brain barrier dysfunction in animal disease models. *Epilepsia* 2012;**53**(Suppl 6):14–21. http://dx.doi.org/10.1111/j.1528-1167.2012.03698.x.
51. Saunders NR, Dziegielewska KM, Mollgard K, Habgood MD. Markers for blood-brain barrier integrity: how appropriate is Evans blue in the twenty-first century and what are the alternatives? *Front Neurosci* 2015;**9**:385. http://dx.doi.org/10.3389/fnins.2015.00385.
52. Brightman MW, Reese TS. Junctions between intimately apposed cell membranes in the vertebrate brain. *J Cell Biol* 1969;**40**:648–77.
53. Davson H. Review lecture. The blood-brain barrier. *J Physiol* 1976;**255**:1–28.
54. Ghosh M, Balbi M, Hellal F, Dichgans M, Lindauer U, Plesnila N. Pericytes are involved in the pathogenesis of cerebral autosomal dominant arteriopathy with subcortical infarcts and leukoencephalopathy. *Ann Neurol* 2015;**78**:887–900.
55. Sagare AP, Bell RD, Zhao Z, et al. Pericyte loss influences Alzheimer-like neurodegeneration in mice. *Nat Commun* 2013;**4**:2932. http://dx.doi.org/10.1038/ncomms3932.
56. Sengillo JD, Winkler EA, Walker CT, Sullivan JS, Johnson M, Zlokovic BV. Deficiency in mural vascular cells coincides with blood-brain barrier disruption in Alzheimer's disease. *Brain Pathol* 2013;**23**:303–10.
57. Bell RD, Winkler EA, Sagare AP, et al. Pericytes control key neurovascular functions and neuronal phenotype in the adult brain and during brain aging. *Neuron* 2010;**68**:409–27.
58. Hofman HJ, Olszewski J. Spread of sodium fluorescein in normal brain tissue. A study of the mechanism of the blood-brain barrier. *Neurology* 1961;**11**:1081–5.
59. Piper S, Bahmani P, Klohs J, et al. Non-invasive surface-stripping for epifluorescence small animal imaging. *Biomed Opt Express* 2010;**1**:97–105.
60. Schwarzmaier SM, Gallozzi M, Plesnila N. Identification of the vascular source of vasogenic brain edema following traumatic brain injury using in vivo 2-photon microscopy in mice. *J Neurotrauma* 2015;**32**:990–1000.
61. Kawamura S, Schurer L, Goetz A, Kempski O, Schmucker B, Baethmann A. An improved closed cranial window technique for investigation of blood-brain barrier function and cerebral vasomotor control in the rat. *Int J Microcirc Clin Exp* 1990(9):369–83.
62. Schurer L, Corvin S, Rohrich F, Abels C, Baethmann A. Leukocyte/endothelial interactions and blood-brain barrier permeability in rats during cerebral superfusion with LTB4. *Acta Neurochir Suppl (Wien)* 1994;**60**:51–4.

CHAPTER

5

In Vitro Models of the Blood–Brain Barrier to Better Understand the Pathophysiology of Brain Edema

Abraham J. Al-Ahmad

Texas Tech University Health Sciences Center, Amarillo, TX, United States

OUTLINE

Brain Edema.
DOI: http://dx.doi.org/10.1016/B978-0-12-803196-4.00005-9

INTRODUCTION

Studies in vivo of the blood–brain barrier (BBB) are a valuable source to better understand the pathophysiology contributing to the formation of cerebral edema in brain disorders. However, a detailed investigation of the cellular and molecular mechanisms such disruption is hampered by two major factors: the location and the accessibility of the site of such condition.

Abbott and coworkers[1] showed that cerebral vessels were different regarding the barrier tightness measured by the electrical resistance of cerebral microvessels in rat. Electrical resistance increased proportionally as the vessel diameter decreased with a cutoff value of 50 μm. Large blood vessels (>50 μm) including arteries and veins exhibited the lowest electrical resistance (300–600 Ω cm^2); however, such value remains largely higher than values reported in peripheral vascular beds with reported values of 3–33 Ω cm^2.[2,3] Vessels with a diameter less than 50 μm showed an average electrical resistance of 1500 Ω cm^2 but did not show differences beyond this diameter. However, the electrical resistance differs between arterioles and venules, with a higher permeability value for venules with an average electrical resistance of 750 Ω cm^2 than arterioles. Such differences in permeabilities between arterioles and venules may partially explain the susceptibility of venules to represent an entry point for leukocytes extravasation.[4,5] Such observation was further confirmed by Pachter and coworkers.[6] The authors also investigated the expression of glucose transporter (GLUT-1), P-gp (drug efflux transporter), and transferrin receptor (receptor-mediated transporter) in freshly isolated rat brain microvessels and fractionated by their diameter size. These proteins are highly enriched in cerebral microvessels with a diameter below 40 μm. Therefore the authors concluded that microvessels with a diameter below 40 μm displayed a "BBB phenotype." These results suggest that any investigation of the BBB in vivo should be mostly accomplished in microvessels displaying a diameter lesser than 50 μm. The accessibility of these vessels is technically challenging in vivo, despite the development of new imaging techniques. In fact the irrigation of the cerebral tissue occurs by the penetration of pial arteries lining on the surface, which developing collateral microvessels to allow blood flow inside the brain parenchyma with an average neuron-capillary distance oscillating between 8 μm and 20 μm.[7] The difficulty to access the deep brain tissue limits our ability to decipher in vivo molecular and cellular events how the "disruption" of the barrier function leads to vasogenic edema.

Therefore, for a better understanding of the cellular and molecular mechanisms underlying the pathophysiological events leading to the BBB disruption, different in vitro models have been developed during the last two decades. These models gave rise to better understand the signaling pathways contributing in the BBB disruption and the onset of vasogenic edema. This chapter will provide the readers an exhaustive list of in vitro BBB models and their usage to model the onset of cerebral edema with a focus on hypoxic/ischemic stress (H/I).

IN VITRO BLOOD–BRAIN BARRIER: EXPERIMENTAL SETUP AND TECHNIQUES

Various experimental models have been described in order to directly assess changes in the barrier properties after stress. Despite the existence of several endothelial cell cultures reproducing BBB fates, the most common experimental method used to monitor changes in the BBB function in vitro is with the usage of the Transwell insert system (Fig. 5.1). The Transwell system is a modification of the original Boyden chamber initially developed by Boyden for assessing leukocytes migration.[8] It consists of a semipermeable membrane made of polycarbonate or polyester with pores with a fixed diameter of 0.4 μm diameter. This diameter is allowing the formation of a continuous monolayer of cells on the insert; however, the presence of a semipermeable membrane facilitates the diffusion of molecules. The transwell reproduces the physiological compartimentation observed in vivo with the separation of an apical (upper) and a basolateral (lower) chamber. In most of the experimental design, brain microvascular endothelial cells (BMECs) are seeded on the upper side of the semiporous membrane after coating either with type-I collagen[9] or with a mixture of type-IV collagen/fibronectin.[10] The coating is critical to improve BMECs cell adhesion on the membrane and allow them to form a confluent monolayer. Once the cells reach confluence the BMECs exhibit a polarization with a differentiation with apical and basolateral

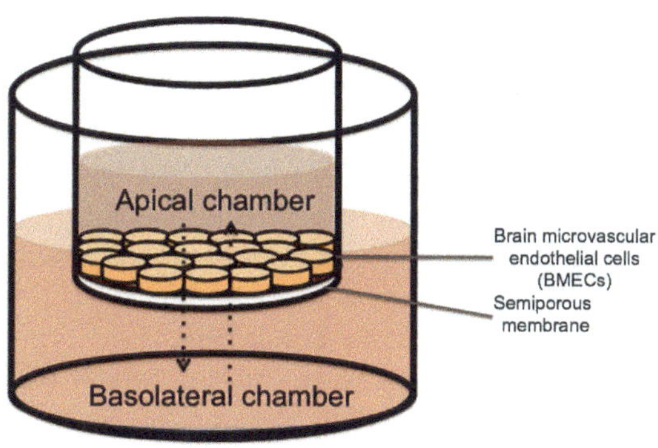

FIGURE 5.1 In vitro model of the blood–brain barrier using the Transwell system.

sides respectively. In fact the presence BMECS show a polarized transport as observed in vivo (see Chapter 3: Noninvasive Imaging Techniques for Brain Edema: From Basic Science to the Clinic) and which constitutes an important landmark in the validation of in vitro models of the BBB. The cellular polarity of the BMECs is evaluated by comparing the apical-to-basolateral (AB) versus the basolateral-to-apical (BA) fluxes of a known compound. The presence of an active uptake transporter at the luminal side will translate into an AB flux higher than the BA flux (e.g., glucose and GLUT-1 transporter). On the other hand, the presence of an active efflux pump at the luminal surface will result into a BA flux higher than the AB flux (e.g., colchicine and P-gp efflux pump). In average, BMECs monolayers are cultivated between 2 days and 10 days to allow formation of tight monolayers and the proper development of cell polarity.[11–15]

In vivo, tight junctions complexes "disruption/relaxation" following injury may result in increased paracellular permeability with eventually vasogenic edema. Thus techniques assessing the diffusion of small ions and solutes diffusing through the paracellular pathway can provide valuable tools to model vasogenic edema in vitro. Two major techniques are commonly used to measure changes in paracellular permeability: the transendothelial electrical resistance (TEER) and the usage of small-polarized paracellular tracers.

The TEER is an electro-physical method relying on the biology on epithelial cell membranes. Epithelial cells monolayers can be visualized as biological resistances providing a resistance by limiting the diffusion of ion fluxes between the apical and basolateral membranes. Such resistance can be measured by different methods including the EVOM2 (World Precision Instruments, Sarasota, FL) or the CellZScope (NanoAnalytics, Muenster, Germany). In this system a tight monolayer will be reflected by a high TEER value. In contrast an increase in paracellular permeability will be reflected by a decrease in TEER. The advantage of TEER compared to other analytical methods is its nonchemical, noninvasive approach allowing the experimenter to monitor changes in paracellular permeability in vitro. Although TEER values show a good correlation with paracellular permeability (Fig. 5.2) such method used alone cannot provide an accurate assessment of barrier mechanical tightness, as variable factors such as temperature or medium composition influence the reading accuracy.[16] In addition to TEER measurements, paracellular flux tracers are used to assess changes in BMECs monolayers tightness. These water-soluble tracers have in common to harbor a small molecular size (<0.5 kDa) and have a very low permeability across the intact BBB. Two types of tracers are commonly used: fluorescent dyes (e.g., Lucifer Yellow (LY), sodium fluorescein) and radiolabeled dyes (e.g., ^{14}C-sucrose, ^{3}H-inulin). These dyes are disposed on the apical chamber whereas sampling of the basolateral chamber at regular timepoints is operated to determine the diffusion rate

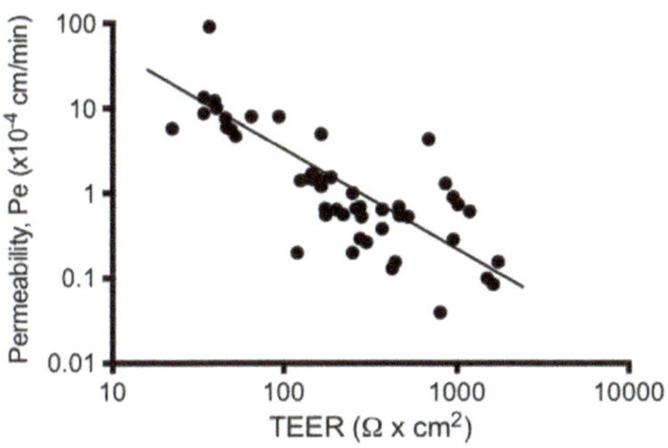

FIGURE 5.2 Correlation between paracellular permeability and TEER in BEMCs monolayers.

of the dye across the monolayer. At the difference of other methods relying on Fick's first law to determine the permeability of these dyes across the monolayers, permeability values (Pe) for compounds diffusing across BMECs monolayers is obtained by a modified iteration of Fick's first law that integrate the calculation of clearance slopes and the surface from BMECs monolayers and from blank inserts.[17]

PRIMARY AND IMMORTALIZED BRAIN MICROVASCULAR ENDOTHELIAL CELLS

Current isolation protocols of primary BMECs from various species originates from the pioneering publication of Ferenc Joo in 1973.[18] BMECs isolation protocols follow a canonical methods summarized in Fig. 5.1. In brief, freshly isolated cortices are firstly depleted from superficial pial vessels and meninges followed by a mechanical tissue homogenization and an ultracentrifugation using a gradient density to separate microvessels from the tissue homogenate. Purified microvessels are further digested using tissue-dissociating enzymes (e.g., collagenase/dispase) that allow the degradation of the basement membrane surrounding such microvessels followed by a final centrifugation step that allows BMECs purification from nonvascular cells (astrocytes end-feet processes, pericytes, etc.). Established protocols have been established for primary BMECs isolation from various mammal brains including, bovine,[19] human,[20] murine,[18,21,22] and porcine[23] origin. Primary cultures are preferentially appreciated compared to other established in vitro models as these cells display a better BBB phenotype than immortalized cell lines.

In particular, primary porcine and bovine BMECs are highly appreciated by their ability to display higher barrier tightness (TEER>800 Ωcm^2; $Pe_{14C\text{-sucrose}}$: 3.5×10^{-4} cm/min). Therefore such bovine and porcine-based in vitro models are commonly used to evaluate the BBB permeability in drug discovery.[24–26] Despite their robust BBB phenotype, primary BMECs suffer from several caveats inherent to their isolation. Firstly, primary BMECs cultures are particularly susceptible to the loss of their barrier phenotype following their isolation and expansion. For instance, comparative transcriptomic studies revealed a significant down-regulation in several genes encoding for BBB genes in both bovine and murine BMECs primary cultures.[27,28] Such dedifferentiation can directly impact the reliability of experimental data obtained in vitro to the in vivo situation. Moreover, primary BMECs are known to have a limited number of passaging, with a maximum of 2–3 passages following isolation. Beyond this number of cell passaging, primary BMECs usually stop to divide.

In addition to limited passaging, BMECs primary cultures are hampered by the low yield of microvessels. For instance, an adult rat brain will yield approximately provide enough cells to seed six 12-well transwells that may suffice for covering the need for a small-scale study. However, the usage of such system for high-throughput screening (>10,000 assays per day)[29] would require an important source of starting material in order to obtain an appreciable yield for large-scale applications yet it also raises important financial, technical and ethical challenges that cannot be solved with the current approach.

Because primary BMECs isolation constitutes an important technical challenge for large-scale studies various protocols have been published in the literature to induce and propagate immortalized BMECs cell lines. The first reported study discussing the establishment of immortalized BMEC lines was published by Montesano and coworkers and isolated from hemangiomas induced by the insertion of the polyoma T oncogene in mice, resulting in the formation of two cell lines: bEnd.3[30,31] and presumably bEnd.5. Following such publication the utilization of the T oncogene (also described as T antigen) had become the standard to obtain various rat brain endothelial cell lines,[32] including the RBE4 cell line.[33,34] Among the different immortalized rodents BMEC lines, it is important to note that two cell lines have been exhaustively used in the literature: bEnd.3 (209 referenced articles in Pubmed) and RBE4 (162 articles referenced).

In addition to rodent established cell lines, different immortalized cell lines have been established and characterized in the literature[35]: hCMEC/D3[12], TY08,[36] and BB19[37] immortalized human BMEC lines. However, among all these different human cell lines, the hCMEC/D3 line developed by Couraud and coworkers[12] constitutes certainly the most common human brain endothelial cell line cited in the scientific literature, with over 150 referenced studies citing this cell line. This cell line retains the expression of key BBB markers, including the expression of vascular markers (PECAM-1; VE-cadherin), tight junction proteins (claudin-5, JAM-A, occludin, ZO-1) as well as drug and nutrient transporters (GLUT-1, MRPs, and P-gp). Despite the advantage of such cells in terms of yield and ease of use, it is important to note these cell lines often yield to monolayers with poor barrier tightness and high paracellular flux (TEER<200 Ωcm^2; $Pe_{\text{Lucifer Yellow}}$: 15×10^{-4} cm/min).[12,38] In summary, immortalized brain endothelial cell liens may represent interesting models.

BRAIN MICROVASCULAR ENDOTHELIAL CELLS DERIVED FROM ENDOTHELIAL PROGENITOR AND STEM CELL SOURCES

During the last decade, human pluripotent stem cells have gained attention by their two important features: a virtual infinite number of cell passaging due to their self-renewal feature and their ability to be differentiate into any cell type of interest including neurons and microvascular endothelial cells.[39–41] Human pluripotent stem cells are classically divided into four major groups based on their pluripotency (from the most to the least pluripotent): embryonic stem cells (ESCs, isolated from embryonic blastocysts),[42] induced pluripotent stem cells (iPSCs, reprogrammed adult somatic cells),[43] mesenchymal stem cells (MSCs, isolated from bone marrow or from adipose tissue)[44,45] and hematopoietic stem cells (HSCs, isolated from umbilical cord blood).[46] Because stem cells share several features often aimed in the establishment of a human in vitro *model* of the BBB, several approaches have been used in order to generate in vitro models of the human BBB from stem cell sources.[47]

Shusta and coworkers pioneered such approach by describing a serum-free differentiation protocol allowing the differentiation of human stem cell-derived BMECs from both established ESCs and iPSCs lines.[14,15] Following the differentiation and purification, such BMECs displayed the expression of key vascular (PECAM-1, VE-cadherin, von Willebrand factor) and BBB (GLUT-1, claudin-5, occludin) markers, reasonable barrier tightness (TEER>200 Ωcm^2, $Pe_{14C\text{-sucrose}}<10^{-4}$ cm/min), an appreciable in vitro-to-in vivo correlation regarding the permeability profile of several known compounds capable to cross the BBB and the presence of a polarized drug transport. Furthermore, these cells were capable to up-regulate their barrier tightness following coculture with primary rat brain astrocytes, further confirming their BBB identity. However, the most striking feature of its model was its ability to respond to retinoic acid (RA) treatment. RA signaling pathway has been recently identified as one of the most highly enriched signaling pathway in freshly isolated mouse BMECs in comparison to other microvascular endothelial cells.[48] Furthermore, treatment of hCMEC/D3 monolayers with RA resulted in an increase barrier tightness,[49] suggesting the importance of such pathway in the onset and the maintenance of the BBB. Following such study, Shusta and coworkers published an updated revision of their in vitro model, by adding RA treatment during the maturation stage.[14,41]

Notably, RA treatment resulted in a significant increase in the stem cell-derived BMEC monolayers tightness (TEER>1500 Ωcm^2, $Pe_{14Csucrose}<10^{-5}$ cm/min) bringing them in par with primary bovine and porcine BMECs monolayers in terms of barrier tightness. Furthermore, the authors reported TEER values ranging from 3500 to 5000 Ωcm^2 when these BMECs where cocultured with differentiated astrocytes and neurons from neural progenitor cells, allowing such coculture system to reach TEER values reported in vivo.[1,50] Similar differentiation approach was used to generate BMECs from HSCs derived from umbilical cord blood.[51] Despite showing a BBB cellular identity the barrier tightness of these cells remained in the range of other in vitro models, with a reported $Pe_{Lucifer\ Yellow} = 6 \times 10^{-4}$ cm/min.[51]

Stem cell-derived BMECs show a substantial advantage to existing models. They provide a scalable and reproducible source of human cells, show a phenotype similar to primary BMECs and display appreciable barrier tightness. Furthermore, iPSCs-derived cells provide a patient-specific source of stem cells, allowing scientists to include population polymorphism and gender differences in their studies, as well as modeling the influence of neurological disorders with a genetic background. Despite the novelty of this model, we still need to demonstrate the ability to transpose such differentiation protocol from established stem cell lines into patient-derived lines. In addition, we still have no conclusive comparative study assessing the genomic and proteomic profile of such stem cell-derived BMECs to freshly isolated human BMECs. Such validation step remains important, as it will allow us to better appreciate how such cells.

Among the different models developed during these last 20 years, iPSC-derived BMECs appears among the best candidate as it represent a scalable, reproducible and patient-specific source of cells. iPSC-derived BMECs, at least from established iPSCs lines, have displayed the highest barrier tightness, making them suitable for modeling the cellular and molecular mechanisms leading to vasogenic edema formation. Yet the ability to these cells to respond to stimuli leading to vasogenic edema formation remains to be documented.

IN VITRO COCULTURES MODEL OF THE BLOOD–BRAIN BARRIER

As described before there are various endothelial cells, which have been used to mimic the BBB in a petri dish (Fig. 5.3), with each of them displaying advantages and disadvantages (see Table 5.1). The simplest model consists of BMECs monolayers cultured alone (monocultures). Although

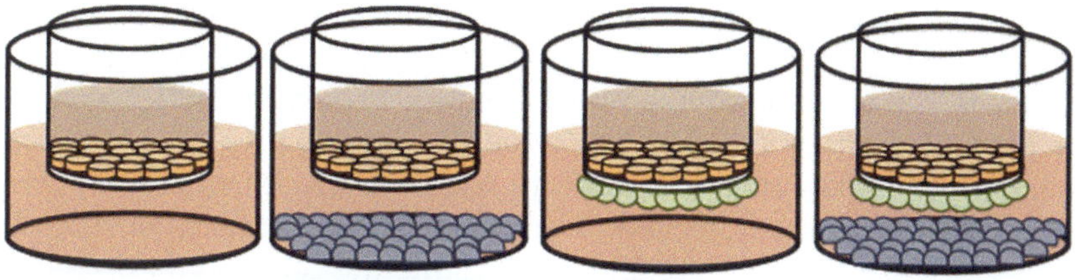

FIGURE 5.3 In vitro models of the blood–brain barrier using monocultures and cocultures.

TABLE 5.1 Summary of the different types of BMECs commonly used in in vitro models of the blood-brain barrier

In Vitro Blood–Brain Barrier	Advantages	Disadvantages	Use
Immortalized cell lines (human, rodents)	Scalable and high passaging possible, ease of access	Poor barrier properties, dedifferentiation	Cellular and molecular physiology and pathophysiology
Primary cells (human, rodents, bovine, porcine, and ovine)	Acceptable barrier properties, lesser dedifferentiation issues	Hardly scalable, isolation challenging	Cellular and molecular physiology and pathophysiology, drug permeability screening
Stem-cell-based models (human embryonic and induced pluripotent)	Scalable, acceptable barrier properties, patient-specific	Phenotype comparable to adult BBB remains to be demonstrated. Dedifferentiation, variability in differentiation between stem cell lines	Cellular and molecular physiology and pathophysiology, drug permeability screening

such iteration constitutes an affordable, transferrable and scalable model for modeling BBB dysfunction with for consequences cerebral edema, the detrimental aspect of such model is the absence of the other cell types of the neurovascular unit including astrocytes, pericytes, and neurons. As an alternative approach to provide an in vitro model that better reflect the in vivo situation, by establishing coculture systems in which BMECs are grown in presence of one or other several cell types of the neurovascular unit astrocytes, pericytes, or neurons.[11,14,15,25,47,52–61] Different permutations have been described in the literature but the most integrated platform includes a three-cells coculture (e.g., astrocytes/pericytes; astrocytes/neurons). In these cocultures, astrocytes and/or pericytes may be seeded on the basolateral side of the insert, allowing physical contact between BMECs and these cells through formation of gap junctions. Cocultures provide a substantial advantage to monocultures as they provide a more integrative approach and bring in vitro models closer from in vivo counterparts. Furthermore the presence of cocultures can partially restore the cellular communication observed at the neurovascular unit. Such restoration in cellular communications is observed by an increase in the barrier tightness (e.g., increased TEER) when BMECs are cocultured with astrocytes and pericytes compared to monocultures (Fig. 5.4). Furthermore the presence of astrocytes and pericytes in cocultures can strikingly affect how the BMEC monolayer responds to injury, in particular to hypoxia/ischemia (H/I). Ogunshola and coworkers showed that the presence of astrocytes or pericytes partially restored the barrier function following hypoxic injury[11,62,63] by modulating BMECs response to oxidative stress and vascular endothelial growth factor (VEGF). However, such model suffers from two caveats: first, it adds a layer of complexity to the model that can be very challenging in high-throughput screening assays. Second, this model limits the ability to investigate the response of astrocytes and pericytes to such type of injury.

In regard of the importance of cocultures in BMECs monolayers response to injury, in particular to H/I injury,[11,62,63] future in vitro models will have to take the contribution of the different cell types of the neurovascular unit during the pathophysiological processes leading to cerebral edema formation. Although such integrated intercellular communications are important to better model the in vivo situation, it also add a layer of complexity that can make the development of such model challenging and unfit for high-throughput screening platforms.

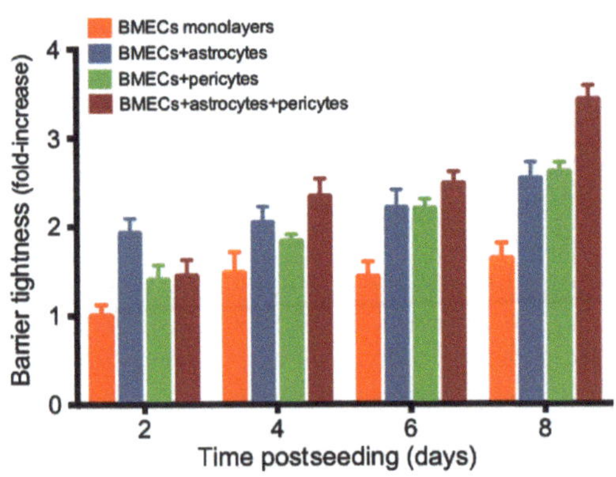

FIGURE 5.4 Impact of the coculture system on BMECs monolayers barrier tightness.

FLOW-BASED IN VITRO MODELS OF THE BLOOD–BRAIN BARRIER

Although in vitro models of the BBB can provide crucial information regarding the cellular and molecular mechanisms underlying the pathophysiology underlying vasogenic edema formation, a major caveat of such models is the absence of a hydrodynamic pressure. In the vast majority of in vitro models of the BBB, fluid on the apical side exercises an hydrostatic pressure, whereas in vivo these cells are exposed to a hydrodynamic pressure resulting in an exposure of BMECs to a luminal shear stress.

Shear-stress stimuli play an important function in endothelial cell biology.[64] Endothelial cells respond to shear-stress stimuli by various mechanosignaling pathways. The nature of such mechanosensors remains unclear but several candidates have been identified as potential candidates: PECAM-1 and the glycocalyx lining the apical surface of microvascular endothelial cells (including BMECs). The presence of a hydrodynamic force on cultured BMECs physiology have been extensively investigated by Janigro and coworkers,[53,65–69] using their DIV-BBB model. Under such model, bovine aortic endothelial cells showed an increase in glucose metabolism and glycolytic activity, such increased metabolism was accompanied by an increase in TEER from 50 to 500 $\Omega\ cm^2$ and a decrease in paracellular flux. Despite the advantage of this dynamic model to hydrostatic ones the utilization of such system remained until now discrete due to the difficulty to use for high-throughput screening.

An alternative model may be coming from the field of the microfluidics engineering. Recently, Searson and coworkers[70] have introduced a microfluidic system to study the rheological properties of primary human BMECs compared to primary human umbilical vein endothelial cells. Under such shear-stress HBMECs showed changes in actin distribution, with the formation of actin stress-fibers, an distribution of ZO-1 at the cell borders and an increased β-catenin and ZO-1 expression suggesting an enhanced barrier function.

The presence of a shear stress on endothelial cells is critical in the onset and undergoing development of the barrier phenotype. Due to its technical challenge to transpose into other research labs, this approach is very discretely used so far. However, the recent emergence of microfluidics in biomedical engineering opens a new vista for fluid-based systems that are easier to setup and transfer into other research groups. It also represents a better alternative to hydrostatic models as it brings current in vitro models closer from the in vivo situation and have a better view of the importance of the flow in the formation of the vasogenic edema.

IN VITRO MODELS FOR MODELING VASOGENIC CEREBRAL EDEMA

Among the different in vitro models of the BBB developed to assess changes in the barrier function during disease stage, in vitro models designed to investigate the effect of H/I on the barrier function are probably the best suited to model vasogenic cerebral edema, in vitro. Hypoxia is defined by an impaired oxygen supply versus tissue oxygenation need, whereas ischemia is defined by the impairment of both oxygen and nutrients (in particular glucose) supply. Hypoxia is well-known to trigger BBB disruption by dislocating the tight junction complexes.[71] Such disruptive effect of hypoxia at the BBB ultimately leads to its opening and eventually onset in vivo cerebral edema by entrance of blood products into the brain parenchyma.

In practice, cells are incubated in a sealed hypoxic chamber with a custom atmosphere containing carbon dioxide (5% gas phase concentration) nitrogen (94%), and oxygen (1%). Hypoxic stress is achieved by incubating cells within such chambers, whereas ischemic stress is commonly modeled using an oxygen-glucose deprivation (OGD) stress, in which cell medium is replaced by a glucose-free equivalent prior hypoxic treatment. Two types of chambers are commonly used in the literature: a static chamber and a dynamic chamber (including the so-called glove-box chambers).

1. Static chambers constitute the most affordable option; with two types of chambers: anaerobic chambers (or anaerobic jars) and static hypoxic chambers. For anaerobic chambers an anaerobic pouch will sequestrate atmospheric oxygen present inside the chamber and in cell culture medium. Although such chambers constitute the most affordable option to induce hypoxic stress, the major caveat is the absence of control of hypoxic environment. In such system, cells rapidly undergo severe anoxia (absence of oxygen) not representative of a pathological situation. The static chambers is using premixed gas to achieve a defined hypoxic environment. In this system, premixed hypoxic gas is injected inside the chamber and allowed to flush out the normoxic air trapped inside the chamber. Upon equilibration, such chamber is placed inside a conventional incubator upon the end of the experiment. One of the major issues of this system is the inability to accurately measure the chamber atmospheric equilibration prior experiment, with for direct consequence a substantial variability in the experimental outcome.
2. Dynamic chambers provide a better control of gas levels using a gas sensor-controller, allowing a fine-tuning in gas phase concentrations. Although such chambers are well suited for hypoxic environment, the cost of such equipment and their maintenance may represent a substantial investment.

Different in vitro models of the BBB have been used to assess changes in the barrier function following hypoxia/OGD stress. In all these models, changes in the barrier function are commonly assessed using the TEER and by measuring changes in paracellular permeability using labeled solutes (radioactive or fluorescent-based tracers). Monoculture-based models constitute probably the most common model by its simplicity, allowing to elucidate the endothelial-specific response to hypoxia.[11,72–77] H/I induce the opening of the BBB with entrance of small ionic polarized compounds, contributing to changes of the osmotic pressure and entry of water. However, such model omits the contribution of the different cell types present at the BBB and their possible influence on the BBB response to hypoxic/ischemic stress. As an alternative, coculture models integrating astrocytes and/or pericytes have been developed and provide evidence of an active cell–cell communication between brain endothelial cells and its surrounding environment. In particular, astrocytes and pericytes cocultures studies documented a protective effect of these cells on the BBB function during hypoxia/OGD stress.[60,61,78–82] The relevance of a coculture-based model versus monocultures models has recently been highlighted by Kuntz and coworkers, using both a in vivo and an in vitro model of stroke injury.[83] In their coculture model based on primary mouse BMECs and astrocytes the authors demonstrated the detrimental effect of astrocytes on the recovery of the barrier function during reperfusion stage. In their study the presence of astrocytes subjected to OGD stress had a detrimental resulted in a higher permeability of the BMEC monolayer to LY compared to BMECs monocultures exposed to OGD stress or to BMECs. Furthermore the authors also demonstrated that the cocultures of BMECs with normoxic astrocytes during the reperfusion stage resulted in a better barrier restoration, as noted by a significant reduction in LY as early as 2h after reperfusion. The changes observed in their in vitro models parallel the in vivo stroke pathophysiology with loss of BBB functions and edema formation. Taken together, such studies as others highlight the importance of in vitro models integrative of the different cell types of the neurovascular unit, including astrocytes and neurons. In vivo models provide us a simplified situation to have a better understanding in molecular and cellular process of the BBB pathophysiology associated with the edema process.

CONCLUDING REMARKS

As reviewed in the chapter during the last two decades an important effort have been accomplished to develop better in vitro models of the BBB, using primary cell cultures, immortalized cell lines and even more recently introducing stem cells as a novel source of BBB cells. Despite the important progress, we are still challenged by having a scalable and representative model of the BBB to investigate the cellular and molecular events leading to the BBB changes leading in the formation of vasogenic edema.

We can expect in the near future to see the emergence of newer in vitro models that are closer from the in vivo situation and allow us to directly model the cellular and molecular response of the BBB to H/I injury. Eventually, we will able to measure and quantify discrete water movement across BMECs monolayers and better elucidate how the different cell types of the neurovascular unit respond to H/I injury resulting in the disruption of the barrier function and formation of the vasogenic edema.

References

1. Butt AM, Jones HC, Abbott NJ. Electrical resistance across the blood–brain barrier in anaesthetized rats: a developmental study. *J Physiol* 1990;**429**:47–62.
2. O'Donnell MP, Vargas FF. Electrical conductivity and its use in estimating an equivalent pore size for arterial endothelium. *Am J Physiol* 1986;**250**(1 Pt 2):H16–21.
3. Crone C, Christensen O. Electrical resistance of a capillary endothelium. *J Gen Physiol* 1981;**77**(4):349–71.
4. McMenamin PG, Forrester JV, Steptoe RJ, Dua HS. Ultrastructural pathology of experimental autoimmune uveitis. Quantitative evidence of activation and possible high endothelial venule-like changes in retinal vascular endothelium. *Lab Invest* 1992;**67**(1):42–55.
5. Wolburg H, Wolburg-Buchholz K, Engelhardt B. Diapedesis of mononuclear cells across cerebral venules during experimental autoimmune encephalomyelitis leaves tight junctions intact. *Acta Neuropathol* 2005;**109**(2):181–90.
6. Ge S, Song L, Pachter JS. Where is the blood-brain barrier…really? *J Neurosci Res* 2005;**79**:421–7.
7. Schlageter KE, Molnar P, Lapin GD, Groothuis DR. Microvessel organization and structure in experimental brain tumors: microvessel populations with distinctive structural and functional properties. *Microvasc Res* 1999;**58**(3):312–28.
8. Boyden S. The chemotactic effect of mixtures of antibody and antigen on polymorphonuclear leucocytes. *J Exp Med* 1962;**115**:453–66.
9. Bowman PD, Ennis SR, Rarey KE, Betz AL, Goldstein GW. Brain microvessel endothelial cells in tissue culture: a model for study of blood-brain barrier permeability. *Ann Neurol* 1983;**14**(4):396–402.
10. Meyer J, Mischeck U, Veyhl M, Henzel K, Galla HJ. Blood-brain barrier characteristic enzymatic properties in cultured brain capillary endothelial cells. *Brain Res* 1990;**514**(2):305–9.
11. Al Ahmad A, Gassmann M, Ogunshola OO. Maintaining blood-brain barrier integrity: pericytes perform better than astrocytes during prolonged oxygen deprivation. *J Cell Physiol* 2009;**218**(3):612–22.
12. Weksler BB, Subileau EA, Perriere N, et al. Blood-brain barrier-specific properties of a human adult brain endothelial cell line. *FASEB J* 2005;**19**:1872–4.
13. Patabendige A, Skinner RA, Abbott NJ. Establishment of a simplified in vitro porcine blood-brain barrier model with high transendothelial electrical resistance. *Brain Res* 2012;**1521**:1–15.
14. Lippmann ES, Al-Ahmad A, Azarin SM, Palecek SP, Shusta EV. A retinoic acid-enhanced, multicellular human blood-brain barrier model derived from stem cell sources. *Sci Rep* 2014;**4**:4160.

15. Lippmann ES, Azarin SM, Kay JE, et al. Derivation of blood-brain barrier endothelial cells from human pluripotent stem cells. *Nat Biotechnol.* 2012;**30**(8):783–91.
16. Srinivasan B, Kolli AR, Esch MB, Abaci HE, Shuler ML, Hickman JJ. TEER measurement techniques for in vitro barrier model systems. *J Lab Autom* 2015;**20**(2):107–26.
17. Perriere N, Demeuse P, Garcia E, et al. Puromycin-based purification of rat brain capillary endothelial cell cultures. Effect on the expression of blood-brain barrier-specific properties. *J Neurochem* 2005;**93**(2):279–89.
18. Joo F, Karnushina I. A procedure for the isolation of capillaries from rat brain. *Cytobios* 1973;**8**(29):41–8.
19. Audus KL, Borchardt RT. Bovine brain microvessel endothelial cell monolayers as a model system for the blood-brain barrier. *Ann N Y Acad Sci* 1987;**507**:9–18.
20. Drewes LR, Broderius MA, Gerhart DZ. Phorbol ester stimulates hexose uptake by brain microvessel endothelial cells. *Brain Res Bull* 1988;**21**(5):771–6.
21. DeBault LE, Henriquez E, Hart MN, Cancilla PA. Cerebral microvessels and derived cells in tissue culture: II. Establishment, identification, and preliminary characterization of an endothelial cell line. *In Vitro* 1981;**17**(6):480–94.
22. Diglio CA, Grammas P, Giacomelli F, Wiener J. Primary culture of rat cerebral microvascular endothelial cells. Isolation, growth, and characterization. *Lab Invest* 1982;**46**(6):554–63.
23. Mischeck U, Meyer J, Galla HJ. Characterization of gamma-glutamyl transpeptidase activity of cultured endothelial cells from porcine brain capillaries. *Cell Tissue Res* 1989;**256**(1):221–6.
24. Yusof SR, Avdeef A, Abbott NJ. In vitro porcine blood-brain barrier model for permeability studies: pCEL-X software pKa(FLUX) method for aqueous boundary layer correction and detailed data analysis. *Eur J Pharm Sci* 2014;**65**:98–111.
25. Patabendige A, Abbott NJ. Primary porcine brain microvessel endothelial cell isolation and culture. *Curr Protoc Neurosci* 2014;**69**:32721–232717.
26. Culot M, Lundquist S, Vanuxeem D, et al. An in vitro blood-brain barrier model for high throughput (HTS) toxicological screening. *Toxicol In Vitro* 2008;**22**(3):799–811.
27. Calabria AR, Shusta EV. A genomic comparison of in vivo and in vitro brain microvascular endothelial cells. *J Cerebral Blood Flow Metabol* 2008;**28**(1):135–48.
28. Lyck R, Ruderisch N, Moll AG, et al. Culture-induced changes in blood-brain barrier transcriptome: implications for amino-acid transporters in vivo. *J Cerebral Blood Flow Metabol* 2009;**29**:1491–502.
29. Sharma P, Ando DM, Daub A, Kaye JA, Finkbeiner S. High-throughput screening in primary neurons. *Methods Enzymol* 2012;**506**:331–60.
30. Montesano R, Pepper MS, Mohle-Steinlein U, Risau W, Wagner EF, Orci L. Increased proteolytic activity is responsible for the aberrant morphogenetic behavior of endothelial cells expressing the middle T oncogene. *Cell* 1990;**62**(3):435–45.
31. Brown RC, Morris AP, O'Neil RG. Tight junction protein expression and barrier properties of immortalized mouse brain microvessel endothelial cells. *Brain Res* 2007;**1130**(1):17–30.
32. Roux F, Couraud PO. Rat brain endothelial cell lines for the study of blood-brain barrier permeability and transport functions. *Cell Mol Neurobiol* 2005;**25**:41–58.
33. Roux F, Durieu-Trautmann O, Chaverot N, et al. Regulation of gamma-glutamyl transpeptidase and alkaline phosphatase activities in immortalized rat brain microvessel endothelial cells. *J Cell Physiol* 1994;**159**:101–13.
34. Begley DJ, Lechardeur D, Chen ZD, et al. Functional expression of P-glycoprotein in an immortalised cell line of rat brain endothelial cells, RBE4. *J Neurochem* 1996;**67**: 988–95.
35. Eigenmann DE, Xue G, Kim KS, Moses AV, Hamburger M, Oufir M. Comparative study of four immortalized human brain capillary endothelial cell lines, hCMEC/D3, hBMEC, TY10, and BB19, and optimization of culture conditions, for an in vitro blood-brain barrier model for drug permeability studies. *Fluids Barrier CNS* 2013;**10**(1):33.

36. Sano Y, Shimizu F, Abe M, et al. Establishment of a new conditionally immortalized human brain microvascular endothelial cell line retaining an in vivo blood-brain barrier function. *J Cell Physiol* 2010;**225**(2):519–28.
37. Prudhomme JG, Sherman IW, Land KM, Moses AV, Stenglein S, Nelson JA. Studies of *Plasmodium falciparum* cytoadherence using immortalized human brain capillary endothelial cells. *Int J Parasitol* 1996;**26**(6):647–55.
38. Poller B, Gutmann H, Krahenbuhl S, et al. The human brain endothelial cell line hCMEC/D3 as a human blood-brain barrier model for drug transport studies. *J Neurochem* 2008;**107**:1358–68.
39. Wilson HK, Canfield SG, Shusta EV, Palecek SP. Concise review: tissue-specific microvascular endothelial cells derived from human pluripotent stem cells. *Stem Cells.* 2014;**32**(12):3037–45.
40. Chuang JH, Tung LC, Lin Y. Neural differentiation from embryonic stem cells in vitro: an overview of the signaling pathways. *World J Stem Cells* 2015;**7**(2):437–47.
41. Broccoli V, Giannelli SG, Mazzara PG. Modeling physiological and pathological human neurogenesis in the dish. *Front Neurosci* 2014;**8**:183.
42. Thomson JA. Embryonic stem cell lines derived from human blastocysts. *Science* 1998;**282**(5391):1145–7.
43. Yu J, Vodyanik MA, Smuga-Otto K, et al. Induced pluripotent stem cell lines derived from human somatic cells. *Science* 2007;**318**(5858):1917–20.
44. Bruder SP, Fink DJ, Caplan AI. Mesenchymal stem cells in bone development, bone repair, and skeletal regeneration therapy. *J Cell Biochem* 1994;**56**(3):283–94.
45. Nicoletti GF, De Francesco F, D'Andrea F, Ferraro GA. Methods and procedures in adipose stem cells: state of the art and perspective for translation medicine. *J Cell Physiol* 2015;**230**(3):489–95.
46. Mendez-Ferrer S, Scadden DT, Sanchez-Aguilera A. Bone marrow stem cells: current and emerging concepts. *Ann N Y Acad Sci* 2015;**1335**:32–44.
47. Lippmann ES, Al-Ahmad A, Palecek SP, Shusta EV. Modeling the blood-brain barrier using stem cell sources. *Fluids Barriers CNS* 2013;**10**(1):2.
48. Daneman R, Zhou L, Agalliu D, Cahoy JD, Kaushal A, Barres BA. The mouse blood-brain barrier transcriptome: a new resource for understanding the development and function of brain endothelial cells. *PLoS One.* 2010;**5**(10):e13741.
49. Mizee MR, Wooldrik D, Lakeman KA, et al. Retinoic acid induces blood-brain barrier development. *J Neurosci* 2013;**33**(4):1660–71.
50. Crone C, Olesen SP. Electrical resistance of brain microvascular endothelium. *Brain Res.* 1982;**241**:49–55.
51. Cecchelli R, Aday S, Sevin E, et al. A stable and reproducible human blood-brain barrier model derived from hematopoietic stem cells. *PLoS One.* 2014;**9**(6):e99733.
52. Lippmann ES, Weidenfeller C, Svendsen CN, Shusta EV. Blood-brain barrier modeling with co-cultured neural progenitor cell-derived astrocytes and neurons. *J Neurochem* 2011;**119**(3):507–20.
53. Cucullo L, Couraud PO, Weksler B, et al. Immortalized human brain endothelial cells and flow-based vascular modeling: a marriage of convenience for rational neurovascular studies. *J Cerebral Blood Flow Metabol* 2008;**28**(2):312–28.
54. Dehouck MP, Meresse S, Delorme P, Fruchart JC, Cecchelli R. An easier, reproducible, and mass-production method to study the blood-brain barrier in vitro. *J Neurochem* 1990;**54**:1798–801.
55. Weidenfeller C, Svendsen CN, Shusta EV. Differentiating embryonic neural progenitor cells induce blood-brain barrier properties. *J Neurochem* 2007;**101**:555–65.
56. Schiera G, Sala S, Gallo A, et al. Permeability properties of a three-cell type in vitro model of blood-brain barrier. *J Cell Mol Med* 2005;**9**:373–9.
57. Schiera G, Bono E, Raffa MP, et al. Synergistic effects of neurons and astrocytes on the differentiation of brain capillary endothelial cells in culture. *J Cell Mol Med* 2003;**7**:165–70.

58. Cestelli A, Catania C, D'Agostino S, et al. Functional feature of a novel model of blood brain barrier: studies on permeation of test compounds. *J Control Release.* 2001;**76**:139–47.
59. Savettieri G, Di Liegro I, Catania C, et al. Neurons and ECM regulate occludin localization in brain endothelial cells. *Neuroreport* 2000;**11**:1081–4.
60. Hayashi K, Nakao S, Nakaoke R, Nakagawa S, Kitagawa N, Niwa M. Effects of hypoxia on endothelial/pericytic co-culture model of the blood-brain barrier. *Regul Pept* 2004;**123**:77–83.
61. Thanabalasundaram G, Schneidewind J, Pieper C, Galla HJ. The impact of pericytes on the blood-brain barrier integrity depends critically on the pericyte differentiation stage. *Int J Biochem Cell Biol* 2011;**43**:1284–93.
62. Lee B, Clarke D, Al Ahmad A, et al. Perlecan domain V is neuroprotective and proangiogenic following ischemic stroke in rodents. *J Clin Invest* 2011;**121**(8):3005–23.
63. Al Ahmad A, Gassmann M, Ogunshola OO. Involvement of oxidative stress in hypoxia-induced blood-brain barrier breakdown. *Microvasc Res* 2012;**84**(2):222–5.
64. Chatterjee S, Fujiwara K, Perez NG, Ushio-Fukai M, Fisher AB. Mechanosignaling in the vasculature: emerging concepts in sensing, transduction and physiological responses. *Am J Physiol Heart Circ Physiol* 2015;**308**(12):H1451–62.
65. Santaguida S, Janigro D, Hossain M, Oby E, Rapp E, Cucullo L. Side by side comparison between dynamic versus static models of blood-brain barrier in vitro: a permeability study. *Brain Res* 2006;**1109**(1):1–13.
66. Marroni M, Kight KM, Hossain M, Cucullo L, Desai SY, Janigro D. Dynamic in vitro model of the blood-brain barrier. Gene profiling using cDNA microarray analysis. *Methods Mol Med* 2003;**89**:419–34.
67. Krizanac-Bengez L, Kapural M, Parkinson F, et al. Effects of transient loss of shear stress on blood-brain barrier endothelium: role of nitric oxide and IL-6. *Brain Res* 2003;**977**(2):239–46.
68. Desai SY, Marroni M, Cucullo L, et al. Mechanisms of endothelial survival under shear stress. *Endothelium* 2002;**9**(2):89–102.
69. Cucullo L, McAllister MS, Kight K, et al. A new dynamic in vitro model for the multidimensional study of astrocyte-endothelial cell interactions at the blood-brain barrier. *Brain Res* 2002;**951**(2):243–54.
70. Reinitz A, DeStefano J, Ye M, Wong AD, Searson PC. Human brain microvascular endothelial cells resist elongation due to shear stress. *Microvasc Res* 2015;**99**:8–18.
71. Ogunshola OO, Al-ahmad A. HIF-1 at the blood-brain barrier: a mediator of permeability? Special topic: the many faces of hypoxia HIF-1 at the blood-brain barrier. *High Altitude Med Biol* 2012;**13**:153–61.
72. Jacob A, Potin S, Saubamea B, et al. Hypoxia interferes with aryl hydrocarbon receptor pathway in hCMEC/D3 human cerebral microvascular endothelial cells. *J Neurochem* 2015;**132**(4):373–83.
73. Lee SR, Lo EH. Induction of caspase-mediated cell death by matrix metalloproteinases in cerebral endothelial cells after hypoxia-reoxygenation. *J Cerebral Blood Flow Metabol* 2004;**24**:720–7.
74. Cucullo L, Hallene K, Dini G, Dal Toso R, Janigro D. Glycerophosphoinositol and dexamethasone improve transendothelial electrical resistance in an in vitro study of the blood-brain barrier. *Brain Res* 2004;**997**(2):147–51.
75. Fischer S, Wobben M, Marti HH, Renz D, Schaper W. Hypoxia-induced hyperpermeability in brain microvessel endothelial cells involves VEGF-mediated changes in the expression of zonula occludens-1. *Microvasc Res* 2002;**63**:70–80.
76. Fischer S, Clauss M, Wiesnet M, Renz D, Schaper W, Karliczek GF. Hypoxia induces permeability in brain microvessel endothelial cells via VEGF and NO. *Am J Physiol* 1999;**276**:C812–820.
77. Abbruscato TJ, Davis TP. Combination of hypoxia/aglycemia compromises in vitro blood-brain barrier integrity. *J Pharmacol Exp Ther* 1999;**289**:668–75.

78. Neuhaus W, Gaiser F, Mahringer A, Franz J, Riethmuller C, Forster C. The pivotal role of astrocytes in an in vitro stroke model of the blood-brain barrier. *Frontiers Cell Neurosci* 2014;**8**:352.
79. Vandenhaute E, Culot M, Gosselet F, et al. Brain pericytes from stress-susceptible pigs increase blood-brain barrier permeability in vitro. *Fluids Barriers CNS* 2012;**9**:11.
80. Lochhead JJ, McCaffrey G, Quigley CE, et al. Oxidative stress increases blood-brain barrier permeability and induces alterations in occludin during hypoxia-reoxygenation. *J Cerebral Blood Flow Metabol* 2010;**30**:1625–36.
81. Azevedo FA, Carvalho LR, Grinberg LT, et al. Equal numbers of neuronal and non-neuronal cells make the human brain an isometrically scaled-up primate brain. *J Comp Neurol* 2009;**513**(5):532–41.
82. Roitbak T, Li L, Cunningham LA. Neural stem/progenitor cells promote endothelial cell morphogenesis and protect endothelial cells against ischemia via HIF-1alpha-regulated VEGF signaling. *J Cerebral Blood Flow Metabol* 2008;**28**(9):1530–42.
83. Kuntz M, Mysiorek C, Petrault O, et al. Stroke-induced brain parenchymal injury drives blood-brain barrier early leakage kinetics: a combined in vivo/in vitro study. *J Cerebral Blood Flow Metabol* 2014;**34**(1):95–107.

SECTION III

MOLECULAR BASIS AND CONCEPTS IN BRAIN EDEMA FORMATION FOR NEW TREATMENT DEVELOPMENT

CHAPTER

6

The Extracellular and Perivascular Spaces of the Brain

Michelle E. Pizzo and Robert G. Thorne
University of Wisconsin-Madison, Madison, WI, United States

OUTLINE

COMPONENTS OF THE CENTRAL NERVOUS SYSTEM (CNS)

The central nervous system (CNS) can be generally divided into three main components—the cells, fluids, and vasculature. First, there are the cells of the brain and spinal cord parenchyma, or tissue, that consist of two main types—neurons and glia. Approximately 86 billion neurons are present in the human brain (with 80%, or 69 billion neurons, in the cerebellum, 16 billion neurons in the cerebral cortex, and slightly less than 1 billion neurons in other regions) along with about 85 billion glial cells comprised of astrocytes, microglia, and oligodendroctyes (with 72% of glial cells located in the cerebral cortex).[1–5] Between all of these cells and their processes is an extracellular space (ECS) compartment that is filled with interstitial fluid

Brain Edema.
DOI: http://dx.doi.org/10.1016/B978-0-12-803196-4.00006-0

(ISF); ISF is carefully regulated because its ionic composition and volume are critical parameters for neurotransmission. Cerebrospinal fluid (CSF) is also present within and surrounding the brain and spinal cord. CSF composition and volume are carefully regulated and optimized to provide mechanical support (making the brain buoyant so it does not deform under the forces of gravity) and mechanical protection (preventing the brain from striking the skull with regular movement), as well as to maintain ion homeostasis.[6,7] The third component is the cerebral vasculature. Cerebral blood vessels are uniquely designed to separate the blood and its contents from the cells and fluids of the CNS (the CSF and ISF), providing highly selective blood–brain and blood–CSF barriers (BCSFB) (for details, see Chapter 3: Noninvasive Imaging Techniques for Brain Edema: From Basic Science to the Clinic). The blood–brain barrier (BBB) is primarily formed by microvessel endothelial cells, which lack fenestrations, possess tight junctions that prevent paracellular transport, have very low rates of transcytosis, and also have a variety of influx and efflux transporters to allow selective entry of some molecules (e.g., glucose) while selectively rejecting others (e.g., toxic substances). In this fashion, the BBB functions to supply the cells of the brain with nutrients, allow for removal of waste products, maintain the ionic homeostasis of the brain (by preventing the alteration of brain ions when plasma levels fluctuate), and to selectively regulate the entry and exit of molecules (e.g., proteins and neurotransmitters) between the blood and brain compartments.[8,9] An important feature is the proximity of the vasculature to the brain cells—neurons are usually not located further than ~10–20 μm from capillaries,[10,11] a short distance likely optimized for oxygen, glucose, and other essential molecules to reach all brain cells efficiently by diffusive transport.

The brain and spinal cord are contained within the meninges, which include (external to internal) the dura mater, arachnoid mater, and pia mater (Fig. 6.1). The outermost external layer is the dura (a tough membrane composed largely of collagen), which lies between the skull and the underlying leptomeninges. The leptomeninges, or "thin" meninges, include the arachnoid and pia, between which lie the CSF-filled subarachnoid space (SAS). Many large blood vessels travel in the SAS and trabeculae (strings or sheets of leptomeningeal cells) also cross the SAS connecting the arachnoid and pia. Beneath all these meningeal layers lies the brain; at the surface of the brain as well as along parenchymal blood vessels, astrocyte (glial) endfeet and their basement membrane form the glia limitans, partly partitioning the brain from the subpial and perivascular spaces.[6,12,13]

THE VENTRICULAR SYSTEM AND CSF

The CSF of mammals is contained within several cavities (ventricles) within the brain along with the SAS that surrounds the brain and spinal

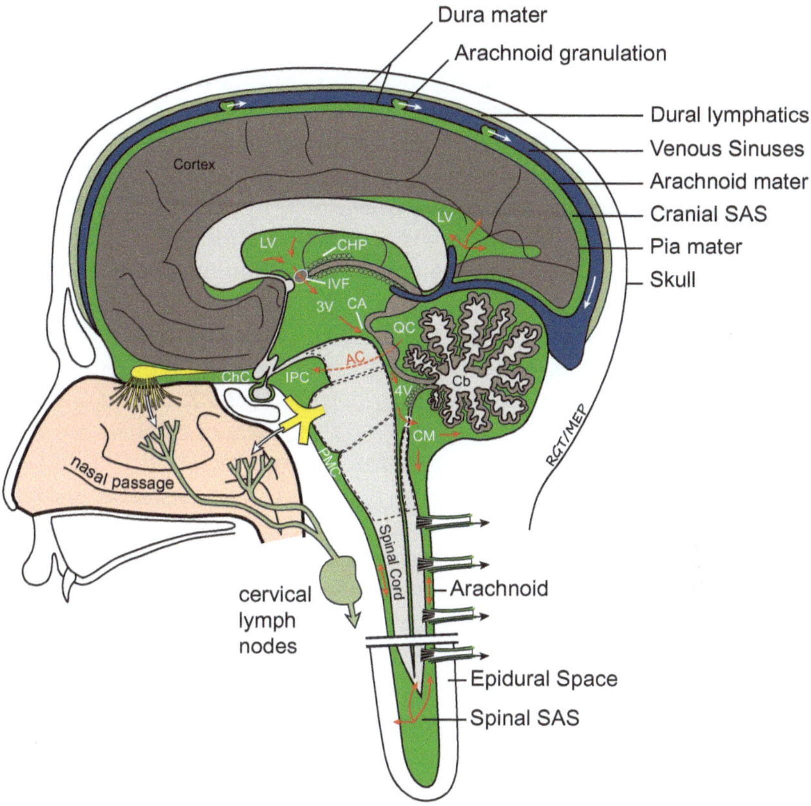

FIGURE 6.1 **Cerebrospinal fluid compartments, flow pattern, and drainage routes.** The CSF (green) flows through the ventricular system (red arrows) and into the subarachnoid space surrounding the brain and spinal cord. The CSF exit pathways (white or black arrows) include drainage into the venous sinuses of the dura via arachnoid granulations, drainage along cranial nerves (e.g., olfactory nerves and possibly others such as the trigeminal nerves) into the nasal mucosa and deep cervical lymph nodes, and drainage along spinal nerves to intercostal and other lymph nodes. In addition, lymphatic vessels in the dura may be a route for CSF to drain to the cervical lymph nodes. Abbreviations: *SAS*, subarachnoid space; *LV*, lateral ventricle; *CHP*, choroid plexus; *IVF*, interventricular foramen; *3V*, third ventricle; *CA*, cerebral aqueduct; *4V*, fourth ventricle; *CM*, cisterna magna (cerebellomedullary cistern); *QC*, quadrigeminal cistern; *AC*, ambient cistern; *ChC*, chiasmatic cistern; *IPC*, interpeduncular cistern; *PMC*, pontomedullary cistern; *Cb*, cerebellum.

cord (Fig. 6.1). Each of the brain's four ventricles contains a choroid plexus, consisting of fenestrated vasculature and specialized secretory epithelia that produce CSF by active transport/secretion[6] (for details, see Chapter 3: Noninvasive Imaging Techniques for Brain Edema: From Basic Science to the Clinic). In addition, CSF is produced in the brain parenchyma, by fluid entering across the BBB.[6,14,15] The ependymal cells lining the ventricles possess cilia, which, in addition to the respiratory and cardiac pulsations, aid in CSF movement through the ventricular system.[6,16,17] The conventional CSF flow pattern is generally described as follows: CSF is produced by the choroid plexuses of two paired lateral ventricles, in which it flows through two interventricular foramina (of Monro) to enter the third ventricle; the third ventricle choroid plexus produces additional CSF, which flows through a narrow cerebral aqueduct (of Sylvius) to enter the fourth ventricle; the fourth ventricle choroid plexus produces additional CSF, which exits via three sites, two lateral foramina (of Luschka), and one medial foramen (of Magendie), to enter the SAS (though some CSF may continue down the central canal at the center of the spinal cord). Once CSF enters the SAS, it fills the SAS surrounding the brain parenchyma and spinal cord. In some brain areas, there is a thin layer of SAS between the brain and the dura/skull, whereas in other areas there are large pockets of SAS referred to as cisterns.[6,18] From the SAS, CSF may drain along arachnoid (Pacchionian) granulations or villi into the blood of the dural venous sinuses, or along cranial nerves (e.g., along olfactory nerves to the nasal lymphatics) to the deep cervical lymph nodes[14,19–23] (reviewed in Ref. [24]); or spinal nerves to the deep cervical, intercostal, and other lymph nodes (reviewed in Ref. [24]), which ultimately drain into the blood. Recently, it has been shown that CSF may also drain to dural lymphatic vessels that also appear to drain to the deep cervical lymph nodes.[22,25,26] It is thought that all these routes to the lymphatic system are important for CNS antigen drainage and immune function.[27–32] Key questions going forward concern how all the different CSF/lymphatic drainage pathways work in concert with each other, their relative importance, species differences, and alterations in their function/relative importance during different disease processes.

In addition to the BBB, there is a BCSFB preventing free communication between the blood and CSF. One region of the BCSFB is created by tight junctions between arachnoid epithelial cells,[33] which prevents plasma contents from fenestrated blood vessels in the dura from entering the CSF of the SAS. Tight junctions between choroid plexus epithelial cells create the other region of the BCSFB.[34] As already mentioned, choroid plexus blood vessels are fenestrated, which aids in CSF production from plasma water. Thus, like the BBB, the BCSFB has transporters to allow selective entry of ions and molecules from the blood compartment and also plays a role in immune regulation/response[6,35–37] (for details, see Chapter 3: Noninvasive Imaging Techniques for Brain Edema: From Basic Science to the Clinic).

THE EXTRACELLULAR SPACE

The ECS refers to the ISF-filled environment that separates brain cells and their processes (Fig. 6.2A). Much work suggests that the ECS accounts

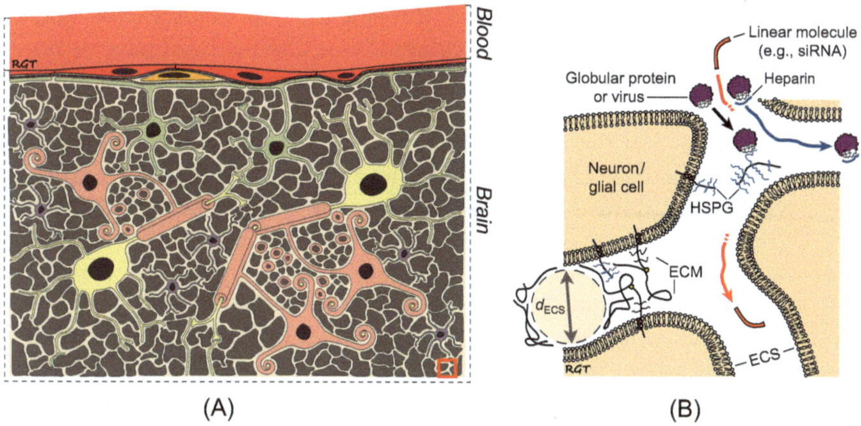

FIGURE 6.2 **The brain microenvironment—blood–brain barrier and extracellular space.** (A) The brain parenchyma is composed of several different cell types including neurons (yellow), oligodendrocytes (pink), microglia (blue), and astrocytes (green); these cells and their processes (gray) create a very narrow and tortuous extracellular space (ECS). Brain microvessels are composed of endothelial cells (dark pink) and their basement membrane (dashed dark gray line) and associated pericytes (light orange). Astrocyte foot processes and their basement membrane (dashed light gray line) further separate the blood vessels from the brain parenchyma and the extracellular spaces of the brain. (B) A magnified view of the ECS (red box) illustrates how linear molecules (e.g., siRNA) or globular macromolecules and macromolecular assemblies (e.g., proteins or viral vectors) are believed to diffuse through the ECS, and how the extracellular matrix (ECM) components (e.g., heparan sulfate proteoglycans) and the ECS width (d_{ECS}) may influence their diffusion. It is possible to improve transport of substances through the ECS by reducing binding to ECM components, e.g., complexing heparin to certain proteins and viruses is expected to reduce their binding to heparan sulfate proteoglycans and enhance their distribution (blue line).

for approximately 20% of the total tissue volume in the normal, adult brain in vivo.[38,39] This ECS volume fraction ($\alpha \sim 0.2$) appears highly conserved between most brain regions and across most species that have been studied to date, including mammals (e.g., rat, mouse, guinea pig, rabbit, dog, and human tissue), reptiles (e.g., turtle), amphibians (e.g., frog), and fish (e.g., skate).[40–43] Careful study of the rodent brain during postnatal development has revealed that the ECS in both the gray and white matter is about twice its adult value in the newborn ($\alpha \sim 0.4$[44,45]) and that the gray matter ECS appears to decline in aged animals at the end of the lifespan (e.g., $\alpha = 0.13$–0.16 in 17–25-month-old mice[46]). It is worth noting that the above information about ECS volume comes primarily from the method of real-time iontophoresis performed in either anesthetized animals in vivo or in acute brain slice preparations (e.g., see Refs. [41] or [43] for review). A recent report utilizing this same method has suggested that the ECS volume may be reduced in the superficial parietal association cortex of awake mice as compared to measurements in the same area of either sleeping or anesthetized adult mice ($\alpha \sim 0.14$ in awake versus $\alpha \sim 0.23$ in sleeping or ketamine/xylazine anesthetized animals[47]); however, this very interesting finding awaits replication and further study across other brain regions and species.

Another important parameter describing the brain ECS is its width (Fig. 6.2B), as this will impose constraints on how larger macromolecules, nanoparticles, and viruses diffuse through the tortuous paths between brain cells. Although early estimates of ECS width from electron microscopy suggested a narrow ECS width of ~20nm or less, it was long appreciated that these estimates were likely confounded by redistribution of extracellular water due to tissue ischemia and processing (associated with animal sacrifice and preparation of samples for microscopy) because the ECS volume in this material often appeared much smaller than physiological estimates had clearly provided.[41,48] More recent estimates from in vivo diffusion measurements in anesthetized rat neocortex suggest the actual ECS width is approximately 40–60nm in adult animals.[48] At least for the gray matter, a large body of evidence from numerous tracer diffusion and distribution studies suggests that these narrow spaces are unlikely to accommodate bulk flow (convection) because basic diffusion theory has been sufficient to fully describe such transport under a variety of experimental conditions in normal animals (e.g., Refs. [49–51]; for review, see Refs. [9,41,43]). ECS transport is most likely governed by diffusion because the hydraulic conductivity (ease of flow) is likely too low (and the resistance to flow too high) to allow for appreciable convective flow.[52,53] In simple terms, diffusion is a size-dependent, slow random-walk migration of molecules that results in net transfer of molecules down a concentration gradient.[54] Studies employing tracers of varying size/chemistry and using a variety of methods have demonstrated that molecular transport in the brain tissue is consistent with Fick's second law relating changes in concentration with time to a diffusion coefficient.[41,43,48,52,55]

An important aspect of the ECS that can further affect transport concerns its structural organization. Brain regions that lack homogeneity (e.g., inhomogeneity is evident in tissue with complex patterns of different layers such as the cerebellum and olfactory bulb) and/or isotropy (e.g., anisotropic aligned fibers within the white matter) display regional variations in diffusion characteristics.[43] Most neocortical areas that have been examined to date, e.g., somatosensory cortex,[51,55–58] have been characterized by isotropic diffusion (diffusion coefficients are the same along any axis so that point source injections result in spherically symmetrical diffusion clouds). The white matter of the corpus callosum provides an example of a brain area where diffusion anisotropy exists, e.g., extracellular diffusion of the small tetramethylammonium ion tracer is significantly greater along the long cylindrical axis of myelinated axons in this region.[59]

Convincing and interesting experimental evidence suggests that convective transport (flow) within and around the brain and spinal cord is limited to the ventricular and SAS CSF compartments, perivascular spaces, and possibly also the white matter; all of these areas are expected to exhibit substantially higher hydraulic conductivity (i.e., lower resistance to flow) than the neuropil of the gray matter.[41,52] Although the white matter has a normal ECS volume similar to the gray matter in adult animals and diffusive transport likely dominates under normal conditions,[44,45] experimental evidence suggests that the white matter ECS paths oriented along fibers likely can undergo significant expansion when exposed to higher than normal pressures or osmotic challenge such that measureable flow may sometimes occur.[50,60] This preferential poroelastic expansion of the white matter ECS may help to explain the marked periventricular white matter edema that occurs with hydrocephalus, whereas the central gray matter and neocortex remain relatively unaffected, even in severe conditions.[61]

The ECS is not simply an empty fluid space; it also contains hyaluronic acid, heparan sulfate and chondroitin sulfate proteoglycans, glycoproteins, laminins, and collagen, all of which may be attached to cells or free floating.[62] All of these ECM components, as well as the cell membranes and associated cell surface molecules together, comprise the ECS boundaries and can impact the transport of endogenous and exogenous molecules (Fig. 6.2B). Molecules that interact (e.g., transiently bind) with ECM components and/or fixed elements on cell surfaces may therefore diffuse more slowly through brain ECS than comparably sized inert molecules that lack such interactions.[41,43,51] Examples in which alterations in measured diffusive transport parameters due to binding to cell surface receptors or extracellular matrix components have recently been described. First, a direct ECM effect in which rapid, reversible binding to endogenous heparan sulfate proteoglycans slowed the extracellular diffusion of the iron-binding protein lactoferrin was shown in vivo.[63] This study took advantage of a comparison between lactoferrin and the closely related transferrin protein to isolate the heparan sulfate proteoglycan binding effect on measured extracellular diffusion parameters; both lactoferrin and transferrin are bilobal ~80-kDa proteins, but only lactoferrin contains a region of basic amino acids near its amino terminus allowing it to bind polyanions, such as heparin, heparan sulfate, and DNA. The study showed that lactoferrin's effective diffusion coefficient was reduced by ~60% due to its binding the relatively fixed, endogenous heparan sulfate binding sites arrayed in brain ECS.[63] Second, IgG antibody binding to endogenous Fc receptors on cell surfaces has also been shown to significantly hinder extracellular diffusion in the somatosensory cortex in vivo;[51] administering an excess of unlabeled Fc (competing away the binding of the labeled IgG domain that specifically binds the endogenous Fc receptors) facilitated improved IgG diffusion.[51]

Although it is thought that diffusion governs transport in the ECS, it is well agreed upon that there are also pathways in the brain for convective, or bulk fluid flow, in which molecules in a fluid, independent of size, move with the fluid at a certain velocity. Convective flow is driven by a pressure gradient and can be substantially faster than diffusion, particularly over longer

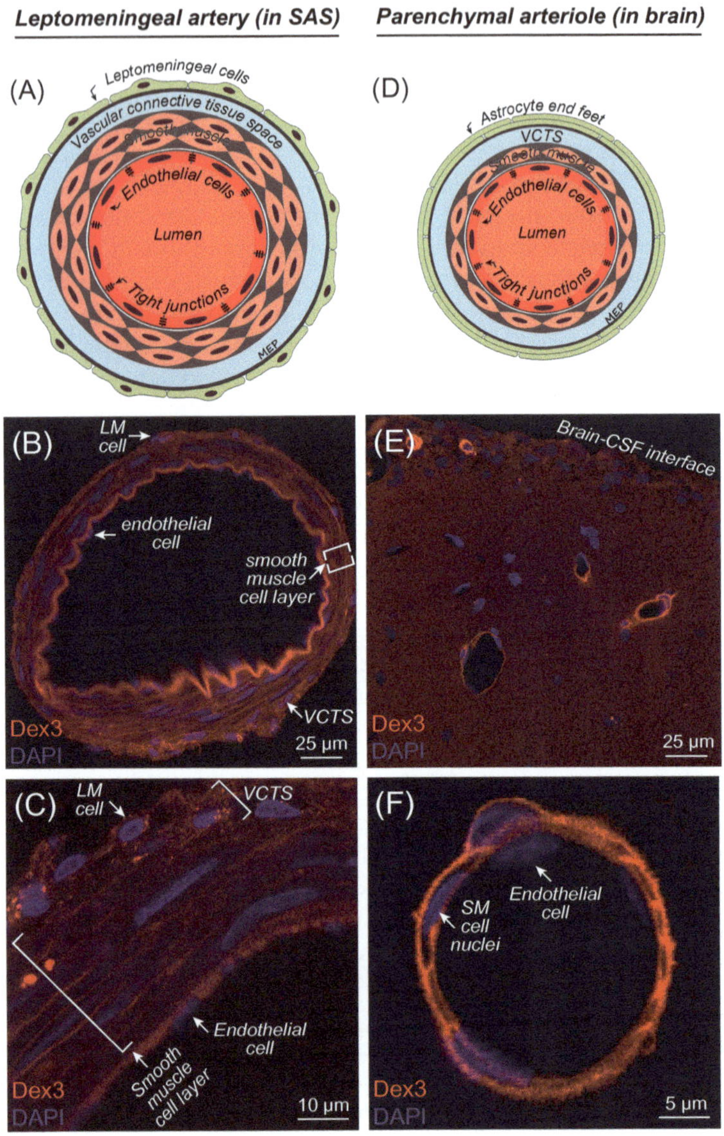

FIGURE 6.3 **The perivascular space of cerebral blood vessels.** (A) Schematic of the layers of a leptomeningeal artery in the subarachnoid space on the brain surface. The blood vessel is ensheathed by a layer of leptomeningeal cells (light green) and a vascular connective tissue space (VCTS), generally referred to as the perivascular space (light blue). The artery wall consists of (outer to inner): circumferentially arranged smooth muscle (SM) cells (light pink) that may be several layers thick, and their surrounding SM basement membrane (gray); an internal elastic lamina (white line); and the endothelial cell basement membrane (gray) and endothelial cells (dark pink), joined by tight junctions. (B) Texas Red labeled 3-kDa dextran was infused for 50 minutes into the cisterna magna in rats (for method details, see Ref. [51]) and the distribution in brain slices observed using ex vivo confocal microscopy. A cross section of the intracranial internal carotid artery, a large leptomeningeal artery, shows 3-kDa dextran around leptomeningeal (LM) cells, in the perivascular/vascular connective tissue space (adventitia), and around smooth muscle cells (tunica media), most likely in the smooth muscle basement membrane, demonstrating that CSF can access the perivascular space and the arterial wall. (C) A magnified image of the wall of the artery of a branch of the anterior cerebral artery also

distances.[9,41] Transport of mass within a fluid may be described in terms of a diffusive flux component (always present as long as the diffusing molecules are mobile) and a convective flux component (present when a volume of fluid containing the molecules is transported at some velocity). Diffusion is caused by Brownian motion (random walks), resulting in a net flux down a concentration gradient, and is directly proportional to a molecule's diffusion coefficient according to Fick's first law. In contrast, convection most often refers to a flow driven by a pressure difference. Darcy's law states that convective flux is directly proportional to the hydraulic permeability (a property influenced by both the material's pores for fluid flow and the fluid itself). Intrinsic hydraulic permeability is a useful component of the hydraulic permeability—It varies directly with the square of the mean pore diameter through which fluid must flow and can be used to understand how convection is dramatically favored within the brain along lower resistance pathways (i.e., those with much larger effective "pores" than the 40–60 nm wide neuropil;[48] such as the ventricular and subarachnoid CSF spaces, the perivascular spaces of the cerebrovasculature, and potentially the white matter). For example, measurements of perivascular space widths in mammals (e.g., arteriole perivascular spaces) commonly appear $\sim 5-10\,\mu m$ or larger in rodents and humans;[64–66] indeed, for a PVS dimension of $7.5\,\mu m$ and an ECS width of 50 nm, it can be determined that the expected resistance to fluid flow would be over 20,000-fold higher in the ECS of the neuropil compared to the larger PVS.[41] Several tracer experiments have now demonstrated that molecules administered into the CNS spread quickly, much faster than diffusion would allow, along preferential pathways in the brain that include the perivascular spaces and extracellular compartments associated with white matter tracts.[14,20,23,49,60,64,67–75] Most research suggests that the convective flow in the PVS is driven by the pulsations of the arteries,[69,76,77] as reducing the pulsations with drugs, physical ligation of vessels, or euthanization reduces transport in the PVS.

There is now great interest in how perivascular spaces, fluid-filled channels surrounding arteries, arterioles, veins, venules, and possibly even microvessels, may offer potential pathways for rapid convective flows into and out of the brain parenchyma. In light of this, we shall

◄ demonstrates 3-kDa dextran in and around the arterial wall. (D) Schematic of an arteriole in the brain parenchyma. The arteriole has fewer smooth muscle cell layers and is separated from the brain parenchyma by astrocyte endfeet (light green) and their basement membrane (gray). (E) After infusion of 3-kDa dextran into the CSF, the dextran has entered into the perivascular spaces of various-sized blood vessels in the cortical parenchyma near the brain surface. A putative arteriole in the frontal cortex shows dextran distributed throughout the arteriole wall, surrounding SM cells (empty pockets, some show DAPI staining of SM cell nuclei), likely distributed within the SM and endothelial basement membranes. Note, the spaces in the schematic drawings are not necessarily drawn to scale, but the vascular connective tissue space/perivascular space is not generally visible in light microscopy (i.e., around the arteriole in F) due to changes in the tissue with aldehyde fixation and processing.

briefly summarize the work of several groups (see Fig. 6.3) to generally define the perivascular compartment.[78–80] The perivascular space exists around vessels in the SAS and follows the vessels (at least for some distance) as they penetrate the brain tissue, narrowing as the vessel diameter narrows. The vessel continues until the fluid compartment lies between the astrocyte (glial) basement membrane that walls off the brain parenchyma and the vessel wall. At the level of the capillaries, in which the blood vessel wall and its components consist of the endothelial cell, the endothelial cell basement membrane, and pericytes, it has been unclear if a perivascular fluid space still remains. Indeed, whether a pericapillary space exists, or if the endothelial BM and the glial BM fuse to obliterate such a space, has been a matter of debate for some time.[2,81–83]

As a pathway, the perivascular space likely has several functions in normal physiology, many of which are still emerging or have only been recently proposed. It may play a role in the distribution and delivery of nutrients and neuromodulators[69,84] to specific brain regions or the entire brain, which therefore may make it an excellent medium/route for drug delivery. For example, it was once proposed that the perivascular space may provide nutrients to the cells in the thick vessel walls, as cerebral vessels lack vasa vasorum, which supply the walls of large vessels in the periphery.[85] It may be involved in solute and waste clearance from the brain, e.g., transporting metabolic products like amyloid beta out of the ISF;[47,64] indeed, it has been suggested that impairment of this system could be related to several pathologies such as Alzheimer's disease (AD), cerebral amyloid angiopathy, and traumatic brain injury.[86–89] In addition, many researchers[14,64,69,90–92] have suggested that the perivascular spaces may very well be functioning as lymphatics (or prelymphatics) for the brain (this was in fact first suggested by His in 1865![67]), partly because the brain noticeably lacks traditional lymphatic vessels for the drainage of ISF and antigens. As mentioned already, the CSF and ISF have been shown to drain along cranial nerves and via dural lymphatic vessels to the deep cervical lymph nodes, and it has been demonstrated that antigens in the CSF and ISF can induce a specific immune response via the cervical lymph nodes.[27,28,93] It is possible that drainage along the perivascular space is an initial step in the CSF/ISF to nasal lymphatic/dural lymphatic to cervical lymph node pathway; surprisingly, the precise direct and/or indirect connections between the ISF and cervical lymph nodes have yet to be firmly established.

Although the perivascular space likely plays a role in normal physiology, it also is linked to several neuropathologies. It has been suggested that in AD, a failure of the perivascular drainage pathway may result in reduced drainage of amyloid-beta from the ISF, forming amyloid aggregates in the perivascular space and blocking drainage further, worsening the disease.[86–88] In cerebral amyloid angiopathy, amyloid-beta is found in the basement membrane of the vessel walls, perhaps also indicating a failure in the perivascular drainage of amyloid-beta.[89]

THE EXTRACELLULAR SPACE AND PERIVASCULAR SPACE IN EDEMA

Although there are three principal classifications of brain edema (vasogenic, cytotoxic, ionic edema; for details see Chapter 12: Edema and BBB Breakdown in Stroke and Ref. [94]), we will focus our brief consideration here on the two general types of brain edema in which excess fluid accumulates in either (1) the intracellular compartment or (2) the extracellular compartment. It is important to note that these two types of edema often occur in succession and are rarely independent. Simplistically, cellular damage can lead to excessive accumulation of fluid in the intracellular compartment resulting in cytotoxic edema. With cytotoxic (or cellular) edema, an ionic imbalance may be induced by toxic/poisonous agents, ischemia, or hypoxia, causing the cellular uptake of water. This causes swelling of the intracellular compartment, shrinkage of the ECS and associated further hindrance of molecular movement in the ECS, potentially worsening edema by preventing drainage of various molecules diffusing extracellularly. Alternatively, excessive extravasation of fluid into the extracellular compartment due to BBB disruption is referred to as vasogenic edema. With vasogenic edema, solutes and proteins, along with water, cross a compromised BBB into the ECS. This BBB disruption may be the result of a traumatic incident, brain tumor, infection, hypertension, altered osmotic homeostasis, and even high altitudes.[94–96] The influx of fluid into the ECS causes the ECS to expand and an associated pressure increase because the accumulating fluid cannot easily escape through the small ECS of surrounding healthy tissue. As the white matter appears to have a higher hydraulic conductivity (lower resistance to flow) compared to the gray matter (discussed above), this may explain why the white matter is often preferentially expanded by fluid during cerebral edema. The white matter may also be a route for excess fluid to move from one brain area to another, as the expansion of the spaces between ordered fibers of the white matter offers a path-of-lesser-resistance compared to the more disordered and tortuous gray matter ECS.[52,97,98]

The most common cause of cytotoxic edema is cerebral infarction or ischemia. One example of cytotoxic cerebral edema is provided by the work of Hossmann,[99] who studied changes in the brain during ischemia. Hossmann used a cat model of ischemia in which blood flow to the brain was obstructed for 1 hour followed by 2–3 hours of reperfusion; cortical impedance measurements were used to infer that the brain ECS volume % dropped from 19% to about 9% at the end of ischemia, returning toward baseline after about 2 hours. Brain osmolality increased during ischemia, but brain water content rose only slightly (as water could not come from the blood, this likely came from the CSF). During reperfusion, the elevated brain osmolality likely caused water to enter the brain from the blood, as the brain was shown to swell and water content increased more than 10% above normal.

Substantial work modeling vasogenic brain edema has described the pathways of edema fluid resolution, and the involvement of the white matter and perivascular spaces, summarized briefly here. Kalimo et al.[100] used stroke prone spontaneously hypertensive rats as a naturally occurring (not induced) model of vasogenic edema. Fluid entry into the brain was inferred from extravasation of systemically administered Evans blue, a dye which binds tightly to albumin. Although the spread of Evans blue (representing albumin and likely other plasma proteins) appeared limited around leaky vessels in the gray matter ECS, Evans blue traveled great distances along the white matter of the corpus callosum (even extending from one hemisphere to the other) and ultimately appeared to drain into ventricular CSF. They also described a role for the perivascular spaces in vasogenic edema fluid resolution. Importantly Kalimo et al. also described several instances in which limited diffuse Evans blue complexed albumin extended around vessels all the way from the apparent site of entry/hemorrhage toward the cortical surface; this tracer seemed to accumulate locally in the subpial space in which the vessel exited the brain, where it was assumed to then access the SAS CSF compartment. Kalimo et al. suggested that they were the first to have described this route of perivascular edema fluid exit, which may have gone unnoticed in previous edema models.

Ohata et al.[101] used a rat infusion model of vasogenic edema, in which albumin is slowly infused into the caudoputamen, mimicking the leakage of plasma proteins and fluid across a damaged BBB. To better understand both the acute phase (where pressure differentials may be greater) and long-term resolution pathways (after pressures have subsided), they followed the infused albumin over several days. Immediately after the infusion (15 minutes), albumin was distributed in the gray matter ECS of the striatum, but upon reaching the white matter it spread great distances throughout the expanded and edematous white matter tracts to the ventricular wall, where it appeared to cross between ependymal cells to reach the CSF. At later time points (1–8 days), albumin distributed further through the white matter of the contralateral side and diffusely into the overlying gray matter cortex, but a full 8-day period was needed to complete removal of the infused albumin. In addition, Ohata et al. also observed perivascular involvement in both the enlarged white matter and gray matter of the cortex; they found that perivascular spaces around arterioles and capillaries were slightly expanded with minimal albumin reaction product, but spaces around venules were particularly large and indicated high levels of albumin, suggesting to the authors that the perivascular spaces of venules were primarily involved in the clearance of the edema fluid. However, Ohata et al. saw no albumin crossing the endothelial cell, indicating that clearance of albumin in the edema fluid was unlikely to occur by drainage into the blood across the BBB, and no intracellular albumin was found, indicating that cellular uptake and metabolism/degradation was also an unlikely resolution pathway (in contrast to conclusions from previous work). The results suggested that the bulk flow pathways described by Cserr[60,68] including the white matter tracts and perivascular spaces (particularly venules, as suggested by Rennels[69]) were likely involved in initial, rapid clearance to the ventricular and subarachnoid CSF but also involved in removal at later time points, and that drainage into the CSF compartment was the predominant resolution pathway.

Using a similar experimental paradigm, Ohata and Marmarou[102] later infused albumin or fluorescent dextran tracers of different molecular sizes into the rat brain; they again demonstrated rapid spreading along white matter tracts (both ipsilateral and contralateral to the infusion) and along the ventricle wall into the ventricular CSF, and some perivascular drainage toward the cortical surface to the subarachnoid CSF. However, the spread of tracers in the gray matter was size dependent, such that the smaller dextran spread further than the larger dextran, indicating diffusive transport (not bulk flow) in the normal (nonedematous) ECS of the gray matter, in agreement with previous work.[103,104] The suggested pathway of edema fluid resolution in this study was bulk flow along white matter tracts and perivascular spaces to the ventricular or subarachnoid CSF, respectively, as well as by diffusion across the nonedematous gray matter to the CSF.

Geer and Grossman[105] used a similar vasogenic edema infusion model to Ohata et al.,[102] infusing albumin with tracers Evans blue and radiolabeled inulin into the rat brain at gray–white matter junctions in the frontal and temporal lobes. Similar to the experiments above, they saw rapid, bulk flow of the tracers along white matter tracts, but the area of spread was very different depending on the infusion site. The authors suggested these white matter tract pathways may contribute to dissemination of cancer cells from tumors with BBB leakage (and thus vasogenic edema) and postulated that the location of the main tumor (and its access to various distribution pathways) may necessitate more or less aggressive therapies to combat local regrowth and distant metastasis.

Reulen et al.[106] used a cold-injury lesion model in cats to induce vasogenic edema. They systemically administered tracers of different molecular sizes to act as plasma markers they could visualize leaking across the damaged BBB. They found that even though these tracers had significantly different molecular sizes and thus different diffusion coefficients, they were able to quickly spread through the edematous white matter over similar distances and at similar rates. In addition, the spread of both tracers far exceeded the calculated distances achievable by diffusion, strongly suggestive of bulk flow within the white matter. However, if cytotoxic edema was induced before cold-injury, which resulted in decreased ECS size due to increased intracellular water content (see the work of Hossmann discussed above), the spread of edema fluid and tracers appeared to be reduced, presumably because the smaller ECS size increased the resistance to flow. They also found increased water content and elevated ISF pressure in edematous white matter as compared to normal tissue and CSF. Reulen et al. suggested that this pressure differential allowed for edema fluid to drain from areas of high pressure—the site of injury and edematous tissue—to areas of lower pressure—the normal tissue and CSF. They ultimately suggested that lower CSF pressure could aid in the clearance of the fluid by increasing the driving force (the pressure differential) and further postulated that it could explain the beneficial effect observed when CSF production and pressure reducing drugs are used in patients with brain edema.

The Reulen group followed up on this work (see Ref. [107]) to further investigate the effect of CSF pressure on edema resolution. After the cold-injury lesion and administration of systemic tracers, animals were surgically prepared for ventriculocisternal perfusion, in which artificial CSF was infused into the ventricles and simultaneously withdrawn from the cisterna magna. This allowed them to adjust CSF pressure by modulating outflow from the cisterna, creating low or high pressure conditions (as well as physiological controls), and also allowed for CSF sampling to determine tracer drainage into the ventricular CSF from the brain tissue. CSF levels of the albumin tracer were much higher in animals with low CSF pressure compared to high CSF pressure, indicating that far more edema fluid/protein was cleared into the CSF when there was a greater pressure drop between the edematous tissue and the ventricular CSF. Evans blue tracer, albumin tracer, and water content in the white matter near the ventricles was much less for the low pressure group than the high pressure or controls, suggesting more tracer/edema fluid was cleared from the tissue into the CSF in the presence of a greater pressure difference. This study therefore also highlights the importance of the CSF resolution pathway and suggested that alteration of CSF pressure can impact edema fluid removal from the brain. Similar experiments performed by Shima and Marmarou,[108] however, saw a trend but no significant difference between low and high intracranial pressure groups; perhaps larger pressure changes were needed to resolve the difference. They also commented on the multiple effects that CSF pressure alterations could have on the balance between edema fluid formation and removal. For example, if the CSF pressure is lowered, this may allow fluid and plasma proteins to more easily cross the damaged BBB into the brain, increasing edema fluid, but lower CSF pressure may increase the pressure differential between brain and CSF increasing edema fluid removal from the brain. If the CSF pressure is high, the driving force between brain ISF to CSF may be reduced, but the higher pressure may keep blood from crossing into the brain as easily. However, the brain tissue between the blood and CSF may also provide sufficient buffering to prevent moderate alterations from impacting edema formation or removal.

Blaumanis et al.[109] induced vasogenic edema unilaterally in cats using a closed skull cold lesion and found edematous white matter on the ipsilateral side. They additionally injected horseradish peroxidase (HRP) into the CSF (at the cisterna magna) to monitor disruption in their proposed paravascular circulation (where CSF enters along the perivascular space of arteries, circulates in the brain along capillary basement membrane, and exits into CSF along the perivascular space of veins) during edema. On the contralateral uninjured side, HRP accessed white matter and to a lesser extent perivascular spaces (perivascular distribution was somewhat similar to infusions in normal animals[69]). However, on the lesioned side the white matter was edematous and there was little tracer present in the white matter or perivascular spaces, even in normal tissue at more distant sites ipsilateral to the lesion. Blaumanis et al. suggested a disruption of the perivascular circulation had occurred with edema.

Although the above studies have demonstrated how edema fluid initiating in the brain is drained, additional studies have also demonstrated how edema can be caused by a failure of these drainage pathways or other downstream drainage. Ligation of the cervical lymph vessels (to which CSF/ISF has been shown to drain) in dogs resulted in enlarged and edematous white matter,[110] as well as enlargement of the perivascular spaces, which also showed accumulation of injected ink and endogenous protein. In hydrocephalus, white matter edema[111,112] and perivascular expansion[113] have also been reported. This suggests that both the white matter and perivascular spaces may be important physiological routes for normal or edematous fluid to drain from the brain to the CSF and cervical lymph nodes, as obstruction of these proposed drainage pathways leads to detrimental accumulation in the upstream white matter and perivascular spaces.

Finally, many studies have implicated a role for aquaporin (AQP) water channel proteins in the regulation of brain water homeostasis (see Chapter 11: Role of Matrix Metalloproteinases in Brain Edema; reviewed in Ref.[114]). Briefly, 3 of the at least 13 AQP family members appear to be prominently expressed in the CNS: AQP1, AQP4, and AQP9.[114–118] AQP4 is highly expressed on astrocyte processes forming the glia limitans at the pial brain surface (i.e., the brain interface with the SAS CSF) as well as on astrocyte endfeet contacting blood vessels (i.e., forming part of the wall separating the perivascular space from the surrounding parenchyma in large vessels and facing the basal lamina of the neurovascular unit in microvessels); it is the most abundant water channel in the CNS and is thought to play a major role in water homeostasis.[114] AQP1 is prominently expressed in the choroid plexus epithelia, in which it likely plays a role in CSF formation,[114] whereas AQP9, an aquaglyceroporin permeable to water as well as glycerol and urea, is widely expressed in astrocytes and more selectively expressed in neurons of the catecholaminergic system.[118] The prominent expression of AQP4 at brain tissue interfaces with the CSF, and blood vessels has long suggested a role for AQP4 in the regulation of brain ECS water. As mentioned above, brain ECS volume fraction declines significantly from birth ($\alpha \sim 0.4$) to reach its smaller adult volume ($\alpha \sim 0.2$); this change occurs in parallel with glial proliferation and increasing AQP4 expression over the same period.[119] Furthermore, careful measurements of somatosensory cortex ECS volume in AQP4-deficient adult mice have revealed a small but significant increase in ECS volume (α change from 0.18 to 0.23[120]), further suggesting a prominent role for AQP4 in regulating extracellular water volume. Although measurements of PVS width in AQP4-null adult mice have suggested that PVS volume is not significantly affected by a lack of AQP4 at the perivascular astrocytic endfeet, tracer influx from the CSF to the parenchyma appeared noticeably reduced.[64] Further research is needed to better define the different roles that AQPs may play in normal brain water circulation as well as in pathological processes resulting in cerebral edema.

CONCLUSIONS

Much evidence exists to suggest that molecular distribution in the brain microenvironment is normally governed by diffusion in the narrow, tortuous ECS of the neuropil in combination with convection or bulk flow within the CSF compartments (ventricular and SASs), and cerebral perivascular fluid spaces and perhaps also within the white matter. ECS and PVS volumes and physiological attributes may be altered in a variety of neurological disease states, including those conditions accompanied by brain edema. Three possible routes of edema fluid resolution have been suggested: (1) cellular uptake and metabolism, (2) transport across the BBB into the blood, and (3) drainage into the CSF. Much of the work described above suggests that the CSF is the predominant efflux pathway in models of vasogenic edema. Edematous white matter tracts show enlarged ECS, with size-independent bulk flow of tracers occurring along the white matter tracts primarily to the ventricular CSF. It would appear that the white matter is uniquely suited for this expansion along the aligned fibers as compared to the random cellular orientation in the gray matter. The narrow ECS of the gray matter in adult animals appears well suited to resist volume expansion in the face of increasing pressure, providing a possible explanation for numerous observations that transport in the gray matter ECS appears limited to size-dependent diffusion and thus the gray matter ECS would be better capable of maintaining a stable fluid environment for the ionic gradients that underlie neurotransmission. The cerebral perivascular spaces are now recognized as a path-of-less-resistance with a role in clearing edema constituents from the brain to the CSF via bulk flow. Both of these pathways show a connection to the downstream CSF drainage route to the cervical lymph nodes, but the precise details remain unclear. Understanding the formation of edema fluid, the mechanisms and pathways of edema fluid removal, and the balance between them will allow for improved, physiologically based treatment of cerebral edema. Central to this endeavor is developing a more complete understanding of just how the fluid compartments of the brain ECS and PVS are regulated under normal conditions as well as during pathological processes accompanied by edema.

References

1. Herculano-Houzel S. *The human advantage: a new understanding of how our brain became remarkable*. The MIT Press; 2016.
2. Peters A, Palay SSL, Webster H deF. *The fine structure of the nervous system: neurons and their supporting cells*. 3rd ed. New York: Oxford University Press; 1991.
3. Kandel ER, Schwartz JH, Jessell TM, Siegelbaum SA, Hudspeth AJ, editors. *Principles of neural science*. 5th ed. New York: McGraw Hill Medical; 2013. http://dx.doi.org/10.1036/0838577016.
4. Paxinos G, Mai JK. *The human nervous system*. Amsterdam: Elsevier Inc; 2004.
5. Thorne RG. Primer on central nervous system structure/function and the vasculature, ventricular system, and fluids of the brain. In: Hammarlund-Udenaes M, de Lange E, Thorne RG, editors. *Drug delivery to the brain: physiological concepts, methodologies, and approaches*. New York: Springer-Verlag; 2014. p. 685–707.
6. Davson H, Segal MB. *Phyioslogy of the CSF and blood–brain barriers*. 1st ed. Boca Raton, FL: CRC Press, Inc; 1995.
7. Sakka L, Coll G, Chazal J. Anatomy and physiology of cerebrospinal fluid. *Eur Ann Otorhinolaryngol Head Neck Dis* 2011;**128**(6):309–16. http://dx.doi.org/10.1016/j.anorl.2011.03.002.
8. Abbott NJ, Rönnbäck L, Hansson E. Astrocyte-endothelial interactions at the blood–brain barrier. *Nat Rev Neurosci* 2006;**7**(1):41–53. http://dx.doi.org/10.1038/nrn1824.
9. Abbott NJ. Evidence for bulk flow of brain interstitial fluid: significance for physiology and pathology. *Neurochem Int* 2004;**45**(4):545–52. http://dx.doi.org/10.1016/j.neuint.2003.11.006.
10. Schlageter KE, Molnar P, Lapin GD, Groothuis DR. Microvessel organization and structure in experimental brain tumors: microvessel populations with distinctive structural and functional properties. *Microvasc Res* 1999;**58**(3):312–28. http://dx.doi.org/10.1006/mvre.1999.2188.
11. Mabuchi T, Lucero J, Feng A, Koziol JA, del Zoppo GJ. Focal cerebral ischemia preferentially affects neurons distant from their neighboring microvessels. *J Cereb Blood Flow Metab* 2005;**25**(2):257–66. http://dx.doi.org/10.1038/sj.jcbfm.9600027.
12. Zhang ET, Inman CB, Weller RO. Interrelationships of the pia mater and the perivascular (Virchow-Robin) spaces in the human cerebrum. *J Anat* 1990;**170**:111–23.
13. Nolte J. *The human brain: an introduction to its functional anatomy*. 6th ed. Amsterdam: Elsevier; 2009.

14. Szentistványi I, Patlak CS, Ellis RA, Cserr HF. Drainage of interstitial fluid from different regions of rat brain. *Am J Physiol* 1984;**246**(6):35–44.
15. Cserr HF. Physiology of the choroid plexus. *Physiol Rev* 1971;**51**(2):273–311.
16. Brinker T, Stopa E, Morrison J, Klinge P. A new look at cerebrospinal fluid circulation. *Fluids Barriers CNS* 2014;**11**(1):10. http://dx.doi.org/10.1186/2045-8118-11-10.
17. Faubel R, Westendorf C, Bodenschatz E, Eichele G. Cilia-based flow network in the brain ventricles. *Science* 2016;**353**(6295):176–8. http://dx.doi.org/10.1126/science.aae0450.
18. Schuenke M, Schulte E, Schumacher U. *Thieme atlas of anatomy: head and neuroanatomy*. 1st ed. New York: Thieme; 2010.
19. Bradbury MW, Westrop RJ. Factors influencing exit of substances from cerebrospinal fluid into deep cervical lymph of the rabbit. *J Physiol* 1983;**339**:519–34.
20. Bradbury MW, Cserr HF, Westrop RJ. Drainage of cerebral interstitial fluid into deep cervical lymph of the rabbit. *Am J Physiol* 1981;**240**(4):F329–36.
21. Faber WM. The nasal mucosa and the subarachnoid space. *Am J Anat* 1937;**62**(1):121–48. http://dx.doi.org/10.1002/aja.1000620106.
22. Kida S, Pantazis A, Weller RO. CSF drains directly from the subarachnoid space into nasal lymphatics in the rat. Anatomy, histology and immunological significance. *Neuropathol Appl Neurobiol* 1993;**19**(6):480–8. http://dx.doi.org/10.1111/j.1365-2990.1993.tb00476.x.
23. Weller RO, Kida S, Zhang ET. Pathways of fluid drainage from the brain—morphological aspects and immunological significance in rat and man. *Brain Pathol* 1992;**2**(4):277–84. http://dx.doi.org/10.1111/j.1750-3639.1992.tb00704.x.
24. Koh L, Zakharov A, Johnston M. Integration of the subarachnoid space and lymphatics: is it time to embrace a new concept of cerebrospinal fluid absorption? *Cerebrospinal Fluid Res* 2005;**2**:6. http://dx.doi.org/10.1186/1743-8454-2-6.
25. Aspelund A, Antila S, Proulx ST, et al. A dural lymphatic vascular system that drains brain interstitial fluid and macromolecules. *J Exp Med* 2015;**212**(7):991–9. http://dx.doi.org/10.1084/jem.20142290.
26. Louveau A, Smirnov I, Keyes TJ, et al. Structural and functional features of central nervous system lymphatic vessels. *Nature* 2015;**523**(7560):337–41. http://dx.doi.org/10.1038/nature14432.
27. Harling-Berg C, Knopf PM, Merriam J, Cserr HF. Role of cervical lymph nodes in the systemic humoral immune response to human serum albumin microinfused into rat cerebrospinal fluid. *J Neuroimmunol* 1989;**25**(2-3):185–93.
28. Walter BA, Valera VA, Takahashi S, Matsuno K, Ushiki T. Evidence of antibody production in the rat cervical lymph nodes after antigen administration into the cerebrospinal fluid. *Arch Histol Cytol* 2006;**69**(1):37–47. http://dx.doi.org/10.1679/aohc.69.37.
29. Cserr HF, DePasquale M, Harling-Berg CJ, Park JT, Knopf PM. Afferent and efferent arms of the humoral immune response to CSF-administered albumins in a rat model with normal blood–brain barrier permeability. *J Neuroimmunol* 1992;**41**(2):195–202.
30. Knopf PM, Harling-Berg CJ, Cserr HF, et al. Antigen-dependent intrathecal antibody synthesis in the normal rat brain: tissue entry and local retention of antigen-specific B cells. *J Immunol* 1998;**161**(2):692–701.
31. Kipnis J. Multifaceted interactions between adaptive immunity and the central nervous system. *Science* 2016;**353**(6301):766–71. http://dx.doi.org/10.1126/science.aag2638.
32. Louveau A, Da Mesquita S, Kipnis J. Lymphatics in neurological disorders: a neuro-lympho-vascular component of multiple sclerosis and Alzheimer's disease? *Neuron* 2016;**91**(5):957–73. http://dx.doi.org/10.1016/j.neuron.2016.08.027.
33. Nabeshima S, Reese TS, Landis DM, Brightman MW. Junctions in the meninges and marginal glia. *J Comp Neurol* 1975;**164**(2):127–69. http://dx.doi.org/10.1002/cne.901640202.
34. Becker NH, Novikoff AB, Zimmerman HM. Fine structure observations of the uptake of intravenously injected peroxidase by the rat choroid plexus. *J Histochem Cytochem* 1967;**15**(3):160–5.

35. Strazielle N, Ghersi-Egea JF. Choroid plexus in the central nervous system: biology and physiopathology. *J Neuropathol Exp Neurol* 2000;**59**(7):561–74.
36. Damkier HH, Brown PD, Praetorius J. Cerebrospinal fluid secretion by the choroid plexus. *Physiol Rev* 2013;**93**(4):1847–92. http://dx.doi.org/10.1152/physrev.00004.2013.
37. Engelhardt B, Sorokin L. The blood–brain and the blood–cerebrospinal fluid barriers: function and dysfunction. *Semin Immunopathol* 2009;**31**(4):497–511. http://dx.doi.org/10.1007/s00281-009-0177-0.
38. Nicholson C, Phillips JM. Ion diffusion modified by tortuosity and volume fraction in the extracellular microenvironment of the rat cerebellum. *J Physiol* 1981;**321**(1981):225–57.
39. Nicholson C, Syková E. Extracellular space structure revealed by diffusion analysis. *Trends Neurosci* 1998;**21**(5):207–15. http://dx.doi.org/10.1016/S0166-2236(98)01261-2.
40. Levin VA, Fenstermacher JD, Patlak CS. Sucrose and inulin space measurements of cerebral cortex in four mammalian species. *Am J Physiol* 1970;**219**(5):1528–33.
41. Wolak DJ, Thorne RG. Diffusion of macromolecules in the brain: implications for drug delivery. *Mol Pharm* 2013;**10**(5):1492–504. http://dx.doi.org/10.1021/mp300495e.
42. Nicholson C. Diffusion and related transport mechanisms in brain tissue. *Rep Prog Phys* 2001;**64**(7):815–84. http://dx.doi.org/10.1088/0034-4885/64/7/202.
43. Syková E, Nicholson C. Diffusion in brain extracellular space. *Physiol Rev* 2008;**88**(4):1277–340. http://dx.doi.org/10.1152/physrev.00027.2007.
44. Lehmenkühler A, Syková E, Svoboda J, Zilles K, Nicholson C. Extracellular space parameters in the rat neocortex and subcortical white matter during postnatal development determined by diffusion analysis. *Neuroscience* 1993;**55**(2):339–51. http://dx.doi.org/10.1016/0306-4522(93)90503-8.
45. Vorisek I, Sykova E. Ischemia-induced changes in the extracellular space diffusion parameters, K +, and pH in the developing rat cortex and corpus callosum. *J Cereb Blood Flow Metab* 1997;**17**(2):191–203. http://dx.doi.org/10.1097/00004647-199702000-00009.
46. Syková E, Vorísek I, Antonova T, et al. Changes in extracellular space size and geometry in APP23 transgenic mice: a model of Alzheimer's disease. *Proc Natl Acad Sci USA* 2005;**102**(2):479–84. http://dx.doi.org/10.1073/pnas.0408235102.
47. Xie L, Kang H, Xu Q, et al. Sleep drives metabolite clearance from the adult brain. *Science* 2013;**342**(6156):373–7. http://dx.doi.org/10.1126/science.1241224.
48. Thorne RG, Nicholson C. In vivo diffusion analysis with quantum dots and dextrans predicts the width of brain extracellular space. *Proc Natl Acad Sci USA* 2006;**103**(14):5567–72. http://dx.doi.org/10.1073/pnas.0509425103.
49. Ichimura T, Fraser PA, Cserr HF. Distribution of extracellular tracers in perivascular spaces of the rat brain. *Brain Res* 1991;**545**(1–2):103–13. http://dx.doi.org/10.1016/0006-8993(91)91275-6.
50. Rosenberg GA, Kyner WT, Estrada E. Bulk flow of brain interstitial fluid under normal and hyperosmolar conditions. *Am J Physiol* 1980;**238**(1):F42–9.
51. Wolak DJ, Pizzo ME, Thorne RG. Probing the extracellular diffusion of antibodies in brain using in vivo integrative optical imaging and ex vivo fluorescence imaging. *J Control Release* 2015;**197**:78–86. http://dx.doi.org/10.1016/j.jconrel.2014.10.034.
52. Fenstermacher JD, Patlak CS. The movements of water and solutes in the brains of mammals. In: Pappius HM, Feindel W, editors. *Dynamics of brain edema*. Berlin Heidelberg: Springer; 1976. p. 87–94. http://dx.doi.org/10.1007/978-3-642-66524-0.
53. Bear J. *Dynamics of fluids in porous media*. New York: American Elsevier Pub. Co; 1972.
54. Berg HC. *Random walks in biology*. Princeton: Princeton University Press; 1993.
55. Thorne RG, Hrabetová S, Nicholson C. Diffusion of epidermal growth factor in rat brain extracellular space measured by integrative optical imaging. *J Neurophysiol* 2004;**92**(6):3471–81. http://dx.doi.org/10.1152/jn.00352.2004.
56. Nicholson C, Tao L. Hindered diffusion of high molecular weight compounds in brain extracellular microenvironment measured with integrative optical imaging. *Biophys J* 1993;**65**(6):2277–90. http://dx.doi.org/10.1016/S0006-3495(93)81324-9.

57. Mazel T, Simonová Z, Syková E. Diffusion heterogeneity and anisotropy in rat hippocampus. *Neuroreport* 1998;**9**(7):1299–304. http://dx.doi.org/10.1097/00001756-199805110-00008.
58. Tao L, Nicholson C. Diffusion of albumins in rat cortical slices and relevance to volume transmission. *Neuroscience* 1996;**75**(3):839–47. http://dx.doi.org/10.1016/0306-4522(96)00303-X.
59. Vorísek I, Syková E. Evolution of anisotropic diffusion in the developing rat corpus callosum. *J Neurophysiol* 1997;**78**(2):912–19.
60. Cserr HF, Cooper DN, Milhorat TH. Flow of cerebral interstitial fluid as indicated by the removal of extracellular markers from rat caudate nucleus. *Exp Eye Res* 1977;**25**(Suppl.):461–73. http://dx.doi.org/10.1016/S0014-4835(77)80041-9.
61. Weller RO. Pathology of cerebrospinal fluid and interstitial fluid of the CNS: significance for Alzheimer disease, prion disorders and multiple sclerosis. *J Neuropathol Exp Neurol* 1998;**57**(10):885–94. http://dx.doi.org/10.1097/00005072-199810000-00001.
62. Novak U, Kaye AH. Extracellular matrix and the brain: components and function. *J Clin Neurosci* 2000;**7**(4):280–90. http://dx.doi.org/10.1054/jocn.1999.0212.
63. Thorne RG, Lakkaraju A, Rodriguez-Boulan E, Nicholson C. In vivo diffusion of lactoferrin in brain extracellular space is regulated by interactions with heparan sulfate. *Proc Natl Acad Sci USA* 2008;**105**(24):8416–21. http://dx.doi.org/10.1073/pnas.0711345105.
64. Iliff JJ, Wang M, Liao Y, et al. A paravascular pathway facilitates CSF flow through the brain parenchyma and the clearance of interstitial solutes, including amyloid β. *Sci Transl Med* 2012;**4**(147):147ra111. http://dx.doi.org/10.1126/scitranslmed.3003748.
65. Pollock H, Hutchings M, Weller RO, Zhang ET. Perivascular spaces in the basal ganglia of the human brain: their relationship to lacunes. *J Anat* 1997;**191**(Pt 3)):337–46.
66. Foley CP, Nishimura N, Neeves KB, Schaffer CB, Olbricht WL. Real-time imaging of perivascular transport of nanoparticles during convection-enhanced delivery in the rat cortex. *Ann Biomed Eng* 2012;**40**(2):292–303. http://dx.doi.org/10.1007/s10439-011-0440-0.
67. Cserr HF, Patlak CS. Secretion and bulk flow of interstitial fluid. In: Bradbury MW, editor. *Physiology and pharmacology of the blood–brain barrier*. Springer-Verlag; 1992.
68. Cserr HF, Cooper DN, Suri PK, Patlak CS. Efflux of radiolabeled polyethylene glycols and albumin from rat brain. *Am J Physiol* 1981;**240**(4):F319–28.
69. Rennels ML, Gregory TF, Blaumanis OR, Fujimoto K, Grady PA. Evidence for a "paravascular" fluid circulation in the mammalian central nervous system, provided by the rapid distribution of tracer protein throughout the brain from the subarachnoid space. *Brain Res.* 1985;**326**(1):47–63. http://dx.doi.org/10.1016/0006-8993(85)91383-6.
70. Rennels ML, Blaumanis OR, Grady PA. Rapid solute transport throughout the brain via paravascular fluid pathways. *Adv Neurol* 1990;**52**(9):431–9.
71. Iliff JJ, Lee H, Yu M, et al. Brain-wide pathway for waste clearance captured by contrast-enhanced MRI. *J Clin Invest* 2013;**123**(3):1299–309. http://dx.doi.org/10.1172/JCI67677.
72. Cserr HF, Ostrach LH. Bulk flow of interstitial fluid after intracranial injection of blue dextran 2000. *Exp Neurol* 1974;**45**(1):50–60. http://dx.doi.org/10.1016/0014-4886(74)90099-5.
73. Fenstermacher JD, Ghersi-Egea JF, Finnegan W, Chen JL. The rapid flow of cerebrospinal fluid from ventricles to cisterns via subarachnoid velae in the normal rat. *Acta Neurochir Suppl* 1997;**70**:285–7.
74. Ghersi-Egea JF, Finnegan W, Chen JL, Fenstermacher JD, Hospital HF. Rapid distribution of intraventricularly administered sucrose into cerebrospinal fluid cisterns via subarachnoid velae in rat. *Neuroscience* 1996;**75**(4):1271–88. http://dx.doi.org/10.1016/0306-4522(96)00281-3.
75. Proescholdt MG, Hutto B, Brady LS, Herkenham M. Studies of cerebrospinal fluid flow and penetration into brain following lateral ventricle and cisterna magna injections of the tracer [14C]inulin in rat. *Neuroscience* 2000;**95**(2):577–92. http://dx.doi.org/10.1016/S0306-4522(99)00417-0.

76. Iliff JJ, Wang M, Zeppenfeld DM, et al. Cerebral arterial pulsation drives paravascular CSF-interstitial fluid exchange in the murine brain. *J Neurosci* 2013;**33**(46):18190–9. http://dx.doi.org/10.1523/JNEUROSCI.1592-13.2013.
77. Hadaczek P, Yamashita Y, Mirek H, et al. The "perivascular pump" driven by arterial pulsation is a powerful mechanism for the distribution of therapeutic molecules within the brain. *Mol Ther* 2006;**14**(1):69–78. http://dx.doi.org/10.1016/j.ymthe.2006.02.018.
78. Frederickson RG, Low FN. Blood vessels and tissue space associated with the brain of the rat. *Am J Anat* 1969;**125**(2):123–45. http://dx.doi.org/10.1002/aja.1001250202.
79. Wagner HJ, Pilgrim C, Brandl J. Penetration and removal of horseradish peroxidase injected into the cerebrospinal fluid: role of cerebral perivascular spaces, endothelium and microglia. *Acta Neuropathol* 1974;**27**(4):299–315. http://dx.doi.org/10.1007/BF00690695.
80. Krahn V. The pia mater at the site of the entry of blood vessels into the central nervous system. *Anat Embryol (Berl)* 1982;**164**(2):257–63.
81. Maynard EA, Schultz RL, Pease DC. Electron microscopy of the vascular bed of rat cerebral cortex. *Am J Anat* 1957;**100**(3):409–33. http://dx.doi.org/10.1002/aja.1001000306.
82. Cervós-Navarro J, Ferszt R. Connective tissue in pericapillary spaces of the human spinal cord. *Acta Neuropathol* 1973;**24**(2):178–83. http://dx.doi.org/10.1007/BF00684839.
83. Jones EG. On the mode of entry of blood vessels into the cerebral cortex. *J Anat* 1970;**106**(Pt 3):507–20.
84. Jessen NA, Munk ASF, Lundgaard I, Nedergaard M. The glymphatic system: a beginner's guide. *Neurochem Res* 2015:1–17. http://dx.doi.org/10.1007/s11064-015-1581-6.
85. Zervas NT, Liszczak TM, Mayberg MR, Black PM. Cerebrospinal fluid may nourish cerebral vessels through pathways in the adventitia that may be analogous to systemic vasa vasorum. *J Neurosurg* 1982;**56**(4):475–81. http://dx.doi.org/10.3171/jns.1982.56.4.0475.
86. Kress BT, Iliff JJ, Xia M, et al. Impairment of paravascular clearance pathways in the aging brain. *Ann Neurol* 2014;**76**(6):845–61. http://dx.doi.org/10.1002/ana.24271.
87. Iliff JJ, Chen MJ, Plog BA, et al. Impairment of glymphatic pathway function promotes tau pathology after traumatic brain injury. *J Neurosci* 2014;**34**(49):16180–93. http://dx.doi.org/10.1523/JNEUROSCI.3020-14.2014.
88. Weller RO, Subash M, Preston SD, Mazanti I, Carare RO. Perivascular drainage of amyloid-beta peptides from the brain and its failure in cerebral amyloid angiopathy and Alzheimer's disease. *Brain Pathol* 2008;**18**(2):253–66. http://dx.doi.org/10.1111/j.1750-3639.2008.00133.x.
89. Preston SD, Steart PV, Wilkinson A, Nicoll JA, Weller RO. Capillary and arterial cerebral amyloid angiopathy in Alzheimer's disease: defining the perivascular route for the elimination of amyloid beta from the human brain. *Neuropathol Appl Neurobiol* 2003;**29**(2):106–17. http://dx.doi.org/10.1046/j.1365-2990.2003.00424.x.
90. Casley-Smith JR, Földi-Börsök E, Földi M. The prelymphatic pathways of the brain as revealed by cervical lymphatic obstruction and the passage of particles. *Br J Exp Pathol* 1976;**57**(2):179–88.
91. Földi M. The brain and the lymphatic system (I). *Lymphology* 1996;**29**(1):1–9.
92. Weller RO, Djuanda E, Yow H-Y, Carare RO. Lymphatic drainage of the brain and the pathophysiology of neurological disease. *Acta Neuropathol* 2009;**117**(1):1–14. http://dx.doi.org/10.1007/s00401-008-0457-0.
93. Walter BA, Valera VA, Takahashi S, Ushiki T. The olfactory route for cerebrospinal fluid drainage into the peripheral lymphatic system. *Neuropathol Appl Neurobiol* 2006;**32**(4):388–96. http://dx.doi.org/10.1111/j.1365-2990.2006.00737.x.
94. Nag S, Manias JL, Stewart DJ. Pathology and new players in the pathogenesis of brain edema. *Acta Neuropathol* 2009;**118**(2):197–217. http://dx.doi.org/10.1007/s00401-009-0541-0.
95. Verbalis JG. Brain volume regulation in response to changes in osmolality. *Neuroscience* 2010;**168**(4):862–70. http://dx.doi.org/10.1016/j.neuroscience.2010.03.042.

96. Wilson MH, Newman S, Imray CH. The cerebral effects of ascent to high altitudes. *Lancet Neurol* 2009;**8**(2):175–91. http://dx.doi.org/10.1016/S1474-4422(09)70014-6.
97. Klatzo I. Pathophysiological aspects of brain edema. *Acta Neuropathol* 1987;**72**(3):236–9. http://dx.doi.org/10.1007/BF00691095.
98. Klatzo I. Presidental address. Neuropathological aspects of brain edema. *J Neuropathol Exp Neurol* 1967;**26**(1):1–14.
99. Hossmann KA. Development and resolution of ischemic brain swelling. In: Pappius HM, Feindel W, editors. *Dynamics of brain edema*. Berlin Heidelberg: Springer; 1976. p. 219–27.
100. Kalimo H, Fredriksson K, Nordborg C, Auer RN, Olsson Y, Johansson B. The spread of brain oedema in hypertensive brain injury. *Med Biol* 1986;**64**(2–3):133–7.
101. Ohata K, Marmarou A, Povlishock JT. An immunocytochemical study of protein clearance in brain infusion edema. *Acta Neuropathol* 1990;**81**(2):162–77. http://dx.doi.org/10.1007/BF00334505.
102. Ohata K, Marmarou A. Clearance of brain edema and macromolecules through the cortical extracellular space. *J Neurosurg* 1992;**77**(3):387–96. http://dx.doi.org/10.3171/jns.1992.77.3.0387.
103. Fenstermacher JD, Kaye T. Drug "diffusion" within the brain. *Ann N Y Acad Sci* 1988;**531**:29–39. http://dx.doi.org/10.1111/j.1749-6632.1988.tb31809.x.
104. Fenstermacher JD, Patlak CS, Blasberg RG. Transport of material between brain extracellular fluid, brain cells and blood. *Fed Proc* 1974;**33**(9):2070–4.
105. Geer CP, Grossman SA. Interstitial fluid flow along white matter tracts: a potentially important mechanism for the dissemination of primary brain tumors. *J Neurooncol* 1997;**32**(3):193–201. http://dx.doi.org/10.1023/A:1005761031077.
106. Reulen HJ, Graham R, Spatz M, Klatzo I. Role of pressure gradients and bulk flow in dynamics of vasogenic brain edema. *J Neurosurg* 1977;**46**(1):24–35. http://dx.doi.org/10.3171/jns.1977.46.1.0024.
107. Reulen HJ, Tsuyumu M, Tack A, Fenske AR, Prioleau GR. Clearance of edema fluid into cerebrospinal fluid: A mechanism for resolution of vasogenic brain edema. *J Neurosurg.* 1978;**48**(5):754–64. http://dx.doi.org/10.3171/jns.1978.48.5.0754.
108. Shima K, Marmarou A. Clearance of infusion edema fluid at high and low levels of intracranial pressure. *Adv Neurol* 1990;**52**:285–7.
109. Blaumanis OR, Rennels ML, Grady PA. Focal cerebral edema impedes convective fluid/tracer movement through paravascular pathways in cat brain. *Adv Neurol* 1990;**52**:385–9.
110. Csanda E, Földi M, Obál F, Zoltán OT. Cerebral oedema as a consequence of experimental cervical lymphatic blockage. *Angiologica* 1968;**5**(1):55–63. http://dx.doi.org/10.1159/000157734.
111. Ebisu T, Naruse S, Horikawa Y, et al. Discrimination between different types of white matter edema with diffusion-weighted MR imaging. *J Magn Reson Imaging* 1993;**3**(6):863–8.
112. McLone DG, Bondareff W, Raimondi AJ. Brain edema in the hydrocephalic hy-3 mouse: submicroscopic morphology. *J Neuropathol Exp Neurol* 1971;**30**(4):627–37.
113. Del Bigio MR, da Silva MC, Drake JM, Tuor UI. Acute and chronic cerebral white matter damage in neonatal hydrocephalus. *Can J Neurol Sci* 1994;**21**(4):299–305.
114. Badaut J, Fukuda AM, Jullienne A, Petry KG. Aquaporin and brain diseases. *Biochim Biophys Acta—Gen Subj* 2014;**1840**(5):1554–65. http://dx.doi.org/10.1016/j.bbagen.2013.10.032.
115. Frigeri A, Gropper MA, Umenishi F, Kawashima M, Brown D, Verkman AS. Localization of MIWC and GLIP water channel homologs in neuromuscular, epithelial and glandular tissues. *J Cell Sci* 1995;**108**(Pt 9)):2993–3002. PMID:8537439.

116. Nielsen S, Nagelhus EA, Amiry-Moghaddam M, Bourque C, Agre P, Ottersen OP. Specialized membrane domains for water transport in glial cells: high-resolution immunogold cytochemistry of aquaporin-4 in rat brain. *J Neurosci* 1997;**17**(1):171–80.
117. Nagelhus EA, Mathiisen TM, Ottersen OP. Aquaporin-4 in the central nervous system: cellular and subcellular distribution and coexpression with KIR4.1. *Neuroscience* 2004;**129**(4):905–13. http://dx.doi.org/10.1016/j.neuroscience.2004.08.053.
118. Badaut J. Aquaglyceroporin 9 in brain pathologies. *Neuroscience* 2010;**168**(4):1047–57. http://dx.doi.org/10.1016/j.neuroscience.2009.10.030.
119. Wen H. Ontogeny of water transport in rat brain: postnatal expression of the aquaporin-4 water channel. *Eur J Neurosci* 1999;**11**(3):935–45. http://dx.doi.org/10.1046/j.1460-9568.1999.00502.x.
120. Yao X, Hrabetova S, Nicholson C, Manley GT. Aquaporin-4-deficient mice have increased extracellular space without tortuosity change. *J Neurosci* 2008;**28**(21):5460–4. http://dx.doi.org/10.1523/JNEUROSCI.0257-08.2008.

CHAPTER

7

Blood-Brain Barrier Mechanisms of Edema Formation: The Role of Ion Transporters and Channels

Martha E. O'Donnell, Heike Wulff and Yi-Je Chen

University of California, Davis, CA, United States

OUTLINE

Brain Edema.
DOI: http://dx.doi.org/10.1016/B978-0-12-803196-4.00007-2

Abbreviations

BBB blood–brain barrier
NVU neurovascular unit
NKCC Na–K–Cl cotransport
NHE Na/H exchange
NBC Na-HCO_3 cotransport
KCa3.1 Calcium-activated K channel KCa 3.1
TRPC transient receptor potential cation
CRAC calcium release activated channel
NCX Na/Ca exchange
[Caɪ] intracellular Ca concentration
AVP arginine vasopressin
NMR nuclear magnetic resonance
MRI magnetic resonance imaging
DWI diffusion weighted imaging
MRS magnetic resonance spectroscopy
Na CSI Na chemical shift imaging
CBF cerebral blood flow
TTC 2,3,5-triphenyltetrazolium chloride
PKC protein kinase C
CaMK Ca-calmodulin-dependent protein kinase
ERK1/2 extracellular signal-related kinase
MLCK myosin light chain kinase

This chapter addresses the roles of blood-brain barrier (BBB) ion transporters and channels in brain edema formation with an emphasis on edema that forms during ischemic stroke. Many stroke studies have focused on understanding how ischemia affects BBB tight junctions as well as cellular metabolic and signaling pathways. However, despite the fact that the BBB plays a critical role in regulating the volume and composition of brain interstitial fluid in healthy brain, little has been known about the role of BBB ion transporters and channels in ischemia-induced edema and brain infarct. Many BBB ion transporters and channels are asymmetrically distributed between luminal and abluminal BBB plasma membranes, giving these cells the ability to perform vectorial transport of ions between blood and brain (for details, see Fig. 4, Chapter 3: Noninvasive Imaging Techniques for Brain Edema: From Basic Science to the Clinic) in a manner similar to vectorial transport of ions across renal and epithelial cells.[1–4] Evidence has now emerged that luminally located BBB Na^+ transporters are stimulated by ischemic conditions to drive increased secretion of ions and water from blood into brain, contributing to formation of cytotoxic edema as astrocytes take up the ions and water presented to them. Recent studies also show that BBB Ca^{2+}-dependent K^+ channels also participate in this process. Details of these findings will be discussed in the following sections of this chapter.

We will first consider cytotoxic and vasogenic forms of ischemia-induced brain edema and then focus on what we know about how BBB

Na^+ transporters and K^+ channels participate in cytotoxic/ionic edema formation. Finally, the manner in which these BBB ion transporters and channels may contribute to changes in BBB tight junction structure and function leading to vasogenic edema will be presented.

CYTOTOXIC VERSUS VASOGENIC EDEMA

An all too commonly held assumption is that during cerebral ischemia, a decrease in available ATP causes immediate BBB disruption with vasogenic edema formation occurring as plasma constituents enter the brain paracellularly through compromised tight junctions. However, stroke-induced brain edema forms by a sequence of events that is considerably more complex than simple breakdown of the barrier. It has been established, in fact, that the BBB remains intact during the early hours of ischemic stroke during which time transcellular, rather than paracellular, mechanisms account for production of edema.[1,3,5,6,7–10] These BBB transcellular mechanisms involving blood to brain transport of ions play an essential role in promoting cytotoxic edema formation through rapid swelling of astrocytes and other brain parenchymal cells as they take up ions and water presented to them by BBB endothelial cells.[1,3,5,11] This molecular process has also been called "ionic" edema (see Chapter 4: Experimental Techniques to Investigate the Formation of Brain Edema In Vivo and Chapter 12: Edema and BBB Breakdown in Stroke). As ischemia continues, over time BBB endothelial cells themselves begin to undergo changes leading to increased paracellular permeability and vasogenic edema formation[1] (Fig. 7.1).

In focal ischemic stroke a core region with the greatest decrease in cerebral blood flow and most severe ischemia is surrounded by a penumbral region that receives some collateral blood flow and experiences only moderate ischemia.[12] The more severe the ischemia, the more quickly the BBB is compromised to the point that vasogenic edema begins to form and the sooner cellular processes leading to necrosis, and/or apoptosis are initiated, resulting in permanent tissue damage. It is the penumbral region, therefore, that is generally the target for therapies aimed at reducing the extent of brain damage following stroke. The sooner the blood flow is restored the greater the eventual recovery of penumbral tissue. If blood flow is not restored the penumbra accrues increasing damage and gradually converts to core, resulting in greater final infarct volume.[5]

Many potential therapeutic targets for reducing brain injury in stroke have been identified throughout cells of the neurovascular unit (NVU), comprised of endothelial cells, astrocytes, neurons, pericytes, microglia, and vascular smooth muscle cells.[4,13,14] However, a great many putative therapies have failed because of inability to reach their targets. This is

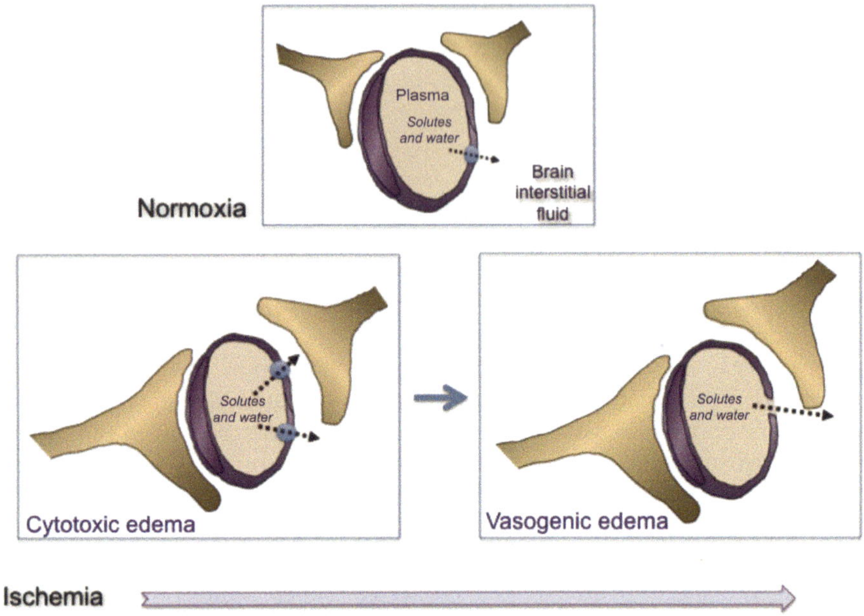

FIGURE 7.1 **Solute movement across the BBB during normoxia and ischemia.** In healthy normoxic brain endothelial tight junctions severely restrict paracellular movement of solutes such that ions and other solutes must be transported across the barrier by specific plasma membrane transporters. Ions cross the BBB by specific ion transporters and channels (top panel). In the early hours of stroke, ischemic conditions stimulate BBB transporters causing a "hypersecretion" of ions and water from blood into brain. Similar ion transporters in astrocyte perivascular endfeet are also stimulated during ischemia, causing the astrocytes to quickly take up ions and water presented to them by the BBB, resulting in cytotoxic edema as the astrocytes swell (bottom left panel). Mechanisms of water movement across the barrier and into the astrocytes are covered in other chapters. As ischemia progresses, metabolic changes in the BBB endothelial cells cause the cells to begin accumulating Na which can lead to elevated intracellular Ca via reversed Na^+/Ca^{2+} exchange. Elevated intracellular Ca^{2+}, in turn, leads to alteration of tight junctions with increased paracellular permeability, resulting in vasogenic edema (bottom right panel).

largely due to BBB drug transporters that quite effectively prevent entry of the therapeutics into the brain, even small molecule lipophilic.[4,13,15–22] A distinct advantage of putative stroke therapies designed to target luminal BBB membrane processes is that they can simply be administered intravenously and do not need to cross the BBB to have effect.

The following sections will first present what we know about BBB cellular mechanisms that contribute to cytotoxic edema formation, in particular the ion transporters and channels of BBB endothelial cells. The extent to which these transporters and channels constitute therapeutic targets for reducing edema and infarct will also be discussed. Finally, we will present evidence supporting the hypothesis that BBB ion transporters and channels also contribute to alteration of BBB permeability and vasogenic edema formation occurring as ischemic conditions persist over time.

BLOOD-BRAIN BARRIER ION TRANSPORTERS AND CHANNELS IN CYTOTOXIC EDEMA

BBB endothelial cells possess an impressive array of plasma membrane transport proteins. This is not surprising given that the barrier is highly restrictive to paracellular solute movement such that solutes enter and leave the brain via transcellular pathways mediated by the transport proteins.[23] These include transporters for nutrients, metabolites, and waste products as well as a variety of xenobiotics, both toxic environmental substances and therapeutic agents.[23,4,13,15,17–21] A great deal of attention has been given to the drug/xenobiotic transporters in large part because of the problem that many therapeutic are prevented from entering the brain by the BBB drug transporters (for details see Chapter 3: Noninvasive Imaging Techniques for Brain Edema: From Basic Science to the Clinic). Thus, often BBB transporter studies focus on drug transporters. However, it is important to not overlook the fact that BBB endothelial cells also have a rich complement of ion transporters and channels. One of the important functions of the BBB is regulation of the composition and volume of brain interstitial fluid (for details see Fig. 5, Chapter 3: Noninvasive Imaging Techniques for Brain Edema: From Basic Science to the Clinic) by transporting ions, including Na^+, K^+, Cl^-, and HCO_3^- between blood and brain as needed.[23–27] Despite this, little has been known about the specific ion transporters and channels involved in this process and how they are affected by disease states, including ischemic stroke. Several BBB ion transporters have been identified in the cells and shown to have an asymmetrical distribution between luminal and abluminal membranes[11,27–31] (Fig. 7.2). This arrangement is observed in all epithelia that perform absorption and/or secretion of ions between the lumen and parenchyma of the organ, such as renal and gastrointestinal (GI) epithelia, and is consistent with vectorial ion transport mediated by BBB endothelial cells as well.[3]

The BBB has been reported to produce up to 30% of brain interstitial fluid in healthy normoxic brain by blood to brain secretion of ions,[32,33] in particular Na^+ and Cl^-, the major ionic constituents of brain extracellular fluid. More recent studies have suggested that the absolute percentage is difficult to assess and may be even greater given the very large total surface area of brain capillaries.[27] During ischemic stroke, BBB ion transporters are in some manner stimulated to greatly increase secretion of Na^+ and Cl^- into the brain with osmotically obliged water following.[1,7,9,34] At the same time, ischemia causes the perivascular astrocytes to swell as they

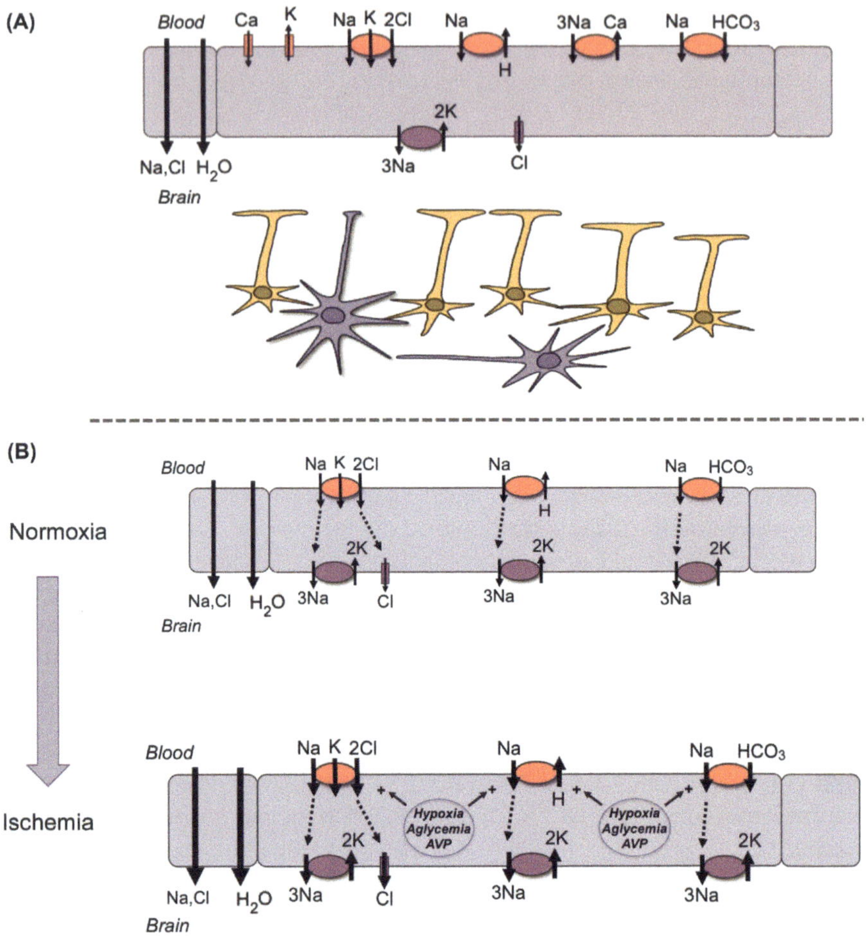

FIGURE 7.2 **Many BBB endothelial cell ion transporters and channels are asymmetrically distributed between luminal and abluminal plasma membranes.** The BBB has a rich complement of ion transporters and channels, only some of which are depicted in this figure (panel (A)). Among these are the luminal Na^+ transporters Na^+-K^+-Cl^- cotransport (NKCC) and Na^+/H^+ exchange (NHE) that, together with abluminal Na/K pump and Cl efflux pathway, mediate ischemic factor-induced hypersecretion of Na^+, Cl^-, and water from blood into brain, contributing to cytotoxic edema formation (panel (B)). Na^+/Ca^{2+} exchange (NCX) is also present and may play a role in mediating increased intracellular Ca^{2+} leading to vasogenic edema when metabolic changes begin to occur in the BBB leading to Na^+ accumulation. Also shown are Ca^{2+} and K^+ channels that appear to sustain luminal Na^+ transport activity during ischemia. Recent studies suggest that a luminal Na^+-HCO_3^- cotransporter (NBC) also participates in edema formation. *Source: Figure is modified with permission from O'Donnell ME. Ion and water transport across the blood–brain barrier. In: Alvarez-Leefmans FJ, Delpire E, editors. Physiology and pathophysiology of chloride transporters and channels in the nervous system. From molecules to diseases. London: Elsevier; 2009. p. 585–606.*

take up the ions and water presented to them by the BBB.[1,10,35–41] In this regard the BBB facilitates astrocyte swelling and formation of astrocytic and then neuronal cytotoxic edema during the early hours of ischemic stroke. This is consistent with the observation that a luminal BBB Na^+ transporter appears to be the rate-limiting factor in ischemia-induced edema formation.[7–9,42,43] Others and we have found evidence that the BBB NKCC and NHE are two ion transporters involved in this process[1] and that the calcium-activated K channel KCa3.1 may also contribute.[44]

Na^+–K^+–Cl^- Cotransport

Na^+–K^+–Cl^- cotransport (NKCC) is an electroneutral bidirectional obligate symporter that carries $1Na^+$, $1K^+$, and $2Cl^-$ ions and is specifically inhibited by bumetanide. NKCC is well known for its role in Na^+ absorption/secretion across kidney, gut, and other epithelia.[45–48] It also plays an important role in regulating intracellular volume, restoring volume following hypertonicity-induced cell shrinkage, and even driving increases in cell volume under some conditions,[48–50] a feature that will be discussed in the third section of this chapter. There are two isoforms of NKCC, the nearly ubiquitously expressed NKCC1,[5,51] and NKCC2, found in renal epithelia.[45,52,53] Early studies by others and we showed that NKCC1 is present in BBB endothelial cells of a variety of species including rat, bovine, and human[54–59] and immunoelectron microscopy studies further showed that it is located predominately in the luminal membrane of the cells.[11] A possible role for the transporter in ischemia-induced brain edema was first suggested by the finding that NKCC activity is stimulated by hypoxia,[56,57] aglycemia,[56] and other factors released during ischemia, including arginine vasopressin (AVP)[55] and endothelin.[59–61] Our in vitro studies using cultured bovine cerebral microvascular endothelial cells (CMEC) revealed that NKCC activity is rapidly increased by more than twofold when the cells are exposed to hypoxia, whether moderate or severe.[56] Exposure of CMEC to aglycemia increases NKCC activity more than threefold while hypoxia–aglycemia in combination increases activity more than fivefold.[56] The peptide AVP, released during ischemia by extra-hypothalamic nerve processes that terminate near brain microvessels,[62–65] is also a potent stimulator of BBB NKCC activity.[55] Our in vitro CMEC studies also demonstrated that hypoxia and aglycemia increase activity of the Na^+/K^+ pump (Na^+/K^+ ATPase), known to reside in the abluminal BBB membrane.[31] However, hypoxia stimulation of the Na^+/K^+ pump is blocked by the NKCC inhibitor bumetanide. This is consistent with NKCC and Na^+/K^+ pump working together to increase secretion of Na^+ from blood into brain during ischemia. Previous in vivo studies showed that in rats subjected to cerebral ischemia, total brain Na^+, Cl^-, and water increase while brain K^+ decreases.[7] The abluminal Cl^- efflux mechanism that must couple with luminal NKCC influx of Cl^- has yet to be identified but is likely to be by a Cl^- channel if not also by Cl^-/HCO_3^- exchange.

If luminal BBB NKCC plays a significant role in increasing secretion of Na^+, Cl^- and water during stroke one predicts that inhibiting the cotransporter by intravenous bumetanide would reduce brain edema in vivo. This has now been explored using the rat intraluminal middle cerebral artery occlusion (MCAO) model of permanent focal ischemia[66] with nuclear magnetic resonance (NMR) evaluation of cytotoxic edema and brain Na^+ uptake. In this model, events occurring in the early hours of ischemia when the barrier is still intact can be observed. Rats subjected to permanent MCAO (pMCAO) and administered bumetanide intravenously showed significant attenuation of cytotoxic edema in both core and penumbral regions as assessed by diffusion weighted imaging (DWI)-derived apparent diffusion coefficients (ADC), and as confirmed by gravimetry compared to rats subjected to pMCAO and administered vehicle.[11] The reduction in edema was accompanied by decreased infarct volume, determined by 2,3,5-triphenyltetrazolium chloride (TTC) staining. Evaluating brain Na^+ uptake and NMR Na^+ chemical shift imaging (CSI) revealed a linear increase in brain Na^+ during 3 hours of imaging,[28] consistent with reports evaluating changes in brain Na^+ by spectrophotometry.[7] As with pMCAO-induced edema, intravenous bumetanide markedly reduced brain Na^+ uptake in a manner sustained throughout the 3 hour imaging period. Rats subjected to pMCAO and treated with bumetanide also exhibited significantly improved neurological outcome at 1 and 2 days after induction of ischemia when evaluated by 14-point neurological assessment and Rotarod tests.[28] Whether the neurologic benefit is observed at times later than 2 days remains to be determined. Bumetanide has also been found to reduce edema and infarct in a model of ischemia with reperfusion (transient MCAO) and found to reduce edema and infarct.[67,68] An important point to consider in administering bumetanide as a therapeutic for ischemic stroke is that both its charge at physiological pH and the presence of BBB drug transporters prevent bumetanide from crossing the BBB.[69,70] Thus, studies employing bumetanide as a therapeutic target for NKCC in brain parenchymal cells (beyond the BBB) need to take that into account.

Na^+/H^+ Exchange

Na^+/H^+ exchange (NHE) is another well-studied and widely distributed transport protein. NHE exchanges $1Na^+$ for $1H^+$ and is inhibited by HOE-642 (also known as cariporide) and related NHE inhibitors.[5,28,29,71,72] There are at least nine isoforms of NHE. NHE1 is nearly ubiquitously expressed in cells and is perhaps best known for its ability to regulate intracellular pH (pH_i).[73] Like NKCC, it also participates in regulation of intracellular volume, as will be discussed in Section "Blood-Brain Barrier Ion Transporters and Channels in Cytotoxic Edema" of this chapter. NHEs 1–5 are found in the plasma membrane while NHEs 6–9 appear to reside in organelle membranes.[73,74] In renal, GI and other epithelia, NHEs 1–3 have been found to function in transepithelial transport of Na^+,[75–77] suggesting that one or all of these could be involved in BBB Na^+ transport between blood and brain. Less likely are contributions from NHE4 and 5 since NHE4 requires hyperosmotic conditions for activation[73] and NHE5 appears to be a neuron-specific isoform.[78] Several studies have established the presence of NHE1 in BBB endothelial cells.[29,79] We have shown that NHE2 is also present in BBB of rat, bovine, and human but that NHE3 is not expressed in the cells.[29] Immunoelectron microscopy experiments revealed that, like NKCC1, both NHE1 and NHE2 are distributed predominantly in the luminal BBB plasma membrane.[28,29] Also as with NKCC, there is good evidence that ischemic factors stimulate NHE activity. HOE-642, the potent NHE inhibitor, blocks both NHE1 and NHE2 activities[80,81] and thus, it has not yet been determined whether it is NHE1, NHE2, or both, that are stimulated during ischemia in BBB endothelial cells. In vitro studies using cultured CMEC have revealed that similar to NKCC, NHE is quite sensitive to stimulation by hypoxia, aglycemia, and AVP with activity increasing from 1.5 to 3-fold in a manner sustained over hours.[29,82] BBB NHE activity has also been shown to be stimulated by endothelin.[60,83] These findings led to the hypothesis that, like NKCC, luminal BBB NHE activity couples to abluminal Na^+/K^+ pump such that during cerebral ischemia the two transporters work in tandem to "hypersecrete" Na^+ from blood into the brain.[28]

In vivo studies have been done to evaluate the possibility that BBB NHE contributes to ischemia-induced cytotoxic edema and, in that regard, might present another therapeutic target for reduction of brain damage in stroke. Using the rat intraluminal filament permanent MCAO model of ischemic stroke again with DWI and Na^+ CSI to assess edema and brain Na^+ uptake, respectively, we found that the specific NHE inhibitor cariporide (HOE-642) significantly attenuates edema and brain Na^+ uptake in both core and penumbra as well as reduces hemispheric brain swelling.[28] HOE-642 was also found to reduce TTC-defined brain infarct volume and to improve sensorimotor outcome evaluated by 14-point neurological assessment and Rotarod two days after induction of ischemia.[28] Other studies have provided evidence that HOE-642 reduces edema in the transient MCAO model of ischemia with reperfusion.[84,85] As with bumetanide, the chemical structure of HOE-642 and our HPLC studies suggest that it does not penetrate the BBB.[28] Thus, its primary site of action when administered intravenously is predicted to be the luminal BBB NHE.

While it is not clear from NHE inhibitor studies to what extent NHE1 versus NHE2 contributes to ischemia-induced edema some insight has been obtained through the use of genetically engineered mice. Studies using NHE1 knockout mice and the reversible MCAO model of ischemia/perfusion showed significantly less edema and infarct compared to wild type mice with intact NHE1.[85] This does not answer the question of whether NHE2 may also contribute to ischemic-induced edema and more specifically, the role of NHE1 versus NHE2 in BBB hypersecretion of Na^+, Cl^-, and water from blood to brain. For that, future studies will need to use cell specific NHE knockouts.

KCa3.1 Channel

Much evidence has been provided that both NKCC and NHE activities are stimulated by Ca^{2+}-dependent signaling mechanisms.[5,55] This is true, for example of AVP-stimulated NKCC activity, which involves V1 receptor activation of phospholipase C and subsequent elevation of cytosolic $[Ca^{2+}]$.[55] NHE activity is also stimulated by phospholipase C and other pathways that elevate $[Ca^{2+}{}_i]$.[5,29] With respect to NKCC, the cotransporter is sensitive to inhibition by elevation of intracellular $[Cl^-]$. This is not due to the fact that Cl^- is an ion carried by NKCC but rather, the result of a Cl^--sensitive kinase that provides feedback inhibition of NKCC activity as intracellular $[Cl^-]$ rises.[86–90] This is as predicted for vectorial transport of Na^+ and Cl^- by luminal NKCC working in tandem with abluminal Na^+/K^+ pump and abluminal Cl^- channel or other Cl^- efflux pathway such as Cl^-/HCO_3^- exchange. Thus, factors increasing intracellular $[Ca^{2+}]$ and/or preventing a rise in intracellular $[Cl^-]$ will tend to increase and sustain NKCC activity. One mechanism predicted to accomplish this in BBB endothelial cells is the recently identified presence of the Ca^{2+}-activated K^+ channel KCa3.1. This channel was found to be present in both freshly isolated rat brain microvessels and cultured bovine CMEC by Western blot as well as in rat and human brain microvascular endothelial cells by in situ immunohistochemistry.[44] In that study, KCa3.1 channel activity was demonstrated in cultured CMEC by radioisotopic ion flux methods as a Ca^{2+}-dependent K^+ flux that is blocked by TRAM-34, the specific KCa3.1 inhibitor.[91] Further, a Ca^{2+}-activated, TRAM-34-sensitive K^+ current was revealed in whole-cell patch-clamp analyses of cultured CMEC. Given these findings, it was hypothesized that KCa3.1 contributes to ischemia-stimulation of BBB Na^+, Cl^-, and water secretion in the following manner. Ischemic factor-induced elevation of BBB endothelial intracellular $[Ca^{2+}]$ occurring through a variety of processes stimulates NKCC activity. The increased intracellular $[Ca^{2+}]$ also causes activation of KCa3.1, allowing efflux of K^+ with consequent hyperpolarization of the cell. The more negative membrane potential provides a greater electrochemical driving force for Cl^- efflux from the cell via abluminal Cl^- channels. This reduces intracellular $[Cl^-]$ to prevent feedback inhibition of NKCC. The

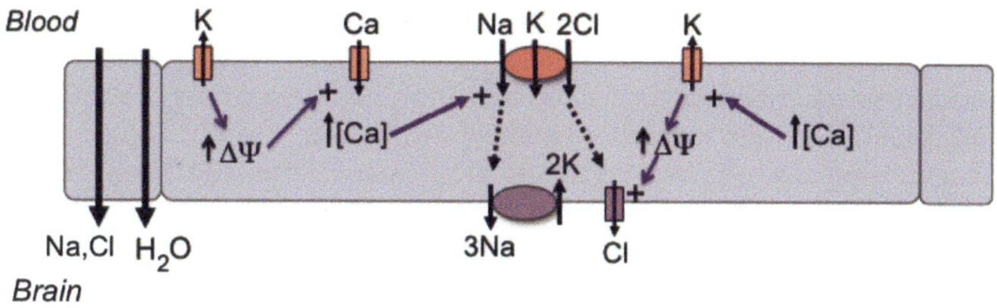

FIGURE 7.3 **Hypothesized role of the BBB KCa3.1 channel in ischemic stroke.** The Na–K–Cl cotransporter (NKCC) is well known to be stimulated by factors that increase intracellular [Ca^{2+}] ([Ca^{2+}]$_i$). It is also very sensitive to intracellular [Cl^-] ([Cl^-]$_i$), with increased levels inhibiting and decreased levels stimulating cotransporter activity. Ischemia causes an initial increase in [Ca^{2+}]$_i$ that increases activity of both NKCC and KCa3.1. Increased NKCC will tend to increase [Cl^-]$_i$ which will cause a feedback inhibition of NKCC (through Cl^--sensitive kinases). However, increased KCa3.1 activity will promote a more negative membrane potential as K^+ exits the cells. This then promotes greater Cl^- efflux from Cl^- channels, avoiding elevated Cl^- levels that would inhibit NKCC activity. The more negative membrane potential will also promote more Ca^{2+} influx via Ca^{2+} channels, which will in turn promote both NKCC and KCa3.1 activity. By this mechanism, these BBB ion transporters and channels work together to cause a robust secretion of Na^+, Cl^-, and water into the brain during ischemic stroke. *Source: Figure is modified with permission from Chen Y-J, Raman G, Bodendiek S, O'Donnell ME, Wulff H. TheKCa3.1 blocker TRAM-34 reduces infarction and neurological deficit in a rat model of ischemia/reperfusion stroke. J Cereb Blood Flow Metab 2011;31:2363–74.*

more negative membrane potential caused by KCa3.1 mediated K^+ efflux also promotes Ca^{2+} influx via inward rectifier plasma membrane Ca^{2+} channels such as the transient receptor potential cation channel (TRPC) or calcium release activated channel (CRAC) which in turn supports the continued elevation of NKCC and KCa3.1 activities. By this hypothesis, KCa3.1 works with luminal NKCC and abluminal Na^+/K^+ pump and Cl^- channel to support and maintain ischemia-induced BBB hypersecretion of Na^+, Cl^-, and water from blood into brain[44,92] (Fig. 7.3). This role of KCa3.1 in supporting Na^+, Cl^-, and water secretion is reminiscent of the channel's well-characterized role in secretory epithelia of the lung and the GI tract where KCa3.1 blockers have long been known to suppress fluid movement.[93–95]

The role of KCa3.1 in ischemia-induced edema was recently evaluated in the rat pMCAO model of stroke using DWI and Na^+ CSI methods. In rats given TRAM-34 at the start of pMCAO brain edema and brain Na^+ uptake were both found to be significantly attenuated in a manner sustained over the 4 hour imaging period following induction of ischemia.[44] These findings are consistent with the hypothesis the BBB KCa3.1 contributes to edema during the early hours of ischemic stroke. HPLC evaluation of blood and brain TRAM-34 has shown that while bumetanide and HOE-642 do not significantly cross the BBB, KCa3.1 is highly brain penetrant.[92,96] Thus, it may be possible that TRAM-34 is improving stroke outcome by inhibiting KCa3.1 in brain cells other than just BBB endothelial cells. This appears to be unlikely, however, since KCa3.1 has not been found in astrocytes or neurons and although it is present in microglia and has been shown to reduce poststroke inflammation by preventing microglial activation that occurs on a time scale of many hours to days rather the minutes to several hours observed for cytotoxic edema.[92,96]

BLOOD-BRAIN BARRIER ION TRANSPORTERS AND CHANNELS IN VASOGENIC EDEMA

As stroke progresses an increasing number of BBB cellular processes emerge in addition to the transendothelial transport of ions and water from blood into brain that began in the earliest stages of ischemia. In particular, changes in tight junctions emerge that lead to increased paracellular permeability.[97] The time of onset and degree of permeability change will of course depend on the severity of the ischemia. It is important to note the transendothelial and paracellular flux of solutes and water from blood into brain are not mutually exclusive. That is, transendothelial flux can continue across the BBB even as paracellular flux begins and eventually even dominates. With regard to the effects of prolonged NKCC and NHE stimulation we have found that exposing cultured BBB endothelial cells to hypoxia for periods of –5 hours causes the cells to swell.[56,98] Thus, while the asymmetric distribution of ion transporters and channels allows for BBB vectorial transport without a net accumulation of ions in the cells at the start of stroke, after a time [ATP] falls to a level that no longer enables the abluminal Na^+/K^+ pump to keep pace with luminal Na^+ transporters and the BBB endothelial cells themselves begin to swell as the ion content of the cells increases. This will be elaborated upon in the following paragraphs.

In addition to alteration of tight junctions and cell swelling, another well established event is the alteration of intracellular [Ca^{2+}]. The role of intracellular [Ca^{2+}] in altered BBB permeability appears to be complex and incompletely understood at best.[97,99] However, there is good evidence that a rise in intracellular [Ca^{2+}] occurring during ischemia through a variety of pathways leads to disruption of the barrier.[97,99,100] Much effort has been focused on identifying the mechanisms by which intracellular Ca^{2+} may affect BBB tight junctions. However, in the interest of developing therapies to improve stroke outcome, given the pronounced rapid and sustained response of BBB ion transporters and channels to ischemia it is important to consider whether and how these may also contribute to development of vasogenic edema as stroke progresses. The possible links between ischemia-stimulated BBB NKCC, NHE and KCa3.1 activities, elevation of intracellular [Ca^{2+}], and barrier disruption will be explored below.

Effects of Elevated Intracellular [Ca^{2+}] on BBB Tight Junctions and Permeability

Calcium plays a central role in regulating paracellular permeability of the BBB. It is well known to be required for maintenance of tight functions. One example of this is that reducing extracellular [Ca^{2+}] promotes disruption of tight junctions in brain microvessels.[101] At the same time, however, elevation of intracellular [Ca^{2+}] is associated with compromised BBB permeability.[97,99,101,102]

Of pertinence to barrier changes occurring in ischemic stroke, a number of studies have established that ischemic conditions cause an increase in intracellular [Ca^{2+}] of BBB endothelial cells.[55,97,100,103] In particular, in vitro studies using a variety of cultured brain microvascular endothelial cell types have demonstrated that exposing the cells to varying levels of hypoxia, with or without aglycemia, triggers an increase in intracellular [Ca^{2+}].[100,103–106] Other ischemic factors, including AVP and endothelin, are also known to elevate intracellular [Ca^{2+}] of brain microvascular endothelial cells.[55,107,108] The source of the increased [Ca_i^{2+}] is not fully understood but there is evidence that both release of Ca^{2+} from intracellular stores as well as influx via plasma membrane Ca^{2+} channels can contribute.[99,100,104,109,110] One important mechanism of Ca^{2+} influx during ischemic conditions is the TRPC channels.[100,109] Regardless of the source, the resulting elevated [Ca_i^{2+}] is an important factor leading to increased BBB permeability. The mechanisms have yet to be fully elucidated but appear to involve direct interaction with junctional proteins as well as activation of several kinase signaling cascades that cause changes in tight junction structure with consequent increased paracellular permeability.[97,99] Signaling pathways that have been shown to be triggered by elevated [Ca_i^{2+}] and that are associated in increased BBB permeability include protein kinase C (PKC), Ca^{2+}-calmodulin-dependent protein kinase (CaMK), extracellular signal-related kinase, ERK1/2 and myosin light chain kinase (MLCK).[99,109]

Role of BBB Ion Transporters and Channels in Alteration of Intracellular [Ca^{2+}] and Barrier Disruption

While ischemia-induced stimulation of BBB NKCC, NHE, and KCa3.1 promotes hypersecretion of ions and water into the brain, contributing to cytotoxic edema formation during the early hours of stroke, there is evidence that this also contributes to alteration of BBB junctional permeability and development of vasogenic edema as stroke progresses. This is because activity of NKCC, NHE, and KCa3.1 can also promote increases in intracellular [Ca^{2+}]. In the simplest sense, as illustrated in Fig. 7.3, continued activation of KCa3.1 will promote hyperpolarization of the cells or at least prevent depolarization of the cells, supporting continued influx of Ca^{2+} through plasma membrane Ca^{2+} channels thereby elevating ($[Ca^{2+}]_i$).[44,92] However, ischemia stimulation of BBB NKCC and NHE may also cause elevation of $[Ca^{2+}]_i$ in the following manner. BBB endothelial cells have been found to be more resistant to hypoxia than other NVU cells, maintaining ATP levels for a longer time compared to their neighbors. In this regard, subjecting cultured CMEC to 4 hours of severe hypoxia with or without aglycemia does not produce a significant drop in ATP content of the cells.[56] This is consistent with the ability of the BBB Na^+/K^+ pump to continue working during the early hours of ischemia, functionally coupling to luminal NKCC and NHE to perform vectorial transport of Na^+ from blood into brain. However, as ischemia progresses, luminal influx of Na^+ via NKCC and NHE activity begins to outpace Na^+/K^+ pump-mediated Na^+ efflux with the result that the Na^+ content of the cells increases. This is accompanied by swelling of the endothelial cells themselves that emerges by 4 hours, whether the hypoxia is moderate or severe.[56,98] At the same time, intracellular Na^+ content increases, nearly doubling in the presence of moderate hypoxia.[98] The hypoxia-induced swelling is blocked by bumetanide and HOE-642, consistent with BBB NKCC and NHE contributing not only to vectorial transport in early stroke but also to swelling of the endothelial cells themselves over time.

In addition to swelling that occurs because of increased Na^+ content (with accompanying Cl^- and osmotically obliged water), the NKCC and NHE-mediated increase in Na^+ content can drive an increase in $[Ca^{2+}]_i$ through reverse operation of Na^+/Ca^{2+} exchange. The Na^+/Ca^{2+} exchanger (NCX) , documented to be expressed in BBB endothelial cells,[111] exchanges $2Na^+$ for $1Ca^{2+}$.[112,113] Under normoxic conditions a large inwardly directed Na^+ gradient drives Ca^{2+} efflux from the cell. However, when intracellular $[Na^+]$ increases, the inwardly directed gradient is decreased or even reverses such that the exchanger begins to mediate Ca^{2+} influx as Na^+ exits the cell.[113] This is a well-observed phenomenon in many types of cells.[113–116] Consistent with a role of BBB Na^+ transporters in contributing to ischemia-driven elevation of $[Ca^{2+}]_i$ and barrier disruption, the highly potent NHE inhibitor sabiporide has been found to reduce hypoxia-evoked increases in $[Ca^{2+}]_i$ and barrier permeability in the bEnd.3 endothelial cell in vitro BBB model.[105] In MCAO studies sabiporide also reduced ischemia-evoked increases in permeability and disruption of tight junctions.[105] Less is known about bumetanide effects on ischemia-induced $[Ca^{2+}]_i$ increases in BBB endothelial cells . However, hypoxia activation of nitric oxide synthase (NOS), a Ca^{2+}-dependent event, has been reported to be both NKCC- and NCX-dependent as well.[117] In addition NKCC has been shown to work in concert with NCX to increase $[Ca_i]$ in cultured astrocytes.[115,118]

CONCLUSIONS

Edema formation occurring in ischemic stroke is complex, involving many cellular processes, and evolving over time from primarily cell swelling-based cytotoxic edema to increasing dominance of BBB tight junction disruption-based vasogenic edema. Of the many cellular components that contribute to edema a prominent role of BBB ion transporters and channels has now become apparent. Luminally predominant NKCC and NHE functionally couple with abluminal Na^+/K^+ pump and Cl^- efflux transporters and/or channels to mediate hypersecretion of Na^+, Cl^-, and water into the brain during early stroke, contributing to cytotoxic edema formation as perivascular astrocytes take up the ions and water presented to them. The KCa3.1 channel also contributes to cytotoxic edema formation through mechanisms that may involve maintenance of elevated Na^+ transporter activity during ischemia. Additional channels have been shown to be upregulated, such as the sulfonylurea receptor 1-transient receptor potential M4 cation (Sur1-Trpm4) channel, in endothelium after stroke and to contribute to Na^+ movements (for details, see Chapter 11: Role of Matrix Metalloproteinases in Brain Edema). As ischemia progresses, metabolic changes in the BBB promote NKCC- and NHE-induced Na^+ accumulation in the cells, leading to increased $[Ca^{2+}]_i$ through reversed NCX. This BBB ion transporter-induced elevation of $[Ca^{2+}]_i$ may significantly contribute to the development of vasogenic edema through direct Ca^{2+} interactions with tight junction proteins and through activation of kinase signaling pathways. In this regard, targeting BBB ion transporters and channels appears to be a promising approach to attenuate both cytotoxic and vasogenic edema occurring in ischemic stroke.

Acknowledgments

The authors' studies presented in this chapter were supported by National Institutes of Health (NIH) R01 NS039953 (to M.E.O.) and R01 GM076063 (to H.W.) and American Heart Association Fellowship 09POST2260973 (to Y.J.C.). The studies were conducted in part in a facility constructed with support from National Center for Research Resources Research Facilities Improvement Program Grant C06 RR-17348-01. Funding for magnetic resonance equipment was provided in part by NIH grant RR02511 and by National Science Foundation grant OSTI 97-24412.

References

1. O'Donnell ME. Ion and water transport across the blood–brain barrier. In: Alvarez-Leefmans FJ, Delpire E, editors. *Physiology and pathophysiology of chloride transporters and channels in the nervous system. From molecules to diseases*. London: Elsevier; 2009. p. 585–606.
2. O'Donnell ME, Daneman R, Fisher MJ, Noble LJ, Ransom BR, Spencer BJ, et al. Molecular physiology of the brain and brain barriers working group report. *Nat Rev Neurosci* 2011;**12**:169–82.

3. O'Donnell ME. Blood–brain barrier Na transporters in ischemic stroke. *Adv Pharmacol* 2014;**71**:113–46.
4. O'Donnell ME. The neurovascular unit. In: Dorovini-Zis K, editor. *The blood–brain barrier in health and disease: morphology, biology and immune function*. CRC Press, Taylor & Francis Group; 2015. p. 86–118.
5. Pedersen SF, O'Donnell ME, Anderson SE, Cala PM. Physiology and pathophysiology of Na^+/H^+ exchange and Na^+–K^+–$2Cl^-$ cotransporter in the heart, brain and blood. *Am J Physiol Regul Integr Comp Physiol* 2006;**291**:R1–R25.
6. Betz AL. Alterations in cerebral endothelial cell function in ischemia. *Adv Neurol* 1996;**71**:301–13.
7. Menzies SA, Betz AL, Hoff JT. Contributions of ions and albumin to the formation and resolution of ischemic brain edema. *J Neurosurg* 1993;**78**:257–66.
8. Menzies SA, Hoff JT, Betz AL. Extravasation of albumin in ischaemic brain oedema. *Acta Neurochir* 1990;**51**:220–2.
9. Schielke GP, Moises HC, Betz AL. Blood to brain sodium transport and interstitial fluid potassium concentration during focal ischemia in the rat. *J Cereb Blood Flow Metab* 1991;**11**:466–71.
10. Kimelberg HK. Current concepts of brain edema. Review of laboratory investigations. *J Neurosurg* 1995;**83**:1051–9.
11. O'Donnell ME, Tran L, Lam T, Liu XB, Anderson SE. Bumetanide inhibition of the blood–brain barrier Na–K–Cl cotransporter reduces edema formation in the rat middle cerebral artery occlusion model of stroke. *J Cereb Blood Flow Metab* 2004;**24**:1046–56.
12. Ginsberg MD, Busto R. Rodent models of cerebral ischemia. *Stroke* 1989;**20**:1627–42.
13. Neuwelt EA, Bauer B, Fahlke C, Fricker G, Iadecola C, Janigro D, et al. Engaging neuroscience to advance translational research in brain barrier biology. *Nat Rev Neurosci* 2011;**12**:169–82.
14. Abbott NJ, Rönnbäck L, Hansson E. Astrocyte-endothelial interactions at the blood–brain barrier. *Nat Rev Neurosci* 2006;**7**:41–53.
15. Löscher W, Potschka H. Blood–brain barrier active efflux transporters: ATP-binding cassette gene family. *J Am Soc Exp Neurotherap* 2005;**2**:86–98.
16. Hagenbuch B, Gao B, Meier PJ. Transport of xenobiotics across the blood–brain barrier. *News Physiol Sci* 2002;**17**:231–4.
17. Kusuhara H, Sugiyama Y. Active efflux across the blood–brain barrier: role of the solute carrier family. *J Am Soc Exp Neurotherap* 2005;**2**:73–85.
18. Dallas S, Miller DS, Bendayan R. Multidrug resistance-associated proteins: expression and function in the central nervous system. *Pharmacol Rev* 2006;**58**:140–61.
19. Ohtsuki S, Terasaki T. Contribution of carrier-mediated transport systems to the blood–brain barrier as a supporting and protecting interface for the brain; importance for CNS drug delivery and development. *Pharm Res* 2007;**24**:1745–58.
20. Miller DS, Bauer B, Hartz AM. Modulation of p-glycoprotein at the blood–brain barrier: opportunities to improve central nervous system pharmacology. *Pharmacol Res* 2008;**60**:196–209.
21. Demeule M, Régina A, Jodoin J, Laplante A, Dagenais C, Berthelet F, et al. Drug transport to the brain: key roles for the efflux pump p-glycoprotein in the blood–brain barrier. *Vasc Pharmacol* 2002;**38**:339–48.
22. Scherrmann J-M. Drug delivery to brain via the blood–brain barrier. *Vasc Pharmacol* 2002;**38**:349–54.
23. Abbott NJ, Patabendige AAK, Dolman DEM, Yusof SR, Begley DJ. Structure and function of the blood–brain barrier. *Neurobiol Dis* 2010;**37**:13–25.
24. Betz AL. Transport of ions across the blood–brain barrier. *Fed Proc* 1986;**45**:2050–4.
25. Betz AL, Goldstein GW. Specialized properties and solute transport in brain capillaries. *Ann Rev Physiol* 1986;**48**:241–50.

26. Bradbury MWB. The structure and function of the blood–brain barrier. *Fed Proc* 1984;**43**:186–90.
27. Hladky SB, Barrand MA. Mechanisms of fluid movement into, through and out of the brain: evaluation of the evidence. *Fluids Barriers CNS* 2014;**11**:26–32.
28. O'Donnell ME, Chen Y-J, Lam TI, Taylor KC, Walton JH, Anderson SE. Intravenous HOE-642 reduces brain edema and Na uptake in the rat permanent middle cerebral artery occlusion model of stroke: evidence for participation of the blood–brain barrier Na/H exchanger. *J Cereb Blood Flow Metab* 2013;**33**:225–34.
29. Lam TI, Wise PM, O'Donnell ME. Cerebral microvascular endothelial cell Na/H exchange: evidence for the presence of NHE1 and NHE2 isofroms and regulation by arginine vasopressin. *Am J Physiol Cell Physiol* 2009;**297**:C278–89.
30. Betz AL, Firth JA, Goldstein GW. Polarity of the blood–brain barrier: distribution of enzymes between the luminal and antiluminal membranes of the brain capillary endothelial cells. *Brain Res* 1980;**192**:17–28.
31. Zlokovic BV, Wand L, Mackic JB, Saraj AJ, McComb JG, McDonough A. *Expression of Na,K-ATPase at the blood–brain interface*. New York: Plenum Press; 1993.
32. Cserr HF, DePasquale M, Patlak CS. Volume regulatory influx of electrolytes from plasma to brain during acute hyperosmolality. *Am J Physiol* 1987;**253**:F530–7.
33. Keep RF. Potassium transport at the blood–brain and blood–csf barriers. *Front Cereb Vasc Biol: Transp Regul* 1993:43–54.
34. Betz AL. Sodium transport from blood to brain: inhibition by furosemide and amiloride. *J Neurochem* 1983;**41**:1158–64.
35. Jayakumar AR, Panickar KS, Curtis KM, Tong XY, Moriyama M, Norenberg MD. Na–K–Cl cotransport-1 in the mechanisms of cell swelling in cultured astrocytes after fluid percussion injury. *J Neurochem* 2011;**117**:437–48.
36. Kintner DB, Su G, Lenart B, Ballard AJ, Meyer JW, Ng LL, et al. Increased tolerance to oxygen and glucose deprivation in astrocytes from Na^+/H^+ exchanger isoform 1 null mice. *Am J Physiol Cell Physiol* 2004;**287**:C12–21.
37. Rutkowsky JM, Wallace BK, Wise PM, O'Donnell ME. Effects of estradiol on ischemic factor-induced astrocyte swelling and AQP4 protein abundance. *Am J Physiol Cell Physiol* 2011;**301**:C204–12.
38. Bourke RS, Kimelberg HK, Nelson LR, Barron KD, Auen EL, Popp AJ, et al. Biology of glial swelling in experimental brain edema. *Adv Neurol: Brain Edema.* 1980;**28**:99–109.
39. Iadecola C. Mechanisms of cerebral ischemic damage. In: Walz W, editor. *Cerebral ischemia: molecular and cellular pathophysiology*. Totowa, NJ: Humana Press; 1999. p. 3–34.
40. Gotoh O, Asano T, Koide T, Takakura K. Ischemic brain edema following occlusion of the middle cerebral artery in the rat. I: the time course of the brain water, sodium and potassium contents and blood–brain barrier permeability to ^{125}I-albumin. *Stroke* 1985;**16**:101–9.
41. Kempski O, Rosen Sv, Weight H, Staub F, Peters J, Baethmann A. Glial ion transport and volume control. In: Abbott NJ, editor. *Glial-neuronal interaction*. New York: The New York Academy of Sciences; 1991. p. 306–17.
42. Betz AL, Keep RF, Beer ME, Ren X. Blood–brain barrier permeability and brain concentration of sodium, potassium, and chloride during focal ischemia. *J Cereb Blood Flow Metab* 1994;**14**:29–37.
43. Kato H, Kogure K, Sakamoto N, Watanabe T. Greater disturbance of water and ion homeostasis in the periphery of experimental focal cerebral ischemia. *Exp Neurol* 1987;**96**:118–26.
44. Chen Y-J, Wallace BK, Yuen N, Jenkins DP, Wulff H, O'Donnell ME. Blood–brain barrier KCa3.1 channels. Evidence for a role in brain Na uptake and edema in ischemic stroke. *Stroke* 2015;**46**:237–44.

45. Haas M. Properties and diversity of (Na–K–Cl) cotransporters. *Ann Rev Physiol* 1989;**51**:443–57.
46. Chipperfield AR. The (Na^+–K^+–Cl^-) co-transport system. *Clin Sci* 1996;**71**:465–76.
47. Haas M. The Na–K–Cl cotransporters. *Am J Physiol* 1994;**267**:C869–85.
48. O'Grady SM, Palfrey HC, Field M. Characteristics and functions of NaK–Cl cotransport in epithelial tissues. *Am J Physiol* 1987;**253**:C177–92.
49. O'Donnell ME. Role of Na–K–Cl cotransport in vascular endothelial cell volume regulation. *Am J Physiol* 1993;**264**:C1316–26.
50. O'Donnell ME, Brandt JD, Curry F-RE. Na–K–Cl cotransport regulates intracellular volume and monolayer permeability of trabecular meshwork cells. *Am J Physiol* 1995;**268**:C1067–74.
51. Yerby TR, Vibat CRT, Sun D, Payne JA, O'Donnell ME. Molecular characterization of the Na–K–Cl cotransporter of bovine aortic endothelial cells. *Am J Physiol* 1997;**273**:C188–97.
52. Payne JA, Forbush III B. Alternatively spliced isoforms of the putative renal NaK–Cl cotransporter are differentially distributed within the rabbit kidney. *Proc Natl Acad Sci USA* 1994;**91**:4544–8.
53. Payne JA, Forbush III B. Molecular characterization of the epithelial Na–K–Cl cotransporter isoforms. *Curr Opin Cell Biol* 1995;**7**:493–503.
54. O'Donnell ME, Martinez A, Sun D. Cerebral microvascular endothelial cell Na–K–Cl cotransport: regulation by astrocyte-conditioned medium. *Am J Physiol* 1995;**268**:C747–54.
55. O'Donnell ME, Duong V, Suvatne S, Foroutan S, Johnson DM. Arginine vasopressin stimulation of cerebral microvascular endothelial cell Na–K–Cl cotransport activity is V1 receptor- and [Ca]-dependent. *Am J Physiol Cell Physiol* 2005;**289**:C283–92.
56. Foroutan S, Brillault J, Forbush B, O'Donnell ME. Moderate to severe ischemic conditions increase activity and phosphorylation of the cerebral microvascular endothelial cell Na–K–Cl cotransporter. *Am J Physiol Cell Physiol* 2005;**289**:C1492–501.
57. Kawai N, McCarron RM, Spatz M. Effect of hypoxia on Na^+–K^+–Cl^- cotransport in cultured brain capillary endothelial cells of the rat. *J Neurochem* 1996;**66**:2572–9.
58. Spatz M, Merkel KN, Bembry J, McCarron RM. Functional properties of cultured endothelial cells derived from large microvessels of human brain. *Am J Physiol* 1997;**272**:C231–9.
59. Vigne P, Farre AL, Frelin C. Na^+–K^+–Cl^- cotransporter of brain capillary endothelial cells. Properties and regulation by endothelins, hyperosmolar solutions, calyculin A, and interleukin-1. *J Biol Chem* 1994;**269**:19925–30.
60. Kawai N, McCarron RM, Spatz M. Endothelins stimulate sodium uptake into rat brain capillary endothelial cells through endothelin a-like receptors. *Neurosci Lett* 1995;**190**:85–8.
61. Kawai N, McCarron RM, Spatz M. The effect of endothelins on ion transport systems in cultured rat brain capillary endothelial cells. *Acta Neurochir* 1997;**70**:138–40.
62. Dóczi T. Volume regulation of the brain tissue—a survey. *Acta Neurochir* 1993;**121**:1–8.
63. Jójárt I, Joó F, Siklós L, László FA. Immunoelectronhistochemical evidence for innervation of brain microvessels by vasopressin-immunoreactive neurons in the rat. *Neurosci Lett* 1984;**51**:259–64.
64. Landgraf R. Central release of vasopressin: stimuli, dynamics, consequences. *Prog Brain Res* 1992;**91**:29–39.
65. Sorensen PS, Gjerris A, Hammer M. Cerebrospinal fluid vasopressin in neurological and psychiatric disorders. *J Neurol Neurosurg Psychiatry* 1985;**48**:50–7.
66. Zea Longa E, Weinstein PR, Carlson S, Cummins R. Reversible middle cerebral artery occlusion without craniectomy in rats. *Stroke* 1989;**20**:84–91.
67. Yan Y, Dempsey RJ, Flemmer A, Forbush B, Sun D. Inhibition of Na^+–K^+–Cl^- cotransporter during focal cerebral ischemia decreases edema and neuronal damage. *Brain Res* 2003;**961**:22–31.
68. Yan Y, Dempsey RJ, Sun D. Na–K–Cl cotransporter in rat focal cerebral ischemia. *J Cereb Blood Flow Metab* 2001;**21**:711–21.

69. Donovan MD, Schellekens H, Boylan GB, Cryan JF, Griffin BT. In vitro bidirectional permeability studies identify pharmacokinetic limitations of NKCC1 inhibitor bumetanide. *Eur J Pharmacol* 2016;**770**:117–25.
70. Tollner K, Brandt C, Romermann K, Loscher W. The organic anion transport inhibitor probenicid increases brain concentrations of the NKCC1 inhibitor bumetanide. *Eur J Pharmacol* 2015;**746**:167–73.
71. Mahnensmith RL, Aronson PS. The plasma membrane sodium–hydrogen exchanger and its role in physiological and pathophysiological processes. *Circ Res* 1985;**56**: 773–88.
72. Aronson PS. Kinetic properties of the plasma membrane Na^+/H^+ exchanger. *Ann Rev Physiol* 1985;**47**:545–60.
73. Chesler M. Regulation and modulation of ph in the brain. *Physiol Rev* 2003;**83**:1183–221.
74. Nakamura N, Tanaka S, Teko Y, Mitsui K, Kanazawa H. Four Na^+/H^+ exchanger isoforms are distributed to golgi and post golgi compartments and are involved in organelle ph regulation. *J Biol Chem* 2005;**280**:1561–72.
75. Wang XF, Yu MK, Lam SY, Leung KM, Jiang JL, Leung PS, et al. Expression, immunolocalization, and functional activity of Na^+/H^+ exchanger isoforms in mouse endometrial epithelium. *Biol Reprod* 2003;**68**:302–8.
76. Noel J, Roux D, Pouyssegur J. Differential localization of Na^+/H^+ exchanger isoforms (NHE1 and NHE3) in polarized epithelial cell lines. *J Cell Sci* 1996;**109**:929–39.
77. Hoogerwerf WA, Tsao SC, Vevuyst O, Levine SA, Yun CHC, Yip JW, et al. NHE2 and NHE3 are human and rabbit intestinal brush-border proteins. *Am J Physiol Gastrointest Liver Physiol* 1996;**270**:G29–41.
78. Bobulescu IA, Sole FD, Moe OW. Na^+/H^+ exchangers: physiology and link to hypertension and organ ischemia. *Curr Opin Nephrol Hypertens* 2005;**14**:485–94.
79. Kalaria RN, Premkumar DR, Lin CW, Kroon SN, Bae JY, Sayre LM, et al. Identification and expression of the Na^+/H^+ exchanger in mammalian cerebrovascular and choroidal tissues: characterization by amiloride-sensitive [3H]MIA binding and RT-PCR analysis. *Brain Res Mol Brain Res* 1998;**58**:178–87.
80. Masereel B, Pochet L, Laeckmann D. An overview of inhibitors of Na^+/H^+ exchanger. *Eur J Med Chem* 2003;**38**:547–54.
81. Scholz M, Wimmer-Greinecker G, Kleine P, Dzemali O, Martens S, Moritz A, et al. Cariporide (HOE642) limits S-100β release during cardiac surgery. *J Cardiovasc Pharmacol* 2003;**41**:468–73.
82. Yuen N, Lam TI, Wallace BK, Klug NR, Anderson SE, O'Donnell ME. Ischemic factor-induced increases in cerebral microvascular endothelial cell Na/H exchange activity and abuncance: evidence for involvement of ERK1/2 MAP kinase. *Am J Physiol Cell Physiol* 2014;**306**:C931–42.
83. Vigne P, Ladoux A, Frelin C. Endothelins activate Na^+/H^+ exchange in brain capillary endothelial cells via a high affinity endothelin-3 receptor that is not coupled to phospholipase C. *J Biol Chem* 1991;**266**:5925–8.
84. Ferrazzano P, Shi Y, Manhas N, Wang Y, Hutchinson B, Chen X, et al. Inhibiting the Na^+/H^+ exchanger reduces reperfusion injury: a small animal mri study. *Front Biosci* 2011;**3**:81–8.
85. Luo J, Chen H, Kintner DB, Shull GE, Sun D. Decreased neuronal death in Na^+/H^+ exchanger isoform 1-null mice after in vitro and in vivo ischemia. *J Neurosci* 2005;**25**:11256–68.
86. Russell JM. Cation-coupled chloride influx in squid axon. Role of potassium and stoichiometry of the transport process. *J Gen Physiol* 1983;**81**:909–25.
87. Lytle C, McManus TJ. Coordinate modulation of NaK–2Cl cotransport and K–Cl cotransport by cell volume and chloride. *Am J Physiol Cell Physiol* 2002;**283**: C1422–31.
88. Delpire E, Gagnon KB. SPAK and OSR1:Ste20 kinases involved in the regulation of ion homeostasis and volume control in mammalian cells. *Biochem J* 2008;**409**:321–31.

89. Gamba G. Molecular physiology and pathophysiology of electroneutral cation-chloride cotransporters. *Physiol Rev* 2005;**85**:423–93.
90. O'Donnell ME, Martinez A, Sun D. Endothelial Na–K–Cl cotransport regulation by tonicity and hormones: phosphorylation of cotransport protein. *Am J Physiol* 1995;**269**:C1513–23.
91. Wulff H, Miller MJ, Hansel W, Grissmer S, Cahalan MD, Chandy KG. Design of a potent and selective inhibitor of the intermediate-conductance $Ca2^+$-activated K channel, IKCa1: a potential immunosuppressant. *Proc Natl Acad Sci* 2000;**97**:8151–6.
92. Kohler R, Oliván-Viguera A, Wulff H. Endothelial small- and intermediate-conductance K channels and endothelium-dependent hyperpolarization as drug targets in cardiovascular disease. *Adv Pharmacol* 2016;**77**:65–104.
93. Rufo PA, Merlin D, Riegler M, Ferguson-Maltzman MH, Dickinson BL, Brugnara C, et al. The antifungal antibiotic, clotrimazole, inhibits chloride secretion by human intestinal T84 cells via blockade of distinct basolateral K^+ conductances. Demonstration of efficacy in intact rabbit colon and in an in vivo mouse model of cholera. *J Clin Investig* 1997;**100**:3111–20.
94. McCann JD, Welsh MJ. Basolateral K^+ channels in airway epithelia. II. Role in Cl^- secretion and evidence for two types of K^+ channel. *Am J Physiol Lung Mol Physiol* 1990;**258**:L343–8.
95. Warth R, Barhanin J. Function of K^+ channels in the intestinal epithelium. *J Membr Biol* 2003;**193**:67–78.
96. Chen Y-J, Raman G, Bodendiek S, O'Donnell ME, Wulff H. TheKCa3.1 blocker TRAM-34 reduces infarction and neurological deficit in a rat model of ischemia/reperfusion stroke. *J Cereb Blood Flow Metab* 2011;**31**:2363–74.
97. Hawkins BT, Davis TP. The blood–brain barrier/neurovascular unit in health and disease. *Pharmacol Rev* 2005;**57**:173–85.
98. Brillault J, Lam TI, Rutkowsky JM, Foroutan S, O'Donnell ME. Hypoxia effects on cell volume and ion uptake of cerebral microvascular endothelial cells. *Am J Physiol Cell Physiol* 2008;**294**:C88–96.
99. Brown RC, Davis TP. Calcium modulation of adherens and tight junction function. *Stroke* 2002;**33**:1706–11.
100. Brown RC, Mark KS, Egleton RD, Davis TP. Protection against hypoxia-induced blood–brain barrier disruption: changes in intracellular calcium. *Am J Physiol Cell Physiol* 2004;**286**:C1045–52.
101. Nagy Z, Goehlert UG, Wolfe LS, Huttner I. $Ca2^+$ depletion-induced disconnection of tight junctions in isolated rat brain microvessels. *Acta Neuropathol* 1985;**68**:48–52.
102. Abbott NJ, Revest PA. Control of brain endothelial permeability. *Cerebrovasc Brain Metab Rev* 1991;**3**:39–72.
103. Rakker K, Bayraktutan U. Increases in intracellular calcium perturb blood–brain barrier via protein kinase C-alpha and apoptosis. *Biochim Biophys Acta* 2016;**1862**:56–71.
104. Kimura C, Oike M, Ito Y. Hypoxia-induced alterations in $Ca2^+$ mobilization in brain microvascular endothelial cells. *Am J Physiol Heart and Circ Physiol* 2000;**279**:H2310–8.
105. Park SL, Lee DH, Yoo S, Yung Y-S. The effect of Na^+/H^+ exchanger-1 inhibition by sabiporide on blood–brain barrier dysfunction after ischemia/hypoxia in vivo and in vitro. *Brain Res* 2010;**1366**:189–96.
106. Park JH, Okayama N, Gute D, Krsmanovic A, Battarbee H, Alexander JS. Hypoxia/aglycemia increases endothelial permeability: role of second messengers and cytoskeleton. *Am J Physiol Cell Physiol* 1999;**277**:C1066–74.
107. Chen Y, McCarron RM, Azzam N, Bembry J, Reutzler C, Lenz FA, et al. Endothelin-1 and nitric oxide affect human cerebromicrovascular endothelial responses and signal transduction. *Acta Neurochir* 2000;**76**:131–5.

108. Chen Y, McCarron RM, Golech S, Bembry J, Ford B, Lenz FA, et al. Et-1- and NO-mediated signal transduction pathway in human brain capillary endothelial cells. *Am J Physiol Cell Physiol* 2003;**284**:C243–9.
109. Hicks K, O'Neil RG, Dubinsky WS, Brown RC. TRPC-mediated actin-myosin is critical for BBB disruption following hypoxic stress. *Am J Physiol Cell Physiol* 2010;**298**:C1583–93.
110. Huber JD, Egleton RD, Davis TP. Molecular physiology and pathophysiology of tight junctions in the blood–brain barrier. *Trends in Neurosci* 2001;**24**:719–25.
111. Dömötör E, Abbott NJ, Adam-Vizi V. Na^+-Ca^{2+} exchange and its implications for calcium homeostasis in primary cultured rat brain microvascular endothelial cells. *J Physiol* 1999;**515**(Pt 1):147–55.
112. O'Donnell ME, Owen NE. Regulation of pumps and carriers in vascular smooth muscle. *Physiol Rev* 1994;**74**:683–721.
113. Blaustein MP, Lederer WJ. Sodium/calcium exchange: its physiological implications. *Physiol Rev* 1999;**79**:763–854.
114. Luo J, Wang Y, Chen H, Kintner DB, Cramer SW, Gerdts JK, et al. A concerted role of Na^+–K^+–Cl^- cotransporter and Na^+/Ca^{2+} exchanger in ischemic damage. *J Cereb Blood Flow Metab* 2007 http://dx.doi.org/10.1038/sj/jcbfm.9600561.
115. Lenart B, Kintner DB, Shull GE, Sun D. Na–K–Cl cotransporter-mediated intracellular Na^+ accumulation affects Ca^{2+} signaling in astrocytes in an in vitro ischemic model. *J Neurosci* 2004;**24**:9585–97.
116. Khananshvili D. Sodium-calcium exchangers (NCX): molecular hallmarks underlying the tissue-specific and systemic functions. *Pflugers Arch—Eur J Physiol* 2014;**466**:43–60.
117. Bauser-Heaton HD, Song J, Bohlen HG. Cerebral microvascular nnos responds to lowered oxygen tension through a bumetanide-sensitive cotransporter and sodium–calcium exchanger. *Am J Physiol Heart Circ Physiol* 2008;**294**:H2166–73.
118. Kintner DB, Luo J, Gerdts J, Ballard AJ, Shull GE, Sun D. Role of Na^+–K^+–Cl^- cotransport and Na^+/Ca^{2+} exchange in mitochondrial dysfunction in astrocytes following in vitro ischemia. *Am J Physiol Cell Physiol* 2007;**292**:C1113–22.

CHAPTER

8

Mechanisms of Cell-Volume Regulation in the Central Nervous System

Nikolaus Plesnila[1] *and Oliver Kempski*[2]

[1]University of Munich Medical Center & Munich Cluster for Systems Neurology (Synergy), Munich, Germany [2]University of Mainz, Mainz, Germany

OUTLINE

KEY PHRASES

- Neuronal activity alters the homeostasis of the brain's extracellular space by the release of neurotransmitters, potassium ions, and protons.
- Astrocytes, which are responsible for the maintenance of the homeostasis of the brain's extracellular space, take up these solutes using the extra-/intracellular sodium gradient. Thereby sodium ions

Brain Edema.
DOI: http://dx.doi.org/10.1016/B978-0-12-803196-4.00008-4

enter the cytoplasm, increase the intracellular osmolarity, trigger the osmotic influx of water, and cause cell swelling. Hence, astrocytic cell swelling is foremost a physiological process.

- Under pathophysiological conditions, e.g., hypoxia or ischemia, neurons depolarize and release high concentrations of neurotransmitters and potassium ions into the extracellular space. Uptake of these solutes by astrocytes causes persistent swelling. In addition, aerobic metabolisms ceases thereby causing tissue acidosis. The uptake of protons by astrocytes causes additional sodium uptake and cell swelling.
- Accordingly, astrocytic cell swelling, clinically known as "cytototxic edema" is the physiological attempt of astrocytes to maintain the homeostasis of the extracellular space of the brain. As specific transporters are involved in this process, therapeutic interventions are possible but need to be addressed with care not to interfere with possible protective function of astrocytes.

INTRODUCTION

Brain edema is defined as the accumulation of water in the brain.[1] This accumulation may occur in cerebral vessels, as in vascular engorgement, in the extracellular space of the brain parenchyma, as in vasogenic edema, or within cells, as in cytotoxic edema. Vascular engorgement is caused by pathological vasodilation as a consequence of disturbed autoregulation of large cerebral vessels, and vasogenic edema is the consequence of altered permeability of the blood–brain barrier. Accordingly, water accumulation within cerebral vessels and in the brain's extracellular space is caused by active pathophysiological processes but is per se passive. In contrast, accumulation of water inside cells is an active event driven by highly specialized and interconnected transport processes for ions located in the cell membrane that then drive water flux trough the cell membrane, a process that can be further facilitated by aquaporins, a family of specialized water channels. These transport processes are not only responsible for regulation of cell volume, but foremost for the regulation of intracellular osmolarity and intracellular pH and are therefore tightly linked to vital cell functions such as enzyme activity and cell metabolism.[2]

MECHANISMS OF CELL-VOLUME REGULATION IN VERTEBRATES

In order to better understand the principles of cell-volume regulation in the brain, it is necessary to first understand the principles of cell-volume regulation in vertebrate cells in general. The volume of cells is determined by two of their most fundamental properties: the semipermeability of the cell membrane, i.e., the cell membrane is several orders of magnitude more permeable for water than for Na^+-, K^+-, and Cl^--ions and the presence of negatively charged large molecular weight proteins and organic phosphates within the cytoplasm, which cannot cross the plasma membrane due to their size and charge.[3] Accordingly, positively charged ions, such as Na^+ and K^+, will enter the cell together with water until according to Gibbs and Donnan an electrostatic equilibrium is reached and a potential across the cell membrane is created. At this point, the intracellular osmolarity is slightly higher than the extracellular one and water will enter the cell until its membrane ruptures. In order to prevent rupture of the cell membrane, a $3Na^+/2K^+$-ATPase pumps sodium out of the cell thereby rendering the cell membrane functionally impermeable for sodium. Thereby, an equilibrium between electrostatic and osmotic pressures becomes possible, and membrane potential and cell volume remain constant.[2] This concept, also called the "pump-leak" model, was proven experimentally by demonstrating that inhibition of the $3Na^+/2K^+$-ATPase indeed caused cell swelling in a whole range of vertebrate cells.[4]

As the intracellular osmolarity usually remains constant, challenges toward the maintenance of cell volume mainly derive from changes of extracellular osmolarity. Therefore, cells are equipped with a plethora of mechanisms to keep their cell volume constant upon extracellular hypo- or hyperosmolarity.

Cells exposed to a hypo-osmotic environment take up water and swell almost immediately as expected on the basis of the very rapid nature of fluxes of ions and water across plasma membranes. Within seconds, however, cells start releasing K^+ and Cl^- ions and organic osmolytes such as taurine and reduce their cell volume to the initial value within 1–2 minutes, a process termed regulatory volume decrease (RVD).[5] The molecular mechanisms behind RVD are (1) a volume-sensitive outwardly rectifying anion channel responsible for Cl^- efflux from the cell,[6] (2) about half of all known 4- and 5-membrane spanning potassium channels for K^+ efflux, (3) the K^+/Cl^--cotransporter family (1–4) for concomitant K^+ and Cl^- efflux, (4) a so far only functionally identified channel or transporter responsible for taurine efflux,[2] and (5) the recently described transient receptor potential isoform 4 (TRPV4), a polymodal swelling-sensitive cation channel[7] (Fig. 8.1). All these mechanisms are not energy dependent and should therefore, at least in part, be active during hypoxia and ischemia and counteract cell swelling under these conditions.

Cells suspended in hypertonic media shrink immediately, i.e., lose intracellular water but recover their initial cell volume within 1–2 minutes by taking up Na^+-, K^+-, and Cl^--ions from the extracellular space. This process is called regulatory volume increase (RVI) and shows the same

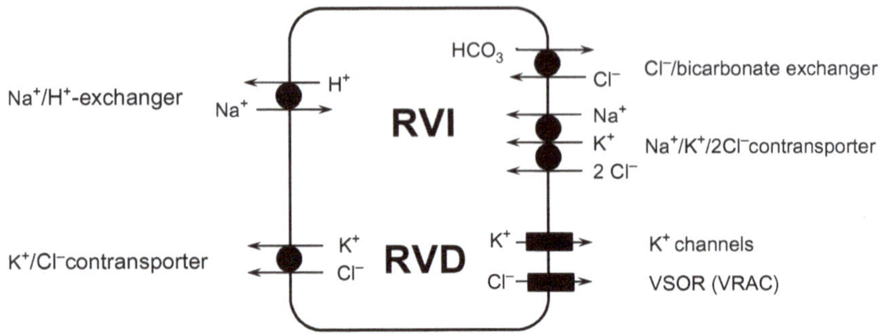

FIGURE 8.1 Ion pumps, transporters, and channels involved in regulatory volume increase (RVI) and regulatory volume decrease (RVD) in mammalian cells.

rapid temporal profile as RVD. Although RVI and RVD seem to be inverse, but otherwise very similar processes, the underlying mechanisms of RVI are very different from those described above for RVD. In the case of RVI, cells take up osmotically active ions by activation of (1) Na^+/H^+ exchangers (NHE 1 and possibly 2) together with the Cl^-/HCO_3^- exchanger, and (2) $Na^+/K^+/2Cl^-$-cotransporters[2] (Fig. 8.1).

In summary, mammalian cells have multiple and effective mechanisms to maintain their cell volume in face of significant changes in extracellular osmolarity or lack of energy. Based on these considerations, it does not become immediately clear why cells in the central nervous system (CNS) nevertheless swell and remain swollen for prolonged periods of time under pathological conditions. In order to approach this issue, it is essential to understand the specific mechanism responsible for cell-volume regulation in the CNS.

MECHANISMS OF PHYSIOLOGICAL CELL-VOLUME REGULATION IN ASTROCYTES

Conventionally it was (and is partially still) believed that astrocytes are the "glue" of the CNS, i.e., that astrocytes are static cells with a low metabolism that have a similar function as fibroblasts in connective tissue. Due to intensive research in the past 25 years, it became more than clear that astrocytes are exactly the opposite. Astrocytes carry most neurotransmitter receptors present on neurons, take up glucose from vessels, preprocess it to lactate and deliver it to neurons, participate in the regulation of cerebral blood flow, and maintain the homeostasis of the brain's extracellular space by scavenging neurotransmitters, protons, potassium ions, and other substances released by neurons during activation, just to name the most important functions.[8] Accordingly, astrocytes are actively and dynamically involved in brain metabolism and function.

In respect to cell-volume regulation, astrocytes carry almost all essential transporters described for non-neuronal cells[9–12] and are therefore able to perform rapid RVD and RVI when investigated in suspension in vitro.[10,13–15] More recent investigations on brain slices and in the brain in vivo using deep-tissue 2-photon imaging, however, demonstrate that astrocytes indeed swell upon exposure to global hypo-osmotic stress induced by an intraperitoneal injection of distilled water but remain swollen for at least 30 minutes, i.e., do not perform rapid RVD under physiological conditions in vivo.[16] The reason for this discrepancy between in vitro and in vivo findings has not been elucidated experimentally so far but may be explained by a completely different expression of volume-regulating channels and transporters in cultured cells in vitro as compared to the situation in vivo making all published results on astrocytic cell-volume regulation obtained on cultured cells difficult to interpret.

Astrocytes also express cotransporters in their cell membrane that take up solutes released by neurons during depolarization/activation, e.g., glutamate, K^+ ions, lactate, and protons (H^+), along the steep extra-/intracellular Na^+ gradient. Accordingly, exposure of astrocytes to these solutes in vitro results in Na^+ uptake and cell swelling.[11,17–22] Such highly dynamic cell-volume changes occur not only in isolated cells in vitro, but also during normal neuronal activity in vivo suggesting that swelling induced by clearance of substances from the extracellular space of the brain is a physiological function of astrocytes.[23] How astrocytes recover their cell volume under these conditions is, however, not clear yet. As RVD does not seem to occur in vivo, spatial buffering of the scavenged solutes within the gap-junction connected astrocytic network and finally into the vascular compartment seem to be the only other viable explanation.[24]

Another important feature of astrocytes regarding their ability to regulate their volume is the expression of water channels, aquaporins, which greatly facilitate the diffusion of water across cell membranes.[25] Among the aquaporin family of water channels, AQP4 is the subtype most abundantly expressed in the brain.[26] High-resolution immunogold analysis using electron microscopy identified AQP4 to be restricted to glial and ependymal cells and to be particularly strongly expressed on glial membranes that are in direct contact with vessels and pial membranes, i.e., mostly on glial end-feet, suggesting that AQP4 is involved in directed water transport from glial cells to these compartments.[27] The function of AQP4 in astrocytes is linked to the expression of TRPV4, a polymodal swelling-sensitive cation channel.[28] TRPV4 senses membrane stretch and induces intracellular Ca^{2+} signaling thereby inducing upregulation of AQP4 and other transporters involved in cell-volume regulation. Hence, TRPV4 is essential for the gene regulation involved in cell-volume regulation in astrocytes but does per se not mediate ion or water fluxes across the cell membrane.[29]

Taken together, scavenging osmotically active solutes from the extracellular space released by neurons during neuronal activity is a fundamental physiological function of astrocytes. This process is associated with Na^+ and subsequent water uptake and, hence, results in rapid astrocytic swelling. Astrocytes regain their normal cell volume quickly by distributing the excess osmotic load and water within the astrocytic network and finally through their end-feet into the cerebral microvasculature.

MECHANISMS OF PHYSIOLOGICAL CELL-VOLUME REGULATION IN NEURONS

Mechanisms of cell-volume regulation in neuronal cells are far less well investigated than in astrocytes. Nevertheless, neurons are believed to have the same ability and to carry the same membrane transporters and channels to regulate their cell volume as all other vertebrate cells.[30,31] Exposure of suspended neuronal cells to hypertonic or hypotonic stress in vitro results in rapid cell shrinkage and cell swelling, respectively. Accordingly, cultured neurons seem to have membranes with a high permeability for water suggestive of expression of aquaporins. Significantly, suspended neurons readily perform RVD, whereas they do not seem able to perform RVI, i.e., they remain shrunken after a hyperosmotic challenge.[31]

Similar to astrocytes also for neurons, the situation between volume regulation in cultured cells and in vivo seems to be completely different. One striking difference regarding the molecular mechanism necessary for cell-volume regulation is that neurons within the brain do not seem to express aquaporins in vivo as evidenced by immunogold analysis using electron microscopy for AQP4 and by conventional immunohistochemistry for all other known aquaporin isoforms.[32] AQP9 and AQP1 are the exception to the rule, as both were found in neurons by immunohistochemistry.[33,34] However, AQP9 is only present on noncortical, catecholaminergic neurons and does not seem to play a role for volume regulation but rather for facilitated transport of metabolites such as glycerol and lactate[33] and AQP1 is expressed in neurons from the septum and seems to be specifically involved in pain processing.[35] The lack of facilitated water transport in neurons is also supported by dynamic imaging of neuronal cell volume during a hypo-osmotic challenge in living brain slices using 2-photon microscopy. Under these conditions, astrocytes swell readily,[16] but neurons do not show any change in cell volume suggesting that in intact tissue water does indeed not cross the neuronal cell membrane.[36] Accordingly, regulation of cell volume under physiological conditions seems to be fundamentally different in astrocytes and neurons. Although astrocytes swell while maintaining the homeostasis of the extracellular space of the brain and in response to osmotic challenges in slices and in vivo, the neuronal cell volume does not change neither during neuronal activity nor during physiological fluctuations of extracellular osmolarity. The molecular basis of this behavior is most likely a lack of functional water channels in neurons.[36]

CELL-VOLUME CHANGES IN THE CNS UNDER PATHOLOGICAL CONDITIONS

Besides the interest in fundamental functions of a cell, most studies on swelling of glial and neuronal cells were triggered by the fact that the brain swells following pathological events, such as hypoglycemia, anoxia, ischemia, or brain trauma.[37–39] Based on neuropathological studies from the 1970s and 1980s, it was largely believed that brain swelling following ischemia is mainly mediated by swelling of glial cells, whereas traumatic brain swelling is caused by rupture of the blood–brain barrier and the formation of vasogenic edema.[37,39–41] Therefore, most efforts regarding the understanding of brain swelling after cerebral ischemia were directed toward the mechanisms of glial cell swelling.

The first consequence of cerebral ischemia is the disruption of glucose and oxygen delivery to the brain parenchyma. As the brain has only very limited storage capacity for energy substrates, the most metabolically active cells of the brain, i.e., the neurons, first lose their ability to maintain their membrane potential and depolarize.[42] Loss of resting membrane potential causes degranulation of axon terminals and massive release of glutamate and other neurotransmitters as well as K^+ ions into the extracellular space of the brain.[43] This process is accompanied by tissue acidosis and accumulation of lactate due to anaerobic glycolysis and by glial swelling as demonstrated in experimental studies performed in cats and nonhuman primates.[37,43] More recent studies suggest that possibly also neurons swell during focal cerebral ischemia.[16] In order to dissect which of these events are particularly responsible for cell swelling, neurons and glial cells were cultured, and subjected to each single ischemia-related event, i.e., anoxia, energy depletion, high extracellular glutamate, or high extracellular K^+, separately. The first question that needed to be addressed was if anoxia and/or energy depletion causes cell swelling. In a series of elegant experiments, Kempski et al. subjected suspended glial cells to anoxia or to anoxia in combination with iodoacetate, an inhibitor of glycolysis.[13] According the pump-leak theory of cell-volume regulation, lack of ATP by anoxia or by shutdown of glucose metabolism should have resulted in an almost immediate stop of the Na^+/K^+ ATPase and subsequent cell swelling. Surprisingly, unlike in many other cell types, glial cells did not change their volume at all and did not die for more than 2 hours of anoxia with or without inhibition of glucose metabolism.

The ionic mechanism of this phenomenon was an efflux of K^+ ions and an influx of Na^+ ions so that despite an inactive Na^+/K^+ ATPase, cell volume remained constant.[13] Similar experiments were performed later in cultured neurons, and the same results were obtained, i.e., also neurons did not swell in anoxia.[31]

Based on these findings, the hypothesis was put forward that not anoxia or ischemia per se cause tissue swelling, but their secondary consequences, such as high extracellular glutamate, potassium concentrations, or tissue acidosis. Indeed, experiments performed mainly during the 1990s demonstrated that this hypothesis may be correct. Neuronal and glial cells exposed to concentrations of glutamate expected to be present in the ischemic brain swelled by more than 40% of their baseline volume due to uptake of glutamate by the Na^+-dependent electrogenic glutamate carrier.[19,22,44–51] Incubation of neurons, glial cell lines, or primary cultured astrocytes in media with a pH similar to that found in ischemic tissue, i.e., 6.2–6.8, resulted in marked cell swelling and intracellular acidosis.[17,52–55] Undelaying mechanisms were the export of protons and the import of bicarbonate along the intra-/extracellular Na^+ and Cl^- gradient by the Na^+/H^+ and Cl^-/HCO_3^- exchanger, respectively, in an attempt to maintain an intracellular pH compatible with cell viability at the expense of an abnormal cell volume.[17,56] Later on it was shown by concomitant measurements of cell volume and intracellular pH that this attempt was futile, as intracellular pH remained acidic even when cell swelling was completely blunted by inhibition of the Na^+/H^+ exchanger.[57] More importantly, this finding demonstrated that acidosis-induced cell swelling could be completely inhibited without having any negative effect on intracellular pH and may therefore represent a viable therapeutic strategy for the treatment of ischemic cell swelling. Next to glutamate and acidosis also a high extracellular concentration of K^+ ions was shown to induce cell swelling of astrocytes.[53,58–60] As for the other swelling inducing substances, i.e., glutamate and protons, K^+-induced swelling of astrocytes also involves specific transport of ions across the cell membrane along the intra-/extracellular Na^+ and Cl^- gradient as K^+ ions have to be imported against a preexisting concentration gradient. The main two transporters involved under this circumstance are the $Na^+/K^+/2Cl^-$cotransporter and the K^+/Cl^- cotransporter, which both cause an increase in intracellular osmolarity and, hence, cell swelling.[53,58,59,61]

The findings of acidosis-, glutamate-, and potassium-induced astrocytic cell swelling suggest that during ischemia astrocytes do exactly what they do during normal brain function: They remove excess (and possibly toxic) solutes released by neurons from the extracellular space in order to maintain brain homeostasis and to protect neurons from damage. As this process is coupled to uptake of Na^+ and Cl^- ions from the extracellular space by specific transporter proteins, astrocytes swell. This explanation of ischemia-induced cell swelling has far reaching consequences, as cell swelling should be regarded as a neuroprotective process and not as a process that needs to be inhibited by all means. One notable exception may be acidosis-induced cell swelling, which does not result in maintenance of a normal intracellular pH and can therefore be inhibited without damaging the cells.

The concept that major parts of ischemic cell swelling are rather beneficial than damaging is hard to prove, specifically in vivo; however, a very elegant recent study in a model of very defined axonal injury suggests that this concept may indeed be correct: acute posttraumatic swelling of perinodal astrocytes helps prevent or postpone axonal degradation by maintaining physiologically relevant levels of extracellular K^+.[62]

Taken together, it cannot be emphasized enough that based on the concept that ischemic cell swelling is a protective reaction of the brain toward lack of energy, it is of utmost importance to fully understand the molecular mechanisms of this process before beginning to develop therapeutic strategies. Despite our quite detailed knowledge on cell-volume regulation in healthy cells and in the uninjured brain, further research into the mechanisms of pathological cell swelling is urgently needed in order to fully understand these processes in detail and to be able to develop viable therapeutic strategies for patients suffering from brain edema.

References

1. Pappius HM. The distribution of water in brain tissues swollen in vitro and in vivo. *Prog Brain Res* 1965;**15**:135–54.
2. Hoffmann EK, Lambert IH, Pedersen SF. Physiology of cell volume regulation in vertebrates. *Physiol Rev* 2009;**89**:193–277.
3. Gianazza E, Righetti PG. Size and charge distribution of macromolecules in living systems. *J Chromatogr* 1980;**193**:1–8.
4. Macknight AD, Leaf A. Regulation of cellular volume. *Physiol Rev* 1977;**57**:510–73.
5. Strange K. Cellular volume homeostasis. *Adv Physiol Educ* 2004;**28**:155–9.
6. Akita T, Okada Y. Characteristics and roles of the volume-sensitive outwardly rectifying (VSOR) anion channel in the central nervous system. *Neuroscience* 2014;**275**:211–31.
7. Jo AO, Ryskamp DA, Phuong TT, et al. TRPV4 and AQP4 channels synergistically regulate cell volume and calcium homeostasis in retinal Muller glia. *J Neurosci* 2015;**35**:13525–37.
8. Ransom BR, Ransom CB. Astrocytes: multitalented stars of the central nervous system. *Methods Mol Biol* 2012;**814**:3–7.
9. Kimelberg HK, Frangakis MV. Furosemide- and bumetanide-sensitive ion transport and volume control in primary astrocyte cultures from rat brain. *Brain Res* 1985;**361**:125–34.
10. Kimelberg HK. Anisotonic media and glutamate-induced ion transport and volume responses in primary astrocyte cultures. *J Physiol (Paris)* 1987;**82**:294–303.
11. Walz W. Swelling and potassium uptake in cultured astrocytes. *Can J Physiol Pharmacol* 1987;**65**:1051–7.
12. Eriksson PS, Nilsson M, Wagberg M, Ronnback L, Hansson E. Volume regulation of single astroglial cells in primary culture. *Neurosci Lett* 1992;**143**:195–9.

13. Kempski O, Zimmer M, Neu A, von RF, Jansen M, Baethmann A. Control of glial cell volume in anoxia. In vitro studies on ischemic cell swelling. *Stroke* 1987;**18**:623–8.
14. Pasantes-Morales H, Murray RA, Lilja L, Moran J. Regulatory volume decrease in cultured astrocytes. I. Potassium- and chloride-activated permeability. *Am J Physiol* 1994;**266**:C165–71.
15. Pasantes-Morales H, Murray RA, Sanchez-Olea R, Moran J. Regulatory volume decrease in cultured astrocytes. II. Permeability pathway to amino acids and polyols. *Am J Physiol* 1994;**266**:C172–8.
16. Risher WC, Andrew RD, Kirov SA. Real-time passive volume responses of astrocytes to acute osmotic and ischemic stress in cortical slices and in vivo revealed by two-photon microscopy. *Glia* 2009;**57**:207–21.
17. Kempski O, Staub F, Jansen M, Schodel F, Baethmann A. Glial swelling during extracellular acidosis in vitro. *Stroke* 1988;**19**:385–92.
18. Jakubovicz DE, Klip A. Lactic acid-induced swelling in C6 glial cells via Na^+/H^+ exchange. *Brain Res* 1989;**485**:215–24.
19. Chan PH, Chu L. Mechanisms underlying glutamate-induced swelling of astrocytes in primary culture. *Acta Neurochir Suppl (Wien)* 1990;**51**:7–10.
20. Lomneth R, Medrano S, Gruenstein EI. The role of transmembrane pH gradients in the lactic acid induced swelling of astrocytes. *Brain Res* 1990;**523**:69–77.
21. Kempski O, von RS, Weigt H, Staub F, Peters J, Baethmann A. Glial ion transport and volume control. *Ann N Y Acad Sci* 1991;**633**:306–17.
22. Hansson E, Johansson BB, Westergren I, Ronnback L. Mechanisms of glutamate induced swelling in astroglial cells. *Acta Neurochir Suppl (Wien)* 1994;**60**:12–14.
23. Sykova E. Extrasynaptic volume transmission and diffusion parameters of the extracellular space. *Neuroscience* 2004;**129**:861–76.
24. Kofuji P, Connors NC. Molecular substrates of potassium spatial buffering in glial cells. *Mol Neurobiol* 2003;**28**:195–208.
25. Verkman AS. Aquaporins at a glance. *J Cell Sci* 2011;**124**:2107–12.
26. Amiry-Moghaddam M, Ottersen OP. The molecular basis of water transport in the brain. *Nat Rev Neurosci* 2003;**4**:991–1001.
27. Nielsen S, Nagelhus EA, miry-Moghaddam M, Bourque C, Agre P, Ottersen OP. Specialized membrane domains for water transport in glial cells: high-resolution immunogold cytochemistry of aquaporin-4 in rat brain. *J Neurosci* 1997;**17**:171–80.
28. Benfenati V, Caprini M, Dovizio M, et al. An aquaporin-4/transient receptor potential vanilloid 4 (AQP4/TRPV4) complex is essential for cell-volume control in astrocytes. *Proc Natl Acad Sci U S A* 2011;**108**:2563–8.
29. Mola MG, Sparaneo A, Gargano CD, et al. The speed of swelling kinetics modulates cell volume regulation and calcium signaling in astrocytes: a different point of view on the role of aquaporins. *Glia* 2016;**64**:139–54.
30. Basavappa S, Ellory JC. The role of swelling-induced anion channels during neuronal volume regulation. *Mol Neurobiol* 1996;**13**:137–53.
31. Mackert BM, Staub F, Peters J, Baethmann A, Kempski O. Anoxia in vitro does not induce neuronal swelling or death. *J Neurol Sci* 1996;**139**:39–47.
32. Papadopoulos MC, Verkman AS. Aquaporin water channels in the nervous system. *Nat Rev Neurosci* 2013;**14**:265–77.
33. Badaut J, Petit JM, Brunet JF, Magistretti PJ, Charriaut-Marlangue C, Regli L. Distribution of Aquaporin 9 in the adult rat brain: preferential expression in catecholaminergic neurons and in glial cells. *Neuroscience* 2004;**128**:27–38.
34. Fukuda AM, Pop V, Spagnoli D, Ashwal S, Obenaus A, Badaut J. Delayed increase of astrocytic aquaporin 4 after juvenile traumatic brain injury: possible role in edema resolution? *Neuroscience* 2012;**222**:366–78.

35. Oshio K, Watanabe H, Yan D, Verkman AS, Manley GT. Impaired pain sensation in mice lacking Aquaporin-1 water channels. *Biochem Biophys Res Commun* 2006;**341**:1022–8.
36. Andrew RD, Labron MW, Boehnke SE, Carnduff L, Kirov SA. Physiological evidence that pyramidal neurons lack functional water channels. *Cereb Cortex* 2007;**17**:787–802.
37. Garcia JH, Kalimo H, Kamijyo Y, Trump BF. Cellular events during partial cerebral ischemia. I. Electron microscopy of feline cerebral cortex after middle-cerebral-artery occlusion. *Virchows Arch B Cell Pathol* 1977;**25**:191–206.
38. Gisselsson L, Smith ML, Siesjo BK. Influence of hypoglycemic coma on brain water and osmolality. *Exp Brain Res* 1998;**120**:461–9.
39. Jenkins LW, Becker DP, Coburn TH. A quantitative analysis of glial swelling and ischemic neuronal injury following complete cerebral ischemia. In: Go KG, Baethmann A, editors. *Recent progress in the study and therapy of brain edema*. Berlin: Springer; 1984. p. 523–37.
40. Fishman RA. Brain edema. *N Engl J Med* 1975;**293**:706–11.
41. Mendelow AD, Teasdale GM. Pathophysiology of head injuries. *Br J Surg* 1983;**70**:641–50.
42. Hossmann KA, Sakaki S, Zimmerman V. Cation activities in reversible ischemia of the cat brain. *Stroke* 1977;**8**:77–81.
43. Astrup J, Symon L, Branston NM, Lassen NA. Cortical evoked potential and extracellular K+ and H+ at critical levels of brain ischemia. *Stroke* 1977;**8**:51–7.
44. Chan PH, Chu L. Ketamine protects cultured astrocytes from glutamate-induced swelling. *Brain Res* 1989;**487**:380–3.
45. Kimelberg HK, Rose JW, Barron KD, Waniewski RA, Cragoe EJ. Astrocytic swelling in traumatic-hypoxic brain injury. Beneficial effects of an inhibitor of anion exchange transport and glutamate uptake in glial cells. *Mol Chem Neuropathol* 1989;**11**:1–31.
46. Koyama Y, Sugimoto T, Shigenaga Y, Baba A, Iwata H. A morphological study on glutamate-induced swelling of cultured astrocytes: involvement of calcium and chloride ion mechanisms. *Neurosci Lett* 1991;**124**:235–8.
47. Koyama Y, Baba A, Iwata H. L-glutamate-induced swelling of cultured astrocytes is dependent on extracellular Ca2+. *Neurosci Lett* 1991;**122**:210–12.
48. Koyama Y, Ishibashi T, Baba A. L-glutamate-induced swelling of cultured astrocytes. *Adv Exp Med Biol* 1992;**315**:375–80.
49. Schneider GH, Baethmann A, Kempski O. Mechanisms of glial swelling induced by glutamate. *Can J Physiol Pharmacol* 1992;**70**:S334–43.
50. Hansson E, Johansson BB, Westergren I, Ronnback L. Glutamate-induced swelling of single astroglial cells in primary culture. *Neuroscience* 1994;**63**:1057–66.
51. Bender AS, Schousboe A, Reichelt W, Norenberg MD. Ionic mechanisms in glutamate-induced astrocyte swelling: role of K+ influx. *J Neurosci Res* 1998;**52**:307–21.
52. Kempski O, Staub F, Jansen M, Baethmann A. Molecular mechanisms of glial cell swelling in acidosis. *Adv Neurol* 1990;**52**:39–45.
53. Kempski O, Staub F, Schneider GH, Weigt H, Baethmann A. Swelling of C6 glioma cells and astrocytes from glutamate, high K+ concentrations or acidosis. *Prog Brain Res* 1992;**94**:69–75.
54. Plesnila N, Haberstok J, Peters J, Kolbl I, Baethmann A, Staub F. Effect of lactacidosis on cell volume and intracellular pH of astrocytes. *J Neurotrauma* 1999;**16**:831–41.
55. Nabekura T, Morishima S, Cover TL, et al. Recovery from lactacidosis-induced glial cell swelling with the aid of exogenous anion channels. *Glia* 2003;**41**:247–59.
56. Staub F, Peters J, Kempski O, Schneider GH, Schurer L, Baethmann A. Swelling of glial cells in lactacidosis and by glutamate: significance of Cl(−)-transport. *Brain Res* 1993;**610**:69–74.
57. Plesnila N, Ringel F, Chang RC, Peters J, Staub F, Baethmann A. Effect of mild and moderate hypothermia on the acidosis-induced swelling of glial cells. *Acta Neurochir Suppl* 1997;**70**:262–4.

58. Walz W, Mukerji S. KCl movements during potassium-induced cytotoxic swelling of cultured astrocytes. *Exp Neurol* 1988;**99**:17–29.
59. Walz W. Mechanism of rapid K(+)-induced swelling of mouse astrocytes. *Neurosci Lett* 1992;**135**:243–6.
60. Su G, Kintner DB, Sun D. Contribution of Na(+)-K(+)-Cl(−) cotransporter to high-[K(+)](o)- induced swelling and EAA release in astrocytes. *Am J Physiol Cell Physiol* 2002;**282**:C1136–46.
61. MacAulay N, Zeuthen T. Glial K(+) clearance and cell swelling: key roles for cotransporters and pumps. *Neurochem Res* 2012;**37**:2299–309.
62. Volman V, Ng LJ. Perinodal glial swelling mitigates axonal degradation in a model of axonal injury. *J Neurophysiol* 2016;**115**:1003–17.

CHAPTER

9

Role of Aquaporins in the Formation and Resolution of Brain Edema

Beatriz Rodriguez-Grande[1], *Jan-Pieter Konsman*[1] *and Jérôme Badaut*[1,2]

[1]CNRS-University of Bordeaux, Bordeaux, France [2]Loma Linda University School of Medicine, Loma Linda, CA, United States

OUTLINE

Brain Edema.
DOI: http://dx.doi.org/10.1016/B978-0-12-803196-4.00009-6

Abbreviations

ADC	apparent diffusion coefficient
AQP	aquaporin
BBB	blood–brain barrier
CNS	central nervous system
CSF	cerebrospinal fluid
Cx43	connexin-43
DTI	diffusion tensor imaging
DWI	diffusion weighted imaging
EAE	experimental autoimmune encephalomyelitis
ISF	interstitial fluid
jTBI	juvenile traumatic brain injury
Kir4.1	inwardly rectifying potassium channel 4.1
MRI	magnetic resonance imaging
MS	multiple sclerosis
OAPs	orthogonal array of particles
SAH	subarachnoid hemorrhage
SCI	spinal cord injury
siAQP4	siRNA targeting AQP4
T2WI	T2-weighted imaging
TBI	traumatic brain injury
VP	vasopressin

As detailed in other chapters, edema mainly arises from direct damage to blood vessels or energy failure, which impairs ion transporters leading to an imbalance in intracellular ionic homeostasis that triggers a cascade of events starting with swelling of astrocyte end feet and neuronal dendrites (anoxic edema) followed by endothelial cell (EC) swelling (ionic edema), and finally by blood–brain barrier (BBB) disruption. The latter allows further water accumulation in the brain parenchyma (vasogenic edema) (Fig. 9.1). Aquaporins (AQPs) are water channels, crucial in regulation of water movement through cell membranes. Thus, they represent key elements in edema formation and resolution. Asides from directly regulating water homeostasis, AQPs have been associated with other physiological functions, which will not be covered in this book, such as cell migration[1] and gas diffusion.[2]

This chapter gives an overview of the AQP family and current knowledge on how AQPs are altered in brain injury, how those changes in AQPs affect the edema process, and what makes AQPs interesting therapeutic targets.

AQUAPORINS

Family Members

AQPs are transmembrane proteins with a common structure constituted by a consensus motif, Asn-Pro-Ala, which is implicated in pore

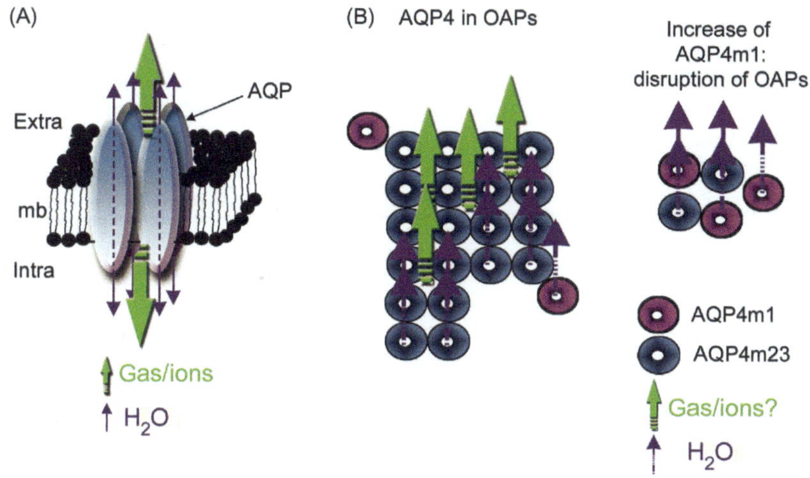

FIGURE 9.1 **Aquaporin structure and organization in the lipid membrane.** (A) Schematic drawing of the AQP homo–tetramer assembly within the lipid membrane resulting in a central pore permeable to cations and gases (green arrows). Each individual AQP facilitates bi-directional water movement depending on the osmotic gradient (blue arrows). (B) AQP4-m1 (purple circles) and AQP4-m23 (blue circles) isoforms contribute together to form OAPs in astrocyte end feet in contact to the blood vessels. It was previously shown that higher expression of AQP4-m23 contributes to the formation of large OAPs. However, increase of AQP4-m1 induced disruption of OAPs with a reduction of the size. This modification is observed in pathological conditions such as stroke. Recent knowledge on AQP leads us to hypothesize that the large OAPs contribute to gas and cation diffusion in the astrocyte membranes through central pores (green arrows). *Source: For A and B: modified from Badaut J, Fukuda AM, Jullienne A, Petry KG. Aquaporin and brain diseases.* Biochim Biophys Acta *2014;***1840***(5):1554–65.*

formation.[3] The AQP family has 13 members, which have been divided into three subgroups based on their permeability: (1) *AQPs* (AQP0, 1, 2, 4, 5, 6, and 8), the "pure water channel family," are primarily only permeable to water; (2) *Aquaglyceroporins* (AQP3, 7, 9, and 10) contribute to water diffusion, as well as glycerol, urea, and some monocarboxylates, which can facilitate lactate diffusion;[3] (3) *Super-AQPs* (AQP11 and 12) are localized in the cytoplasm, and they have been proposed to be involved in regulation of intracellular water transport, organelle volume, and intravesicular homeostasis.[3] It is important to remember that the water movement is dependent of the difference of osmotic pressure between intra and extracellular space.

From Genes to Arrays of Pores

Most AQP genes consist of a long exon 1, encoding one half of the protein that contains the N-terminal, followed by exons 2–4, encoding the

C-terminal half of the protein. The degree of homology of the N-terminal half between different AQPs is higher than that of the C-terminal half, which suggests that the first exon encodes common basic aspects of all AQPs, while the rest encodes other characteristics more specific to each AQP subtype.[4] One AQP gene can give rise to different transcripts, enhancing AQP diversity even more.[5] As we will describe later, transcriptional activation of AQP genes in certain situations (e.g., inflammation) can lead to increased water permeability and facilitate edema formation.[6] However, so far, transcriptional regulation is not well known for all AQP subtypes.

AQPs are highly conserved through evolution, since AQP or AQP-like proteins are found in eubacteria, yeast, plants and mammals.[7] This reinforces the idea that regulation of water trafficking is a fundamental mechanism present in all organisms. The typical AQP protein weighs around 30 kDa and has six membrane spanning domains with its carboxyl (C) and amino (N) termini presented in the intracellular side. Some AQP proteins assemble as homo-tetramers, giving rise to a central pore (Fig. 9.1), which can be permeable to other substrates such as cations and gas.[8]

In the case of AQP4, the tetrameric pores are organized as highly dense suprastructures (orthogonal arrays of particles, OAPs)[9] (Fig. 9.1). The ratio of long (AQP4-m1) to short (AQP4-m23) splice variants of AQP4 determines the size of these super structures.[9] AQP4-m23 stabilizes OAPs, facilitating the formation of bigger arrays (Fig. 9.1).[9,10] These super structures are abundant in processes of vascular astrocytes and in astrocytes of the supraoptic nucleus and have been proposed to be involved in diffusion of gases such as CO_2 along with water (Fig. 9.1).[2,11–13]

Distribution in the CNS

AQPs are ubiquitously distributed throughout mammalian tissues including the brain and the spinal cord.[8,14] AQP1, 3, 4, 5, 6, 8, 9, and 12 have been described in brain cells;[15] however, in the rest of the chapter, we will focus only on AQP1, 4, and 9, since they are the most studied in relation with edema formation.

AQP1 is primarily found in epithelial cells of the choroid plexus in primate and rodent brains (Fig. 9.2). As previously discussed (see Chapter 3: Noninvasive Imaging Techniques for Brain Edema: From Basic Science to the Clinic, for details), AQP1 seems to play a role in cerebrospinal fluid (CSF) formation, a process that may indirectly contribute to an increase in intracranial pressure.[16] AQP1 has also been found in the dorsal horn of the spinal cord and in the trigeminal sensory ganglia, possibly involved in nociception,[17] and in some neuronal processes in the brain septum, with increased expression after traumatic brain injury (TBI).[18] It is interesting to note that the distribution of AQP1 seems to differ between species,

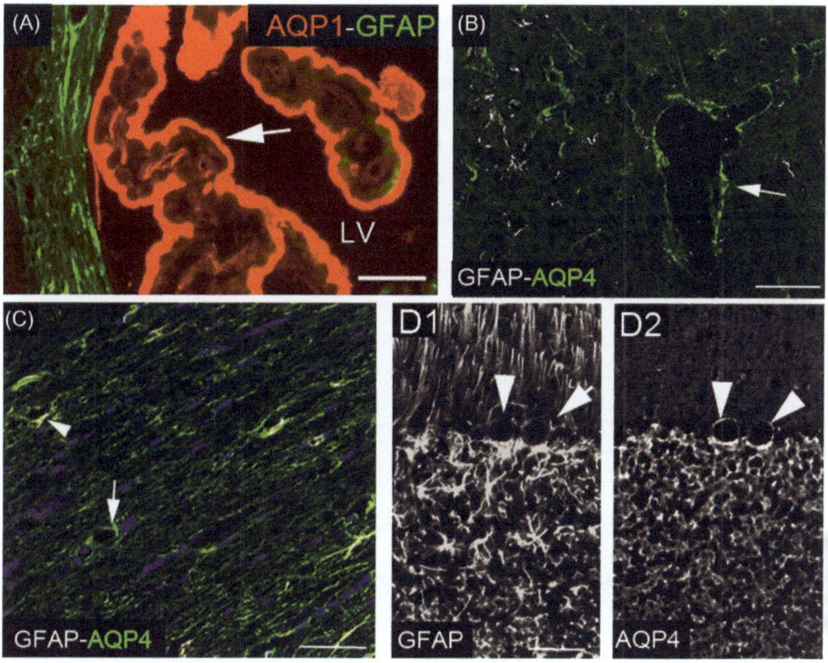

FIGURE 9.2 **Variety in astrocyte AQPs distribution.** (A) AQP1 immunostaining (red) in choroid plexus epithelium (arrow) located in the lateral ventricle (LV). The border of the LV is outlined by the GFAP staining (green), specific marker of the astrocytes. (B) AQP4 distribution in the cortex layer showed the "polarization" of the AQP4 labeling around the blood vessels (arrows). The double staining AQP4 and GFAP exhibits its presence on the astrocyte end feet (arrows). (C) AQP4 staining (green) in the corpus callosum (CC) is abundant around the blood vessel (arrows) and at distance from the perivascular space (arrowheads). In the white matter structure, AQP4 exhibits a different pattern of staining, with distribution of the protein on the astrocyte processes (gray, GFAP staining), possibly in association with node of Ranvier. (D1, D2) GFAP (D1) and AQP4 (D2) staining in the cerebellum differs from the cortex, with an abundant staining around all the neurons of the granular layer as well as around the Purkinje cell bodies (P, arrowheads). AQP4 is observed around the blood vessels in the molecular layer, where the radial glia is located. Bar A, B, C = 50 μm; D = 40 μm. *Source: Modified from Badaut J, Fukuda AM, Jullienne A, Petry KG. Aquaporin and brain diseases.* Biochim Biophys Acta *2014;***1840***(5):1554–65.*

for example, as it is expressed in a subpopulation of astrocytes in non-human primates,[19] but not in rodents. This suggests that AQP1 may have an additional role in brain water homeostasis in non-human primates.[19]

AQP4 messenger ribonucleic acid was first found in the hypothalamic supraoptic nuclei and perivascular brain regions,[20] and AQP4 has been the most studied AQP in the central nervous system (CNS) ever since. AQP4 is abundant in all brain structures including the gray and white matter of the spinal cord.[14,21] AQP4 is expressed exclusively by astrocytes, mostly concentrated on perivascular astrocyte end feet (Fig. 9.2),[22,23] which makes it a key element for regulation of cerebrovascular interactions. This typical perivascular distribution of AQP4 is very well described in the cerebral cortex and the striatum (Fig. 9.2),[22] and also in the spinal cord.[14] Nevertheless, AQP4 is also found in cell bodies and astrocytic processes in contact with neurons and possibly with nodes of Ranvier.[22,23] Some of these patterns of staining have been observed in the hippocampus (Fig. 9.2),[24] the corpus callosum (Fig. 9.2), the cerebellum (Fig. 9.2),[25] the magnocellular nuclei of the hypothalamus,[22,26] the brainstem (Fig. 9.2) and the spinal cord.[14]

AQP9, an aquaglyceroporin, has been identified in both rodent and primate brains.[19] AQP9 is present in astrocytes,[27,28] brain ECs,[28] and catecholaminergic neurons.[19,28,29] Our studies suggest that this channel is probably involved in energy metabolism.[31]

ROLE OF AQUAPORINS IN EDEMA

Mechanisms of Regulation of Water Movement

The obvious mechanism of regulation of water movement by AQPs is the water-permeable pore itself. Therefore, increase or decrease in the level of AQP expression would lead to increase or decrease water permeability of the cell membrane. However, AQPs also interact with each other and with other proteins to further regulate water homeostasis.

AQP1, which is present in peripheral blood vessels, has never been observed in brain ECs.[8] In vitro experiments suggest that AQP1 can be expressed by brain ECs, but this expression is inhibited when they are in contact with astrocytes.[32] This suggests that astrocytic AQP4 controls water homeostasis not only by regulating influx of water in and out of astrocytes, but also by limiting the permeability of brain ECs to water through AQP1 inhibition.

In theory, AQP4 can also regulate permeability by changing its suprastructural organization. For instance, the ratio between AQP4-m1 and AQP4-m23, which is critical to determine the size of the OAPs[9] (Fig. 9.1) is altered in a stroke model. AQP4-m1 is more highly expressed than AQP4-m23 in the ischemic hemisphere after stroke,[33] suggesting a disorganization of the OAPs in accordance with previous rat stroke data.[34] From what is known about isoforms and OAPs formation (see Section "From Genes to Arrays of Pores" of this chapter), one would hypothesize that AQP4 is assembled in smaller arrays in the ischemic hemisphere. Although this could potentially alter water movement, the functional consequences of AQP4-m1 increase and OAP disorganization is not yet elucidated (Fig. 9.1).

AQP4 also works together with other astrocytic proteins such as the gap junction connexin-43 (Cx43) proteins and the inwardly rectifying potassium channel 4.1 (Kir4.1) to clear excess of water. Cx43 forms gap-junctions between astrocytes and contributes to the diffusion of water and solutes within the astrocytic network. Interestingly, in astrocyte cultures, Cx43 is downregulated after treatment with silencing RNA against AQP4,[35] and it may decrease the connectivity between astrocytes, thus limiting water diffusion. Kir4.1 colocalizes with AQP4 in astrocyte end feet[36] and potassium has been linked with water flux during astrocyte swelling in a spinal cord injury (SCI) model.[37] In addition, cellular potassium reuptake is delayed in AQP4 knockout ($AQP4^{-/-}$) mice in an epilepsy model.[38] Water homeostasis is thus associated with spatial potassium buffering by astrocytes after neuronal activity and the presence of AQP4 in astrocyte end feet has been proposed to be part of this process[39] although a clear functional relationship between AQP4 and Kir4.1 has not been demonstrated yet.[40] Therefore, we can hypothesize that high expression of AQP4 in astrocytes in the corpus callosum may also be associated with buffering of osmotic pressure at the node of Ranvier as a prerequisite for efficient axonal conductance. Finally, AQP4 has also been proposed to play a role in the detection of variations in plasma osmotic pressure within the hypothalamic nuclei.[22]

Astrocytic AQP4 has also been recently associated with a general bulk flow occurring through the perivascular astrocytic network. This perivascular flow seems to be involved in the clearance of waste proteins such as beta amyloid from the cerebral cortex.[41] Although the mechanism is unclear, these recent results confirm the importance of AQP4 in the clearance of potentially harmful substances in the central nervous system.[41] Mechanisms of CSF and interstitial fluid (ISF) exchange are associated with AQP4, and the movement of CSF along the arteries is dependent on arterial pulsatility.[42] Regarding these new results, it is interesting to pinpoint that AQP4 knockout mice have a wider extracellular space as compared to wild-type (WT) mice.[43] Therefore, it seems that the presence of AQP4 is critical for the regulation of the volume of the extracellular space, by facilitating water movement.[8] Recent observations from Nedergaard's group suggest a similarity in function between the peripheral lymphatic system and the astrocytic network, the so-called "glymphatic system."[41] The proposed model seems to fit well with the cortical pattern of AQP4 expression, where AQP4 is mostly found in astrocyte end feet; but not necessarily with the pattern of AQP4 expression in other brain structures where AQP4 is not only limited to end feet.

Aquaporin Expression in Edema

Various studies have shown changes in the level of AQP1, 4, and 9 expression in several brain disorders in rodent models as well as in human samples.[18,27,44]

AQP1 expression increases in human astrocytes after subarachnoid hemorrhage (SAH),[44] in rodent cortical ECs 24 hours after intracerebral hemorrhage[45] and in murine astrocytes and neurons several months after SCI.[46] Since AQP1 is one of the classical "pure water channels," these changes of expression suggest that it could affect water movement after CNS injury; however, these changes of expression have not been directly linked with the edema process.

AQP9 protein expression is gradually upregulated over time in reactive astrocytes along the infarct border after ischemic injury. However, this upregulation does not correlate with the degree of brain swelling.[27,29] Moreover, species-dependent changes in AQP9 have also been reported. AQP9 expression in mice is increased 7 days after stroke;[29] however, in gerbils a differential AQP9 expression is found after occlusion within hippocampal pyramidal neurons (CA1: 6 hours and CA2 and CA3: 12 hours).[47] As pyramidal neurons do not express this channel under physiological conditions[27,29] it is likely that metabolic stressors induce expression of this "energy" channel.[31] All together, these results suggest that AQP9 does not play a key role in edema formation after transient cerebral ischemia. The functional consequences of increased AQP9 expression in neurons and astrocytes need further investigations.

Most studies on AQPs and edema have focused on AQP4 because changes in AQP4 expression seem to parallel changes in the edema process.[18,29,30] The complexity of the role of AQP4 in edema formation is demonstrated by the variety of changes in AQP4 expression observed after brain injury. Increases[48–52] as well as decreases in AQP4 expression[53–55] were detected and seem to be related to differences in injury type, rodent strains, and age at impact.[18] The following sections deal with the role of AQP4 in edema formation and resolution in different pathologies such as traumatic brain injury,[18,54] cerebral ischemia,[29,56] SAH,[44] and experimental multiple sclerosis (MS).[57]

Aquaporin-4 in Edema Formation

The role of AQP4 in the edema process was first studied by characterizing changes in AQP4 expression. Later on, the field benefited from the generation of $AQP4^{-/-}$ mice by Verkman and colleagues.[58] $AQP4^{-/-}$ mice do not exhibit any major structural or physiological phenotype, except that the extracellular space of $AQP4^{-/-}$ mice is expanded by about 20% as compared to WT mice.[43,59,60] The recent development of siRNA targeting AQP4 (siAQP4) now allows us to study the functions of this protein in more detail.[61,62] So far, the available data strongly suggest that AQP4 promotes the formation of cytotoxic edema by facilitating the entry of water into astrocytes.

In stroke models, early increases in expression of AQP4 seem to be associated with the development of ionic brain edema (Fig. 9.3) and

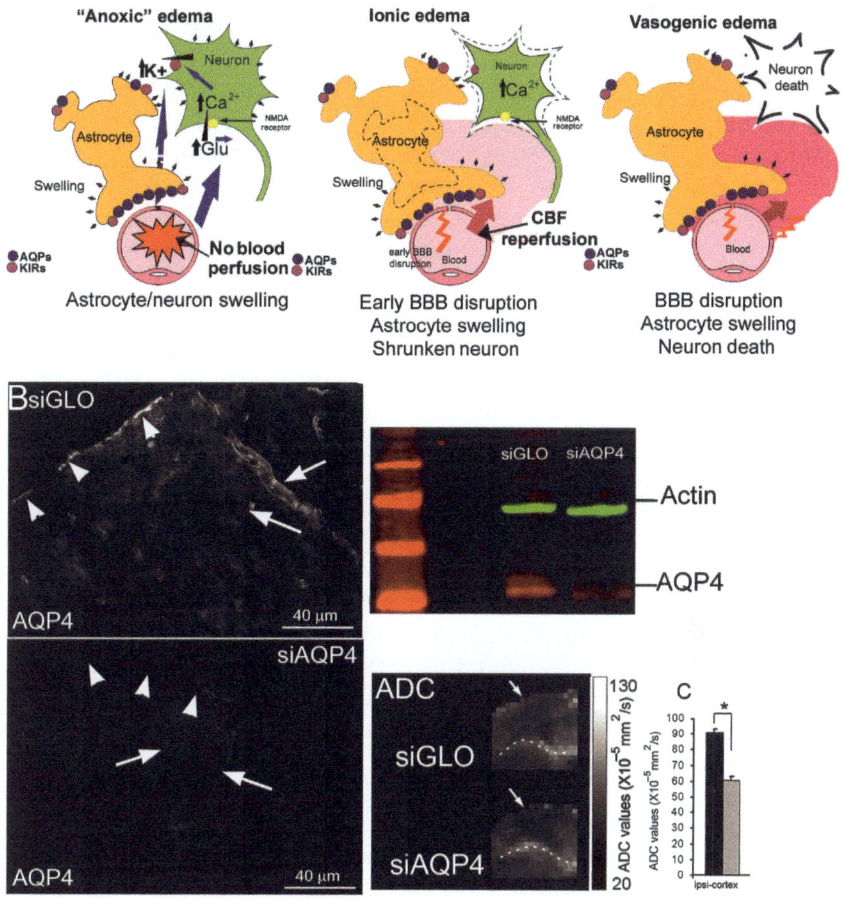

FIGURE 9.3 **AQP4 water diffusion in brain.** (A) Schematic drawing of the events happening during edema formation with three different edema phases: anoxic, ionic, and vasogenic edema. During the injury with decrease of brain perfusion, the first minutes are characterized by anoxic edema. Anoxic edema is characterized as a swelling of the astrocytes and the neuronal dendrites caused by a disruption of the cellular ionic gradients and the entry of ions followed by water entry and leading to cellular swelling. During the ionic edema, astrocytes become swollen and neuronal death starts to occur resulting in shrinkage of the neurons, shear stress, and endothelial dysfunctions on the nonperfused vascular tree, which results in early transient leakage of the BBB. Vasogenic edema is a result of disruption of the tight junctions between the ECs, leading to increased permeability of the cerebral blood vessels to albumin and other plasma proteins, further contributing to swelling of astrocytes and subsequent neuronal cell death. (B) AQP4 labeling in siGLO and siAQP4 treated rats showed a significant decrease in the intensity of AQP4 staining in the glia limitans and perivascular astrocyte in the cortex of siAQP4 treated rats. AQP4 western blot (red) showed a significant decrease of the intensity in siAQP4 compared to siGLO treated rats. Decrease of the AQP4 is associated with a decrease of the water diffusion. Enlarged ADC images focusing on the contralateral cortex showed the decrease in ADC in siAQP4 compared to siGLO treated rats. The quantification showed a 50% decrease of the ADC signals where the AQP4 expression is decreased by 30%. *Source: For (A): Adapted from Badaut J, Fukuda AM, Jullienne A, Petry KG. Aquaporin and brain diseases.* Biochim Biophys Acta *2014;***1840***(5):1554–65. For B Badaut J, Ashwal S, Adami A, et al. Brain water mobility decreases after astrocytic aquaporin-4 inhibition using RNA interference.* J Cereb Blood Flow Metab *2011;***31***(3):819–31.*

swelling of astrocyte processes.[63] In a mouse model of stroke with transient occlusion of the middle cerebral artery, AQP4 expression was rapidly upregulated in perivascular astrocyte end feet, peaking at 1 hour after stroke onset.[29,33] This increase in AQP4 was observed in the infarct core and in the ischemic penumbra and correlated with degree and temporal profile of brain swelling.[29,33] It is interesting to note that increased AQP4 expression is not observed in more severe stroke models,[64] prompting the hypothesis that under severe ischemia the brain is not able to synthesize AQP4 during the early phase of reperfusion.

Experimental autoimmune encephalomyelitis (EAE) is used to model the pathophysiology of MS. In this model, upregulation of AQP4 started 10 days before the onset of cerebellar swelling and correlated with BBB disruption in the cerebellum and the decrease of tight junction proteins such as occludin.[57] Accordingly, $AQP4^{-/-}$ mice showed a less severe clinical score and tissue inflammation in various MS models as compared to WT animals.[65] These results are in line with a deleterious role of AQP4 in MS.

In a model of juvenile traumatic brain injury (jTBI) we showed that AQP4 expression is not changed during the development of edema within the first 24 hours after injury.[18] Our results therefore suggest that stable AQP4 levels may promote water entry leading to cellular swelling (lower apparent diffusion coefficient, ADC) and increased edema (increased T2) near the site of injury.[18] Reduction of cortical AQP4 expression by intracortical injection of siAQP4 reduced edema formation and improved cognitive outcome 2 months postinjury, suggesting that siAQP4 could be used as a specific drug to prevent edema formation by transiently decreasing the level of AQP4 expression in vivo.[62]

Following SCI, $AQP4^{-/-}$ mice have less brain edema and smaller lesions early after injury.[66] However, these beneficial effects do not seem to last long term.[67] Several other contradictory results using these $AQP4^{-/-}$ mice in combination with several injury models[33,67,68] indicate that embryonic deletion of AQP4 may not be the ideal tool to solve the role of AQP4 in edema formation. The apparently contradictory effects of AQP4 led some authors to propose a dual role for AQP4 in the edema process: deleterious during edema formation and beneficial during the edema resolution phase.[69]

Aquaporins in Edema Resolution

Although most studies have focused on the role of AQP4 in the edema build-up phase, some studies have started to explore the role of AQP4 at later stages. So far, several results support the hypothesis that AQP4 is important in edema resolution through facilitation of water clearance.

The first evidence for a role of AQP4 in the resolution of brain edema was that the infusion of saline solution into the brain parenchyma of

AQP4$^{-/-}$ mice induced a significantly higher increase in intracranial pressure as compared to WT mice.[70] In addition, the temporal profile of increased AQP4 expression was rather associated with edema resolution than edema formation in vivo,[18,56,61,62,71,72] that is the peak of AQP4 is found 48 hours after the insult in most ischemic, traumatic, or inflammatory brain injury models.[29,30,33,18,48–52] Most of the time, the increase of AQP4 is observed near the lesion site in perivascular astrocyte end feet, astrocyte processes, and glia limitans.[18,29] These changes may indicate that at this time point excess AQP4 could facilitate edematous fluid elimination through the subarachnoid space.[30,73–74] As observed in a model of jTBI, increased AQP4 in the glia limitans may compensate for water accumulation at 1 and 3 days (higher magnetic resonance imaging, MRI, T2 values), with a gradual increase of AQP4 at 3 days and normalization of both AQP4 and T2 values by 7 days.[18] The same was observed in a rat neuroinflammatory lesion model where the ADC, a MRI parameter indicating cytotoxic brain edema, was most significantly increased when the expression of AQP4 peaked.[30]

These results support the idea that AQP4 plays a dual role in the edema process with a deleterious role during edema build-up and a beneficial role during edema resolution.[30,66,70]

Chronic Changes in Aquaporin-4 in Relation to Edema

Long-term changes of AQP4 have been observed as well: AQP4 expression remained increased up to 28 days in the border region of the injury site in a stroke model[75] and after SCI.[21] In fact, AQP4 is known to be involved in astrocyte migration during glial scar formation by facilitating water entry necessary for filopodia formation,[1] and to facilitate cell adhesion between astrocytes.[76] It is likely that the high expression of AQP4 in astrocytes could contribute to glial scar formation by facilitating astrocyte migration and cell–cell interactions.

In chronic brain diseases such as Alzheimer's disease or chronic TBI, decreases of AQP4 expression were observed in blood vessels surrounded by beta amyloid deposits.[77,78] This reduction of AQP4 expression may lower perivascular flow, thus limiting the clearance of beta amyloid from the brain.[41]

AQPs as a Potential Link Between Neuroinflammation and Edema

Neuroinflammation, the inflammatory response which occurs within the brain, encompasses activation of astrocytes, microglia, and blood vessels and secretion of cytokines and chemokines, soluble molecules which mediate and regulate this activation. Activation of brain ECs at the BBB is associated with dysregulation of water homeostasis and may therefore result in brain edema formation. AQP4 is frequently upregulated during neuroinflammation and has therefore been proposed as one of the key mediators of neuroinflammation-induced edema.[6]

The pro-inflammatory cytokine IL-1 has been shown to induce AQP4 expression in a concentration-dependent manner.[79] This IL-1-mediated increase of AQP4 is expected to occur in all types of brain injury, since high IL-1 is a hallmark of inflammation in all tissues, but particularly important in neuroinflammation.[80] Regarding the mechanism of action, IL-1β has been shown to upregulate AQP4 at a transcriptional level: downstream of the IL-1 receptor, NFκB is translocated to the nucleus and binds the *AQP4* gene promoter.[79,81] Another pro-edematous substance, the high mobility group box protein-1 released by necrotic neurons, induces a neuroinflammatory response together with an increase in AQP4 possibly triggered at least in part by the pro-inflammatory cytokine IL-6.[82] Interestingly, the upregulation of AQP4 takes place through microglia–astrocyte interactions. Following microglial activation, factors secreted by these cells (including the above mentioned cytokines), induce AQP4 expression in astrocytes.[81,82] In fact, inhibition of microglial activation has been shown to reduce neuroinflammation and edema after TBI.[82]

The relationship between AQPs and inflammation is bi-directional. In fact, inflammation alters expression levels of AQPs and changes in AQPs modulate the inflammatory response. Changes in AQPs have been shown to be associated with alterations in cytokine secretion and activation of astrocytes, microglia, and macrophages. Absence of AQP4 in astrocyte end feet is associated with reduced astrocyte migration toward the site of injury and reduced astrocytic hypertrophy.[1,83] Thus, astrogliosis may require the presence of AQP4 to facilitate the water movement that is required for migration and hypertrophy. In addition, recent work suggests a potential relationship between AQP4 and microglial activation. First, $AQP4^{-/-}$ mice show more microglial activation and less astrogliosis one month after TBI.[84] A similar increase in microglial activation and reduction in astrogliosis has been reported in $AQP4^{-/-}$ mice after TBI[85] and in rats treated with siAQP4 in a model of jTBI.[86] The siAQP4 treatment after jTBI showed a decrease in AQP4 associated with less edema formation and better behavior outcomes up to 2 months postinjury, although differences in glial activation were no longer seen at that time point.[62] Alterations in early glial responses could thus be having an effect in long-term edema resolution. With regard to the mechanism, AQP4 could be regulating microglial activation through stretch-activated chloride (Cl^-) channels expressed in microglia and known to be activated by osmotic stress.[87,88] The activation of these channels contributes to the maintenance of the nonactivated (ramified) phenotype of microglia.[87] Since AQP4 is responsible for water transport and osmotic pressure, we hypothesize that inhibition of AQP4 modifies the osmotic stress within the extracellular space surrounding microglia. This would change the activation status of the swelling-activated chloride channels, resulting in microglial activation. Another possibility is the crosstalk between astrogliosis and microglial activation.[89] The decrease of reactive astrogliosis as a result of the absence of AQP4 may cause an increase in microglial activity. Regarding cytokine release, reduced production of the pro-inflammatory cytokines tumor necrosis factor-alpha and IL-6 was observed in $AQP4^{-/-}$ compared to WT mice astrocyte cultures whereas, in those same cultures, adenovirus-mediated increases in expression of AQP4 or AQP1 lead to an increased release of those cytokines.[65] Additionally, it has been shown that water transport, mediated by AQP1, induces IL-1 release from macrophages,[90] although this phenomena has not been studied in the CNS.

In summary, the regulation between AQPs and inflammation works in both directions: inflammatory signals promote the expression of AQP4, and AQP4 facilitates astrocytic activation but seems to limit microglial activation. Given the time-dependent actions of AQP4, the use of anti-inflammatory treatments at early or late time points after acute injuries could differently affect edema formation and resolution through regulation of AQP4 expression.

TREATMENT OF EDEMA: DRUGS AGAINST AQPS

As mentioned previously, there is no specific inhibitor to block the AQP4 channel and such a compound is critical for evaluating the utility of AQP4 as a treatment for edema. Although nonspecific, a range of compounds already commercially available that may block AQP4 have been tested. Bumetanide blocks the AQP4 channel and impairs water permeability in oocytes.[91] Bumetanide decreases AQP4 expression[91] and prevents edema formation after brain ischemia.[91,92] However, bumetanide is also an inhibitor of Na–Cl–K cotransporter expressed in ECs and astrocytes. These multiple sites of action complicate the interpretation of the mechanisms of action of bumetanide, and makes it difficult to tease out the specific effects of AQP4 on edema formation.[92] Acetazolamide (AZA), a sulfonamide carbonic anhydrase inhibitor, was also proposed as an inhibitor of water permeability associated with AQP1 and AQP4.[93,94] However, it was reported that AZA has no effect on water permeability.[95,96] Similarly, two other inhibitors belonging to sulfonamide carbonic anhydrase inhibitor class, methazolamide and valproic acid, have been tested without clear effects on water permeability.[94,97] Recently, Nakada's group developed a PET ligand for AQP4 imaging, which could also be a potent AQP4 inhibitor.[98]

In between, the siRNA strategy has been used with success in normal brain to silence AQP4 expression.[61] The use of siRNA against AQP4 has shown its critical role in the context of jTBI and the contribution of astrocytic AQP4 to changes observed in neuroimaging.[62] Decrease of AQP4 is associated with a decrease of ADC values (Fig. 9.3).[61] siAQP4 was used as treatment in jTBI and resulted in a decrease of edema formation post-TBI and functional improvement up to 2 months postinjury.[62] These results encourage the use of siAQP4 as a specific drug to prevent edema formation; however, given the heterogeneous alterations in AQP4 expression at different time points and in different injury models, further preclinical studies are required to determine in which cases it has beneficial effects and in which ones it is ineffective.

Various cytokines and hormones regulate AQPs, and it is very well established that vasopressin (VP) plays on the level of expression of AQPs. Therefore, VP could be a potential treatment of the edema by modulation of AQP expression (see Chapter 19: Experimental therapies for brain edema and intracranial hypertension). As a more general note, it is important to remember that the division of the edema process in anoxic, ionic, and vasogenic has been proposed in the context of acute brain injuries and may not be adequately adapted for other brain disorders such as brain tumors.[99] In this regard, it is important to underline that clinical treatments solely focusing on osmotic challenges may not be efficient or sufficient for treatment of cerebral edema, because of the complexity and diversity of molecular mechanisms underlying edema formation.

CONCLUSIONS

AQPs are essential for water movement and thus they can be very useful for targeting edema. Current research suggests that AQP1, AQP9 and, especially, AQP4 are the most relevant AQPs for the formation of brain edema. The time-dependent effects of AQP4 highlight the importance of choosing the right time for the use of anti-AQP4 therapeutic approaches. In addition, recent discoveries suggest that AQPs are involved in several other physiological processes such as cell migration and transport of gases. A better understanding of the roles of AQPs is required to make sure that interventions against AQP-mediated edema do not have unwanted side effects. Despite the lack of specific anti-AQP4 channel blockers, the development of siAQP4 provides a valuable tool for targeting the functions of this channel at different time points and anatomical locations.

Statement of Financial Support

This study was supported by grants from the National Institute of child disorders (NICHD) R01HD061946 (J.B.), Fondation des gueules cassées, and eraNET neuron CNSaFlame. ANR-TRAIL (J.B).

References

1. Saadoun S, Papadopoulos MC, Watanabe H, Yan D, Manley GT, Verkman AS. Involvement of aquaporin-4 in astroglial cell migration and glial scar formation. *J Cell Sci* 2005;**118**(Pt 24):5691–8.
2. Herrera M, Hong NJ, Garvin JL. Aquaporin-1 transports NO across cell membranes. *Hypertension* 2006;**48**(1):157–64.
3. Gonen T, Walz T. The structure of aquaporins. *Q Rev Biophys* 2006;**39**(4):361–96.
4. Lee MD, King LS, Agre P. The aquaporin family of water channel proteins in clinical medicine. *Medicine (Baltimore)* 1997;**76**(3):141–56.
5. Isokpehi RD, Rajnarayanan RV, Jeffries CD, Oyeleye TO, Cohly HH. Integrative sequence and tissue expression profiling of chicken and mammalian aquaporins. *BMC Genom* 2009;**10**(Suppl 2):S7.
6. Fukuda AM, Badaut J. Aquaporin 4: a player in cerebral edema and neuroinflammation. *J Neuroinflammation* 2012;**9**:279.
7. Kruse E, Uehlein N, Kaldenhoff R. The aquaporins. *Genome Biol* 2006;**7**(2):206.
8. Badaut J, Fukuda AM, Jullienne A, Petry KG. Aquaporin and brain diseases. *Biochim Biophys Acta* 2014;**1840**(5):1554–65.
9. Rash JE, Davidson KGV, Yasumura T, Furman CS. Freeze-fracture and immunogold analysis of aquaporin-4 (AQP4) square arrays, with models of AQP4 lattice assembly. *Neuroscience* 2004;**129**(4):915–34.
10. Furman CS, Gorelick-Feldman DA, Davidson KG, et al. Aquaporin-4 square array assembly: opposing actions of M1 and M23 isoforms. *Proc Natl Acad Sci USA* 2003;**100**(23):13609–14.
11. Baeten KM, Akassoglou K. Extracellular matrix and matrix receptors in blood–brain barrier formation and stroke. *Dev Neurobiol* 2011;**71**(11):1018–39.
12. Musa-Aziz R, Chen LM, Pelletier MF, Boron WF. Relative CO_2/NH_3 selectivities of AQP1, AQP4, AQP5, AmtB, and RhAG. *Proc Natl Acad Sci USA* 2009;**106**(13):5406–11.
13. Yu J, Yool AJ, Schulten K, Tajkhorshid E. Mechanism of gating and ion conductivity of a possible tetrameric pore in aquaporin-1. *Structure* 2006;**14**(9):1411–23.
14. Oshio K, Binder DK, Yang B, Schecter S, Verkman AS, Manley GT. Expression of aquaporin water channels in mouse spinal cord. *Neuroscience* 2004;**127**(3):685–93.
15. Badaut J, Lasbennes F, Magistretti PJ, Regli L. Aquaporins in brain: distribution, physiology, and pathophysiology. *J Cereb Blood Flow Metab* 2002;**22**(4):367–78.
16. Brown PD, Davies SL, Speake T, Millar ID. Molecular mechanisms of cerebrospinal fluid production. *Neuroscience* 2004;**129**(4):955–68.
17. Shields SD, Mazario J, Skinner K, Basbaum AI. Anatomical and functional analysis of aquaporin 1, a water channel in primary afferent neurons. *Pain* 2007;**131**(1-2):8–20.
18. Fukuda AM, Pop V, Spagnoli D, Ashwal S, Obenaus A, Badaut J. Delayed increase of astrocytic aquaporin 4 after juvenile traumatic brain injury: possible role in edema resolution? *Neuroscience* 2012;**222**:366–78.
19. Arcienega II, Brunet JF, Bloch J, Badaut J. Cell locations for AQP1, AQP4 and 9 in the non-human primate brain. *Neuroscience* 2010;**167**(4):1103–14.
20. Jung JS, Bhat RV, Preston GM, Guggino WB, Baraban JM, Agre P. Molecular characterization of an aquaporin cDNA from brain: candidate osmoreceptor and regulator of water balance. *Proc Natl Acad Sci USA* 1994;**91**(26):13052–6.
21. Nesic O, Lee J, Ye Z, et al. Acute and chronic changes in aquaporin 4 expression after spinal cord injury. *Neuroscience* 2006;**143**(3):779–92.
22. Badaut J, Nehlig A, Verbavatz J, Stoeckel M, Freund-Mercier MJ, Lasbennes F. Hypervascularization in the magnocellular nuclei of the rat hypothalamus: relationship with the distribution of aquaporin-4 and markers of energy metabolism. *J Neuroendocrinol* 2000;**12**(10):960–9.

23. Badaut J, Verbavatz JM, Freund-Mercier MJ, Lasbennes F. Presence of aquaporin-4 and muscarinic receptors in astrocytes and ependymal cells in rat brain: a clue to a common function? *Neurosci Lett* 2000;**292**(2):75–8.
24. Hsu MS, Seldin M, Lee DJ, Seifert G, Steinhauser C, Binder DK. Laminar-specific and developmental expression of aquaporin-4 in the mouse hippocampus. *Neuroscience* 2011;**178**:21–32.
25. Wen H, Nagelhus EA, Amiry-Moghaddam M, Agre P, Ottersen OP, Nielsen S. Ontogeny of water transport in rat brain: postnatal expression of the aquaporin-4 water channel. *Eur J Neurosci* 1999;**11**(3):935–45.
26. Nielsen S, Nagelhus EA, Amiry-Moghaddam M, Bourque C, Agre P, Ottersen OP. Specialized membrane domains for water transport in glial cells: high-resolution immunogold cytochemistry of aquaporin-4 in rat brain. *J Neurosci* 1997;**17**(1):171–80.
27. Badaut J, Hirt L, Granziera C, Bogousslavsky J, Magistretti PJ, Regli L. Astrocyte-specific expression of aquaporin-9 in mouse brain is increased after transient focal cerebral ischemia. *J Cereb Blood Flow Metab* 2001;**21**(5):477–82.
28. Badaut J, Petit JM, Brunet JF, Magistretti PJ, Charriaut-Marlangue C, Regli L. Distribution of Aquaporin 9 in the adult rat brain: preferential expression in catecholaminergic neurons and in glial cells. *Neuroscience* 2004;**128**(1):27–38.
29. de Castro Ribeiro M, Hirt L, Bogousslavsky J, Regli L, Badaut J. Time course of aquaporin expression after transient focal cerebral ischemia in mice. *J Neurosci Res* 2006;**83**(7):1231–40.
30. Tourdias T, Mori N, Dragonu I, et al. Differential aquaporin 4 expression during edema build-up and resolution phases of brain inflammation. *J Neuroinflammation* 2011;**8**:143.
31. Badaut J. Aquaglyceroporin 9 in brain pathologies. *Neuroscience* 2010;**168**(4):1047–57.
32. Dolman D, Drndarski S, Abbott NJ, Rattray M. Induction of aquaporin 1 but not aquaporin 4 messenger RNA in rat primary brain microvessel endothelial cells in culture. *J Neurochem* 2005;**93**(4):825–33.
33. Hirt L, Ternon B, Price M, Mastour N, Brunet JF, Badaut J. Protective role of early aquaporin 4 induction against postischemic edema formation. *J Cereb Blood Flow Metab* 2009;**29**(2):423–33.
34. Suzuki M, Iwasaki Y, Yamamoto T, Konno H, Yoshimoto T, Suzuki J. Disintegration of orthogonal arrays in perivascular astrocytic processes as an early event in acute global ischemia. *Brain Res* 1984;**300**(1):141–5.
35. Nicchia GP, Srinivas M, Li W, Brosnan CF, Frigeri A, Spray DC. New possible roles for aquaporin-4 in astrocytes: cell cytoskeleton and functional relationship with connexin43. *FASEB J* 2005;**19**(12):1674–6.
36. Nagelhus EA, Mathiisen TM, Ottersen OP. Aquaporin-4 in the central nervous system: cellular and subcellular distribution and coexpression with KIR4.1. *Neuroscience* 2004;**129**(4):905–13.
37. Dibaj P, Kaiser M, Hirrlinger J, Kirchhoff F, Neusch C. Kir4.1 channels regulate swelling of astroglial processes in experimental spinal cord edema. *J Neurochem* 2007;**103**(6):2620–8.
38. Binder DK, Yao X, Zador Z, Sick TJ, Verkman AS, Manley GT. Increased seizure duration and slowed potassium kinetics in mice lacking aquaporin-4 water channels. *Glia* 2006;**53**(6):631–6.
39. Niermann H, Amiry-Moghaddam M, Holthoff K, Witte OW, Ottersen OP. A novel role of vasopressin in the brain: modulation of activity-dependent water flux in the neocortex. *J Neurosci* 2001;**21**(9):3045–51.
40. Zhang H, Verkman AS. Aquaporin-4 independent Kir4.1 K+ channel function in brain glial cells. *Mol Cell Neurosci* 2008;**37**(1):1–10.
41. Iliff JJ, Wang M, Liao Y, et al. A paravascular pathway facilitates CSF flow through the brain parenchyma and the clearance of interstitial solutes, including amyloid beta. *Sci Transl Med* 2012;**4**(147):147ra111.

42. Iliff JJ, Wang M, Zeppenfeld DM, et al. Cerebral arterial pulsation drives paravascular CSF-interstitial fluid exchange in the murine brain. *J Neurosci* 2013;**33**(46): 18190–9.
43. Binder DK, Papadopoulos MC, Haggie PM, Verkman AS. In vivo measurement of brain extracellular space diffusion by cortical surface photobleaching. *J Neurosci* 2004;**24**(37):8049–56.
44. Badaut J, Brunet JF, Grollimund L, et al. Aquaporin 1 and aquaporin 4 expression in human brain after subarachnoid hemorrhage and in peritumoral tissue. *Acta Neurochir Suppl* 2003;**86**:495–8.
45. Yatsushige H, Ostrowski RP, Tsubokawa T, Colohan A, Zhang JH. Role of c-Jun N-terminal kinase in early brain injury after subarachnoid hemorrhage. *J Neurosci Res* 2007;**85**(7):1436–48.
46. Nesic O, Lee J, Unabia GC, et al. Aquaporin 1—a novel player in spinal cord injury. *J Neurochem* 2008;**105**(3):628–40.
47. Hwang IK, Yoo KY, Li H, et al. Aquaporin 9 changes in pyramidal cells before and is expressed in astrocytes after delayed neuronal death in the ischemic hippocampal CA1 region of the gerbil. *J Neurosci Res* 2007;**85**(11):2470–9.
48. Higashida T, Kreipke CW, Rafols JA, et al. The role of hypoxia-inducible factor-1alpha, aquaporin-4, and matrix metalloproteinase-9 in blood–brain barrier disruption and brain edema after traumatic brain injury. *J Neurosurg* 2011;**114**(1):92–101.
49. Tomura S, Nawashiro H, Otani N, et al. Effect of decompressive craniectomy on aquaporin-4 expression after lateral fluid percussion injury in rats. *J Neurotrauma* 2011;**28**(2):237–43.
50. Ding JY, Kreipke CW, Speirs SL, Schafer P, Schafer S, Rafols JA. Hypoxia-inducible factor-1alpha signaling in aquaporin upregulation after traumatic brain injury. *Neurosci Lett* 2009;**453**(1):68–72.
51. Guo Q, Sayeed I, Baronne LM, Hoffman SW, Guennoun R, Stein DG. Progesterone administration modulates AQP4 expression and edema after traumatic brain injury in male rats. *Exp Neurol* 2006;**198**(2):469–78.
52. Sun MC, Honey CR, Berk C, Wong NL, Tsui JK. Regulation of aquaporin-4 in a traumatic brain injury model in rats. *J Neurosurg* 2003;**98**(3):565–9.
53. Zhao J, Moore AN, Clifton GL, Dash PK. Sulforaphane enhances aquaporin-4 expression and decreases cerebral edema following traumatic brain injury. *J Neurosci Res* 2005;**82**(4):499–506.
54. Ke C, Poon WS, Ng HK, Pang JC, Chan Y. Heterogeneous responses of aquaporin-4 in oedema formation in a replicated severe traumatic brain injury model in rats. *Neurosci Lett* 2001;**301**(1):21–4.
55. Kiening KL, van Landeghem FK, Schreiber S, et al. Decreased hemispheric Aquaporin-4 is linked to evolving brain edema following controlled cortical impact injury in rats. *Neurosci Lett* 2002;**324**(2):105–8.
56. Meng S, Qiao M, Lin L, Del Bigio MR, Tomanek B, Tuor UI. Correspondence of AQP4 expression and hypoxic-ischaemic brain oedema monitored by magnetic resonance imaging in the immature and juvenile rat. *Eur J Neurosci* 2004;**19**(8):2261–9.
57. Ge S, Song L, Pachter JS. Where is the blood–brain barrier … really? *J Neurosci Res* 2005;**79**(4):421–7.
58. Manley GT, Fujimura M, Ma T, et al. Aquaporin-4 deletion in mice reduces brain edema after acute water intoxication and ischemic stroke. *Nat Med* 2000;**6**(2):159–63.
59. Yao X, Hrabetova S, Nicholson C, Manley GT. Aquaporin-4-deficient mice have increased extracellular space without tortuosity change. *J Neurosci* 2008;**28**(21):5460–4.
60. Zador Z, Magzoub M, Jin S, Manley GT, Papadopoulos MC, Verkman AS. Microfiberoptic fluorescence photobleaching reveals size-dependent macromolecule diffusion in extracellular space deep in brain. *FASEB J* 2008;**22**(3):870–9.

61. Badaut J, Ashwal S, Adami A, et al. Brain water mobility decreases after astrocytic aquaporin-4 inhibition using RNA interference. *J Cereb Blood Flow Metab* 2011;**31**(3): 819–31.
62. Fukuda AM, Adami A, Pop V, et al. Posttraumatic reduction of edema with aquaporin-4 RNA interference improves acute and chronic functional recovery. *J Cereb Blood Flow Metab* 2013.
63. Risher WC, Andrew RD, Kirov SA. Real-time passive volume responses of astrocytes to acute osmotic and ischemic stress in cortical slices and in vivo revealed by two-photon microscopy. *Glia* 2009;**57**(2):207–21.
64. Friedman B, Schachtrup C, Tsai PS, et al. Acute vascular disruption and aquaporin 4 loss after stroke. *Stroke* 2009;**40**(6):2182–90.
65. Li L, Zhang H, Varrin-Doyer M, Zamvil SS, Verkman AS. Proinflammatory role of aquaporin-4 in autoimmune neuroinflammation. *FASEB J* 2011;**25**(5):1556–66.
66. Saadoun S, Papadopoulos MC. Aquaporin-4 in brain and spinal cord oedema. *Neuroscience* 2010;**168**(4):1036–46.
67. Kimura A, Hsu M, Seldin M, Verkman AS, Scharfman HE, Binder DK. Protective role of aquaporin-4 water channels after contusion spinal cord injury. *Ann Neurol* 2010;**67**(6):794–801.
68. Saadoun S, Bell BA, Verkman AS, Papadopoulos MC. Greatly improved neurological outcome after spinal cord compression injury in AQP4-deficient mice. *Brain* 2008;**131** (Pt 4):1087–98.
69. Verkman AS, Binder DK, Bloch O, Auguste K, Papadopoulos MC. Three distinct roles of aquaporin-4 in brain function revealed by knockout mice. *Biochim Biophys Acta* 2006
70. Papadopoulos MC, Manley GT, Krishna S, Verkman AS. Aquaporin-4 facilitates reabsorption of excess fluid in vasogenic brain edema. *FASEB J* 2004;**18**(11):1291–3.
71. Badaut J, Ashwal S, Obenaus A. Aquaporins in cerebrovascular disease: a target for treatment of brain edema? *Cerebrovasc Dis* 2011;**31**(6):521–31.
72. Tourdias T, Dragonu I, Fushimi Y, et al. Aquaporin 4 correlates with apparent diffusion coefficient and hydrocephalus severity in the rat brain: a combined MRI-histological study. *Neuroimage* 2009;**47**(2):659–66.
73. Nicchia GP, Mastrototaro M, Rossi A, et al. Aquaporin-4 orthogonal arrays of particles are the target for neuromyelitis optica autoantibodies. *Glia* 2009;**57**(13):1363–73.
74. Fernandez-Teruel A, Gimenez-Llort L, Escorihuela RM, et al. Early-life handling stimulation and environmental enrichment: are some of their effects mediated by similar neural mechanisms? *Pharmacol Biochem Behav* 2002;**73**(1):233–45.
75. Badaut J, Ashwal S, Tone B, Regli L, Tian HR, Obenaus A. Temporal and regional evolution of aquaporin-4 expression and magnetic resonance imaging in a rat pup model of neonatal stroke. *Pediatr Res* 2007;**62**(3):248–54.
76. Hiroaki Y, Tani K, Kamegawa A, et al. Implications of the aquaporin-4 structure on array formation and cell adhesion. *J Mol Biol* 2006;**355**(4):628–39.
77. Pop V, Sorensen DW, Kamper JE, et al. Early brain injury alters the blood–brain barrier phenotype in parallel with beta-amyloid and cognitive changes in adulthood. *J Cereb Blood Flow Metab* 2013;**33**(2):205–14.
78. Wilcock DM, Vitek MP, Colton CA. Vascular amyloid alters astrocytic water and potassium channels in mouse models and humans with Alzheimer's disease. *Neuroscience* 2009;**159**(3):1055–69.
79. Ito H, Yamamoto N, Arima H, et al. Interleukin-1beta induces the expression of aquaporin-4 through a nuclear factor-kappaB pathway in rat astrocytes. *J Neurochem* 2006;**99**(1):107–18.
80. Giles JA, Greenhalgh AD, Davies CL, et al. Requirement for interleukin-1 to drive brain inflammation reveals tissue-specific mechanisms of innate immunity. *Eur J Immunol* 2015;**45**(2):525–30.

81. Ohnishi M, Monda A, Takemoto R, et al. High-mobility group box 1 up-regulates aquaporin 4 expression via microglia–astrocyte interaction. *Neurochem Int* 2014;**75**:32–8.
82. Laird MD, Shields JS, Sukumari-Ramesh S, et al. High mobility group box protein-1 promotes cerebral edema after traumatic brain injury via activation of toll-like receptor 4. *Glia* 2014;**62**(1):26–38.
83. Auguste KI, Jin S, Uchida K, et al. Greatly impaired migration of implanted aquaporin-4-deficient astroglial cells in mouse brain toward a site of injury. *FASEB J* 2007;**21**(1):108–16.
84. Lu DC, Zador Z, Yao J, Fazlollahi F. Manley GTMDPD. Aquaporin-4 reduces post-traumatic seizure susceptibility by promoting astrocytic glial scar formation in mice. *J Neurotrauma* 2011.
85. Saubamea B, Cochois-Guegan V, Cisternino S, Scherrmann JM. Heterogeneity in the rat brain vasculature revealed by quantitative confocal analysis of endothelial barrier antigen and P-glycoprotein expression. *J Cereb Blood Flow Metab* 2012;**32**(1):81–92.
86. Fukuda AM, Adami A, Pop V, et al. Posttraumatic reduction of edema with aquaporin-4 RNA interference improves acute and chronic functional recovery. *J Cereb Blood Flow Metab* 2013;**33**(10):1621–32.
87. Eder C, Klee R, Heinemann U. Involvement of stretch-activated Cl-channels in ramification of murine microglia. *J Neurosci* 1998;**18**(18):7127–37.
88. Schlichter LC, Mertens T, Liu B. Swelling activated Cl-channels in microglia: biophysics, pharmacology and role in glutamate release. *Channels* 2011;**5**(2):128–37.
89. Liu P, Jing Y, Collie ND, Campbell SA, Zhang H. Pre-aggregated Abeta (25–35) alters arginine metabolism in the rat hippocampus and prefrontal cortex. *Neuroscience* 2011;**193**:269–82.
90. Rabolli V, Wallemme L, Lo Re S, et al. Critical role of aquaporins in interleukin 1β (IL-1β)-induced inflammation. *J Biol Chem* 2014;**289**(20):13937–47.
91. Migliati E, Meurice N, DuBois P, et al. Inhibition of aquaporin-1 and aquaporin-4 water permeability by a derivative of the loop diuretic bumetanide acting at an internal pore-occluding binding site. *Mol Pharmacol* 2009;**76**(1):105–12.
92. O'Donnell ME, Tran L, Lam TI, Liu XB, Anderson SE. Bumetanide inhibition of the blood-brain barrier Na–K–Cl cotransporter reduces edema formation in the rat middle cerebral artery occlusion model of stroke. *J Cereb Blood Flow Metab* 2004;**24**(9):1046–56.
93. Huber VJ, Tsujita M, Yamazaki M, Sakimura K, Nakada T. Identification of arylsulfonamides as Aquaporin 4 inhibitors. *Bioorg Med Chem Lett* 2007;**17**(5):1270–3.
94. Tanimura Y, Hiroaki Y, Fujiyoshi Y. Acetazolamide reversibly inhibits water conduction by aquaporin-4. *J Struct Biol* 2009;**166**(1):16–21.
95. Yang B, Zhang H, Verkman AS. Lack of aquaporin-4 water transport inhibition by antiepileptics and arylsulfonamides. *Bioorg Med Chem* 2008;**16**(15):7489–93.
96. Sogaard R, Zeuthen T. Test of blockers of AQP1 water permeability by a high-resolution method: no effects of tetraethylammonium ions or acetazolamide. *Pflugers Arch* 2008;**456**(2):285–92.
97. Huber VJ, Tsujita M, Kwee IL, Nakada T. Inhibition of aquaporin 4 by antiepileptic drugs. *Bioorg Med Chem* 2009;**17**(1):418–24.
98. Nakamura Y, Suzuki Y, Tsujita M, Huber VJ, Yamada K, Nakada T. Development of a novel ligand, [C]TGN-020, for aquaporin 4 positron emission tomography imaging. *ACS Chem Neurosci* 2011;**2**(10):568–71.
99. Nico B, Ribatti D. Role of aquaporins in cell migration and edema formation in human brain tumors. *Exp Cell Res* 2011;**317**(17):2391–6.

CHAPTER

10

Sur1–Trpm4-Promising Target for Brain Edema Treatment in Brain Edema: From Molecular Mechanisms to Clinical Practice

Sebastian Urday[1], *Kevin N. Sheth*[1] *and J. Marc Simard*[2]

[1]Yale University School of Medicine, New Haven, CT, United States
[2]University of Maryland School of Medicine, Baltimore, MD, United States

OUTLINE

Brain Edema.
DOI: http://dx.doi.org/10.1016/B978-0-12-803196-4.00010-2

INTRODUCTION

Brain edema is a common clinical problem after acute neurologic injury and is responsible for significant morbidity and mortality.[1,2] Despite the clinical relevance of brain edema, its current treatment is nonspecific and suboptimal.[1] Furthermore, current therapies (e.g., sedation, osmotherapy) are reactive, instituted after swelling becomes symptomatic.[3] A more effective approach is to deliver therapies that reduce or prevent edema formation.[4]

A growing body of evidence indicates that the sulfonylurea receptor 1–transient receptor potential M4 (SUR1–TRPM4) cation channel is a critical mediator of brain edema development across various acute neurocritical care conditions. It contributes to every type of edema (cytotoxic, ionic, and vasogenic edema), neuroinflammation, and to total microvascular failure, which results in the extravasation of blood and the formation of petechial hemorrhages.[3] In this chapter, we first examine the biological properties of SUR1–TRPM4. We then discuss the key role of SUR1–TRPM4 in the pathogenesis of brain edema and microvascular dysfunction following ischemic stroke, intracerebral hemorrhage (ICH), subarachnoid hemorrhage (SAH), traumatic brain injury (TBI), and hemorrhagic encephalopathy of prematurity (EP). Our goal is to highlight that pharmacological inhibition of SUR1–TRPM4 is a highly promising approach to ameliorate brain edema in these conditions.

SUR1–TRPM4

SUR1–TRPM4 is a novel ion channel complex formed from two subunits, a regulatory subunit, SUR1, and a pore-forming subunit, TRPM4 (Fig. 10.1).[5,6] SUR1–TRPM4 is normally absent and is transcriptionally upregulated de novo in all members of the neurovascular unit (astrocytes, oligodendrocytes, neurons, endothelial cells) in the setting of acute brain injury.[7–9] The main function of the SUR1–TRPM4 channel is to protect cells from excessive calcium influx during pathological conditions.[7] An excessive rise in intracellular calcium is one of the hallmarks of neurologic injury, resulting in calcium-dependent cell death cascades.[3,7,10,11] The entry of calcium into a cell is determined by the electrochemical gradient for calcium, and the electrical portion of that gradient is determined by the cell-membrane potential.[7] Depolarization decreases the inward driving force for calcium, thereby decreasing its influx.[7] SUR1–TRPM4 is ideally designed to regulate calcium influx.[7] First, because it conducts monovalent cations nonselectively, its activation results in cell depolarization.[7] Second, because it requires calcium for opening, the cell-membrane potential becomes directly linked to the intracellular calcium concentration.[7]

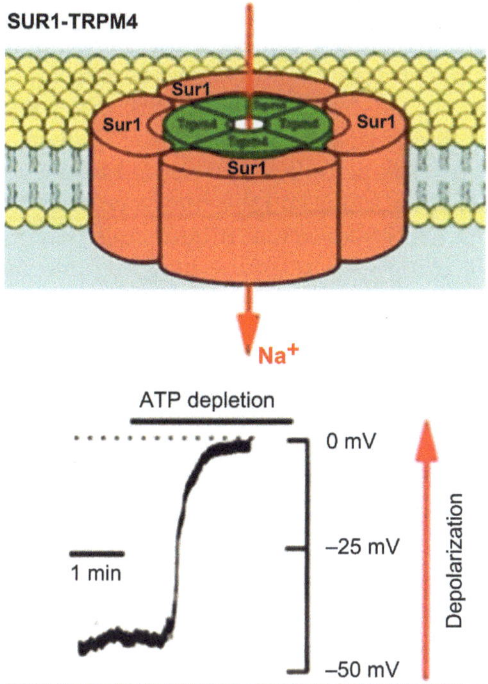

FIGURE 10.1 **Schematic diagram of the SUR1–TRPM4 channel.** SUR1–TRPM4 channels are formed by coassembly of Sur1 and transient receptor potential melastatin 4 (Trpm4). The principal physiologic action of the ion channel complex when activated by ATP depletion is shown: inward flux of Na^+ via the nonselective monovalent cation pore-forming subunit, resulting in depolarization. Abbreviations: *SUR1*, sulfonylurea receptor 1; *TRPM4*, transient receptor potential cation channel subfamily M member 4. *Source: Adapted from Simard JM, Woo SK, Schwartzbauer GT, Gerzanich V. Sulfonylurea receptor 1 in central nervous system injury: a focused review. J Cereb Blood Flow Metab. 2012;32:1699–717, with permission.*

Although initially protective, sustained channel opening in the setting of brain injury is deleterious.[7] Critical depletion of adenosine triphosphate (ATP), which is common after acute brain injury, leads to unchecked channel opening resulting in a strong inward current, widespread cell depolarization, and oncotic cell swelling (cytotoxic edema).[7] When this process occurs in endothelial cells, a process of progressive microvascular dysfunction ensues.[7] This results in brain edema and culminates in the formation of petechial hemorrhages if there is complete structural loss of endothelium.[7–9]

Glibenclamide (US adopted name, glyburide), a second-generation sulfonylurea, blocks the SUR1–TRPM4 channel complex with high affinity and specificity (EC_{50}, 48 nM).[12] The inhibitory effect of glibenclamide is due to prolongation of, and an increase in the probability of long closed states of the channel.[7] Normally, glibenclamide does not accumulate in the brain.[13] However, penetration of glibenclamide into the brain is enhanced after injury.[13] Both blood–brain barrier (BBB) disruption and the relatively low-pH environment induced by tissue damage facilitate uptake of glibenclamide (a weak acid) into the injured brain.[13] As a result, relatively low doses of drug (nonhypoglycemogenic) can be used to achieve a favorable therapeutic effect, while minimizing potential side effects.[13] Selective blockade using glibenclamide has provided investigators with an invaluable tool to understand SUR1–TRPM1 function in preclinical and human studies.

SUR1–TRPM4 expression is upregulated progressively over several hours after the onset of brain injury, which provides an extended time window for therapeutic intervention.[7] For example, in a rat model of cerebral ischemia induced by permanent mechanical middle cerebral artery occlusion (MCAo), 3 hours are required before Abcc8 (which encodes SUR1) is increased by 2.5-fold, and 8 hours are required before Sur1 protein is increased 2.5-fold.[7,14] In a model with 4.5 hours of ischemia followed by a 30-minute infusion of recombinant tissue plasminogen activator (rt-PA), Sur1 upregulation increased 7- to 11-fold over the subsequent 19 hours after infusion.[7,15] Similar observations have been made after other acute brain injuries. For instance, in a rat model of cortical impact TBI, Sur1 upregulation peaked at 12 hours after injury.[7,16]

SUR1–TRPM4 in the Pathogenesis of Ischemic Brain Edema

Our discussion of ischemic brain edema provides a framework to understand the critical role of SUR1–TRPM4 in microvascular failure after neurologic injury. Edema development can be conceptualized as a stepwise temporal progression from cytotoxic edema to ionic edema, and finally to vasogenic edema (Fig. 10.2).[2] If microvascular dysfunction is profound enough, catastrophic structural failure of capillaries occurs (i.e., hemorrhagic transformation, HT).[13] Although several chemical mediators are involved in the pathophysiology of ischemic brain edema, SUR1–TRPM4 plays a key role in every step, as evidenced by the salutary effects of inhibiting this channel with glibenclamide.[13]

The initial type of edema following ischemia is cytotoxic edema.[1] Focal brain ischemia stimulates the upregulation of SUR1–TRPM4 in all cells of the neurovascular unit.[7] ATP depletion is the tipping point, causing this channel to open and leading to the inflow of Na^+ ions from the extracellular to the intracellular space.[17] Chloride follows the influx of sodium through chloride channels, and water follows via aquaporin water channels (see Chapter 11: Role of Matrix Metalloproteinases in Brain Edema) to

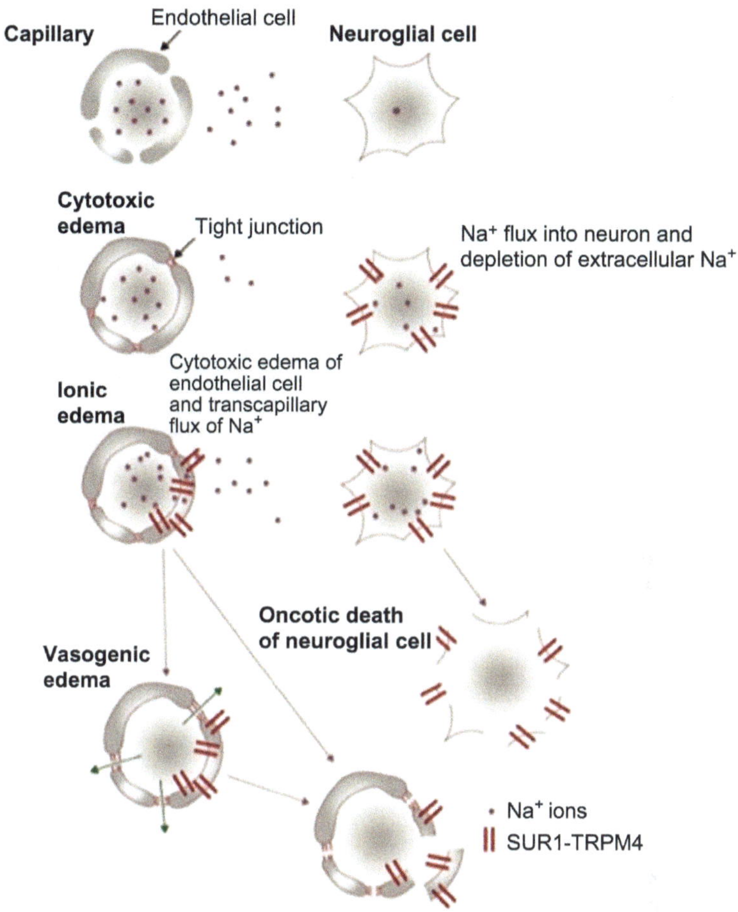

FIGURE 10.2 **Schematic diagram illustrating the role of SUR1–TRPM4 in ischemia-triggered cytotoxic, ionic, vasogenic edema, and hemorrhagic transformation.** Normally, Na^+ concentrations in serum and in extracellular space are the same, and much higher than inside neuroglial cells. Cytotoxic edema of neuroglial cells is due to entry of Na^+ into ischemic neuroglial cells via SUR1–TRPM4, depleting extracellular Na^+ and thereby setting up a concentration gradient between intravascular and extracellular compartments. Ionic edema results from cytotoxic edema of endothelial cells, due to expression of SUR1–TRPM4 on both the apical and basolateral side, allowing Na^+ from the intravascular compartment to traverse the capillary wall and replenish Na^+ in the extracellular space. Vasogenic edema results from degradation of tight junctions between endothelial cells, transforming capillaries into "fenestrated" capillaries that allow extravasation (outward filtration) of proteinacious fluid. Oncotic death of neuroglial cells is the ultimate consequence of cytotoxic edema. Oncotic death of endothelial cells results in complete loss of capillary integrity and in extravasation of blood—i.e., hemorrhagic transformation. Abbreviations: *SUR1*, sulfonylurea receptor 1; *TRPM4*, transient receptor potential cation channel subfamily M member 4. *Source: Adapted from Simard JM, Kent TA, Chen M, Tarasov KV, Gerzanich V. Brain oedema in focal ischaemia: molecular pathophysiology and theoretical implications. Lancet Neurol. 2007;6:258–68, with permission.*

maintain electrical and osmotic neutrality, respectively.[17] Cytotoxic edema does not directly cause tissue swelling, but creates a new transendothelial gradient of sodium that contributes to the osmotic force that drives ionic edema.[17] SUR1–TRPM4 is upregulated in both apical and basolateral endothelial cell membranes, thus allowing the transcapillary flux of sodium (and, consequently, water) down its concentration gradient into the interstitium, causing ionic edema.[18] Ionic edema is the earliest phase of endothelial dysfunction triggered by ischemia and precedes vasogenic edema by ~6 hours.[1] Although multiple mediators are involved (e.g., thrombin) in the progressive breakdown of the BBB that results in vasogenic edema, SUR1–TRPM4 plays a critical role because cytotoxic edema perturbs endothelial cell volume.[18] This deranges the actin cytoskeleton and weakens intercellular tight junctions, leading to an increase in BBB permeability and vasogenic edema.[18,19] Increased BBB permeability and oncotic death of endothelial cells leads to the final step from vasogenic edema to HT.[18] Thus, SUR1–TRPM4 upregulation and activation plays a central role in each stage of endothelial dysfunction from cytotoxic edema to HT. A key point is that the extravasation of blood (HT) represents the end stage in the progressive process of microvascular failure. In some conditions (e.g., contusive TBI), the formation of petechial hemorrhages is the main manifestation that causes neurologic injury. The model described above has been shown to operate in all acute brain injuries discussed in this chapter.

Animal Models

The therapeutic potential of glibenclamide (SUR1–TRPM4 inhibition) for ameliorating ischemic brain edema and HT has been well documented in preclinical studies. Investigators have examined various rodent models of stroke, including nonlethal models of thromboembolic, permanent and temporary occlusion, as well as models of malignant cerebral edema.[20] In these models, SUR1–TRPM4 inhibition results in improvements in edema, brain swelling, lesion size, white matter preservation, neurological function, and mortality.[20] A few selected examples are described below.

In a rodent model of massive ischemic stroke with malignant cerebral edema associated with high mortality (68%), glibenclamide reduced mortality, and cerebral edema by half.[14,21] Notably, in another model of malignant infarction, low-dose glibenclamide administered 6 hours after onset of ischemia led to a reduction in brain swelling sufficient to prevent mortality as effectively as decompressive craniectomy (DC).[4] The fact that glibenclamide was as effective as DC in eliminating mortality illustrates the critical role of SUR1–TRPM4 in brain edema formation.[4] Significantly, neurological function over 2 weeks after treatment was significantly better with glibenclamide compared to DC.[4] This finding demonstrates that preventing swelling is preferable to decompressing the already swollen brain.[4] Gross pathological examination after ischemic injury reveals better functional tissue perfusion and preservation with SUR1–TRPM4 inhibition.[18] For example, significantly more cortical sparing was seen in animals treated with glibenclamide compared to controls.[14,18] It is likely that edema reduction facilitates leptomeningeal collateral blood flow, one of the key determinants of clinical outcome in humans with large vessel occlusion.[14,18]

Independent confirmation that glibenclamide reduces ischemic brain edema was shown in a permanent MCAo rat model of stroke.[22] Wali et al. found that continuous low-dose glibenclamide administration after stroke significantly reduces infarct volume, hemispheric swelling, and neurologic deficits.[22] Treatment with glibenclamide decreased hemispheric swelling dramatically (>50%) compared to the vehicle-treated group.[22]

HT is a major complication of ischemic stroke, particularly in those patients treated with rt-PA.[20] Matrix metalloproteinases (MMPs, see Chapter 13: Brain Edema Formation in Traumatic Brain Injury) are extracellular proteases that are upregulated after cerebral ischemia and play a leading role in HT.[20,23–25] MMP-9 protein is upregulated 10-fold in the infarct core compared to contralateral tissues.[26] There is evidence that MMP-9 activity in cerebral ischemia is increased by rt-PA.[27] For example, with 2- to 3-hour mechanical MCAo, administration of rt-PA at the time of recanalization results in a 2- to 2.5-fold increase in MMP-9 activity above levels observed with ischemia alone.[20,28] Importantly, treatment with glibenclamide reduces both MMP-9 activity and symptomatic HT.[15] In a rat model of malignant cerebral edema, administration of glibenclamide at the time of recanalization significantly reduced brain swelling and symptomatic HT and improved neurological scores both acutely and over the course of 2 weeks of observation.[15] These studies suggest that SUR1–TRPM4 inhibition may complement rt-PA in patients with acute ischemic stroke.

Retrospective Clinical Studies

The exciting results of SUR1–TRPM4 inhibition in preclinical stroke models are supported by retrospective clinical studies of diabetic patients who presented with ischemic stroke. A study by Kunte et al. suggested that patients maintained on sulfonylureas throughout their hospitalization achieved greater functional independence and lower NIH stroke scale scores, compared with controls.[29] Furthermore, a follow-up study revealed that no patients maintained on sulfonylureas experienced symptomatic HT or death, compared with controls (11% and 10%, respectively).[30] Independent investigators using different cohorts have corroborated these findings.[31] In support of the proposed beneficial effect of sulfonylurea therapy, brain biopsy specimens obtained from a patient with symptomatic HT showed prominent upregulation of SUR1 in microvessels and neurons.[30]

Prospective Clinical Studies

The preclinical and retrospective clinical studies documenting the potential for glibenclamide to preempt edema development and improve patient outcome after ischemic stroke have led to prospective trials.[3] These prospective trials are evaluating the therapeutic effects of RP-1127, an IV formulation of glibenclamide.[3]

Recently, a Phase IIa study was completed that evaluated the potential benefit of RP-1127 in 10 patients with a severe anterior circulation ischemic stroke at high risk for malignant cerebral edema (Clinicaltrials.gov Identifier: NCT01268683).[3,32–34] Compared to an untreated comparison cohort, glibenclamide reduced the incidence of malignant edema, and only 2/10 patients required osmotherapy, intubation, or DC.[3] Clinically significant parenchymal hematomas were reduced from an anticipated rate of ~30% to 0%.[3] Compared to a pooled analysis of DC trials, the proportion of patients with 30-day modified Rankin Scale scores ≤ 4 was increased with RP-1127 therapy.[3] Significantly, glibenclamide treatment was associated with a decreased serum MMP-9 level compared with controls.[33] This finding is consistent with preclinical data and supports the notion that glibenclamide reduces vasogenic edema and HT.[33] Currently, a larger randomized, placebo-controlled trial is underway to demonstrate the safety and efficacy of IV RP-1127 following severe anterior circulation ischemic stroke (Clinicaltrials.gov Identifier: NCT01794182).[3]

SUR1–TRPM4 in the Pathogenesis of Perihematomal Edema

Perihematomal edema (PHE) formation starts with the presence of serum from both the primary and secondary hemorrhages in the interstitium of the brain.[35] Initially, the BBB is still intact and ionic edema accumulates due to the transcapillary flux of electrolytes and water from the intravascular to the interstitial compartment.[35] The osmotic gradients that drive transendothelial water flux are generated by clot retraction and cytotoxic edema of neuroglial cells.[35] Clot retraction is a unique aspect of PHE that is absent in other types of edema, such as that caused by cerebral ischemia.[35] Clot retraction results from activation of the coagulation cascade that limits bleeding at the site of injury; during this process, serum proteins are extruded from the hematoma into the interstitium, which raises the interstitial osmotic pressure.[35,36] The inflammatory response to ICH and iron derived from hemolysis then lead to the progressive breakdown of the BBB, resulting in vasogenic edema.[35] PHE progresses fastest in the first 2–3 days, peaks at ~2 weeks, and typically resolves by 1 month.[35,37–39]

There is preliminary evidence that SUR1–TRPM4 also plays a key role in the pathogenesis of PHE by contributing to ionic edema.[35] SUR1–TRPM4

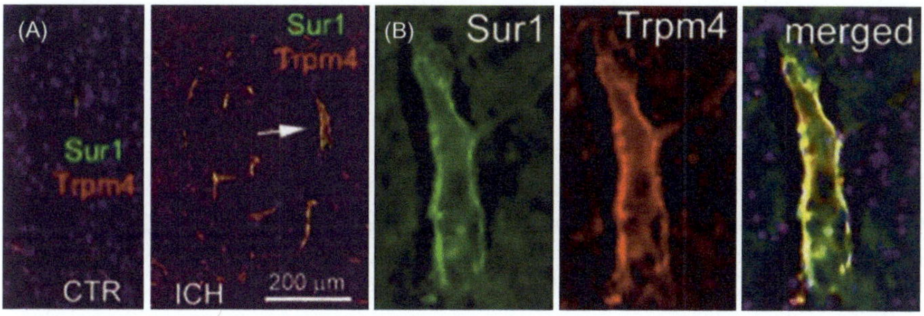

FIGURE 10.3 **Expression of SUR1–TRPM4 in perihematomal tissues following primary spontaneous ICH.** (A) Healthy control (left) and perihematomal tissues obtained by autopsy 3 days after the ictus (right). Tissues are immunolabeled for SUR1 (green) and TRPM4 (red); superimposed images show minimal expression of both proteins in control tissues, but prominent expression of SUR1–TRPM4 after ICH in elongated structures that have the appearance of microvessels. Arrow indicates microvessel that is shown at higher magnification in part B. (B) High-magnification image of microvessel immunolabeled for SUR1 (green, left) and TRPM4 (red, middle). Superimposition of the two images shows coexpression of the two proteins (yellow, right). Abbreviations: *ICH*, intracerebral hemorrhage; *SUR1*, sulfonylurea receptor 1; *TRPM4*, transient receptor potential cation channel subfamily M member 4. *Source: From Urday S, Kimberly WT, Beslow LA, Vortmeyer AO, Selim MH, Rosand J, et al. Targeting secondary injury in intracerebral haemorrhage-perihaematomal oedema. Nat Rev Neurol. 2015;11:111–22, with permission.*

may be upregulated in both apical and basolateral endothelial membranes by hemorrhage-induced inflammation, thus allowing the transcapillary flux of sodium down its concentration gradient into the interstitium if ATP becomes depleted.[12,35,40,41] Nuclear factor κB (NFκB), which mediates the inflammatory response to ICH, induces the transcription of SUR1–TRPM4 (discussed below in the context of SAH).[40] A role for this channel in PHE is further supported by the findings that autopsy specimens from patients with both primary and secondary ICH (due to rupture of an arteriovenous malformation) show upregulation of SUR1 in microvessels (Fig. 10.3). Findings thus far are promising and suggest that SUR1–TRPM4 inhibition could be a therapeutic approach to reduce PHE. This will be explored in future preclinical studies.

SUR1–TRPM4 in the Pathogenesis of Edema After Subarachnoid Hemorrhage

SUR1–TRPM4 channels are newly synthesized in tissues adjacent to experimental SAH and contribute to cytotoxic edema of endothelial cells, vasogenic edema, neuroinflammation, and poor behavioral outcomes.[40,41] Preclinical studies demonstrate that the inflammatory response to SAH induces the upregulation of SUR1–TRPM4 in neighboring cortex and vessels.[40,41] The 5′-flanking region of the Abcc8 promoter from rat and human contains at least 2 consensus NFκB-binding sites.[40] Experiments with cultured brain endothelial cells exposed to tumor necrosis factor α (TNFα) show upregulation of Abcc8 mRNA and SUR1 protein, consistent with the hypothesis that TNFα/NFκB stimulates de novo expression of SUR1 after hemorrhage.[40] Overall, a model emerges whereby inflammation causes perihemorrhagic upregulation of SUR1–TRPM4 channels.[40] Opening of these channels in endothelial cells results in sodium influx, endothelial cell swelling (cytotoxic edema) and actin cytoskeletal rearrangement, loss of endothelial tight junction integrity, and formation of vasogenic edema.[40] A key corollary of this model is that SUR1–TRPM4 expression is not only induced by inflammation, but also potentiates inflammation by mediating BBB breakdown and vasogenic edema.[40] Vasogenic edema is injurious and causes activation of microglia and infiltration of neutrophils.[40]

Several lines of evidence support the critical role of SUR1–TRPM4 in edema formation after SAH. Using a model of mild–moderate SAH, Simard et al. documented prominent protein extravasation in cortex adjacent to the hemorrhage.[40] SUR1–TRPM4 inhibition with glibenclamide resulted in a significant reduction in protein extravasation, reflecting a significant attenuation in the SAH-induced alteration in BBB permeability and resultant vasogenic edema.[40] Furthermore, experimental SAH disrupted the normal junctional localization of the tight junction protein, zona occludens (ZO) 1, and glibenclamide significantly reduced ZO-1 abnormalities, resulting in less edema formation.[13,40] Consistent with the proinflammatory role of SUR1–TRPM4 postulated above, glibenclamide treatment reduces TNFα expression in adjacent cortical cells ~4-fold.[41] Follow-up experiments evaluated the role of SUR1–TRPM4 in a rat model of entorhinal SAH, which is well suited for the investigation of cognitive deficits in this condition.[41] The authors showed that rats treated with glibenclamide for 7 days after SAH exhibited better spatial learning and memory functions compared with vehicle-treated controls.[41] Notably, the investigators also showed evidence of SUR1–TRPM4 upregulation in human autopsy brains with SAH.[41] Altogether, these studies suggest that SUR1–TRPM4 inhibition may be a valuable therapeutic strategy to ameliorate vasogenic edema and improve outcomes in patients with SAH.

SUR1–TRPM4 in the Pathogenesis of Edema and Hemorrhagic Progression of a Contusion After Traumatic Brain Injury

One of the most severe forms of TBI is contusive brain injury.[42] A frequent finding after contusive brain injury is the appearance of new hemorrhagic lesions hours after trauma in regions where there was initially no blood.[42] This phenomenon is termed hemorrhagic progression of a contusion (HPC) and represents the end stage in the cascade of microvascular dysfunction triggered in the region of injury, similar in nature to the progressive capillary dysfunction observed after other acute brain injuries.[9,42] HPC is common, occurring in about half of all cases, even following mild head injury.[42–44] It has great clinical relevance because it causes irreversible hemorrhagic necrosis of brain parenchyma, resulting in increased morbidity and mortality.[42] Allard et al. found that HPC was associated with a 5-fold higher risk of death.[45]

Preclinical studies demonstrate that SUR1–TRPM4 is integrally involved in the development of edema and HPC.[9,46] It is helpful to conceptualize three regions of injury: (1) the volume of initial contusion, (2) a "shell" of penumbra surrounding the contusion, and (3) a "shell" of parapenumbra surrounding the penumbra.[42] After a typical contusion, sufficient kinetic energy is deposited near the epicenter of the impact to shear tissues and microvessels, resulting in an immediate hemorrhagic lesion (the initial contusion).[42] In the penumbra, the amount of energy deposited is not enough to shear tissues and fracture microvessels but is enough to activate two mechanosensitive transcriptional factors (specificity protein 1—Sp1- and NFκB) that upregulate SUR1–TRPM4.[42,47,48] A few hours later, penumbral cells, especially microvascular endothelial cells, newly express SUR1–TRPM4 channels.[9,42] Opening of these channels leads to oncotic swelling and eventually oncotic death of endothelial cells, leading to capillary fragmentation.[42] Capillary fragmentation results in the extravasation of blood and formation of petechial hemorrhages.[42] Over the 12 hours after contusion injury, petechial hemorrhages continue to form, grow in number, and coalesce, resulting in progressive accumulation of extravasated blood, or HPC.[9,42]

In an animal model of contusive TBI, Sur1 expression mirrored the spatial and temporal extent of the hemorrhagic contusion.[9,13] Furthermore, immunolabeling of penumbral capillaries with vimentin (a marker of activated endothelial cells) showed foreshortened, broken segments of capillaries in regions where hemorrhages developed.[9] In glibenclamide-treated rats, penumbral structures that labeled positively for vimentin were elongated and had the normal appearance of capillaries.[9] Together, these findings are consistent with the model that HPC occurs because capillaries fragment in association with newly synthesized SUR1–TRPM4; block of SUR1–TRPM4 with glibenclamide reduces HPC by preserving capillary integrity.[9] In further support of this model, glibenclamide treatment remarkably largely eliminated HPC, as determined by serial quantification of extravasated blood.[9] Preventing expansion of the hemorrhage was associated with significant improvements in neurologic function.[9] Based on these exciting findings, a prospective, multicenter, placebo-controlled,

Phase IIa trial of RP-1127 (IV glibenclamide) is underway to test the effect of this agent in patients with moderate-to-severe TBI.[13] The primary outcome measures are change in edema from baseline within 72 hours, change in hemorrhage from baseline within 72 hours, and safety through 180 days.[13]

SUR1–TRPM4 in the Pathogenesis of Hemorrhagic Encephalopathy of Prematurity

Hemorrhagic EP is a common complication of prematurity, occurring in 15%–45% of cases.[49] The most severe forms involve hemorrhages that originate in periventricular tissues and extend as intraventricular hemorrhage.[8] Infants who survive suffer life-long morbidity from cerebral palsy.[8] Three key risk factors predispose to hemorrhagic EP: (1) prematurity, (2) perinatal ischemia/hypoxia, and (3) early postnatal mechanical ventilation.[8] Prematurity is associated with immaturity of periventricular vessels and renders them selectively vulnerable to rupture.[8] Ischemia/hypoxia alters periventricular vessels, rendering them even more vulnerable to rupture.[8] Mechanical ventilation is the final insult leading to hemorrhage, by causing elevated venous pressures that rupture fragile periventricular veins.[8] Preclinical and early-phase clinical studies suggest that SUR1–TRPM4 is pivotal in the progressive microvascular dysfunction leading to hemorrhage.

In a study of postmortem tissues from premature infants, Sur1 was prominently upregulated in periventricular tissue and veins in infants who sustained periventricular hemorrhage, but not in controls.[13,50] A subsequent study used a rat model of hemorrhagic EP and showed that SUR1–TRPM4 is causally linked to this syndrome.[8] In this novel model, in utero ischemia (IUI) is combined with an episode of postnatal raised intrathoracic pressure, to mimic the effect of mechanical ventilation on central venous pressure.[13,51] The authors found prominent Sur1 upregulation in periventricular vessels after brief IUI.[8] Furthermore, low-dose glibenclamide administered to the mother at the end of pregnancy dramatically reduced brain hemorrhages and inflammation in pups.[8] The mortality of pups born from glibenclamide-treated mothers was reduced by over 30%.[8] Furthermore, neurological outcomes in young adult life were dramatically better in the glibenclamide-treated cohort, as determined by complex vestibulomotor and spatial learning tasks.[8] Most strikingly, glibenclamide treatment significantly attenuated developmental delay.[8] These findings highlight the potential of SUR1–TRPM4 inhibition to ameliorate secondary injury and improve outcomes in this devastating disorder.

CONCLUSIONS

Novel treatment strategies are urgently needed to combat brain edema and progressive secondary hemorrhage after acute neurological injury. We conclude that the SUR1–TRPM4 channel is a highly promising target for brain edema treatment in neurocritical care conditions. First, given its central role in the continuum from cytotoxic edema to total microvascular failure; second, because its transcriptional upregulation provides a broad time-window for clinicians to institute therapy; third, given exciting preclinical and clinical data documenting the beneficial effects of SUR1–TRPM4 inhibition in neurological outcome. Lastly, and crucially, due to our ability to use a safe medication (glibenclamide) to selectively block SUR1–TRPM4 in the injured brain. If studies continue to show promising results, it is possible that glibenclamide may become a therapeutic option for patients with devastating neurological disorders.

References

1. Kahle KT, Simard JM, Staley KJ, Nahed BV, Jones PS, Sun D. Molecular mechanisms of ischemic cerebral edema: role of electroneutral ion transport. *Physiology (Bethesda)* 2009;**24**:257–65.
2. Walcott BP, Kahle KT, Simard JM. Novel treatment targets for cerebral edema. *Neurotherapeutics* 2012;**9**:65–72.
3. Caffes N, Kurland DB, Gerzanich V, Simard JM. Glibenclamide for the treatment of ischemic and hemorrhagic stroke. *Int J Mol Sci* 2015;**16**:4973–84.
4. Simard JM, Tsymbalyuk N, Tsymbalyuk O, Ivanova S, Yurovsky V, Gerzanich V. Glibenclamide is superior to decompressive craniectomy in a rat model of malignant stroke. *Stroke* 2010;**41**:531–7.
5. Woo SK, Kwon MS, Ivanov A, Gerzanich V, Simard JM. The sulfonylurea receptor 1 (sur1)-transient receptor potential melastatin 4 (trpm4) channel. *J Biol Chem* 2013;**288**:3655–67.
6. Simard JM, Woo SK, Gerzanich V. Transient receptor potential melastatin 4 and cell death. *Pflugers Arch.* 2012;**464**:573–82.
7. Simard JM, Woo SK, Schwartzbauer GT, Gerzanich V. Sulfonylurea receptor 1 in central nervous system injury: a focused review. *J Cereb Blood Flow Metab* 2012;**32**:1699–717.
8. Tosun C, Koltz MT, Kurland DB, Ijaz H, Gurakar M, Schwartzbauer G, et al. The protective effect of glibenclamide in a model of hemorrhagic encephalopathy of prematurity. *Brain Sci* 2013;**3**:215–38.
9. Simard JM, Kilbourne M, Tsymbalyuk O, Tosun C, Caridi J, Ivanova S, et al. Key role of sulfonylurea receptor 1 in progressive secondary hemorrhage after brain contusion. *J Neurotrauma* 2009;**26**:2257–67.
10. Arundine M, Tymianski M. Molecular mechanisms of calcium-dependent neurodegeneration in excitotoxicity. *Cell Calcium* 2003;**34**:325–37.
11. Bano D, Nicotera P. Ca2+ signals and neuronal death in brain ischemia. *Stroke* 2007;**38**:674–6.

12. Simard JM, Kahle KT, Gerzanich V. Molecular mechanisms of microvascular failure in central nervous system injury—synergistic roles of nkcc1 and sur1/trpm4. *J Neurosurg* 2010;**113**:622–9.
13. Kurland DB, Tosun C, Pampori A, Karimy JK, Caffes NM, Gerzanich V, et al. Glibenclamide for the treatment of acute cns injury. *Pharmaceuticals (Basel)* 2013;**6**:1287–303.
14. Simard JM, Chen M, Tarasov KV, Bhatta S, Ivanova S, Melnitchenko L, et al. Newly expressed sur1-regulated nc(ca-atp) channel mediates cerebral edema after ischemic stroke. *Nat Med* 2006;**12**:433–40.
15. Simard JM, Woo SK, Tsymbalyuk N, Voloshyn O, Yurovsky V, Ivanova S, et al. Glibenclamide-10-h treatment window in a clinically relevant model of stroke. *Transl Stroke Res* 2012;**3**:286–95.
16. Patel AD, Gerzanich V, Geng Z, Simard JM. Glibenclamide reduces hippocampal injury and preserves rapid spatial learning in a model of traumatic brain injury. *J Neuropathol Exp Neurol* 2010;**69**:1177–90.
17. Simard JM, Kent TA, Chen M, Tarasov KV, Gerzanich V. Brain oedema in focal ischaemia: molecular pathophysiology and theoretical implications. *Lancet Neurol.* 2007;**6**:258–68.
18. Khanna A, Kahle KT, Walcott BP, Gerzanich V, Simard JM. Disruption of ion homeostasis in the neurogliovascular unit underlies the pathogenesis of ischemic cerebral edema. *Transl Stroke Res* 2014;**5**:3–16.
19. Rajasekaran SA, Beyenbach KW, Rajasekaran AK. Interactions of tight junctions with membrane channels and transporters. *Biochim Biophys Acta* 2008;**1778**:757–69.
20. Simard JM, Geng ZH, Silver FL, Sheth KN, Kimberly WT, Stern BJ, et al. Does inhibiting sur1 complement rt-PA in cerebral ischemia? *Thrombolysis Acute Stroke Treat* 2012;**1268**:95–107.
21. Simard JM, Woo SK, Bhatta S, Gerzanich V. Drugs acting on sur1 to treat cns ischemia and trauma. *Curr Opin Pharmacol* 2008;**8**:42–9.
22. Wali B, Ishrat T, Atif F, Hua F, Stein DG, Sayeed I. Glibenclamide administration attenuates infarct volume, hemispheric swelling, and functional impairments following permanent focal cerebral ischemia in rats. *Stroke Res Treat* 2012;**2012**:460909.
23. Castellanos M, Leira R, Serena J, Pumar JM, Lizasoain I, Castillo J, et al. Plasma metalloproteinase-9 concentration predicts hemorrhagic transformation in acute ischemic stroke. *Stroke* 2003;**34**:40–6.
24. Ramos-Fernandez M, Bellolio MF, Stead LG. Matrix metalloproteinase-9 as a marker for acute ischemic stroke: a systematic review. *J Stroke Cerebrovascul Dis* 2011;**20**:47–54.
25. Gidday JM, Gasche YG, Copin JC, Shah AR, Perez RS, Shapiro SD, et al. Leukocyte-derived matrix metalloproteinase-9 mediates blood–brain barrier breakdown and is proinflammatory after transient focal cerebral ischemia. *Am J Physiol-Heart Circul Physiol* 2005;**289**:H558–68.
26. Cuadrado E, Rosell A, Penalba A, Slevin M, Alvarez-Sabin J, Ortega-Aznar A, et al. Vascular mmp-9/timp-2 and neuronal mmp-10 up-regulation in human brain after stroke: A combined laser microdissection and protein array study. *J Proteome Res* 2009;**8**:3191–7.
27. Ning M, Furie KL, Koroshetz WJ, Lee H, Barron M, Lederer M, et al. Association between TPA therapy and raised early matrix metalloproteinase-9 in acute stroke. *Neurology* 2006;**66**:1550–5.
28. Tsuji K, Aoki T, Tejima E, Arai K, Lee SR, Atochin DN, et al. Tissue plasminogen activator promotes matrix metalloproteinase-9 upregulation after focal cerebral ischemia. *Stroke* 2005;**36**:1954–9.
29. Kunte H, Schmidt S, Eliasziw M, del Zoppo GJ, Simard JM, Masuhr F, et al. Sulfonylureas improve outcome in patients with type 2 diabetes and acute ischemic stroke. *Stroke* 2007;**38**:2526–30.

30. Kunte H, Busch MA, Trostdorf K, Vollnberg B, Harms L, Mehta RI, et al. Hemorrhagic transformation of ischemic stroke in diabetics on sulfonylureas. *Ann Neurol* 2012;**72**:799–806.
31. Silver FL, Fang J, Robertson AC, Casaubon L, Kapral MK. Possible neuroprotective effects of sufonylureas in diabetic patients with acute ischemic stroke. *Abstracts from the 2009 International Stroke Conference*. 2009.
32. Sheth KN, Kimberly WT, Elm JJ, Kent TA, Mandava P, Yoo AJ, et al. Pilot study of intravenous glyburide in patients with a large ischemic stroke. *Stroke* 2014;**45**:281–3.
33. Kimberly WT, Battey TWK, Pham L, Wu O, Yoo AJ, Furie KL, et al. Glyburide is associated with attenuated vasogenic edema in stroke patients. *Neurocrit Care*. 2014;**20**:193–201.
34. Sheth KN, Kimberly WT, Elm JJ, Kent TA, Yoo AJ, Thomalla G, et al. Exploratory analysis of glyburide as a novel therapy for preventing brain swelling. *Neurocrit Care* 2014;**21**:43–51.
35. Urday S, Kimberly WT, Beslow LA, Vortmeyer AO, Selim MH, Rosand J, et al. Targeting secondary injury in intracerebral haemorrhage-perihaematomal oedema. *Nat Rev Neurol* 2015;**11**:111–22.
36. Wagner KR, Xi GH, Hua Y, Kleinholz M, deCourtenMyers GM, Myers RE, et al. Lobar intracerebral hemorrhage model in pigs—rapid edema development in perihematomal white matter. *Stroke* 1996;**27**:490–7.
37. Venkatasubramanian C, Mlynash M, Finley-Caulfield A, Eyngorn I, Kalimuthu R, Snider RW, et al. Natural history of perihematomal edema after intracerebral hemorrhage measured by serial magnetic resonance imaging. *Stroke* 2011;**42**:73–80.
38. Inaji M, Tomita H, Tone O, Tamaki M, Suzuki R, Ohno K. Chronological changes of perihematomal edema of human intracerebral hematoma. *Acta Neurochir Suppl* 2003;**86**:445–8.
39. Suga S, Sato S, Yunoki K, Mihara B. Sequential change of brain edema by semiquantitative measurement on mri in patients with hypertensive intracerebral hemorrhage. *Acta Neurochir Suppl (Wien)* 1994;**60**:564–7.
40. Simard JM, Geng Z, Woo SK, Ivanova S, Tosun C, Melnichenko L, et al. Glibenclamide reduces inflammation, vasogenic edema, and caspase-3 activation after subarachnoid hemorrhage. *J Cereb Blood Flow Metab* 2009;**29**:317–30.
41. Tosun C, Kurland DB, Mehta R, Castellani RJ, deJong JL, Kwon MS, et al. Inhibition of the sur1–trpm4 channel reduces neuroinflammation and cognitive impairment in subarachnoid hemorrhage. *Stroke* 2013;**44**:3522–8.
42. Kurland D, Hong C, Aarabi B, Gerzanich V, Simard JM. Hemorrhagic progression of a contusion after traumatic brain injury: a review. *J Neurotrauma* 2012;**29**:19–31.
43. Sifri ZC, Homnick AT, Vaynman A, Lavery R, Liao W, Mohr A, et al. A prospective evaluation of the value of repeat cranial computed tomography in patients with minimal head injury and an intracranial bleed. *J Trauma* 2006;**61**:862–7.
44. Smith JS, Chang EF, Rosenthal G, Meeker M, von Koch C, Manley GT, et al. The role of early follow-up computed tomography imaging in the management of traumatic brain injury patients with intracranial hemorrhage. *J Trauma* 2007;**63**:75–82.
45. Allard CB, Scarpelini S, Rhind SG, Baker AJ, Shek PN, Tien H, et al. Abnormal coagulation tests are associated with progression of traumatic intracranial hemorrhage. *J Trauma* 2009;**67**:959–67.
46. Zweckberger K, Hackenberg K, Jung CS, Hertle DN, Kiening KL, Unterberg AW, et al. Glibenclamide reduces secondary brain damage after experimental traumatic brain injury. *Neuroscience* 2014;**272**:199–206.
47. Abumiya T, Sasaguri T, Taba Y, Miwa Y, Miyagi M. Shear stress induces expression of vascular endothelial growth factor receptor flk-1/kdr through the ct-rich sp1 binding site. *Arterioscler Thromb Vasc Biol* 2002;**22**:907–13.

48. Davis ME, Grumbach IM, Fukai T, Cutchins A, Harrison DG. Shear stress regulates endothelial nitric-oxide synthase promoter activity through nuclear factor kappa b binding. *J Biol Chem* 2004;**279**:163–8.
49. Kadri H, Mawla AA, Kazah J. The incidence, timing, and predisposing factors of germinal matrix and intraventricular hemorrhage (GMH/IVH) in preterm neonates. *Childs Nerv Syst* 2006;**22**:1086–90.
50. Simard JM, Castellani RJ, Ivanova S, Koltz MT, Gerzanich V. Sulfonylurea receptor 1 in the germinal matrix of premature infants. *Pediatr Res* 2008;**64**:648–52.
51. Koltz MT, Tosun C, Kurland DB, Coksaygan T, Castellani RJ, Ivanova S, et al. Tandem insults of prenatal ischemia plus postnatal raised intrathoracic pressure in a novel rat model of encephalopathy of prematurity. *J Neurosurg Pediatr* 2011;**8**:628–39.

CHAPTER

11

Role of Matrix Metalloproteinases in Brain Edema

Changjun Yang and Eduardo Candelario-Jalil
University of Florida, Gainesville, FL, United States

OUTLINE

INTRODUCTION

The blood–brain barrier (BBB) is an interface that restricts the passage of blood-borne cells and substances, including plasma proteins, into the central nervous system (CNS). The brain endothelial cells were initially thought to determine the low permeability of the BBB to blood components. However, it has been recently recognized that not only the brain

Brain Edema.
DOI: http://dx.doi.org/10.1016/B978-0-12-803196-4.00011-4

endothelium, but that also the cell–cell interactions between endothelial cells and neurons, astrocytes, perivascular microglia, pericytes, and smooth muscle cells determine the BBB properties and the maintenance of the cerebral homeostasis. This highly complex functional and structural unit is termed the neurovascular unit.[1–4] A summary schematic of the cellular and molecular components of the neurovascular unit is depicted in Fig. 11.1. The specific role of each cell in maintaining the function of the neurovascular unit has been extensively reviewed in recent years[4–10] and will not be discussed in detail in this chapter (see Chapter 2: Blood–Brain Interfaces Organization in Relation to Inorganic Anion Transport, CSF Secretion, and Circulation; and Chapter 6: The Extracellular and Perivascular Spaces of the Brain).

Unlike peripheral endothelium, brain endothelial cells are unique in that they have a low expression of adhesion molecules, have fewer caveolae (vesicular invaginations of the plasma membrane), have a lack of fenestrations, and are zipped together by tight junction (TJ) proteins (TJPs).[6,10–13] These unique structural characteristics underlie the low paracellular permeability of the BBB and limited infiltration of peripheral immune cells into the perivascular space of healthy CNS. Due to the low number of caveolae in the brain endothelium, the transcellular trafficking of proteins and other solutes via transcytosis between blood and brain is limited under normal conditions.[14,15] Brain endothelial cells express a large number of transporters to facilitate the passage of essential molecules to the brain, as well as to extrude metabolites and xenobiotics from the brain into the blood[16,17] (for details see Chapter 2: Blood–Brain Interfaces Organization in Relation to Inorganic Anion Transport, CSF Secretion, and Circulation).

Brain edema is classically defined as the accumulation of fluid within the brain parenchyma and is a major cause of mortality in many neurological pathologies such as stroke, traumatic brain injury (TBI), seizures, infections, and tumors (see Chapters 14, 15, 18, 19, and 20). Distinct forms of edema have been described including cytotoxic, vasogenic, interstitial, and osmotic/hypostatic edema. In most neurological diseases, these types of edema rarely exist in isolation and depending on the time course of the disease, one particular type of edema could prevail over others. A good example is ischemic stroke, in which cellular swelling/cytotoxic edema is most important during the early stages following vessel occlusion. However, over several hours/days, when oxidative stress and neuroinflammatory processes damage the BBB, fluid accumulation in the brain is mainly due to vasogenic edema.

Vasogenic edema due to dysfunction of the BBB plays a prominent role in fluid accumulation after brain injury by increasing the passage of plasma proteins and followed by water from blood into the brain parenchyma. Many mediators have been implicated in endothelial barrier

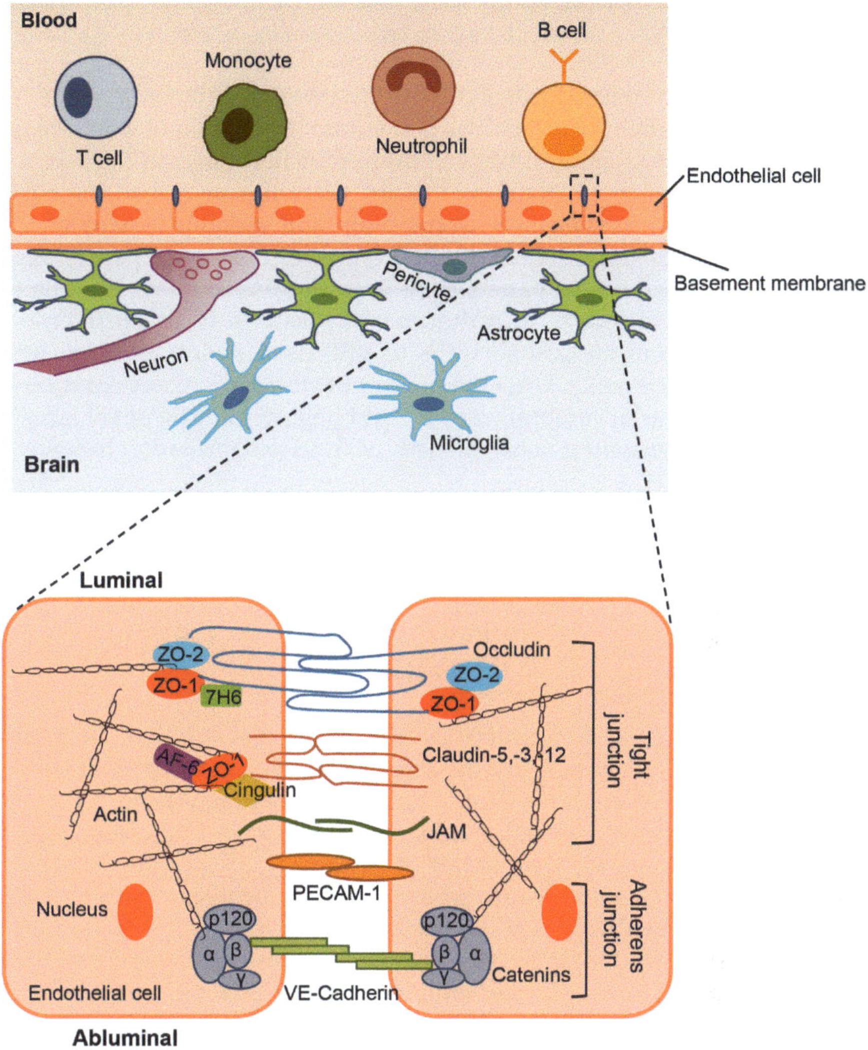

FIGURE 11.1 **Schematic diagram of the neurovascular unit.** The neurovascular unit or blood–brain barrier (BBB) is formed by endothelial cells, which are surrounded by the basement membrane, astrocytic end-feet processes, and pericytes. The basement membrane (BM) mostly comprises extracellular matrix (ECM) proteins such as type IV collagen, laminin, fibronectin, heparan sulfate proteoglycans, dystroglycan, and integrins. The BM surrounds the endothelial cells and pericytes, provides mechanical support, and facilitates cell–cell and cell–ECM interactions at the BBB. The perivascular microglial cells make contact with the cerebral microvessels and modulate the BBB permeability. Blood-borne immune cells such as lymphocytes, monocytes, and neutrophils cannot penetrate the BBB under basal conditions. The tight and adherent junctions connect brain endothelial cells and confer the low paracellular permeability of the BBB. The tight junction proteins (TJPs) form an intricate complex of proteins linked to the actin cytoskeleton. Claudins and occludin have four transmembrane domains with two extracellular loops, which are important in forming the "seal" between two adjacent endothelial cells. These proteins associate with the cytoskeleton via accessory proteins such as zona occludens (ZO)-1, ZO-2, AF6 (afadin), 7H6, cingulin, and others. The junctional adhesion molecule (JAM) family forms part of the TJPs and mediates attachment of cell membranes via homophilic interactions. The most important components of the adherent junctions are vascular endothelial cadherin (VE-cadherin) and catenins, which provide structural integrity and attachment between the endothelial cells, and are necessary for the formation of tight junctions. Platelet endothelial cell adhesion molecule-1 (PECAM-1), also known as CD31, is highly expressed by the endothelium and contributes to the function of the BBB.

failure after brain injury. Proteolytic degradation of BBB components by matrix metalloproteinases (MMPs) is a common effector of BBB breakdown leading to vasogenic edema in numerous neurological disorders.

A large body of work done over many years has described the basic principles of water homeostasis in the brain and identified many of the molecular mechanisms of edema formation (for review see Chapters 8, 9, 11, and 12). Several recent papers have been published about the role of MMPs in cerebral injury.[18–25] In this chapter, we will focus on the most recent studies implicating the MMPs in BBB dysfunction and development of vasogenic edema. Understanding how these proteases contribute to the pathogenesis of cerebral edema could help to identify novel targets to reduce the devastating consequences of fluid accumulation following brain injury.

MOLECULAR ELEMENTS OF THE BBB: PROTEINS COMPOSING THE TIGHT JUNCTIONS, ADHERENS JUNCTIONS, AND EXTRACELLULAR MATRIX

Embedded in the extracellular matrix (ECM), the highly specialized endothelial cells composing the neurovascular unit are tightly attached by a complex arrangement of TJ and adherens junction (AJ) proteins that play an important role in maintaining BBB function.[5,26] TJPs are critical in maintaining the low paracellular permeability of the neurovascular unit. Occludin and claudins are transmembrane proteins that form the primary seal of the BBB by tightly attaching two adjacent endothelial cells. Accessory proteins like zona occludens (ZO)-1 link occludin and claudins to the actin cytoskeleton. In the adult mammalian brain, the most abundant TJPs are occludin, ZO-1, claudin-5, -3, and -12.[11,27] Fig. 11.1 shows the major cellular and molecular elements of the neurovascular unit.

The phosphorylation status of occludin regulates its association with the endothelial cell membrane.[28] Occludin interacts with ZO-1 through its intracellular C-terminal domain.[29–32] Phosphorylation of occludin at tyrosine residues in its intracellular domain results in destabilization of the TJ complex, loss of occludin from the membrane, and increased BBB permeability.[33] Increased phosphorylation of occludin has been associated with BBB dysfunction in models of ischemic stroke.[34,35] Dissociation of ZO-1 from the TJ complexes significantly impairs BBB permeability, indicating that ZO-1 is critical in maintaining TJ stability and function.[4,11,36]

The junctional adhesion molecule (JAM) family forms part of the TJPs and contribute to endothelial cell adhesion and leukocyte–endothelial cell interactions. Several JAM proteins have been described, including JAM-A, -B, and -C. JAM-A, the most abundant JAM at the BBB, maintains endothelial cell–cell interactions and actively participates in the formation of TJs under physiological conditions,[37] but orchestrates leukocyte recruitment under inflammatory conditions.[38]

The AJ proteins (AJPs) are widely expressed in the brain vasculature and mediate many functions, including adhesion of two adjacent endothelial cells, regulation of paracellular permeability, and contact inhibition during vascular remodeling and growth.[11,39] Vascular endothelial cadherin (VE-cadherin) forms AJs between neighboring brain endothelial cells via homophilic molecular interactions between its transmembrane domains. VE-cadherin is linked to the actin cytoskeleton via catenins. There is a dynamic crosstalk between TJPs and AJPs. It has been demonstrated that formation of TJ complexes generally requires the existence of AJs.[39,40] Conversely, recent evidence indicates that the TJP ZO-1 is a key regulator of VE-cadherin-dependent AJs modulating BBB formation and maintenance.[41]

Another important molecular element of the neurovascular unit is platelet endothelial cell adhesion molecule-1 (PECAM-1), also known as CD31. PECAM-1 serves a vital role in regulating endothelial adhesion and BBB permeability through either homophilic binding between two PECAM-1 molecules or via its interaction with other proteins such as integrins and catenins.[42,43] PECAM-1 deficient mice have prolonged and increased BBB permeability following inflammatory insults.[42–44] Passage of leukocytes across the BBB is facilitated by PECAM-1 (reviewed in Ref. [42]).

ECM proteins make up the basement membrane (BM) of the neurovascular unit. Two BMs are located in the CNS: an endothelial/vascular wall BM and a parenchymal BM of the astrocytic glia limitans. Under basal conditions, these two BMs are in very close contact and appear as one sheet-like structure under microscopic examination.[26] The BM directly regulates barrier function, provides structural support, anchors cells into place, and regulates many cellular signaling pathways via interactions with integrins and other ECM receptors.[5,26] All the cellular elements of the BBB contribute to the production of ECM proteins. There is a bidirectional relationship between ECM proteins and cells in that, once produced and secreted, ECM proteins interact with cells to maintain the properties of the neurovascular unit.

The BM of the BBB mostly comprises structural proteins such as type IV collagen, laminin, fibronectin, nidogen-1/entactin, and heparin sulfate proteoglycans (perlecan and agrin). Dystroglycan and integrins are the two main types of receptor/adhesion proteins involved in cell–cell and cell–ECM interactions at the level of the BBB.[5,26] These matrix receptors are present in all cells of the BBB and function as regulators of signaling pathways to allow cells to adapt to changes in the microenvironment. They also form a physical connection between the ECM and the cytoarchitecture, which anchors cells in place and regulate their motility. Integrin-mediated regulation of VE-cadherin modulates vascular permeability. For example, activation of $\alpha v\beta_3$ integrin signaling results in disruption of VE-cadherin localization at the AJ resulting in increased BBB permeability.[45] Other integrins facilitate BBB formation and maturation. Mice deficient in $\alpha v\beta_8$ integrin have significant BBB impairment and increased permeability.[46,47]

MATRIX METALLOPROTEINASES: MOLECULAR ASPECTS AND REGULATION OF THEIR FUNCTION IN BRAIN

The MMPs comprise a large family of 23 zinc-dependent endopeptidases that cleave many protein substrates.[48–50] They belong to the larger family of metzincin metalloproteinases, which also include other members: a disintegrin and metalloproteinases (ADAMs) and ADAMs with thrombospondin repeats. These enzymes have a conserved zinc-binding motif within the catalytic site.[48,51]

Based on substrate specificity and cellular localization, MMPs are classified into different groups: the collagenases (MMP-1, -8, -13, and -18); the gelatinases A and B (MMP-2 and -9); stromelysins (MMP-3, -10, and -11); membrane-bound, anchored by a transmembrane domain (MMP-14, -15, -16, -23, and -24) or by a glycosylphosphatidylinositol linkage (MMP-17 and -25). Other members of the MMPs include metalloelastase (MMP-12), matrylisins (MMP-7 and -26), enamelysin (MMP-20), epilysin (MMP-28), and MMP-19 and -21.[48,49,51]

In the CNS, MMPs are synthesized and secreted by neurons, glia, and endothelial cells. Under basal conditions, transcript levels of *MMP-2, MMP-3, MMP-8, MMP-9, MMP-11, MMP-12, MMP-13, MMP-14, MMP-15*, and *MMP-24* are present in the adult mammalian brain, whereas the mRNA levels of other members of the MMP family are very low or undetectable.[52,53] Neurons produce MMP-9, brain endothelial cells express mainly MMP-9, astrocytes produce MMP-2 and MMP-14, pericytes produce MMP-9 and MMP-3, whereas microglia constitutively produce MMP-12 and show strong induction of MMP-3 and MMP-9 after injury or inflammation.[54–58]

Under normal conditions, MMPs contribute to the remodeling of the ECM associated with important physiological processes such as angiogenesis and neurogenesis.[59–61] Evidence indicates that MMPs, specifically MMP-9, contribute to normal synaptic plasticity/remodeling and long-term potentiation. These processes require MMP-9 activation.[52,62,63] Proteolysis by MMPs occurs primarily extracellularly, but new evidence indicates that some MMPs cleave substrates inside the cell modulating cellular function and survival.[64,65]

MMPs are produced as a latent (inactive) proform, which can be activated by different mechanisms including free radicals and other proteases such as furin, plasminogen activator, and other MMPs. A large body of evidence indicates that MMPs have a conserved zinc-coordinating cysteine residue in the prodomain that masks the catalytic site (the "cysteine switch") rendering the MMPs inactive. Cleavage of the prodomain by proteases or conformational changes due to oxidation or *S*-nitrosylation of the sulfhydryl group on the inhibitory cysteine results in MMP activation.[18,66–69]

Under physiological conditions, endogenous MMP activity is regulated by tissue inhibitors of metalloproteinases (TIMPs).[70] The TIMPs are small proteins (21–28 kDa) that bind reversibly to the MMPs in a 1:1 ratio. The overall proteolytic activity of the MMPs depends on the balance of MMP/TIMP levels. Some of the TIMPs have a preference toward specific MMPs. For example, TIMP-1 mainly inhibits MMP-9. Proteolytic activity of MMP-2 is regulated by TIMP-2, but paradoxically, TIMP-2 participates in the molecular activation of pro-MMP-2. The sheddase activity of MMP-14, MMP-3, and tumor necrosis factor-α (TNF-α) converting enzyme is inhibited by TIMP-3, the only TIMP that is bound to the ECM.[71] Little is known about the function of TIMP-4 in the brain.[70,72]

Activation of MMP-2 involves the interaction between pro-MMP-2, TIMP-2, and MMP-14 (one of the membrane-bound MMPs). The MMP-2 activation process occurs close to the cell surface restricting MMP-2 action and providing local proteolysis of the ECM. Activation of MMP-9 occurs mainly by MMP-3-dependent proteolysis[73] and *S*-nitrosylation.[67] Furin, a proprotein convertase, is an activator of MMP-14.[74] Molecular mechanisms responsible for the activation of MMP-3 in vivo are not well established. Choi et al.[75] suggested that an intracellular serine protease cleaves pro-MMP-3 into the active form. However, the molecular identity of this serine protease has not yet been elucidated.

Many cytokines produced in response to brain injury and/or inflammation are able to induce MMP expression and activation. Interleukin-1β (IL-1β) and TNF-α increase the activation of the transcription factors AP-1 and NF-κB, which directly induce the expression of many MMPs, notably MMP-9, MMP-3, and MMP-14.[19,76]

Oxidative stress due to excessive production of reactive oxygen species (ROS) is an important contributor to brain damage and edema formation in many neurological diseases. ROS attack cells composing the BBB, activate MMPs, and impair organization and cellular localization of TJP complexes.[69,77–80] Treatment with antioxidants or inhibiting cellular sources of ROS significantly reduce the BBB opening and brain edema in experimental models of intracerebral hemorrhage,[81,82] ischemia,[83,84] and TBI,[85,86] which underscores the prominent role of ROS in edema formation.

Oxidative modifications of the thiol group on the zinc-binding cysteine of the MMPs increase their enzymatic activity.[67,69,87,88] Increased nitric oxide (NO) and peroxynitrite formation result in *S*-nitrosylation of MMP-9, a process that dramatically increases MMP-9-dependent proteolysis of structural components of the neurovascular unit after ischemia.[67,89] In a more recent study, increased NO production after stroke was found to depend on the downregulation of caveolin-1, the major component of the caveolae and a negative regulator of endothelial nitric oxide synthase. Ischemia-induced reduction in caveolin-1 correlated with increased MMP-2 and MMP-9 proteolytic activities, loss of TJPs, and BBB breakdown.[90]

TIGHT JUNCTION PROTEINS AND EXTRACELLULAR MATRIX PROTEINS AS MOLECULAR TARGETS OF THE MMPS: IMPLICATIONS FOR BBB DYSFUNCTION AND EDEMA FORMATION

Aberrant MMP activity plays a key role in brain damage in ischemic stroke, TBI, meningitis, multiple sclerosis, seizures, and brain tumors by proteolytically degrading the BM and TJPs, which results in the opening of the BBB and vasogenic edema.[25,91,92] Loss of TJPs and BM proteins is very striking after stroke or TBI and correlates with BBB dysfunction and edema.[93–95] A summary of the potential mechanisms involved in BBB disruption and vasogenic edema following brain injury is presented in Fig. 11.2.

In addition to the well-known effects on BBB leakage, MMP-mediated proteolysis of ECM components significantly interferes with normal matrix signaling via integrins, which is important in functional interactions between cells and the matrix. MMP-dependent matrix disruption has been shown to induce anoikis, a type of apoptotic cell death induced by lack of correct cell/ECM attachment.[67,89,96]

Collagen type IV, laminin, fibronectin, and proteoglycans are substrates of the active MMPs.[76,89,97] By proteolytically degrading the BM components, the MMPs compromise the structural integrity of capillaries, weaken the vessel walls, and increase the risks of rupture and hemorrhage.[98] In addition, TJPs of the BBB are susceptible to MMP-mediated proteolysis and degradation. Occludin, claudin-5, and ZO-1 are vulnerable to the attack by active MMPs in ischemic stroke,[91,97,99–101] neuroinflammation,[102,103] and gliomas.[104]

Cerebral ischemia and the neuroinflammatory events that follow this type of injury activate pro-MMPs and induce de novo synthesis and

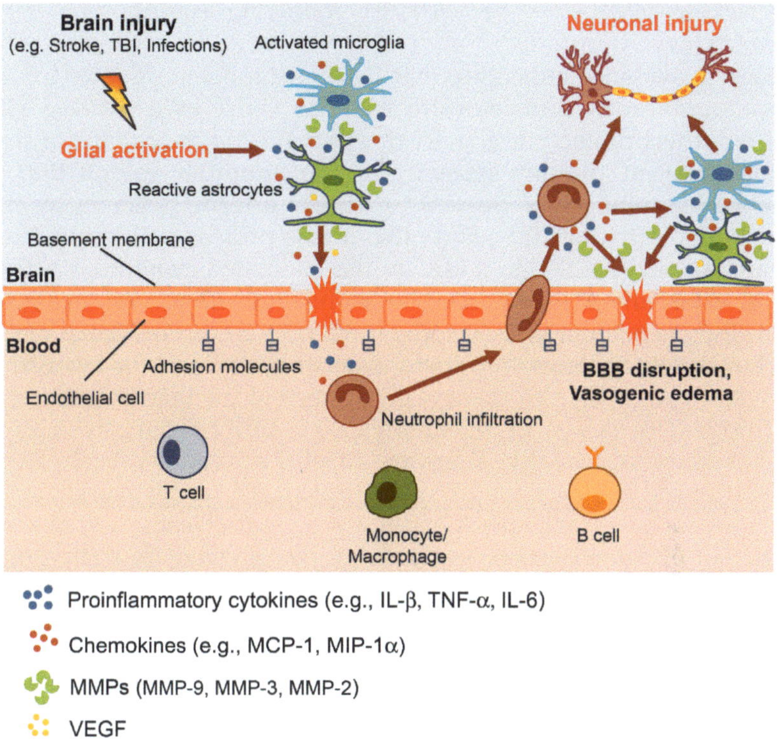

FIGURE 11.2 **Schematic diagram of the potential mechanisms of blood–brain barrier disruption and development of vasogenic edema following brain injury.** Initial injury events such as excitotoxicity, oxidative stress, and hypoxia that follow stroke, traumatic brain injury (TBI) and infections, result in activation of microglia and astrocytes to secret proinflammatory cytokines, chemokines, matrix metalloproteinases (MMPs), and vascular endothelial growth factor (VEGF). These inflammatory mediators and proteases contribute to BBB breakdown and upregulation of cell adhesion molecules on endothelial cells. This promotes infiltration of blood-derived immune cells, mainly neutrophils, into the brain parenchyma through the damaged BBB. Activated neutrophils subsequently release cytokines, chemokines, and active MMPs to further disrupt the BBB. Concurrently, these cytokines and chemokines derived from the activated neutrophils also cause a further activation of glial cells to generate more inflammatory mediators and MMPs, which exacerbate the progression of BBB disruption, vasogenic edema, and neuronal injury.

release of MMPs, especially MMP-2, MMP-3, MMP-9, and MMP-14.[105–109] Convincing evidence from experimental models of ischemic stroke demonstrates that MMP inhibitors significantly reduce vasogenic brain edema following ischemia.[106,110–112] Further evidence implicating the MMPs in stroke-induced BBB injury comes from studies using knockout mice. Degradation of the BM, loss of ZO-1, and BBB breakdown are reduced in mice lacking MMP-9[97,108] or MMP-3[113] following experimental cerebral ischemia.

There is evidence indicating that the plasminogen activator/plasmin system significantly contributes to vascular injury by activating MMPs or by the direct proteolytic activity of plasmin.[114] In the context of stroke, the plasminogen/plasmin system plays an important role in BBB dysfunction. Plasminogen is cleaved to its active form, plasmin, by tissue plasminogen activator (tPA) and other serine proteases. Plasmin directly activates several MMPs. In addition, tPA-mediated proteolysis activates MMPs, especially MMP-9. In order to lyse the fibrin clot and restore blood flow, recombinant tPA is given within 4.5 hour of stroke onset in the clinical setting. However, there is an important risk of hemorrhagic conversion when tPA treatment is given beyond the therapeutic window of 4.5 hour. This is due to ischemia-induced BBB opening, which is the result of increased oxidative stress and MMP-2 activation during the early stages after stroke. When the BBB is disrupted, tPA readily crosses into the brain tissue where it activates MMP-9, a process that leads to further exacerbation of BBB dysfunction, hemorrhagic transformation, and vasogenic edema. Data from ischemic stroke models show that MMP inhibitors (i.e., BB-94) block the opening of the BBB and reduce the risk of hemorrhage after tPA treatment.[110,115–117]

In cerebral ischemia, neutrophils that infiltrate the damaged tissue are a major source of MMP-9.[118] In human stroke, neutrophil extravasation is associated with BBB disruption, hemorrhagic transformation, and collagen IV degradation.[119,120] Clinical findings indicate that increased MMP-9 levels in plasma correlate with poorer neurological outcomes,[121] increased BBB damage, hemorrhage, and vasogenic edema.[122–125]

In addition to ischemic stroke, MMP-dependent disruption of the BBB has been demonstrated in many other CNS disorders. Targeting the MMPs using broad-spectrum pharmacological inhibitors (i.e., BB-94, BB-1101, GM6001) significantly reduces brain edema following intracerebral hemorrhage,[126] cortical impact injury,[127] and bacterial meningitis.[128–130]

CONCLUSIONS

Extensive experimental evidence emphasizes the key role of the MMPs in BBB damage and vasogenic edema formation in a wide range of neuropathological conditions. These proteases significantly contribute to the inflammatory response to injury and are part of a complex and highly interconnected cascade of molecular events resulting in dysfunction of the neurovascular unit and brain tissue damage. In spite of the prominent role of the MMPs in BBB damage and edema, pharmacological targeting of these proteases is challenging because current MMP inhibitors are unspecific and could produce significant side effects, which limit their clinical use.[131] MMPs are also important in tissue repair by promoting neurogenic and angiogenic responses at later times after injury. Therefore, a better understanding of the temporal changes in MMP expression and activation is needed for the development of successful therapeutic strategies targeting these proteases to reduce BBB damage and vasogenic edema.

References

1. Muldoon LL, Alvarez JI, Begley DJ, et al. Immunologic privilege in the central nervous system and the blood–brain barrier. *J Cereb Blood Flow Metab* 2013;**33**(1):13–21.
2. Lok J, Gupta P, Guo S, et al. Cell–cell signaling in the neurovascular unit. *Neurochem Res* 2007;**32**(12):2032–45.
3. Zhang JH, Badaut J, Tang J, Obenaus A, Hartman R, Pearce WJ. The vascular neural network—a new paradigm in stroke pathophysiology. *Nat Rev Neurol* 2012;**8**(12):711–16.
4. Zlokovic BV. The blood–brain barrier in health and chronic neurodegenerative disorders. *Neuron* 2008;**57**(2):178–201.
5. Obermeier B, Daneman R, Ransohoff RM. Development, maintenance and disruption of the blood–brain barrier. *Nat Med* 2013;**19**(12):1584–96.
6. Abbott NJ, Ronnback L, Hansson E. Astrocyte-endothelial interactions at the blood–brain barrier. *Nat Rev Neurosci* 2006;**7**(1):41–53.
7. Alvarez JI, Katayama T, Prat A. Glial influence on the blood brain barrier. *Glia* 2013;**61**(12):1939–58.
8. Cabezas R, Avila M, Gonzalez J, et al. Astrocytic modulation of blood brain barrier: perspectives on Parkinson's disease. *Front Cell Neurosci* 2014;**8**:211.
9. Krueger M, Bechmann I. CNS pericytes: concepts, misconceptions, and a way out. *Glia* 2010;**58**(1):1–10.
10. Blanchette M, Daneman R. Formation and maintenance of the BBB. *Mech Dev* 2015;**138**(1):8–16.
11. Hawkins BT, Davis TP. The blood–brain barrier/neurovascular unit in health and disease. *Pharmacol Rev* 2005;**57**(2):173–85.
12. Sedlakova R, Shivers RR, Del Maestro RF. Ultrastructure of the blood–brain barrier in the rabbit. *J Submicrosc Cytol Pathol* 1999;**31**(1):149–61.
13. Ransohoff RM, Engelhardt B. The anatomical and cellular basis of immune surveillance in the central nervous system. *Nat Rev Immunol* 2012;**12**(9):623–35.
14. Predescu SA, Predescu DN, Malik AB. Molecular determinants of endothelial transcytosis and their role in endothelial permeability. *Am J Physiol Lung Cell Mol Physiol* 2007;**293**(4):L823–42.
15. Stewart PA. Endothelial vesicles in the blood–brain barrier: are they related to permeability? *Cell Mol Neurobiol* 2000;**20**(2):149–63.
16. Pardridge WM. Drug transport across the blood–brain barrier. *J Cereb Blood Flow Metab* 2012;**32**(11):1959–72.
17. Miller DS. Regulation of ABC transporters at the blood–brain barrier. *Clin Pharmacol Ther* 2015;**97**(4):395–403.
18. Candelario-Jalil E, Yang Y, Rosenberg GA. Diverse roles of matrix metalloproteinases and tissue inhibitors of metalloproteinases in neuroinflammation and cerebral ischemia. *Neuroscience* 2009;**158**(3):983–94.
19. Rosenberg GA. Matrix metalloproteinases and their multiple roles in neurodegenerative diseases. *Lancet Neurol* 2009;**8**(2):205–16.
20. Ning M, Wang X, Lo EH. Chapter 6 Reperfusion injury after stroke: neurovascular proteases and the blood–brain barrier. *Handb Clin Neurol* 2008;**92**:117–36.

21. Chaturvedi M, Kaczmarek L. Mmp-9 inhibition: a therapeutic strategy in ischemic stroke. *Mol Neurobiol* 2014;**49**(1):563–73.
22. Lakhan SE, Kirchgessner A, Tepper D, Leonard A. Matrix metalloproteinases and blood–brain barrier disruption in acute ischemic stroke. *Front Neurol* 2013;**4**:32.
23. Rosenberg GA. Neurological diseases in relation to the blood–brain barrier. *J Cereb Blood Flow Metab* 2012;**32**(7):1139–51.
24. Yang Y, Rosenberg GA. Matrix metalloproteinases as therapeutic targets for stroke. *Brain Res* 2015;**1623**:30–8.
25. Seo JH, Guo S, Lok J, et al. Neurovascular matrix metalloproteinases and the blood–brain barrier. *Curr Pharm Des* 2012;**18**(25):3645–8.
26. Baeten KM, Akassoglou K. Extracellular matrix and matrix receptors in blood–brain barrier formation and stroke. *Dev Neurobiol* 2011;**71**(11):1018–39.
27. Ohtsuki S, Yamaguchi H, Katsukura Y, Asashima T, Terasaki T. mRNA expression levels of tight junction protein genes in mouse brain capillary endothelial cells highly purified by magnetic cell sorting. *J Neurochem* 2008;**104**(1):147–54.
28. Hirase T, Kawashima S, Wong EY, et al. Regulation of tight junction permeability and occludin phosphorylation by Rhoa-p160ROCK-dependent and -independent mechanisms. *J Biol Chem* 2001;**276**(13):10423–31.
29. Furuse M, Hirase T, Itoh M, Nagafuchi A, Yonemura S, Tsukita S. Occludin: a novel integral membrane protein localizing at tight junctions. *J Cell Biol* 1993;**123**(6 Pt 2): 1777–88.
30. Hirase T, Staddon JM, Saitou M, et al. Occludin as a possible determinant of tight junction permeability in endothelial cells. *J Cell Sci* 1997;**110**(Pt 14):1603–13.
31. Tsukita S, Furuse M, Itoh M. Molecular architecture of tight junctions: occludin and ZO-1. *Soc Gen Physiol Ser* 1997;**52**:69–76.
32. Tash BR, Bewley MC, Russo M, et al. The occludin and ZO-1 complex, defined by small angle X-ray scattering and NMR, has implications for modulating tight junction permeability. *Proc Natl Acad Sci USA* 2012;**109**(27):10855–60.
33. Elias BC, Suzuki T, Seth A, et al. Phosphorylation of Tyr-398 and Tyr-402 in occludin prevents its interaction with ZO-1 and destabilizes its assembly at the tight junctions. *J Biol Chem* 2009;**284**(3):1559–69.
34. Kago T, Takagi N, Date I, Takenaga Y, Takagi K, Takeo S. Cerebral ischemia enhances tyrosine phosphorylation of occludin in brain capillaries. *Mol Cell Biol Res Commun* 2006;**339**(4):1197–203.
35. Fukumoto K, Takagi N, Yamamoto R, Moriyama Y, Takeo S, Tanonaka K. Prostanoid EP1 receptor antagonist reduces blood–brain barrier leakage after cerebral ischemia. *Eur J Pharmacol* 2010;**640**(1-3):82–6.
36. Fanning AS, Jameson BJ, Jesaitis LA, Anderson JM. The tight junction protein ZO-1 establishes a link between the transmembrane protein occludin and the actin cytoskeleton. *J Biol Chem* 1998;**273**(45):29745–53.
37. Ebnet K, Suzuki A, Ohno S, Vestweber D. Junctional adhesion molecules (JAMs): more molecules with dual functions? *J Cell Sci* 2004;**117**(Pt 1):19–29.
38. Stamatovic SM, Sladojevic N, Keep RF, Andjelkovic AV. Relocalization of junctional adhesion molecule A during inflammatory stimulation of brain endothelial cells. *Mol Cell Biol* 2012;**32**(17):3414–27.
39. Tietz S, Engelhardt B. Brain barriers: crosstalk between complex tight junctions and adherens junctions. *J Cell Biol* 2015;**209**(4):493–506.
40. Taddei A, Giampietro C, Conti A, et al. Endothelial adherens junctions control tight junctions by VE-cadherin-mediated upregulation of claudin-5. *Nat Cell Biol* 2008;**10**(8):923–34.
41. Tornavaca O, Chia M, Dufton N, et al. ZO-1 controls endothelial adherens junctions, cell–cell tension, angiogenesis, and barrier formation. *J Cell Biol* 2015;**208**(6):821–38.

42. Privratsky JR, Newman PJ. PECAM-1: regulator of endothelial junctional integrity. *Cell Tissue Res* 2014;**355**(3):607–19.
43. Biswas P, Canosa S, Schoenfeld D, et al. PECAM-1 affects GSK-3beta-mediated beta-catenin phosphorylation and degradation. *Am J Pathol* 2006;**169**(1):314–24.
44. Graesser D, Solowiej A, Bruckner M, et al. Altered vascular permeability and early onset of experimental autoimmune encephalomyelitis in PECAM-1-deficient mice. *J Clin Invest* 2002;**109**(3):383–92.
45. Alghisi GC, Ponsonnet L, Ruegg C. The integrin antagonist cilengitide activates alphaVbeta3, disrupts VE-cadherin localization at cell junctions and enhances permeability in endothelial cells. *PLoS ONE* 2009;**4**(2):e4449.
46. McCarty JH, Lacy-Hulbert A, Charest A, et al. Selective ablation of alphav integrins in the central nervous system leads to cerebral hemorrhage, seizures, axonal degeneration and premature death. *Development* 2005;**132**(1):165–76.
47. McCarty JH, Monahan-Earley RA, Brown LF, et al. Defective associations between blood vessels and brain parenchyma lead to cerebral hemorrhage in mice lacking alphav integrins. *Mol Cell Biol* 2002;**22**(21):7667–77.
48. Yong VW. Metalloproteinases: mediators of pathology and regeneration in the CNS. *Nat Rev Neurosci* 2005;**6**(12):931–44.
49. Klein T, Bischoff R. Physiology and pathophysiology of matrix metalloproteases. *Amino Acids* 2011;**41**(2):271–90.
50. Hadler-Olsen E, Fadnes B, Sylte I, Uhlin-Hansen L, Winberg JO. Regulation of matrix metalloproteinase activity in health and disease. *FEBS J* 2011;**278**(1):28–45.
51. Yong VW, Power C, Forsyth P, Edwards DR. Metalloproteinases in biology and pathology of the nervous system. *Nat Rev Neurosci* 2001;**2**(7):502–11.
52. Huntley GW. Synaptic circuit remodelling by matrix metalloproteinases in health and disease. *Nat Rev Neurosci* 2012;**13**(11):743–57.
53. Ulrich R, Gerhauser I, Seeliger F, Baumgartner W, Alldinger S. Matrix metalloproteinases and their inhibitors in the developing mouse brain and spinal cord: a reverse transcription quantitative polymerase chain reaction study. *Dev Neurosci* 2005;**27**(6):408–18.
54. Candelario-Jalil E, Yang Y, Rosenberg GA. Diverse roles of matrix metalloproteinases and tissue inhibitors of metalloproteinases in neuroinflammation and cerebral ischemia. *Neuroscience* 2009;**158**(3):983–94.
55. del Zoppo GJ, Frankowski H, Gu YH, et al. Microglial cell activation is a source of metalloproteinase generation during hemorrhagic transformation. *J Cereb Blood Flow Metab* 2012;**32**(5):919–32.
56. del Zoppo GJ, Milner R, Mabuchi T, et al. Microglial activation and matrix protease generation during focal cerebral ischemia. *Stroke* 2007;**38**(2 Suppl):646–51.
57. Rosenberg GA, Cunningham LA, Wallace J, et al. Immunohistochemistry of matrix metalloproteinases in reperfusion injury to rat brain: activation of MMP-9 linked to stromelysin-1 and microglia in cell cultures. *Brain Res* 2001;**893**(1–2):104–12.
58. Crocker SJ, Frausto RF, Whitton JL, Milner R. A novel method to establish microglia-free astrocyte cultures: comparison of matrix metalloproteinase expression profiles in pure cultures of astrocytes and microglia. *Glia* 2008;**56**(11):1187–98.
59. Wang L, Zhang ZG, Zhang RL, et al. Matrix metalloproteinase 2 (MMP2) and MMP9 secreted by erythropoietin-activated endothelial cells promote neural progenitor cell migration. *J Neurosci* 2006;**26**(22):5996–6003.
60. Wojcik-Stanaszek L, Sypecka J, Szymczak P, et al. The potential role of metalloproteinases in neurogenesis in the gerbil hippocampus following global forebrain ischemia. *PLoS ONE* 2011;**6**(7):e22465.
61. Lee SR, Kim HY, Rogowska J, et al. Involvement of matrix metalloproteinase in neuroblast cell migration from the subventricular zone after stroke. *J Neurosci* 2006;**26**(13):3491–5.

62. Nagy V, Bozdagi O, Matynia A, et al. Matrix metalloproteinase-9 is required for hippocampal late-phase long-term potentiation and memory. *J Neurosci* 2006;**26**(7): 1923–34.
63. Nagy V, Bozdagi O, Huntley GW. The extracellular protease matrix metalloproteinase-9 is activated by inhibitory avoidance learning and required for long-term memory. *Learn Mem* 2007;**14**(10):655–64.
64. Yang Y, Candelario-Jalil E, Thompson JF, et al. Increased intranuclear matrix metalloproteinase activity in neurons interferes with oxidative DNA repair in focal cerebral ischemia. *J Neurochem* 2010;**112**(1):134–49.
65. Si-Tayeb K, Monvoisin A, Mazzocco C, et al. Matrix metalloproteinase 3 is present in the cell nucleus and is involved in apoptosis. *Am J Pathol* 2006;**169**(4):1390–401.
66. Van Wart HE, Birkedal-Hansen H. The cysteine switch: a principle of regulation of metalloproteinase activity with potential applicability to the entire matrix metalloproteinase gene family. *Proc Natl Acad Sci USA* 1990;**87**(14):5578–82.
67. Gu Z, Kaul M, Yan B, et al. *S*-Nitrosylation of matrix metalloproteinases: signaling pathway to neuronal cell death. *Science* 2002;**297**(5584):1186–90.
68. Cunningham LA, Wetzel M, Rosenberg GA. Multiple roles for MMPs and TIMPs in cerebral ischemia. *Glia* 2005;**50**(4):329–39.
69. Liu KJ, Rosenberg GA. Matrix metalloproteinases and free radicals in cerebral ischemia. *Free Radic Biol Med* 2005;**39**(1):71–80.
70. Brew K, Dinakarpandian D, Nagase H. Tissue inhibitors of metalloproteinases: evolution, structure and function. *Biochim Biophys Acta* 2000;**1477**(1-2):267–83.
71. Leco KJ, Khokha R, Pavloff N, Hawkes SP, Edwards DR. Tissue inhibitor of metalloproteinases-3 (TIMP-3) is an extracellular matrix-associated protein with a distinctive pattern of expression in mouse cells and tissues. *J Biol Chem* 1994;**269**(12):9352–60.
72. Melendez-Zajgla J, Del Pozo L, Ceballos G, Maldonado V. Tissue inhibitor of metalloproteinases-4. The road less traveled. *Mol Cancer* 2008;**7**:85.
73. Nagase H. Activation mechanisms of matrix metalloproteinases. *Biol Chem* 1997;**378**(3–4):151–60.
74. McMahon S, Grondin F, McDonald PP, Richard DE, Dubois CM. Hypoxia-enhanced expression of the proprotein convertase furin is mediated by hypoxia-inducible factor-1: impact on the bioactivation of proproteins. *J Biol Chem* 2005;**280**(8):6561–9.
75. Choi DH, Kim EM, Son HJ, et al. A novel intracellular role of matrix metalloproteinase-3 during apoptosis of dopaminergic cells. *J Neurochem* 2008;**106**(1):405–15.
76. Rosenberg GA. Matrix metalloproteinases in neuroinflammation. *Glia* 2002;**39**(3): 279–91.
77. Chan PH. Reactive oxygen radicals in signaling and damage in the ischemic brain. *J Cereb Blood Flow Metab* 2001;**21**(1):2–14.
78. Heo JH, Han SW, Lee SK. Free radicals as triggers of brain edema formation after stroke. *Free Radic Biol Med* 2005;**39**(1):51–70.
79. Kamada H, Yu F, Nito C, Chan PH. Influence of hyperglycemia on oxidative stress and matrix metalloproteinase-9 activation after focal cerebral ischemia/reperfusion in rats: relation to blood–brain barrier dysfunction. *Stroke* 2007;**38**(3):1044–9.
80. Kahles T, Luedike P, Endres M, et al. NADPH oxidase plays a central role in blood–brain barrier damage in experimental stroke. *Stroke* 2007;**38**(11):3000–6.
81. Nakamura T, Keep RF, Hua Y, Schallert T, Hoff JT, Xi G. Deferoxamine-induced attenuation of brain edema and neurological deficits in a rat model of intracerebral hemorrhage. *J Neurosurg* 2004;**100**(4):672–8.
82. Nakamura T, Kuroda Y, Yamashita S, et al. Edaravone attenuates brain edema and neurologic deficits in a rat model of acute intracerebral hemorrhage. *Stroke* 2008;**39**(2): 463–9.
83. Ding-Zhou L, Marchand-Verrecchia C, Palmier B, et al. Neuroprotective effects of (S)-*N*-[4-[4-[(3,4-Dihydro-6-hydroxy-2,5,7,8-tetramethyl-2*H*-1-benzopyran-2-yl)

carbonyl]-1-piperazinyl]phenyl]-2-thiophenecarboximid-amide (BN 80933), an inhibitor of neuronal nitric-oxide synthase and an antioxidant, in model of transient focal cerebral ischemia in mice. *J Pharmacol Exp Ther* 2003;**306**(2):588–94.
84. Suda S, Igarashi H, Arai Y, Andou J, Chishiki T, Katayama Y. Effect of edaravone, a free radical scavenger, on ischemic cerebral edema assessed by magnetic resonance imaging. *Neurol Med Chir (Tokyo)* 2007;**47**(5):197–202.
85. Zhang QG, Laird MD, Han D, et al. Critical role of NADPH oxidase in neuronal oxidative damage and microglia activation following traumatic brain injury. *PLoS ONE* 2012;**7**(4):e34504.
86. Terpolilli NA, Zweckberger K, Trabold R, et al. The novel nitric oxide synthase inhibitor 4-amino-tetrahydro-L-biopterine prevents brain edema formation and intracranial hypertension following traumatic brain injury in mice. *J Neurotrauma* 2009;**26**(11):1963–75.
87. Okamoto T, Akaike T, Sawa T, Miyamoto Y, van der Vliet A, Maeda H. Activation of matrix metalloproteinases by peroxynitrite-induced protein *S*-glutathiolation via disulfide S-oxide formation. *J Biol Chem* 2001;**276**(31):29596–602.
88. Fu X, Kassim SY, Parks WC, Heinecke JW. Hypochlorous acid oxygenates the cysteine switch domain of pro-matrilysin (MMP-7). A mechanism for matrix metalloproteinase activation and atherosclerotic plaque rupture by myeloperoxidase. *J Biol Chem* 2001;**276**(44):41279–87.
89. Gu Z, Cui J, Brown S, et al. A highly specific inhibitor of matrix metalloproteinase-9 rescues laminin from proteolysis and neurons from apoptosis in transient focal cerebral ischemia. *J Neurosci* 2005;**25**(27):6401–8.
90. Gu Y, Zheng G, Xu M, et al. Caveolin-1 regulates nitric oxide-mediated matrix metalloproteinases activity and blood–brain barrier permeability in focal cerebral ischemia and reperfusion injury. *J Neurochem* 2012;**120**(1):147–56.
91. Yang Y, Estrada EY, Thompson JF, Liu W, Rosenberg GA. Matrix metalloproteinase-mediated disruption of tight junction proteins in cerebral vessels is reversed by synthetic matrix metalloproteinase inhibitor in focal ischemia in rat. *J Cereb Blood Flow Metab* 2007;**27**(4):697–709.
92. Liu J, Jin X, Liu KJ, Liu W. Matrix metalloproteinase-2-mediated occludin degradation and caveolin-1-mediated claudin-5 redistribution contribute to blood–brain barrier damage in early ischemic stroke stage. *J Neurosci* 2012;**32**(9):3044–57.
93. Del Zoppo GJ, Milner R, Mabuchi T, Hung S, Wang X, Koziol JA. Vascular matrix adhesion and the blood–brain barrier. *Biochem Soc Trans* 2006;**34**(Pt 6):1261–6.
94. Kwon I, Kim EH, del Zoppo GJ, Heo JH. Ultrastructural and temporal changes of the microvascular basement membrane and astrocyte interface following focal cerebral ischemia. *J Neurosci Res* 2009;**87**(3):668–76.
95. Vajtr D, Benada O, Kukacka J, et al. Correlation of ultrastructural changes of endothelial cells and astrocytes occurring during blood brain barrier damage after traumatic brain injury with biochemical markers of BBB leakage and inflammatory response. *Physiol Res* 2009;**58**(2):263–8.
96. Lee SR, Lo EH. Induction of caspase-mediated cell death by matrix metalloproteinases in cerebral endothelial cells after hypoxia-reoxygenation. *J Cereb Blood Flow Metab* 2004;**24**(7):720–7.
97. Asahi M, Wang X, Mori T, et al. Effects of matrix metalloproteinase-9 gene knock-out on the proteolysis of blood–brain barrier and white matter components after cerebral ischemia. *J Neurosci* 2001;**21**(19):7724–32.
98. Mun-Bryce S, Rosenberg GA. Matrix metalloproteinases in cerebrovascular disease. *J Cereb Blood Flow Metab* 1998;**18**(11):1163–72.
99. Liu W, Hendren J, Qin XJ, Shen J, Liu KJ. Normobaric hyperoxia attenuates early blood–brain barrier disruption by inhibiting MMP-9-mediated occludin degradation in focal cerebral ischemia. *J Neurochem* 2009;**108**(3):811–20.

100. Rosenberg GA, Yang Y. Vasogenic edema due to tight junction disruption by matrix metalloproteinases in cerebral ischemia. *Neurosurg Focus* 2007;**22**(5):E4.
101. McColl BW, Rothwell NJ, Allan SM. Systemic inflammation alters the kinetics of cerebrovascular tight junction disruption after experimental stroke in mice. *J Neurosci* 2008;**28**(38):9451–62.
102. Gurney KJ, Estrada EY, Rosenberg GA. Blood–brain barrier disruption by stromelysin-1 facilitates neutrophil infiltration in neuroinflammation. *Neurobiol Dis* 2006;**23**(1):87–96.
103. Reijerkerk A, Kooij G, van der Pol SM, Khazen S, Dijkstra CD, de Vries HE. Diapedesis of monocytes is associated with MMP-mediated occludin disappearance in brain endothelial cells. *FASEB J* 2006;**20**(14):2550–2.
104. Ishihara H, Kubota H, Lindberg RL, et al. Endothelial cell barrier impairment induced by glioblastomas and transforming growth factor beta2 involves matrix metalloproteinases and tight junction proteins. *J Neuropathol Exp Neurol* 2008;**67**(5):435–48.
105. Rosenberg GA, Navratil M, Barone F, Feuerstein G. Proteolytic cascade enzymes increase in focal cerebral ischemia in rat. *J Cereb Blood Flow Metab* 1996;**16**(3):360–6.
106. Rosenberg GA, Estrada EY, Dencoff JE. Matrix metalloproteinases and TIMPs are associated with blood–brain barrier opening after reperfusion in rat brain. *Stroke* 1998;**29**(10):2189–95.
107. Romanic AM, White RF, Arleth AJ, Ohlstein EH, Barone FC. Matrix metalloproteinase expression increases after cerebral focal ischemia in rats: inhibition of matrix metalloproteinase-9 reduces infarct size. *Stroke* 1998;**29**(5):1020–30.
108. Asahi M, Asahi K, Jung JC, del Zoppo GJ, Fini ME, Lo EH. Role for matrix metalloproteinase 9 after focal cerebral ischemia: effects of gene knockout and enzyme inhibition with BB-94. *J Cereb Blood Flow Metab* 2000;**20**(12):1681–9.
109. Heo JH, Lucero J, Abumiya T, Koziol JA, Copeland BR, del Zoppo GJ. Matrix metalloproteinases increase very early during experimental focal cerebral ischemia. *J Cereb Blood Flow Metab* 1999;**19**(6):624–33.
110. Lapchak PA, Chapman DF, Zivin JA. Metalloproteinase inhibition reduces thrombolytic (tissue plasminogen activator)-induced hemorrhage after thromboembolic stroke. *Stroke* 2000;**31**(12):3034–40.
111. Copin JC, Merlani P, Sugawara T, Chan PH, Gasche Y. Delayed matrix metalloproteinase inhibition reduces intracerebral hemorrhage after embolic stroke in rats. *Exp Neurol* 2008;**213**(1):196–201.
112. Gasche Y, Copin JC, Sugawara T, Fujimura M, Chan PH. Matrix metalloproteinase inhibition prevents oxidative stress-associated blood–brain barrier disruption after transient focal cerebral ischemia. *J Cereb Blood Flow Metab* 2001;**21**(12):1393–400.
113. Suzuki Y, Nagai N, Umemura K, Collen D, Lijnen HR. Stromelysin-1 (MMP-3) is critical for intracranial bleeding after t-PA treatment of stroke in mice. *J Thromb Haemost* 2007;**5**(8):1732–9.
114. Cuzner ML, Opdenakker G. Plasminogen activators and matrix metalloproteases, mediators of extracellular proteolysis in inflammatory demyelination of the central nervous system. *J Neuroimmunol* 1999;**94**(1–2):1–14.
115. Pfefferkorn T, Rosenberg GA. Closure of the blood–brain barrier by matrix metalloproteinase inhibition reduces rtPA-mediated mortality in cerebral ischemia with delayed reperfusion. *Stroke* 2003;**34**(8):2025–30.
116. Wang X, Tsuji K, Lee SR, et al. Mechanisms of hemorrhagic transformation after tissue plasminogen activator reperfusion therapy for ischemic stroke. *Stroke* 2004;**35**(11 Suppl 1):2726–30.
117. Murata Y, Rosell A, Scannevin RH, Rhodes KJ, Wang X, Lo EH. Extension of the thrombolytic time window with minocycline in experimental stroke. *Stroke* 2008;**39**(12):3372–7.

118. Gidday JM, Gasche YG, Copin JC, et al. Leukocyte-derived matrix metalloproteinase-9 mediates blood–brain barrier breakdown and is proinflammatory after transient focal cerebral ischemia. *Am J Physiol Heart Circ Physiol* 2005;**289**(2):H558–68.
119. Cuadrado E, Ortega L, Hernandez-Guillamon M, et al. Tissue plasminogen activator (t-PA) promotes neutrophil degranulation and MMP-9 release. *J Leukoc Biol* 2008;**84**(1):207–14.
120. Rosell A, Cuadrado E, Ortega-Aznar A, Hernandez-Guillamon M, Lo EH, Montaner J. MMP-9-positive neutrophil infiltration is associated to blood–brain barrier breakdown and basal lamina type IV collagen degradation during hemorrhagic transformation after human ischemic stroke. *Stroke* 2008;**39**(4):1121–6.
121. Lucivero V, Prontera M, Mezzapesa DM, et al. Different roles of matrix metalloproteinases-2 and -9 after human ischaemic stroke. *Neurol Sci* 2007;**28**(4):165–70.
122. Castellanos M, Leira R, Serena J, et al. Plasma metalloproteinase-9 concentration predicts hemorrhagic transformation in acute ischemic stroke. *Stroke* 2003;**34**(1):40–6.
123. Serena J, Blanco M, Castellanos M, et al. The prediction of malignant cerebral infarction by molecular brain barrier disruption markers. *Stroke* 2005;**36**(9):1921–6.
124. Batra A, Latour LL, Ruetzler CA, et al. Increased plasma and tissue MMP levels are associated with BCSFB and BBB disruption evident on post-contrast FLAIR after experimental stroke. *J Cereb Blood Flow Metab* 2010;**30**(6):1188–99.
125. Jha R, Battey TW, Pham L, et al. Fluid-attenuated inversion recovery hyperintensity correlates with matrix metalloproteinase-9 level and hemorrhagic transformation in acute ischemic stroke. *Stroke* 2014;**45**(4):1040–5.
126. Rosenberg GA, Navratil M. Metalloproteinase inhibition blocks edema in intracerebral hemorrhage in the rat. *Neurology* 1997;**48**(4):921–6.
127. Shigemori Y, Katayama Y, Mori T, Maeda T, Kawamata T. Matrix metalloproteinase-9 is associated with blood–brain barrier opening and brain edema formation after cortical contusion in rats. *Acta Neurochir Suppl* 2006;**96**:130–3.
128. Paul R, Lorenzl S, Koedel U, et al. Matrix metalloproteinases contribute to the blood–brain barrier disruption during bacterial meningitis. *Ann Neurol* 1998;**44**(4):592–600.
129. Leib SL, Clements JM, Lindberg RL, et al. Inhibition of matrix metalloproteinases and tumour necrosis factor alpha converting enzyme as adjuvant therapy in pneumococcal meningitis. *Brain* 2001;**124**(Pt 9):1734–42.
130. Leib SL, Leppert D, Clements J, Tauber MG. Matrix metalloproteinases contribute to brain damage in experimental pneumococcal meningitis. *Infect Immun* 2000;**68**(2):615–20.
131. Vandenbroucke RE, Libert C. Is there new hope for therapeutic matrix metalloproteinase inhibition? *Nat Rev Drug Discov* 2014;**13**(12):904–27.

SECTION IV

BRAIN EDEMA PROCESS IN PRECLINICAL MODELS

CHAPTER

12

Edema and BBB Breakdown in Stroke

Kathleen E. Salmeron, Danielle N. Edwards, Justin F. Fraser and Gregory J. Bix

University of Kentucky, Lexington, KY, United States

OUTLINE

Brain Edema.

DOI: http://dx.doi.org/10.1016/B978-0-12-803196-4.00012-6

INTRODUCTION

Stroke is a leading cause of death and disability affecting 15 million people worldwide. Although rapid recognition and treatment have dropped mortality rates, many are left with permanent disability. Approximately 87% of all strokes result from the occlusion of cerebrovasculature (ischemic strokes).[1] Current research distinguishes between zones of core infarct and areas of metabolically compromised but potentially viable tissue receiving collateral circulation known as *penumbra*.[2] Research has long focused on how endogenous mechanisms of neuroprotection and neurorepair affect penumbral expansion. Such therapeutic approaches have included antiinflammatory interventions, reactive oxygen species (ROS) scavengers, and many other targets with the common goal of mitigating the acute and chronic inflammatory responses typically seen in an ischemic stroke. This chapter will discuss acute and chronic molecular mechanisms underlying edema following ischemic stroke.

Despite advancements in reperfusion and recanalization through intravenous tPA and mechanical thrombectomy, stroke still often results in significant disability. Edema and inflammation compound the injury and can turn a disabling infarct into a deadly lesion causing mass effect. Despite advancements in recanalization, the expansion of ischemic injury occurs regardless of the restoration of blood flow. A key component of such injury expansion is the breakdown of the blood–brain barrier (BBB).[3] Here, we will examine how this breakdown leads to brain swelling (*edema*). To do this, we the first need to understand the composition and physiologic function of the BBB.

PHYSIOLOGIC STRUCTURE AND FUNCTION OF THE BLOOD–BRAIN BARRIER

Under physiologic conditions, the BBB separates the healthy brain from the systemic blood supply. It is a multilayered unit made up of astrocytic endfeet, extracellular matrix (ECM), microglia, pericytes, and nonfenestrated capillaries, which are tightly interconnected in order to protect the brain parenchyma from noxious substances in blood such as immune cells and cytokines. Moreover, the BBB is also responsible for regulating the flow of nutrients into the brain and metabolic by-products out of the brain, thereby helping to maintain a healthy brain environment.[4,5] Specifically, although the vasculature in the majority of the brain is tightly contained, astrocytes with endfeet in close proximity to the vasculature are tasked with filtering nutrients and water into the brain parenchyma and delivering them to neurons.[5,6] Importantly, astrocytes remain unreactive under normal conditions.[7] Astrocytic endfeet are anchored to, and separated from the endothelium by, the **ECM.**[4,5] The same is true with vascular endothelial cells, which are anchored to the matrix. Previously, biochemists believed that the ECM was just a water-based substance that filled the space between cells throughout the body. Today, we know that the matrix provides crucial signal transduction proteins, scaffolding for cell stability and motility, and facilitates homeostasis.[8]

Among the most important cellular signaling receptors that interact with the ECM are integrins. These interactions with ECM components such as collagen, laminin, and fibronectin, aid in cell signaling, activation, and migration.[5,9–13] In addition, brain endothelial cells form intercellular barriers called tight junctions (TJs), made up of TJs proteins including claudin, occludin, and junction adhesion molecules.[5] TJs are responsible for regulating and maintaining the relative impermeability of the brain parenchyma to cells as well as to large molecules.[4,5] Pericytes, while not as well understood, are thought to aid in this regulation of BBB permeability under normal conditions.[14,15] Taken together, the barrier is essential for protecting the brain parenchyma and, if disturbed, can result in or exacerbate brain injury.

THE PHASES OF EDEMA IN STROKE

Stroke occurs in three main phases: acute (can be further split into hyperacute and acute phases), subacute, and chronic. The hyperacute phase, occurring within the first several minutes of stroke onset, is the phase in which clinical symptoms appear but is *not* the phase in which most damage occurs. The second phase, the acute phase, occurs within the first few hours of a stroke and is the phase in which most, if not all, clinical intervention must occur for there to be any positive change in patient outcome.[2] The subacute phase occurs 24–48 hours after onset. This phase is difficult to define because the infarct has not yet reached its maximum volume. It is, however, past the therapeutic window for either IV t-PA or for endovascular thrombectomy.[2,6,16,17] The final or chronic phase can last for 1–2 weeks. This is the phase wherein the brain pathophysiology transitions from neuroprotective to neuroreparative mechanisms.[5,6,16]

HYPERACUTE PHASE: THE FIRST FEW MINUTES

The brain's response to an ischemic event is swift. Because the cells are not receiving the oxygen and energy that they need, ion pumps fail to maintain homeostatic sodium and potassium concentrations, resulting in calcium influx into the cytoplasm. This rapid ion exchange causes excitatory cells (neurons) to depolarize in a process called anoxic depolarization,[18] which releases glutamate into the extracellular space. Astrocytes, in carrying out their physiological function, take up the released glutamate using the glutamate–sodium cotransporter. In addition, the further influx of water through channels, known as aquaporins, causes osmotic swelling of the cell.[2,19] The end result is that neurons and astrocytes within the ischemic core terminally depolarize and will eventually undergo apoptosis. Some studies have shown that blockade or selective knock-down of these aquaporins, particularly aquaporin-4, helps to prevent the formation of metabolic edema after events such as stroke and traumatic brain injury.[6,19]

Meanwhile, the cells of the neurovascular unit release Hypoxia Inducible Factors (HIFs) as well as other signaling molecules, which rapidly activate glial cells. HIFs are typically not released in sufficient quantities to initiate widespread effects. Local astrocytes and microglia; however, will release pro- and antiinflammatory cytokines and other inflammatory markers into the core of the infarcted area.[17] This release of inflammatory mediators results in acute, local swelling, which triggers a decrease in diameter of the vasculature.[5,6] In addition, many of the cytokines released by glial cells cause a breech in the integrity of the BBB.[17] Although this process is just beginning within the first few minutes of ischemia, these early initial events set up a cascade of longer lasting (up to 2 weeks) consequences.

THE ACUTE PHASE: THE FIRST HOURS

At the onset of an ischemic stroke, a patient has very limited time to get treatment to reperfuse brain tissue. On the contrary, reperfusion, whereas necessary, can introduce new cascades that do compound damage. In the hyperacute phase, endothelial cells release pro- and antiinflammatory cytokines, which begin to cause the opening and closing of the BBB.[5] The mechanism by which this opening occurs is relatively unknown. However, the open barrier allows the infiltration of potentially harmful leukocytes and neutrophils.[20] These immune cells attach to the vascular endothelial cell barrier due to an increase in the interaction between selectins, the immunoglobulin (IG) superfamily such as ICAM 1 and 2 and VCAM-1, and integrins.[13,21] This attachment triggers the leukocytes to release many proinflammatory mediators into the brain parenchyma.[17] The members of the IG superfamily are indirectly activated by ischemic injury but are directly activated by the release of cytokines in a biphasic response,[5,22] with the degree of ischemia dictating the amount of cytokine that is released.[22]

If the amount of cytokine reaches a certain threshold, the BBB opens wide enough to allow the attached leukocytes and neutrophils to properly infiltrate through new gaps in the endothelial cell layer and pass through gaps between astrocytic endfeet and endothelial cell adhesions into the brain parenchyma.[13] These inflammatory cells release cytokines into the injured area in excess concentrations than are common in the brain, thus exacerbating astrocyte and microglia activation.[7] Elevated brain cytokine levels also contribute to neuronal apoptosis, thereby expanding the area of permanent ischemic damage.[2] Cellular milieu released during apoptosis causes further activation of astrocytes and microglia in the penumbral tissue.[23,24] Astrocytes and microglia, in turn, also release reactive chemicals as part of their scavenging and damage control mechanisms.[7] Activated astrocytes release inflammatory mediators such as cytokines, chemokines, and inducible nitric oxide synthase (iNOS)[25]; the latter is particularly problematic because it is a mediator of DNA damage through its synthesis of ROS (discussed later in this chapter).[26,27]

Microglia, acting originally as the brain's macrophages, become much more like phagocytes, releasing cytotoxic factors into the surrounding tissue.[17,28] This process persists throughout the subsequent phases of ischemia and induces secondary mechanisms of injury. On the contrary, these mechanisms are not halted once a patient is recanalized (Figs. 12.1 and 12.2).

Electrically, the spread of the ischemic penumbra can be partially explained by phenomena known as periinfarct spreading depression-like depolarizations (PIDs).[2] This largely accepted and widely studied hypothesis states that the expansion of the penumbra is due in part to bioelectrical disturbances caused by the combination of an increased energy demand of penumbral cells and the decreased energy supply due to ischemia. These intermittent depolarizations of neurons proximal to the core cause widespread energy demand by surrounding neurons. An increase in energy demand without adequate energy supply causes further terminal depolarization. Each one of these PIDs may even increase the infarct volume by more than 20%.[29,30] Moreover, PIDs are implicated in many more downstream mechanisms of secondary injury (discussed below) such as the formation of ROS, DNA damage, and mitochondrial dysfunction.[2,29] In fact, many researchers believe that this mechanism holds the key to pharmacological interventions for stroke patients. The PID phenomena is just one of the many excitotoxic pathways by which infarcts develop, and it is by this mechanism that receptor blockers and antagonists (glutamate, NMDA, etc.) act.[29]

THE PATIENT IN THE CLINIC: REPERFUSION AND EDEMA

Most brain injury that has thus far occurred is from ischemia brought on by the clot, but by removing the clot (recanalization) and allowing normal blood flow (reperfusion) the brain is exposed to not only the

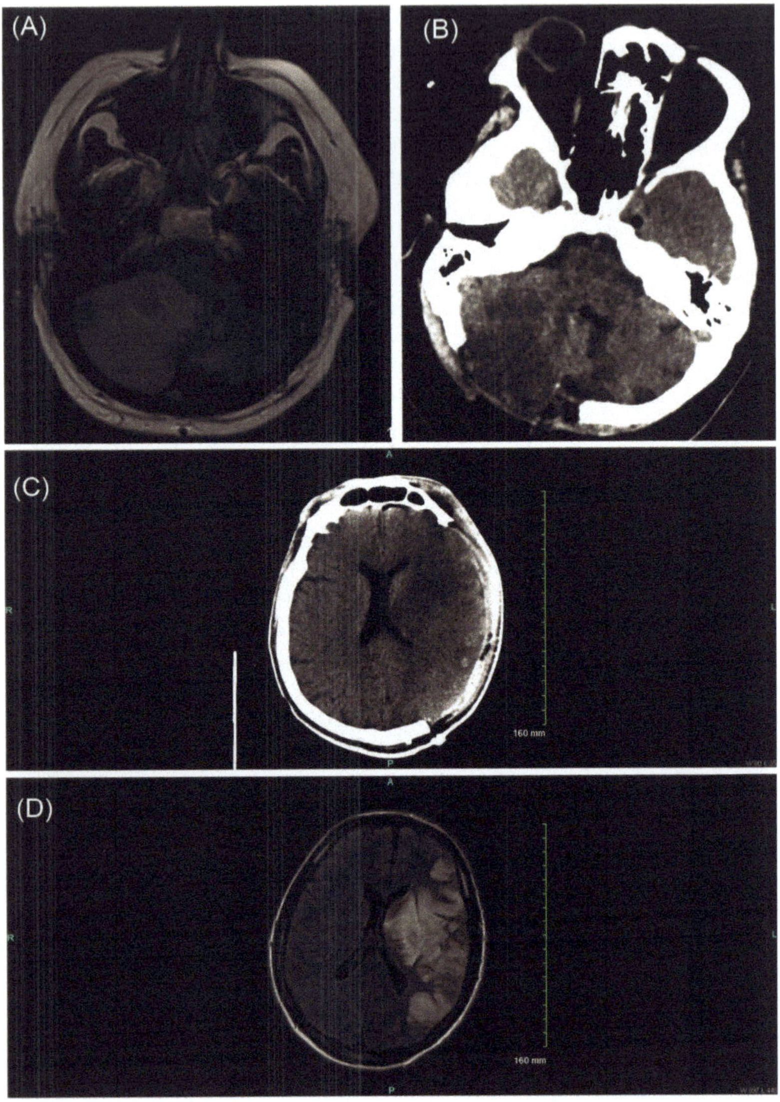

FIGURE 12.1 **A 39-year-old female with underlying coagulopathy presented with a cerebellar stroke with significant edema and underwent decompression.** MRI FLAIR sequence showing large left cerebellar stroke, compression of brainstem, and fourth ventricle (A). Postoperative CT shows craniectomy and partial lobectomy with relief of mass effect on brainstem (B). A 53-year-old male with left acute occlusion of internal carotid and anterior/middle cerebral arteries underwent successful intravenous tPA and mechanical thrombectomy but developed significant edema requiring decompression. MRI FLAIR sequence showing large left MCA territory edema from some infarction (C). Postoperative CT shows craniectomy with significant herniation of edematous brain out of craniectomy defect (D).

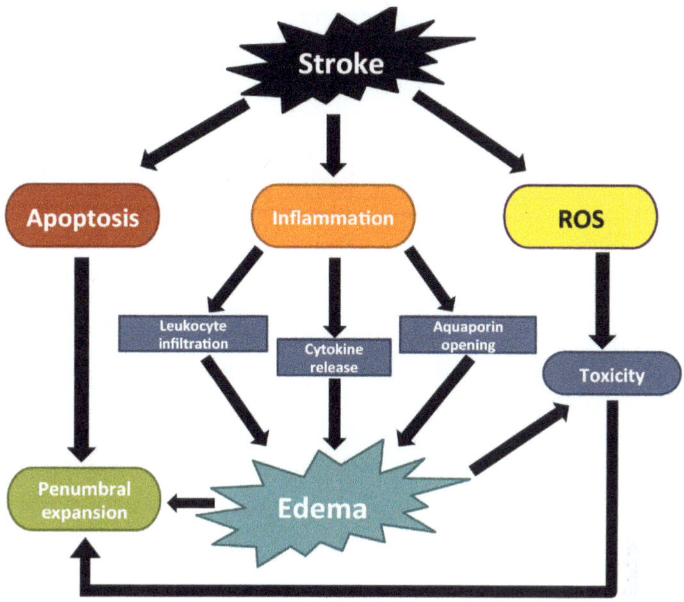

FIGURE 12.2 **Overview of the cyclical and compounding nature of edema and stroke.** Stroke causes apoptosis due to lack of oxygen and nutrients in the brain. The introduction of inflammation contributes to edema through the release of inflammatory cytokines, water, and the infiltration of leukocytes into the brain parenchyma. Reactive oxygen species contribute to glutamate and excitotoxicity, which results in further apoptosis thereby expanding the dead or infarcted area.

oxygen and nutrients that it needs but also to deleterious components of the blood due to the pathologic opening of the BBB.[5] Edema caused by recanalization can be fatal.[31,32] The mechanisms of this deterioration have until recently been poorly understood.[33] This acute worsening condition in some patients can be partially explained by the observation that many of them did not achieve recanalization. Another well-studied hypothesis is that it is due in part to the aforementioned biphasic opening of the BBB.[33] This opening allows the influx of materials from the blood, mainly fibrin and fibrinogen from the blood clot, to enter the infarct.[4,34–38] Once the fibrin and fibrinogen exit the lumen, the rate of translocation of leukocytes increases,[39–41] activating microglia, and thus phagocytosis[42] and binding receptor ICAM-1,[4] the same as described above. In addition, fibrinogen is able to inhibit the growth of new neurites[4,43] as well as inhibit axon regenerations, preventing rejuvenation of the tissue and allowing symptoms to remain and progress.[4] As such, current trends in recanalization therapy, specifically selection of patients for mechanical thrombectomy, seek to use imaging modalities to select patients for recanalization who are most likely to benefit. This continues to lead to avenues in research, as imaging modalities and rapid systems of stroke care seek to avoid recanalizing tissue in which the balance of risk/benefit tilts toward induction of further harm.

EDEMA

All of the above circumstances can lead to edema. Although edema is considered a poor outcome in peripheral tissues, it is potentially deadly in the brain.[44] Of the four specific kinds of edema (intravascular, extracellular, intracellular, and hydrocephalic or CSF edema),[45] clinicians can see all of them in stroke patients. Although edema is typically thought of in the circumstances of wet edema, nearly all of the damage seen in stroke is caused by cytotoxic, or metabolic, edema.[45] Metabolic edema is so named for its characteristic disruption of metabolic homeostatic processes, in particular the sodium pump discussed earlier. This cytotoxic edema causes upregulation of matrix metalloproteases that disrupt TJs and other components of the BBB resulting in further swelling.[46,47] As the barrier continues to open, the influx of proinflammatory proteins and cells lead to a state of secondary inflammation, which will be discussed later. It is important to note, however, that this secondary inflammation pathway forms in the periphery of the ischemic tissue presumably to aid in the protection of healthier tissue.

In addition, edema is nearly impossible to effectively mitigate. The immediate process of ischemic stroke causes primarily cytotoxic edema in the injured tissue.[45] Although this is typically well-tolerated in cases of small strokes, it can compound brain injury and even cause death in two particular types of ischemic strokes. Large anterior circulation strokes involving the middle cerebral artery territory can cause significant compression of the brain and uncal herniation due to edema.[48] In the posterior fossa in which the potential space is smaller, strokes of the cerebellum and brainstem (even relatively small in size) can result in significant edema, which then further compresses the brainstem, resulting in obstruction of outflow of cerebrospinal fluid. Obstruction of the 4th ventricle and its outflow results in hydrocephalus, which further compounds the compressive damage to the brain. Either of these cases can quickly result in neurological injury and death. As such, multiple studies and guidelines support decompression of ischemic strokes in the posterior fossa and in selected patients with MCA strokes. Surgical decompression (Decompressive Craniectomy) involves the removal of the overlying skull, opening of the dura, and, in some cases, resection of the ischemic tissue.[49–51] This provides a space into which the edematous tissue can swell, reducing, and, in most cases, eliminating the compressive forces on the brain and brainstem.[46] It can also improve perfusion to ischemic tissue, most notably to the penumbral areas.[52] Besides decompression, there are tenets of medical therapy designed to minimize the edema, though they are not as effective for the cytotoxic edema present in ischemic stroke.[46,53]

THE SUBACUTE PHASE: 24–48 HOURS

On the contrary, the patients struggle with edema do not end upon recanalization. In fact, some patients worsen during the subacute phase of a stroke. Inflammatory cytokines and ROS continue to cause devastating damage until the infarct achieves maturity. The massive increase in cytokine release occurs cyclically; injury mediates cytokine release, causing further injury, which mediates more cytokine release. In particular, cytokines MCP-1, interleukin-1 (IL-1), and TNF-α are upregulated.[12,54] There is a biphasic upregulation in the brain after ischemia of IL-1β within minutes and of TNF-α within hours of injury.[55–58] MCP-1 is upregulated in astrocytes and brain endothelial cells, allowing for an increase in leukocyte infiltrantion.[10,12] Chemokines are also released after ischemia to recruit inflammatory and immune responses such as increasing leukocyte infiltration to the vascular wall.[59] There are four groups associated with chemokines, C, CC, CXC, and CX3C.[60] Their increase has been shown to occur and localize to neurons in the brain.[60,61] The stimulated immune response by all the above factors also activates the arachidonic acid cascade[62] causing an influx of calcium[2] due to the disruption of the ATP-dependent calcium efflux mechanisms and SERCA on endothelial cells,[11,63] and a conversion of arachidonic acid to prostaglandin.[13] These processes occur over a spatiotemporal gradient meaning that it occurs over an extended period of time and contributes to penumbral expansion.[7]

Inflammatory mediators such as cytokines and chemokines are not the only deleterious molecules to be released in the brain during injury. ROS, such as superoxide, peroxide, and hydroxyl anion,[5] which are the most damaging, are produced as normal by-products of energy producing mechanisms, particularly the electron transport chain occurring in the mitochondria, and are necessary for some normal cell signaling.[64] During injury, these ROS are able to build up in damaging amounts by overcoming the scavenging mechanisms and by intracellular increases in calcium sodium and adenosine as a direct result from the ischemic injury. This results in damage to signaling pathways, TJ proteins, and enzyme activation.[5,65]

Isoforms of nitric oxide (NO), enodothelial-NOS (eNOS), neuronal-NOS (nNOS), and iNOS, indirectly produce ROS through signaling pathways.[5,66] An increase in both the calcium-independent iNOS and calcium-dependent nNOS in both the core and penumbra after 24–48 hours corresponds with the increase in symptoms and the size of the infarct volume.[5,26,27]

THE CHRONIC PHASE: 3 DAYS–2 WEEKS

During the first week after a patient suffers a stroke, there are a multitude of changes that go on in the brain. The lesion volume continues to increase until days 3–8 poststroke and subsequently decreases in size.[67] In fact, the lesion continues to change up to 30 days after stroke onset.[68] Edema may persist, and the patient may remain hospitalized. Despite all of this, the brain is doing remarkable things to repair itself. The vasculature is rebuilding itself by means of angiogenesis, which inadvertently results in a more open BBB due to the generation of leaky neovasculature.[5,35,69] This process serves not only to rebuild and revascularize dead brain but also to serve as a scaffold network for neuronal progenitor cells (NPCs) to migrate along in their efforts to repopulate and regenerate the infarcted area.[69] Indeed, the newly generated periinfarct blood vessels, driven by poststroke increases in angiogenic growth factors such as VEGF and angiopoietin-2, appear to be overall less effective at restoring blood flow to ischemic brain and more linked to forming a protective neurovascular niche for guiding NPCs toward the site of injury. As such, these nascent blood vessels are tortuous and leaky and often support little to no active blood flow as this would only further worsen potential hemorrhage and edema into periinfarct brain.[70]

The brain's repair pathways appear to be largely inadequate and may represent a simple reactivation of dormant developmental pathways. Significant research effort has focused on improving neurorepair pathways.[69,71] For one example, the activated and migrated astrocytes in the infarcted region undergo astrogliosis forming an area of scarring that forms a physical barrier that prevents new neurons from repopulating the dead region. This results in very few neural progenitor cells reaching the infarcted brain region.[72]

Another problem that NPCs face is that even those that reach their termination site often do not differentiate into a neuronal line. In fact, only about 1% of the NPCs born at the subventricular zone mature into neurons.[72] Current research has aimed to encourage the repopulation of the region with mature neurons by administering trophic factors to the site of injury. Thus far, these trials have met little success. Other labs have administered NPCs directly into the infarct. These experiments have met marginally greater success but still not enough to make a clinical impact[69] as the NPCs tend to exacerbate scarring by differentiating in a nonneuronal pathway. Still other labs have worked to reduce astrogliosis, which would, in theory, revitalize repopulation efforts.[7]

SUMMARY

Stroke is a devastating event that affects millions of people every year. Although much research is conducted to improve stroke therapy, more work remains to improve patient outcomes. Although most research to date has focused on mitigating events happening in the hyperacute and acute phases after stroke, a tremendous need persists to prevent or mitigate later events such as edema. Edema, secondary to disruption of the BBB, significantly worsens ischemic stroke brain injury in part via mass effect expansion of the brain in its finite cranial vault, a process that can lead to fatal brain herniation if left unchecked. Indeed, our few current therapeutic options for ischemic stroke that target dislodging or dissolving the at fault blood clot, actually further exacerbate the potential of BBB breakdown and edema via resultant hyperemia. Therefore, our success at targeting the offending blood clot must be paired with additional therapeutic approaches designed to limit or eliminate subsequent edema if we are to make significant gains in ischemic stroke treatment. In addition, a renewed focus on neurorepair strategies may also yield better therapies to minimize long-term disability that is common even with the current standard of care therapy.

References

1. Mozaffarian D, Benjamin EJ, Go AS, Arnett DK, Blaha MJ, Cushman M, et al. Heart disease and stroke statistics—2015 update: a report from the American Heart Association. *Circulation* 2015;**131**(4):e29–322.
2. Heiss WD. The ischemic penumbra: how does tissue injury evolve? *Ann N Y Acad Sci* 2012;**1268**:26–34.
3. Denes A, Ferenczi S, Kovacs K. Systematic inflammatory changes comprise survival after experimental stroke via augmenting brain inflammation, blood–brain barrier damage and brain oedema independently of infarct size. *J Neuroinflammation* 2011;**8**:164.
4. Baeten KM, Akassoglou K. Extracellular matrix and matrix receptors in blood–brain barrier formation and stroke. *Dev Neurobiol* 2011;**71**(11):1018–39.
5. Sandoval KE, Witt KA. Blood–brain barrier tight junction permeability and ischemic stroke. *Neurobiol Dis* 2008;**32**(2):200–19.
6. Yang Y, Rosenberg GA. Blood–brain barrier breakdown in acute and chronic cerebrovascular disease. *Stroke* 2011;**42**(11):3323–8.
7. Choudhury GR, Ding S. Reactive astrocytes and therapeutic potential in focal ischemic stroke. *Neurobiol Dis* 2015;**85**:234–44.
8. Hynes RO. The extracellular matrix: not just pretty fibrils. *Science* 2009;**326**(5957):1216–9.
9. del Zoppo GJ, Milner R. Integrin-matrix interactions in the cerebral microvasculature. *Arterioscler Thromb Vasc Biol* 2006;**26**(9):1966–75.
10. Dimitrijevic OB, Stamatovic SM, Keep RF, Andjelkovic AV. Effects of the chemokine CCL2 on blood–brain barrier permeability during ischemia-reperfusion injury. *J Cereb Blood Flow Metab* 2006;**26**(6):797–810.
11. Brown RC, Davis TP. Calcium modulation of adherens and tight junction function—a potential mechanism for blood–brain barrier disruption after stroke. *Stroke* 2002;**33**(6):1706–11.
12. Huang H, Upadhyay M, Tamargo RJ. Inflammation in stroke and focal cerebral ischemia. *World Neurosurg* 2006;**66**(3):232–45.
13. Wang CX, Shuaib A. Critical role of microvasculature basal lamina in ischemic brain injury. *Prog Neurobiol* 2007;**83**(3):140–8.
14. Mitic LL, Anderson JM. Molecular architecture of tight junctions. *Annu Rev Physiol* 1998;**60**:121–42.

15. von Tell D, Armulik A, Betsholtz C. Pericytes and vascular stability. *Exp Cell Res* 2006;**312**(5):623–9.
16. Jung JE, Kim GS, Chen H, Maier CM, Narasimhan P, Song YS, et al. Reperfusion and neurovascular dysfunction in stroke: from basic mechanisms to potential strategies for neuroprotection. *Mol Neurobiol* 2010;**41**(2–3):172–9.
17. Wang Q, Tang XN, Yenari MA. The inflammatory response in stroke. *J Neuroimmunol* 2007;**18**(1–2):53–68.
18. Revah O, Lasser-Katz E, Fleidervish IA, Gutnick MJ. The earliest neuronal responses to hypoxia in the neocortical circuit are glutamate-dependent. *Neurobiol Dis.* 2016;**95**:158–67. http://dx.doi.org/10.1016/j.nbd.2016.07.019. PubMed PMID: 27443966.
19. Zador Z, Stiver S, Wang V, Manley G. Role of aquaporin-4 in cerebral edema and stroke. *Handb Exp Pharmacol* 2009;**190**:159–70.
20. Hallenbeck J. Significance of the inflammatory response in brain ischemia. *Acta Neurochir* 1996;**66**(Supplement 66):27–31.
21. DeGraba TJ. The role of inflammation after acute stroke—utility of pursuing anti-adhesion molecule therapy. *Neurology* 1998;**51**(3):S62–8.
22. Allan SM, Rothwell NJ. Cytokines and acute neurodegeneration. *Nat Rev Neurosci* 2001;**2**(10):734–44.
23. Lull ME, Block ML. Microglial activation and chronic neurodegeneration. *Neurotherapeutics* 2010;**7**(4):354–65. http://dx.doi.org/10.1016/j.nurt.2010.05.014. PubMed PMID: 20880500; PMCID: PMC2951017.
24. Adle-Biassette H, Chrétien F, Wingertsmann L, Héry C, Ereau F, Scaravilli F, et al. Neuronal apoptosis does not correlate with dementia in HIV infection but is related to microglial activation and axonal damage. *Neuropathol Appl Neurobiol* 1999;**25**:123–33.
25. Dong YS, Benveniste EN. Immune function of astrocytes. *Glia* 2001;**36**(2):180–90.
26. Iadecola C, Zhang FG, Xu S, Casey R, Ross ME. Inducible nitric-oxide synthase gene-expression in brain following cerebral-ischemia. *J Cereb Blood Flow Metab* 1995;**15**(3):378–84.
27. Iadecola C, Zhang FY, Xu XH. Inhibition of inducible nitric-oxide synthase ameliorates cerebral ischemic damage. *Am J Physiol-Regul Integr Comp Physiol* 1995;**268**(1):R286–92.
28. Schwartz M. Macrophages and microglia in central nervous system injury: are they helpful or harmful? *J Cereb Blood Flow Metab* 2003;**23**(4):385–94.
29. Schwartz M, Shaked I, Fisher J, Mizrahi T, Schori H. Protective autoimmunity against the enemy within: fighting glutamate toxicity. *Trends Neurosci* 2003;**26**(6):297–302.
30. Busch E, Gyngell ML, Eis M, Hoehn-Berlage M, Hossmann KA. Potassium-induced cortical spreading depressions during focal cerebral ischemia in rats: contribution to lesion growth assessed by diffusion-weighted NMR and biochemical imaging. *J Cereb Blood Flow Metab* 1996;**16**:1090–9.
31. Fugate JE, Kalimullah EA, Wijdicks EF. Angioedema after tPA: what neurointensivists should know. *Neurocrit Care* 2012;**16**(3):440–3.
32. Strbian D, Meretoja A, Putaala J, Kaste M, Tatlisumak T, Helsinki Stroke Thrombolysis Registry G Cerebral edema in acute ischemic stroke patients treated with intravenous thrombolysis. *Int J Stroke* 2013;**8**(7):529–34.
33. Tisserand M, Seners P, Turc G, Legrand L, Labeyrie MA, Charron S, et al. Mechanisms of unexplained neurological deterioration after intravenous thrombolysis. *Stroke* 2014;**45**(12):3527–34.
34. Baumann E, Preston E, Slinn J, Stanimirovic D. Post-ischemic hypothermia attenuates loss of the vascular basement membrane proteins, agrin and SPARC, and the blood–brain barrier disruption after global cerebral ischemia. *Brain Res* 2009;**1269**:185–97.
35. Schachtrup C, Ryu JK, Helmrick MJ, Vagena E, Galanakis DK, Degen JL, et al. Fibrinogen triggers astrocyte scar formation by promoting the availability of active TGF-beta after vascular damage. *J Neurosci* 2010;**30**(17):5843–54.

36. Okada Y, Copeland BR, Fitridge R, Koziol JA, Delzoppo GJ. Fibrin contributes to microvascular obstructions and parenchymal changes during early focal cerebral-ischemia and reperfusion. *Stroke* 1994;**25**(9):1847–53.
37. Murray V, Norrving B, Sandercock PAG, Terent A, Wardlaw JM, Wester P. The molecular basis of thrombolysis and its clinical application in stroke. *J Intern Med* 2010;**267**(2):191–208.
38. Adams RA, Schachtrup C, Davalos D, Tsigelny I, Akassoglou K. Fibrinogen signal transduction as a mediator and therapeutic target in inflammation: Lessons from multiple sclerosis. *Curr Med Chem* 2007;**14**(27):2925–36.
39. Languino LR, Plescia J, Duperray A, Brian AA, Plow EF, Geltosky JE, et al. Fibrinogen mediates leukocyte adhesion to vascular endothelium through an icam-1-dependent pathway. *Cell* 1993;**73**(7):1423–34.
40. Sahni A, Arevalo MT, Sahni SK, Simpson-Haidaris PJ. The VE-cadherin binding domain of fibrinogen induces endothelial barrier permeability and enhances transendothelial migration of malignant breast epithelial cells. *Int J Cancer* 2009;**125**(3):577–84.
41. Petzelbauer P, Zacharowski PA, Miyazaki Y, Friedl P, Wickenhauser G, Castellino FJ, et al. The fibrin-derived peptide B beta(15-42) protects the myocardium against ischemia-reperfusion injury. *Nat Med* 2005;**11**(3):298–304.
42. Tilling T, Korte D, Hoheisel D, Galla HJ. Basement membrane proteins influence brain capillary endothelial barrier function in vitro. *J Neurochem* 1998;**71**(3):1151–7.
43. Schachtrup C, Lu P, Jones LL, Lee JK, Lu J, Sachs BD, et al. Fibrinogen inhibits neurite outgrowth via beta 3 integrin-mediated phosphorylation of the EGF receptor. *Proc Natl Acad Sci USA* 2007;**104**(28):11814–19.
44. Battey TW, Karki M, Singhal AB, Wu O, Sadaghiani S, Campbell BC, et al. Brain edema predicts outcome after nonlacunar ischemic stroke. *Stroke* 2014;**45**(12):3643–8.
45. O'Brien MD. Ischemic cerebral edema. A review. *Stroke* 1979;**10**(6):623–8.
46. Bardutzky J, Schwab S. Antiedema therapy in ischemic stroke. *Stroke* 2007;**38**(11): 3084–94.
47. Kahle KT, Simard JM, Staley KJ, Nahed BV, Jones PS, Sun D. Molecular mechanisms of ischemic cerebral edema: role of electroneutral ion transport. *Physiology (Bethesda)* 2009;**24**:257–65.
48. Menon BK, Bal S, Modi J, Sohn SI, Watson TW, Hill MD, et al. Anterior temporal artery sign in CT angiography predicts reduced fatal brain edema and mortality in acute M1 middle cerebral artery occlusions. *J Neuroimaging* 2012;**22**(2):145–8.
49. Geurts M, van der Worp HB, Kappelle LJ, Amelink GJ, Algra A, Hofmeijer J, et al. Surgical decompression for space-occupying cerebral infarction: outcomes at 3 years in the randomized HAMLET trial. *Stroke* 2013;**44**(9):2506–8.
50. Hofmeijer J, Kappelle LJ, Algra A, Amelink GJ, van Gijn J, van der Worp HB. Surgical decompression for space-occupying cerebral infarction (the Hemicraniectomy After Middle Cerebral Artery infarction with Life-threatening Edema Trial [HAMLET]): a multicentre, open, randomised trial. *Lancet Neurol* 2009;**8**(4):326–33.
51. Jauch EC, Saver JL, Adams Jr. HP, Bruno A, Connors JJ, Demaerschalk BM, et al. Guidelines for the early management of patients with acute ischemic stroke: a guideline for healthcare professionals from the American Heart Association/American Stroke Association. *Stroke* 2013;**44**(3):870–947.
52. Slotty PJ, Kamp MA, Beez T, Beenen H, Steiger HJ, Turowski B, et al. The influence of decompressive craniectomy for major stroke on early cerebral perfusion. *J Neurosurg* 2015;**123**(1):59–64.
53. Sheth KN. Novel approaches to the primary prevention of edema after ischemia. *Stroke* 2013;**44**(6 Suppl 1):S136.
54. Han HS, Yenari MA. Cellular targets of brain inflammation in stroke. *Curr Opin Investig Drugs* 2003;**4**(5):522–9.

55. Offner H, Subramanian S, Parker SM, Afentoulis ME, Vandenbark AA, Hurn PD. Experimental stroke induces massive, rapid activation of the peripheral immune system. *J Cereb Blood Flow Metab* 2006;**26**(5):654–65.
56. Murakami Y, Saito K, Hara A, Zhu Y, Sudo K, Niwa M, et al. Increases in tumor necrosis factor-alpha following transient global cerebral ischemia do not contribute to neuron death in mouse hippocampus. *J Neurochem* 2005;**93**(6):1616–22.
57. Liu T, Clark RK, McDonnell PC, Young PR, White RF, Barone FC, et al. Tumor-necrosis-factor-alpha expression in ischemic neurons. *Stroke* 1994;**25**(7):1481–8.
58. Haqqani AS, Nesic M, Preston E, Baumann E, Kelly J, Stanimirovic D. Characterization of vascular protein expression patterns in cerebral ischemia/reperfusion using laser capture microdissection and ICAT-nanoLC–MS/MS. *Faseb Journal* 2005;**19**(13):1809–21.
59. Smith CW. Leukocyte-endothelial cell-interactions. *Semin Hematol* 1993;**30**(4):45–55.
60. Bajetto A, Bonavia R, Barbero S, Florio T, Schettini G. Chemokines and their receptors in the central nervous system. *Front Neuroendocrinol* 2001;**22**(3):147–84.
61. Emsley HCA, Tyrrell PJ. Inflammation and infection in clinical stroke. *J Cereb Blood Flow Metab* 2002;**22**(12):1399–419.
62. Schwab JM, Beschorner R, Meyermann R, Gozalan F, Schluesener HJ. Persistent accumulation of cyclooxygenase-1-expressing microglial cells and macrophages and transient upregulation by endothelium in human brain injury. *J Neurosurg* 2002;**96**(5):892–9.
63. Tiruppathi C, Minshall RD, Paria BC, Vogel SM, Malik AB. Role of Ca2+ signaling in the regulation of endothelial permeability. *Vascul Pharmacol* 2002;**39**(4–5):173–85.
64. Chan PH. Reactive oxygen radicals in signaling and damage in the ischemic brain. *J Cereb Blood Flow Metab* 2001;**21**(1):2–14.
65. Kimura C, Oike M, Ito Y. Hypoxia-induced alterations in Ca2+ mobilization in brain microvascular endothelial cells. *Am J Physiol-Heart Circ Physiol* 2000;**279**(5):H2310–8.
66. Banan A, Farhadi A, Fields JZ, Zhang LJ, Shaikh M, Keshavarzian A. The delta-isoform of protein kinase C causes inducible nitric-oxide synthase and nitric oxide up-regulation: key mechanism for oxidant-induced carbonylation, nitration, and disassembly of the microtubule cytoskeleton and hyperpermeability of barrier of intestinal epithelia. *J Pharmacol Exp Therap* 2003;**305**(2):482–94.
67. Lansberg MG, O'Brien MW, Tong DC, Moseley ME, Albers GW. Evolution of cerebral infarct volume assessed by diffusion-weighted magnetic resonance imaging. *Arch Neurol* 2001;**58**(4):613–17.
68. Gaudinski MR, Henning EC, Miracle A, Luby M, Warach S, Latour LL. Establishing final infarct volume: stroke lesion evolution past 30 days is insignificant. *Stroke* 2008;**39**(10):2765–8. http://dx.doi.org/10.1161/STROKEAHA.107.512269. PubMed PMID: 18635854; PMCID: PMC2740989.
69. Skardelly M, Gaber K, Burdack S, Scheidt F, Schuhmann MU, Hilbig H, et al. Transient but not permanent benefit of neuronal progenitor cell therapy after traumatic brain injury: potential causes and translational consequences. *Front Cell Neurosci* 2014;**8**:318.
70. Ohab JJ, Fleming S, Blesch A, Carmichael ST. A neurovascular niche for neurogenesis after stroke. *J Neurosci* 2006;**26**(50):13007–16. http://dx.doi.org/10.1523/JNEUROSCI.4323-06.2006. PubMed PMID: 17167090.
71. Rodriguez-Grande B, Swana M, Nguyen L, Englezou P, Maysami S, Allan SM, et al. The acute-phase protein PTX3 is an essential mediator of glial scar formation and resolution of brain edema after ischemic injury. *J Cereb Blood Flow Metab* 2014;**34**(3):480–8.
72. Kahle MP, Bix GJ. Neuronal restoration following ischemic stroke: influences, barriers, and therapeutic potential. *Neurorehabil Neural Repair* 2013;**27**(5):469–78.

Further Reading

Hacke W, Kaste M, Bluhmki E, Brozman M, Davalos A, Guidetti D, et al. Thrombolysis with alteplase 3 to 4.5 hours after acute ischemic stroke. *N Engl J Med* 2008;**359**(13):1317–29.

Hom J, Dankbaar JW, Soares BP, Schneider T, Cheng SC, Bredno J, et al. Blood–brain barrier permeability assessed by perfusion CT predicts symptomatic hemorrhagic transformation and malignant edema in acute ischemic stroke. *AJNR Am J Neuroradiol* 2011;**32**(1):41–8.

Saqqur M, Uchino K, Demchuk AM, Molina CA, Garami Z, Calleja S, et al. Site of arterial occlusion identified by transcranial Doppler predicts the response to intravenous thrombolysis for stroke. *Stroke* 2007;**38**(3):948–54.

Wardlaw JM, Warlow CP, Counsell C. Systematic review of evidence on thrombolytic therapy for acute ischaemic stroke. *Lancet* 1997;**350**(9078):607–14.

Balasubramaian A, Mitchell P, Dowling R, Yan B. Evolution of endovascular therapy in acute stroke: implications of device development. *J Stroke* 2015;**17**(2):127–37.

CHAPTER

13

Brain Edema Formation in Traumatic Brain Injury

Sighild Lemarchant[1] *and Jérôme Badaut*[1,2]

[1]CNRS-University of Bordeaux, Bordeaux, France [2]Loma Linda University School of Medicine, Loma Linda, CA, United States

OUTLINE

Brain Edema.
DOI: http://dx.doi.org/10.1016/B978-0-12-803196-4.00013-8

Abbreviations

AD	Alzheimer's disease
ADC	apparent diffusion coefficient
AP5	amino-5-phosphonopentanoate
AQP	aquaporin
AQP4	aquaporin 4
AMT	adsorptive-mediated transcytosis
ASIC	acid-sensing ion channel
ATP	adenosine triphosphate
ATPase	adenosine triphosphatase
BBB	blood–brain barrier
CC	corpus callosum
CCI	controlled cortical impact
CHI	closed-head injury
CNS	central nervous system
CT	computerized tomography
DAI	diffuse axonal injury
DWI	diffusion weighted imaging
EAAT	excitatory amino-acid transporter
ET_B-R	endothelin B receptor
FPI	fluid percussion injury
GCS	Glasgow Coma Scale
GLAST	glutamate aspartate transporter
GLT1	glutamate transporter-1
GLUT	glucose transporter
ICP	intracranial pressure
IgG	immunoglobulin G
IL	interleukin
IL-1β	interleukin-1β
Kir4.1	inward-rectifying potassium channel 4.1
MCT	monocarboxylate transporter
MRI	magnetic resonance imaging
MMP	matrix metalloproteinase
NHE	sodium-hydrogen antiporter
NKCC1	$Na^+/K^+/Cl^-$ cotransporter-1
NMDAr	*N*-methyl-D-aspartate receptor
PND	postnatal day
RMT	receptor-mediated transcytosis
ROS	reactive oxygen species
SGLT1	sodium-dependent glucose cotransporter type-1
siAQP4	small interfering RNA targeting aquaporin 4
SUR1-$NC_{Ca\text{-}ATP}$	sulfonylurea receptor 1-regulated nonselective cation channel
TBI	traumatic brain injury
TNF-α	tumor necrosis factor α
VEGF	vascular endothelial growth factor
VEGFR-2	vascular endothelial growth factor receptor-2
WD	weight drop

TRAUMATIC BRAIN INJURY PATHOPHYSIOLOGY

Clinical Aspects of Traumatic Brain Injury

Traumatic brain injury (TBI) is caused by a sudden and unexpected physical insult to the brain and represents one of the leading causes of disability and mortality worldwide.[1] Due to the increased number of road traffic accidents and aging populations the incidence of TBI is rising on a global level.[2] In the United States, the annual incidence of TBI was 1.7 million during 2002–2006 with young children being at the greatest risk (<5 years old), followed by adolescents (15–19 years old), and older adults (>75 years old).[3] It is estimated that 3.2–5.3 million people, or 1.1%–1.7% of the US population, currently suffer from TBI-related disability and long-term complications interfering with daily quality of life.[4–8] Although the TBI-related death rate decreased by 8.2% in the United States during the 1997–2007 decade compared to previous years,[9] the incidence of TBI did not decrease significantly and complications remain a major public health problem. In recent years, it was recognized that TBI is not only an acute event but represents a complex chronic brain disease, sharing common injury pathways with chronic neurodegenerative disorders, such as Alzheimer disease (AD).[10–12]

TBI leads to heterogeneous brain damage resulting from various external mechanical forces, for example direct impact to the skull, rapid acceleration and deceleration, blast waves, or penetration injury. TBI is often followed by short or long-term loss of consciousness[13] and is therefore clinically evaluated by the patient's level of consciousness by the Glasgow Coma Scale (GCS). This evaluation is in most cases combined with a morphological evaluation of the brain using computerized tomography (CT) to classify the injury to the brain as mild, moderate, or severe.[14] However, these severity categories are not fully standardized and may not reflect all forms of injury, especially diffuse axonal injury (DAI) which is not visible in CT scans and particularly frequent among patients with moderate and mild TBI. Generally, brain edema is associated with more severe forms of TBI, that is moderate and severe TBI; however, there is a recent literature on the presence and importance of edema formation following repetitive mild head injury.[15] Specifically under these circumstances massive edema formation may occur. The current chapter focuses on preclinical research investigating the molecular and cellular events contributing to the formation of traumatic brain edema. Refer to Chapter 25, Perspectives on Future Translational Research on Brain Edema, for the clinical aspect and treatment of the edema following TBI.

Pre-clinical Models and Edema Aspects

We will not discuss here which model is the most suitable to study pathophysiological events after TBI since these models have been extensively described and reviewed in various other articles.[16,17] Instead, we will review the most common animal models used to study the molecular and cellular mechanisms behind the edema process. In the animal models, TBI is produced with a quantifiable mechanical force in amplitude, duration, velocity, and acceleration directly applied to the brain. The animals have either the head fixed or free of movement. Depending of the pathophysiology, the animal models can be classified into five major groups in order to reflect the clinical picture: focal, mixed, diffuse, complex, and other.[17] In the current chapter we will present results from three common models mostly: weight drop (WD), fluid percussion injury (FPI), and controlled cortical impact (CCI). Based on the previous classification of TBI, these models belong to the focal (CCI), mixed (WD, lateral FPI) and diffuse (central FPI) subgroups (Fig. 13.1). In these three models, the development and the mechanisms of brain edema formation have been described and studied abundantly.

The WD model reproduces many characteristics of closed-head injury (CHI) in humans. It is a mix between focal and diffuse injuries. The model developed by Marmarou and colleagues[18,19] reproduces several features of DAI observed in TBI patients such as diffuse perivascular edema, small intra-parenchymal hemorrhages, widespread axonal injury, and increase of intracranial pressure (ICP).[19,20] The WD model induces TBI with a weight falling from a designated height on a steel disk glued onto the skull of the experimental animal. The steel disk is necessary to spread the mechanic energy of the weight over the skull in order to prevent fractures while still injuring the brain (Fig. 13.1A). The head of the experimental animal is not fixed and placed on foam with a defined elasticity to allow acceleration and deceleration of the brain.

FPI induces TBI by applying a rapid pressure pulse directly on the intact dural surface by attaching a water filled tube connected to a weight driven piston to the skull (Fig. 13.1B).[17] FPI models can be applied laterally (over parietal cortex) or onto the brain midline (central). Lateral FPI is considered to produce a mixed injury with cortical contusions combined with remote changes in the ipsilateral hemisphere and white matter, while central FPI rather results in diffuse injury with a wide variety of axonal injury, hippocampal damage, and brain stem lesion.[17] Both FPI variants produce cytotoxic and vasogenic edema with quantifiable magnetic resonance imaging (MRI) changes in T2 and diffusion weighted imaging (DWI).[21]

CCI models are used to induce focal cortical contusions. To produce brain injury, a craniotomy is performed without damaging the dura and a pneumatic or an electromagnetic piston is used to deform the exposed cortex with defined velocity and depth (Fig. 13.1C). The impact of the piston on the cortex results in a cortical contusion, hemorrhage, and

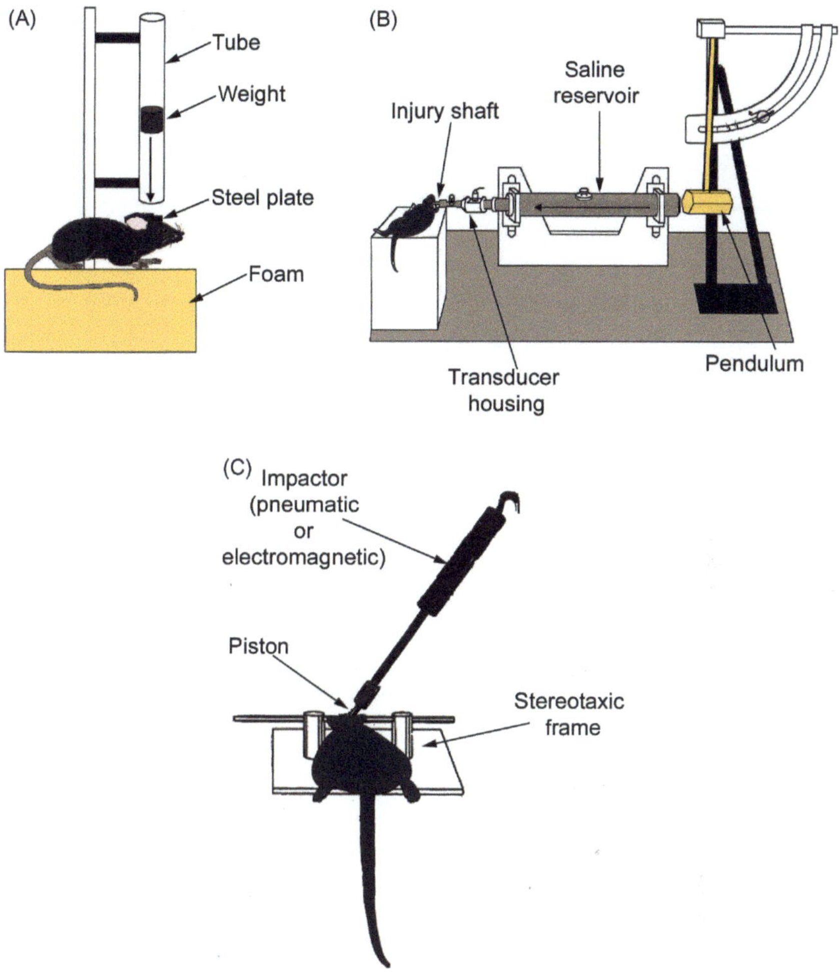

FIGURE 13.1 **Experimental setup of the three classical models for rodent traumatic brain injury: (A) weight drop model, (B) fluid percussion injury model, and (C) controlled cortical impact model.** *Source: Adapted from Marklund N, Hillered L. Animal modelling of traumatic brain injury in preclinical drug development: where do we go from here? Br J Pharmacol 2011;***164***(4):1207–29 and A new model of diffuse brain injury in rats. Part I: pathophysiology and biomechanics. J Neurosurg 1994;***80***(2):291–300.*

blood–brain barrier (BBB) disruption with almost immediate edema formation. After CCI, the craniotomy has to be resealed in order to allow the development of intracranial hypertension and its sequels,[22–24] for example peri-focal ischemia, reduced cerebral perfusion pressure, and microcirculatory dysfunction.[25,26] Vasogenic and cytotoxic edema are observed, which makes CCI a very good model to study edema pathophysiology.[21] This model can easily be adapted to study pediatric TBI by performing the model on rats on postnatal day (PND) 17.[27,28] The main limitation of the model is that it is often not used as originally described by Dixon and colleagues.[22] One major deviation from the original protocol is that the craniotomy is left open. In this case, intracranial hypertension is reduced or blunted and less or no secondary injury occurs.[23,24] A second major violation of the original protocol is the application of the impact to the skull instead of the dura. Thereby skull fractures are produced in a random manner and the model is not controlled anymore, in fact one of the major reasons why the model was initially developed.[22]

Accordingly, a plethora of well-defined and clinically relevant models of TBI are available to study brain edema formation after brain injury.

Pathophysiology of Edema Formation in Traumatic Brain Injury Models

As indicated in the introduction of the chapter, depending on the primary injury, the pathogenesis of TBI is complex, heterogeneous, and multifactorial.[29]

A traumatic cerebral contusion leads to an immediate hemorrhagic core with necrosis and a quick decrease of the cerebral perfusion.[30] The perilesional tissue frequently exhibits a significant decrease of cerebral blood flow possibly contributing to secondary injury events.[26] During the initial 24–72 hours following TBI, a series of vascular, cellular, and molecular processes follow, such as BBB disruption, edema formation, apoptosis, inflammation, and excitotoxicity, which result in additional, that is secondary, tissue damage with a progression from the core to adjacent primarily uninjured tissue (Fig. 13.2).[31,32] For example, CCI in juvenile

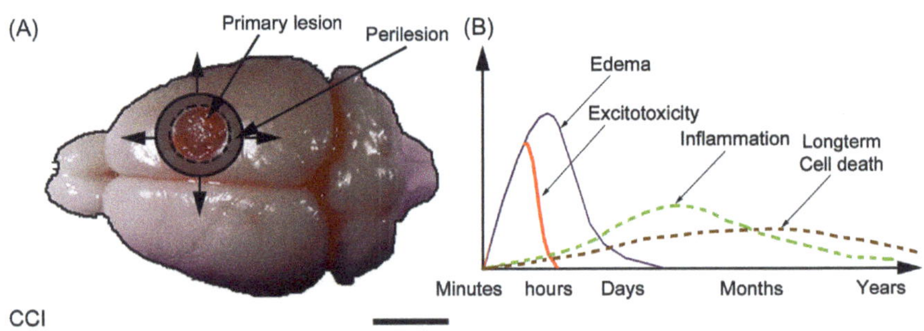

FIGURE 13.2 **(A) Schematic model of secondary injury events spreading after a focal TBI using controlled cortical impact model.** In the center of the primary lesion, there is bleeding and cell death. The primary lesion is spreading over time to the perilesional tissue with pathological processes described in (B). (B) There are chronological pathological processes in the perilesional tissue including excitotoxicity, edema, inflammation, and long-term cell death.

rats induced significant changes in MRI signals in the primary lesion site and perilesional areas at 24 hours postinjury (Fig. 13.3A and B) suggesting ongoing edema formation.[27] At 3 days post-CCI, a cavity filled with cerebrospinal fluid is formed (Fig. 13.3D) which becomes larger over time suggesting a progressive process (Fig. 13.3C and E) starting with edema and then followed by chronic inflammation and neurodegenerative processes.[33] Similar mechanisms are present in moderate and severe diffuse adult TBI. Brain edema formation and intracranial hypertension are critical steps in secondary processes after TBI, since prevention of intracranial hypertension by surgical decompression completely blunts the evolution of secondary injuries (Fig. 13.2).[23,24] Since surgical decompression is not always a clinical option, it is of utmost importance to understand the mechanisms underlying contusion-induced brain edema formation in order to develop medical therapies available for all affected patients.

Classification of the Different Forms of Edema

Brain edema has been recently classified in three subcategories: cytotoxic, ionic, and vasogenic edema. Traditionally, edema following TBI was primarily viewed as being vasogenic.[30] This idea was initially supported by classical neuropathological findings and CT imaging on TBI patients showing hyperdense hemorrhagic lesions early after injury followed by hypo-dense halo around the hemorrhagic lesion.[30] However, later this view was challenged by ultrastructural studies using electron microscopy and MR/CT imaging investigations using intravascular contrast agents. These studies demonstrated massive swelling of astrocytic endfeet and failed to demonstrate BBB leakage acutely after TBI.[34,35] Later on, these findings were confirmed by experimental work performed by Marmarou and colleagues using DWI, which suggested that cytototxic brain edema is also present in the acute phase of TBI.[19] Accordingly, it is very likely that a complex mix between the three brain edema categories with a continuum during the progression of the secondary injuries occurs after TBI. After mechanical injury to the brain, neurons, astrocytes, and—most importantly for the development of brain edema—blood vessels are damaged.[25,36] Vessel injury results in the almost immediate formation of vasogenic edema as a direct consequence of BBB disruption and the subsequent pressure-driven extravasation of fluid from the intravascular into the extracellular space of the brain (Fig. 13.4). Within the first few hours after injury, the resulting edema fluid is rich in serum proteins (but has a lower oncotic pressure than plasma) and has a high potassium but an unchanged sodium concentration.[37,38] These findings suggest that protein-rich fluid from the vascular space extravasated into the brain due to a ruptured BBB and mixes with intracellular, potassium-rich fluid deriving from ruptured cells from lesioned brain tissue. Excess extracellular

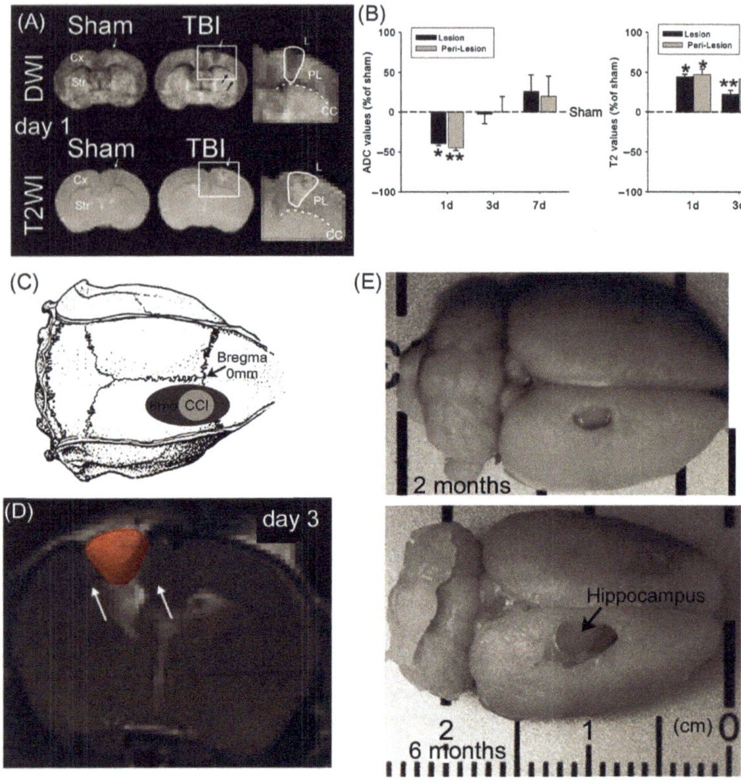

FIGURE 13.3 **(A) Diffusion weighted images (DWI, top) and T2WI (bottom) at 24) hours after juvenile traumatic brain injury (jTBI) shows changes in signal in the lesion (L) and perilesion (PL) over the corpus callosum (CC).** (B) There are relative changes of the apparent diffusion coefficient (ADC) and T2 values in the lesion and perilesion; with a transient initial decrease in the ADC values at 1 day (*$P<0.05$, **$P<0.01$); with an increase of T2 values at 1 and 3 days. These changes in MRI show the ongoing edema process in the controlled cortical impact (CCI) model on juvenile rats. (C) A drawing of the rat skull showing the approximate location of the original CCI done at postnatal day (PND) 17, and the expansion of the lesion observed at 6 months. (D) A T2WI at day 3 post-jTBI shows the lesion in red associated with the presence of a cavity filled of cerebrospinal fluid (CSF). White arrows indicate an increase of T2 signal suggesting ongoing edema, possibly vasogenic edema. (E) Representative jTBI brains photographed postperfusion show the lesion as an open cavity spanning ~3 mm in length at 2 months and then ~6 mm to reveal the hippocampal formation (black arrow) beneath the missing cortex. *Source: (A) and (B) Modified from Fukuda AM, Pop V, Spagnoli D, Ashwal S, Obenaus A, Badaut J. Delayed increase of astrocytic aquaporin 4 after juvenile traumatic brain injury: possible role in edema resolution? Neuroscience 2012;222:366–78. For (E): modified from Kamper JE, Pop V, Fukuda AM, Ajao DO, Hartman RE, Badaut J. Juvenile traumatic brain injury evolves into a chronic brain disorder: behavioral and histological changes over 6months. Exp Neurol 2013;250:8–19.*

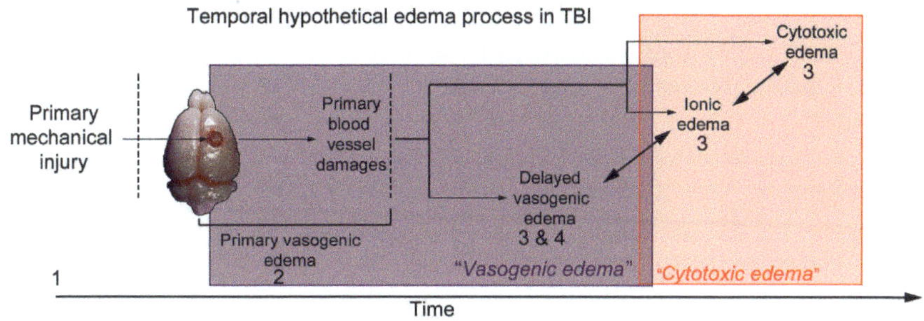

FIGURE 13.4 **The temporal evolution of the edema process in the context of a contusion type of TBI is very likely starting with an initial breakdown of blood vessels conducting to a primary vasogenic edema (step 2).** The initial mechanical impact may induce cell death. Local cell death in combination with primary vasogenic edema will induce hypoperfusion contributing to the formation of ionic edema (step 3) in endothelium, cytotoxic edema (step 3), and delayed vasogenic edema with massive opening of the BBB (steps 3 and 4). These events are likely happening very close in time and the edema process is spreading out of the primary lesion site (see Fig. 13.3).

potassium ions are then taken up by astrocytes along the intra-/extracellular chloride gradient, a process which results in an increase of cytoplasmic osmolarity, osmotic uptake of water, and finally astrocytic cell swelling.[39,40] The consequence of these processes is vasogenic edema in the contusion core and astrocytic swelling in the rim of the contusion which mix with time. These early events promote local ischemia and drive delayed vasogenic edema, which is caused by the physical opening of the BBB due to ischemic vascular damage and the subsequent decreased expression of tight junction proteins (such as claudin-5).[41] In the next sections, we will review the molecular and cellular mechanisms possibly involved in each subcategory of brain edema formation after TBI.

VASOGENIC EDEMA FORMATION

Vasogenic edema viewed to be a landmark of TBI pathophysiology implies BBB disruption. The BBB is a "gate keeper" consisting of a physical barrier established by tight junctions through extracellular adhesion proteins between cerebrovascular endothelial cells (for details see Chapter 2: Blood-Brain Barrier interfaces organization in relation to inorganic anion transport, CSF secretion and circulation and chapter 5: In Vitro Models of the Blood-Brain Barrier to Better Understand the Pathophysiology of Brain Edema). The BBB restricts the paracellular transport thereby forcing circulating solutes to be transported via controlled transcellular mechanisms such as specific transporters or caveolar transcytosis. Similarly, traffic of blood born cells through the BBB is limited and tightly regulated (for more details see Chapter 2: Blood-Brain Barrier interfaces organization in relation to inorganic anion transport, CSF secretion). Currently, the time-course of BBB disruption after TBI is believed to occur in two major phases peaking at 1–3 hours postinjury, as a direct consequence of mechanical disruption of microvascular vessels at the site of the impact, and at 3–7 days postinjury due to secondary injury events with the presence of pro-permeability factors, called delayed vasogenic edema. Considering the recent interests for mild TBI and concussion, it is important to highlight that the relative contribution of mechanical and molecular pathways in BBB disruption depends on the severity of injury.[42] BBB disruption has various devastating consequences including vasogenic edema, hemorrhage, and leukocyte infiltration followed by inflammation.[43,44]

Primary Vasogenic Edema After Mechanical Disruption of Cerebral Vessels

The initial injury causes, among others, a sudden mechanical rupture of cerebral blood vessels resulting in hemorrhage and focal ischemia immediately after the impact.[36] In this area of the brain, for example in the center of a cortical contusion, edema formation is minimal due to the lack or severe reduction of perfusion. Further away from this area of severe and almost complete mechanical injury, vessels are less damaged and are able to maintain their barrier function toward red blood cells and leukocytes. However, due to stretch injury, the BBB is impaired and immediate extravasation of a protein-rich plasma transudate occurs into the brain parenchyma[25] (Fig. 13.5). Very similar events occur in rats exposed to intense blast shock waves, where acute BBB disruption leads to a diffuse extravasation of immunoglobulin G (IgG) around capillaries along with red blood cells at the time of injury.[45] Deep-cortical imaging using two-photon microscopy demonstrates that vascular leakage primarily starts from capillaries, arterioles, and venules a couple of hours postinjury and increases over time after CCI in mice,[25] contributing to local vasogenic edema formation. This early disruption of blood vessels and formation of vasogenic edema contributes to the development and progress of secondary injuries (including delayed vasogenic edema) at the rim of the primary lesion (Figs. 13.2, 13.4, and 13.5).

Delayed Vasogenic Edema After Later Blood–Brain Barrier Dysfunctions: Cellular and Molecular Events

It is important to outline that the changes of the barrier function of the BBB after trauma are described with various expressions in the literature, for example BBB dysfunction, BBB disruption, BBB leakage, and opening

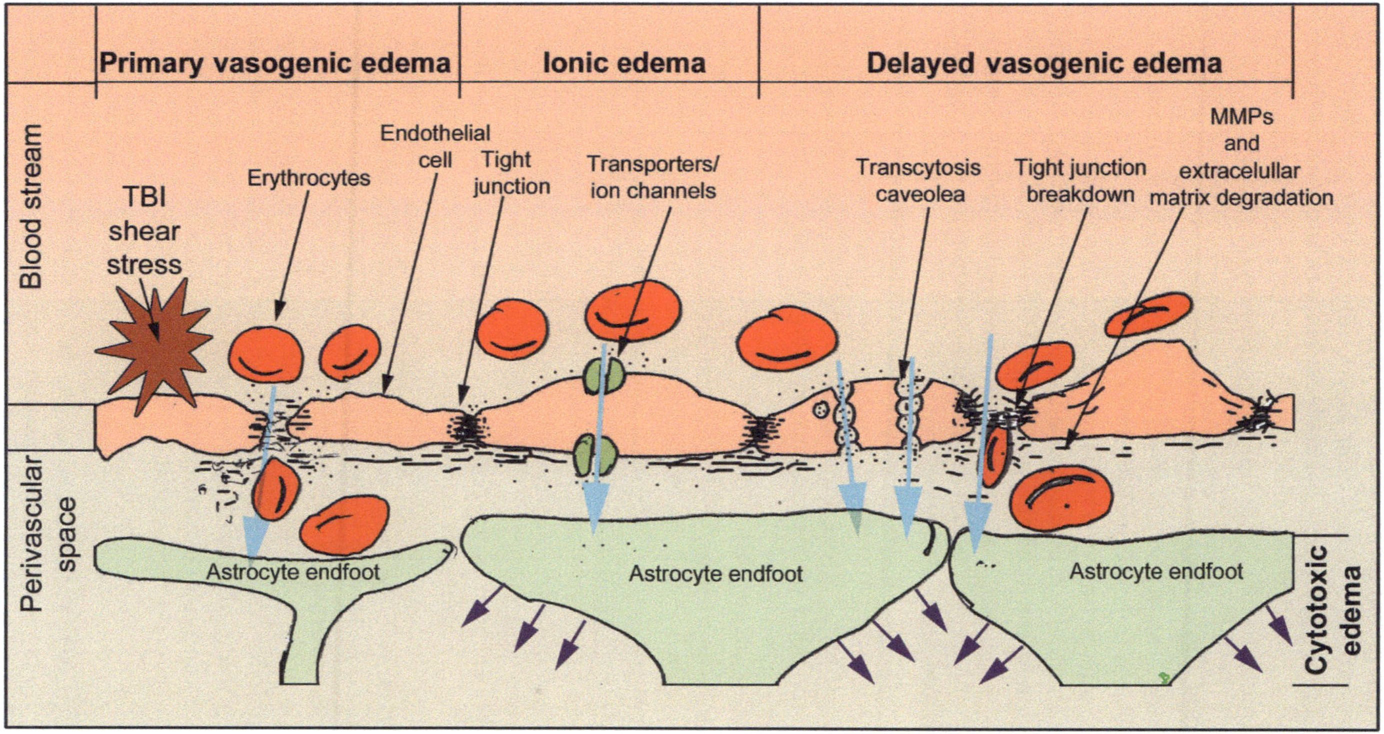

FIGURE 13.5 **This schematic drawing depicts the changes in the neurovascular unit.** Acute vasogenic edema occurs shortly after the impact. The physical rupture of tight junctions contributes to the entry of plasma proteins and potentially erythrocytes, along with some water (light blue arrows). Then, ionic edema in intact endothelial cells is characterized by water entry (light blue arrows) along with ion flux through transporters and ion channels in endothelial cells. At the same time, cytotoxic edema process takes place first in astrocyte end-feet, with a swelling of cells (dark blue arrows). Delayed vasogenic edema is a consequence of the increased transcytosis in the endothelium, rupture of tight junctions, and degradation of the extracellular matrix by metalloproteinases. *Source: Modified from Stokum JA, Gerzanich V, Simard JM. Molecular pathophysiology of cerebral edema. J Cereb Blood Flow Metab 2016;36(3):513–38.*

of the BBB. The problems that authors encounter to find a good expression for changes in BBB function after injury, reflects the complexity of the underlying pathophysiology. Particularly, since primary and secondary edema may be found at the same time after injury depending on the temporal and spatial dynamics of lesion progression (Fig. 13.2).

Transcytosis and Caveolae

Molecules with a large molecular weight, for example, large proteins or peptides, are able to cross the BBB and enter the central nervous system (CNS) via endocytotic mechanisms. Receptor-mediated transcytosis (RMT), as described for insulin, or adsorptive-mediated transcytosis (AMT), as described for albumin, are the two best-known examples.[46] In RMT, macromolecules bind to specific receptors on the luminal membrane of endothelial cells, which is then internalized into a vesicular caveolae and routed across the cytoplasm to be exocytosed at the abluminal side.[46] Ultrastructural analyses reveal that the number of endothelial caveolae increases within minutes in vessels showing BBB disruption.[47,48] For example, in ischemic stroke the increase in transcytotic vesicles was associated with the entry of large molecular weight solutes into the brain and a rise in oncotic pressure.[49]

Caveolae are involved in vesicular trafficking, transcytosis, endocytosis, and potocytosis of proteins.[41,50,51] Caveolae are composed of caveolins. In the brain, caveolin-1 and -2 are expressed in endothelial cells and caveolin-1 and -3 in astrocytes.[41,50,51] Several lines of research show that shear stress at the surface of endothelial cells increases endothelial expression of caveolin-1 during BBB breakdown following TBI.[52–54] Furthermore, phosphorylation of caveolin-1 plays an essential role in mechanisms mediating caveolae formation and motility.[54–56] Interestingly, the increased expression and phosphorylation of caveolin-1 are reduced in matrix metalloproteinase-9 (MMP9) knockout mice after CCI, which partially restores BBB permeability thereby reducing caveolae-transcytosis and vascular protein extravasation to the brain parenchyma.[54] These findings suggest that MMP9 plays a major role in the early phase of vasogenic edema by promoting caveolae formation. The increase of caveolin-1 expression at the lesion site occurs as early as 12 hours postinjury and precedes alterations of tight junction proteins following cortical cold-injury model in rats.[53] It is interesting to note that the effects of caveolin-1 increase is likely to be dual: detrimental during the early phase of the acute injury (increase of caveolae and transcytosis) and then beneficial by facilitating BBB recovery (stabilization of tight junction proteins in caveolae).[41]

Tight Junction Degradation

Tight junctions between cerebrovascular endothelial cells are composed of transmembrane proteins (occludin, claudins, and junctional adhesion molecules) anchored to cytoplasmic protein complexes (zonula occludens and cingulin) (Chapter 2: Blood-Brain Barrier interfaces organization in relation to inorganic anion transport, CSF secretion and circulation and chapter 5: In Vitro Models of the Blood-Brain Barrier to Better Understand the Pathophysiology of Brain Edema). Alterations of tight junctions contribute to BBB disruption and vasogenic edema formation in TBI. A decreased expression of claudin-5, occludin, junctional adhesion molecule-A, and zonula occludens 1 was observed in several rodent models of TBI, and associated with an increased BBB disruption or IgG/Evan's blue extravasation, from 12 hours to 4 days postinjury.[41,57–60] However, the importance of claudin-5 in vasogenic edema formation is doubtful since cerebral edema was not evident in claudin-5 knockout mice, most likely due to the fact that BBB permeability is limited solely to low molecular weight molecules in these mice.[61]

Several molecules such as cytokines, matrix metalloproteinases (MMPs), or vascular endothelial growth factor (VEGF) have been described to alter the structure and integrity of tight junctions; however, MMPs seem to be of particular relevance for edema formation after TBI (Fig. 13.5).[62] MMPs are a family of zinc-dependent endopeptidases capable to remodel the extracellular matrix through the proteolytic breakdown of collagen, laminin, and fibronectin.[63] In the CNS, MMP2, MMP3, and MMP9 are constitutively expressed in astrocytes, microglia, neurons, and/or endothelial cells and participate in tissue growth and development and in wound healing. However, increased expression and activity of MMPs after TBI result in an excessive degradation of vascular basal lamina and tight junction proteins thereby causing BBB disruption and vasogenic brain edema[62,64] (Fig. 13.5). MMPs also promote the infiltration of leukocytes and monocytes into the brain and the activation and motility of astrocytes and microglia. In addition, MMPs degrade endothelial vascular endothelial growth factor receptor-2 (VEGFR-2) which increases VEGF expression levels thereby contributing to apoptosis and neuroinflammation through caspase-1/3 activation and interleukin-1β (IL-1β) release.[65] Inhibition of MMPs with broad-spectrum MMP inhibitors (minocycline, melatonin, GM6001, and MMI270) prevents BBB disruption, brain swelling, and injury in rodent models of TBI.[66–68] Playing on up-stream signaling, injection of selective endothelin B receptor (ET_B-R) antagonists (BQ788 or IRL-2500) prevents the increased expression of MMP9 and VEGF-A in astrocytes and therefore reduces BBB disruption as well as vasogenic brain edema in a cold-injury model in rats.[69,70] Increased expression of endothelin 1, ET_A-R, and ET_B-R have also been observed between 4 and 8 hours postinjury in pericytes.[71] Changes in endothelin expression have been proposed to be related to vasoconstriction[71]; however, it is very likely that an increased expression of endothelin in pericytes contributes to BBB disruption acting on VEGF and MMP9 expressions. Interestingly, it was suggested that the movement of pericytes away from the capillary bed contributes to BBB disruption after TBI.[72]

In summary, several molecules released during inflammation process such as endothelin, as well as bradykinin,[73,74] substance P,[75,76] have direct or indirect effects on MMPs,[62,64] then VEGF, which alter the BBB, leading to the influx of protein-rich fluids and leukocytes inside the brain parenchyma. Leukocytes, pericytes, activated astrocytes, and microglia secrete large amounts of vascular permeability factors, cytokines and chemokines, which promote BBB hyperpermeability and thereby further increase the extravasation of blood-derived protein-rich fluids into the brain parenchyma, all part of the delayed and progressive vasogenic edema post-TBI. This cascade is spreading from the initial injury site to the adjacent regions. It is a feedforward mechanism associated with ionic and cytotoxic edema (Figs. 13.4 and 13.5).

IONIC AND CYTOTOXIC EDEMA FORMATION AFTER TRAUMATIC BRAIN INJURY

Under resting conditions, water movements in the brain are driven by osmotic, oncotic, and hydrostatic pressures.[77] Water flux across cell membranes depends on the osmotic and oncotic pressures between the two compartments separated by the membrane. The osmotic pressure is determined by the difference in electrolyte concentrations between the two compartments, while oncotic pressure is mediated by the presence of macromolecules. Under physiological conditions, the oncotic pressure within cells is maintained by the regular expression of cytoplasmic proteins; the osmotic pressure builds up due to the semipermeable function of the cell membrane and the resulting Donnan forces and is maintained by the Na^+/K^+ adenosine triphosphatase (ATPase).[78]

After TBI, the decrease of cerebral blood flow reduces the delivery of oxygen and glucose to the brain thereby leading to a drop of adenosine triphosphate (ATP) production.[79,80] Under circumstances of reduced cellular energy, the activity of the Na^+/K^+ ATPase is reduced as well and Na^+ accumulates in astrocytes, neurons, and possibly in endothelial cells. Additionally, other cellular cation channels are disturbed, including acid-sensing ion channels (ASIC), sulfonylurea receptor 1 (SUR1)-regulated nonselective cation channels ($NC_{Ca\text{-}ATP}$) (SUR1-$NC_{Ca\text{-}ATP}$), $Na^+/K^+/Cl^-$ cotransporter-1 (NKCC1), and *N*-methyl-D-aspartate receptors (NMDAr). Although a passive efflux of K^+ is considered as an attempt of the cell to prevent the establishment of an osmotic gradient, this compensatory action is annihilated by the increased influx of Cl^- through chloride channels and transporters.[81] Overall, the influx of extracellular ions down their electrochemical gradients increases intracellular osmolarity. Then, there is a compensatory movement of water inside the cells—for the astrocytes through specialized transmembrane water channels called aquaporins (AQPs) resulting in cell swelling.[31,32] It is important to note that at early stages of cytotoxic edema, the BBB is intact and impermeable to ions and fluids, and therefore the total brain volume does not increase (Fig. 13.5). But later on, the inflow of Na^+ accompanied by an extravasation of water from the vascular compartment to the extracellular space leads to ionic edema, thereby increasing total brain volume and ICP. The ionic and vasogenic edema are very closely associated.[81] In secondary injury events, ionic and cytotoxic edema are very likely evolving together and precede the delayed vasogenic edema (Figs. 13.4 and 13.5). Ionic and cytotoxic edema will be discussed with a focus on the contribution of several cation channels, transporter and receptors cited above in brain edema formation and secondary injuries after TBI.

Ionic Edema Formation

It is only recently that ionic and cytotoxic edema have been defined as separated distinct edema by Simard's group.[82] Ionic edema is a subtype of extracellular edema that lacks serum protein extravasation. The depletion of oxygen and energy nutrients induces major changes in the cellular ionic gradients due to the absence of efficient energy-dependent cotransporters present in endothelial cells leading to a transcapillary flux of Na^+[83,84] with extracellular and perivascular water accumulation and overall tissue swelling. In addition to the initial energy depletion, the ionic edema is fed by extracellular Na^+ caused by astrocytic/neuronal cytotoxic edema. The extracellular Na^+ is transported inward along its concentration gradient in the endothelial cells, and therefore followed by water by osmotic gradient in the cells (for review[81] and details see Chapter 10: Sur1-Trpm4 - Promising Target for Brain Edema Treatment). Therefore local *cytotoxic edema* quickly evolves to become *ionic edema*. It has been proposed that ionic edema is the earliest phase of endothelial dysfunction, leading to delayed vasogenic edema (Figs. 13.4 and 13.5).

Constitutive transporters could directly contribute to the ionic edema: sodium-hydrogen antiporter (NHE) family, NKCC1, sodium-dependent glucose cotransporter type-1 (SGLT1), glucose transporter-1 (GLUT1), monocarboxylate transporters (MCT), and many other transporters. They are a potential source of Na^+ entry followed by water. NHE family members contribute to Na^+ influx across the luminal side of endothelium[85]; and NKCC1 controls intracellular Na^+ concentration levels and cell volume.[86] Similarly to NHE, NKCC1 is expressed on the luminal side of endothelial cells; however, its expression is increased following brain injuries.[87] It leads to an excessive influx of Na+ into cells followed by water, which contributes to ionic edema formation.[87,88] Indeed, NKCC1 inhibition with bumetanide or short-hairpin RNA silencing NKCC1 prevents intracellular increase in Na+ concentration level, brain edema, and apoptosis after CCI in mice.[89]

Some channels are up-regulated after acute injury such as SUR1-$NC_{Ca\text{-}ATP}$ (for details see Chapter 10. Sur1-Trpm4 - Promising Target for Brain Edema Treatment). The SUR1-$NC_{Ca\text{-}ATP}$ channel is a nonselective monovalent cation channel that is activated by intracellular Ca^{2+} and ATP. This channel is not constitutively expressed, yet its expression is induced following brain injury in endothelium, astrocytes, and neurons.[90–92] Depletion of ATP leads to the opening of SUR1-$NC_{Ca\text{-}ATP}$ and causes an influx of Na^+ into cells, thereby contributing to ionic and cytotoxic edema (see below) after ischemic stroke and TBI.[92,93] Inhibition of SUR1-$NC_{Ca\text{-}ATP}$ by glibenclamide (a sulfonylurea compound) reduces cytotoxic edema formation and contusion volume after CCI in rats.[93] The ionic edema is contributing to early entry of water in brain tissue through the endothelium before BBB breakdown, leading to brain swelling as observed for example in stroke models within 30 minutes after reperfusion[94,95] associated with a late increase of BBB permeability.[95,96]

Cytotoxic Edema Formation

Cytotoxic edema is a process where the cells swell due to influx of osmolites and water from the extracellular to intracellular space. In contrast with the two others forms of edema, pure cytotoxic edema does not induce tissue swelling because water moves only between cell compartments and no additional water enters the brain. However, cytotoxic edema is a critical step in the feedforward secondary injury process. As indicated earlier, clinical and experimental data suggest that cytotoxic edema precedes vasogenic edema.[30]

As mentioned previously for ionic edema, in absence of ATP, the activity of the Na^+/K^+ ATPase is stopped leading to the accumulation of Na^+ and then water in astrocytes and neurons. As for many brain injuries, astrocytes are the first cells to swell (Fig. 13.5), since one of their main functions is to buffer extracellular K^+ and glutamate.[81] As described for ionic edema, NKCC1 transporters expressed in astrocytes are also involved in the swelling process in addition to the inward-rectifying potassium channel 4.1 (Kir4.1), glutamate transporters (glutamate transporter-1, GLT1 and glutamate aspartate transporter, GLAST).[81]

The cation channels ASIC and SUR1-$NC_{Ca\text{-}ATP}$ are involved in the process of cytotoxic edema formation after TBI. ASICs are proton-gated, non-voltage-gated Na^+ and Ca^{2+} channels that are activated by a reduction of extracellular pH. They belong to the epithelial sodium channel/degenerin family. Six isoforms are described in mammals: ASIC1a, ASIC1b, ASIC2a, ASIC2b, ASIC3, and ASIC4.[97] In the CNS, they are found as heterotrimers composed by ASIC1a and ASIC2 subunits in neurons and astrocytes.[98,99] After TBI, the lack of oxygen switches glucose metabolism to glycolysis in affected cerebral regions, leading to lactate-mediated acidosis.[100,101] Extracellular acidosis activates ASICs resulting in Na^+ and Ca^{2+} influx in neurons, causing neuronal death by excitotoxicity.[102] ASIC1a knockout mice are partly protected against neuronal death and spatial memory impairments compared with wild-type mice after lateral FPI.[103] Similarly, inhibition of ASICs by amiloride decreases neuronal death in the hippocampus after lateral FPI in rats.[104] While no direct evidence of ASICs in brain edema formation is reported so far, the increased influx of Na^+ and Ca^{2+} through ASICs in neurons and astrocytes is most likely contributing to the overall increased intracellular osmolality and therefore to brain edema formation. As discussed for ionic edema, SUR1-$NC_{Ca\text{-}ATP}$ channel is a nonselective monovalent cation channel, activated by intracellular Ca^{2+} and ATP; and its expression is induced following brain injury in astrocytes, neurons, and endothelial cells.[90–92] Inhibition of SUR1-$NC_{Ca\text{-}ATP}$ by glibenclamide (a sulfonylurea compound) reduces edema formation after CCI in rats.[93] Similar results have been observed in the stroke pathophysiology (for details see Chapter 10. Sur1-Trpm4 - Promising Target for Brain Edema Treatment). NKCC1 transporters could also have a critical role in cytotoxic edema in astrocytes and neurons, and its inhibition with bumetanide reduces cytotoxic edema in vitro in neuronal or astrocyte cultures and also in vivo in CCI model in rat and mouse.[89,105–107]

Dysfunctions of ionic gradients create a transmembrane osmotic gradient that drives water movements into astrocytes and causes swelling. Three routes are classically described[108] (1) astrocytic transmembrane water channels, primarily AQP channels, and to a less extent, SGLT1, and glucose transporter types 1 and 2 (GLUT1 and GLUT2), where water movements are driven by an osmotic gradient; (2) ion transporters such as NKCC1 (expressed by endothelial cells, astrocytes and neurons) and excitatory amino-acid transporter (EAAT1, EAAT2 for astrocytes; EAAT3 and EAAT4 for neurons) which also transport few water molecules; (3) transmembrane diffusion through the lipid bilayer. AQP is very likely one of the principal route of water influx into astrocytes after TBI. As described in Chapter 9, Role of Aquaporins for the Formation and Resolution of Brain Edema, AQPs are major water channel proteins for brain water balance. Among the 13 AQPs, aquaporin 4 (AQP4) shows the highest and spatially most heterogeneous expression in the CNS and is predominantly expressed in astrocytic end-feet in cortex and striatum.[109] Water intoxication and stroke experiments strongly suggest a deleterious role of AQP4 in cytotoxic edema.[110] In contrast, AQP4 is important in edema resolution by facilitating water clearance (for details see Chapter 9: Role of Aquaporins for the Formation and Resolution of Brain Edema). In the context of TBI, increased and decreased expression of AQP4 have been observed in various experimental models of TBI.[27,111–114] AQP4 knockout mice are protected from cytotoxic edema and increased ICP in experimental models of TBI.[115,116] Yao et al. showed reduced astrocyte foot-process area in AQP4 knockout mice compared to WT at 24 hours after CCI, suggesting reduced cytotoxic edema and water accumulation.[116] The authors suggest that the moderate improvement observed in AQP4 knockout mice is probably due to the small contribution of cytotoxic edema in CCI models compared to other injuries such as stroke.[116] Beneficial effects were also observed in a pediatric TBI model using small interfering RNA targeting AQP4 (siAQP4) treatment.[28] Fukuda et al. showed that the local siAQP4 treatment prevents edema formation (including vasogenic edema) and increased survival of neurons in perilesional areas. The decrease of AQP4 expression after juvenile TBI using siAQP4 is associated with BBB preservation, suggesting that less osmotic stress on astrocytes may prevent BBB disruption by maintaining physical and mechanical properties of endothelial cells and also the neurovascular unit.[28] A recent in vitro study suggests that application of acetazolamide reduces astrocyte cell swelling and astrocyte survival by decreasing AQP4 expression.[113] Altogether, these results show that AQP4 plays a pivotal role in the cascade injury after TBI.

Neurons are exposed to cytotoxic edema because of ionic gradients alterations, caused by Na^+/K^+ ATPase dysfunction, neuronal depolarization, and subsequent release of glutamate (an excitatory neurotransmitter). The activation of NMDAr induces an excessive influx of Ca^{2+} into neurons which ultimately participates to cytoxic edema and neuronal death in TBI. Inhibition of NMDAr by a selective NMDAr antagonist called amino-5-phosphonopentanoate (AP5) reduces brain edema formation following TBI in rats.[117] Similarly, inhibition of NMDAr by delucemine (NPS1506), a noncompetitive NMDAr antagonist, reduces Ca^{2+} and Na^+ in the injured hemisphere, and decreases brain edema, which improves functional outcome after CHI in rats.[118]

In summary, the pathophysiology of the three edema subtypes defined to date (vasogenic, ionic, and cytotoxic) are most likely interconnected and evolve over time in various parts of the injured brain, a process which was hard to investigate with the methods available so far. The severity of edema formation depends on the severity of the initial TBI, but also on secondary mechanisms.

CONCLUSIONS

Cerebral edema is hard to treat and sometimes uncontrollable, and therefore often worsens morbidity and mortality after TBI. Therefore, there is an urgent need for pharmacological interventions to halt molecular cascades leading to brain swelling. The development of efficient pharmacological strategies is particularly challenging due the existence of three distinct molecular and cellular pathways causing brain edema (cytotoxic, ionic, and vasogenic); however, recent discoveries discussed in this chapter shed new light on the pathophysiology of brain edema formation and may therefore raise hopes for the development of novel strategies to treat brain edema in the near future.

Acknowledgments

The authors thank Dr Obenaus, Loma Linda University, and Aleksandra Ichkova, University of Bordeaux, for commenting the chapter.

Statement of financial support: This study was supported by grants from the National Institute of child disorders (NICHD) R01HD061946 (J.B.), Fondation des gueules cassées (J.B.), eraNet neuron CNSaFlame (J.B.), and eraNet Neuron TRAINS (J.B.).

References

1. Bauer D, Tung ML, Tsao JW. Mechanisms of traumatic brain injury. *Semin Neurol* 2015;35(1):e14–22.
2. Wade SL, Michaud L, Brown TM. Putting the pieces together: preliminary efficacy of a family problem-solving intervention for children with traumatic brain injury. *J Head Trauma Rehabil* 2006;21(1):57–67.
3. Faul M, Xu L, Wald M, Coronado V. *Traumatic brain injury in the United States: emergency department visits, hospitalizations, and deaths, 2002–2006*. Atlanta, GA: National Center for Injury Prevention and Control, CDC; 2010.
4. Zaloshnja E, Miller T, Langlois JA, Selassie AW. Prevalence of long-term disability from traumatic brain injury in the civilian population of the United States, 2005. *J Head Trauma Rehabil* 2008;23(6):394–400.
5. Selassie AW, Zaloshnja E, Langlois JA, Miller T, Jones P, Steiner C. Incidence of long-term disability following traumatic brain injury hospitalization, United States, 2003. *J Head Trauma Rehabil* 2008;23(2):123–31.
6. Thurman DJ, Alverson C, Dunn KA, Guerrero J, Sniezek JE. Traumatic brain injury in the United States: a public health perspective. *J Head Trauma Rehabil* 1999;14(6):602–15.
7. Thurman D, Guerrero J. Trends in hospitalization associated with traumatic brain injury. *JAMA* 1999;282(10):954–7.
8. Langlois JA, Rutland-Brown W, Wald MM. The epidemiology and impact of traumatic brain injury: a brief overview. *J Head Trauma Rehabil* 2006;21(5):375–8.
9. Coronado VG, Xu L, Basavaraju SV, McGuire LC, Wald MM, Faul MD, et al. (CDC) CfDCaP. Surveillance for traumatic brain injury-related deaths—United States, 1997–2007. *MMWR Surveill Summ* 2011;60(5):1–32.
10. Smith DH, Uryu K, Saatman KE, Trojanowski JQ, McIntosh TK. Protein accumulation in traumatic brain injury. *Neuromol Med* 2003;4(1-2):59–72.
11. Gavett BE, Stern RA, Cantu RC, Nowinski CJ, McKee AC. Mild traumatic brain injury: a risk factor for neurodegeneration. *Alzheimers Res Ther* 2010;2(3):18.
12. Johnson VE, Stewart W, Smith DH. Traumatic brain injury and amyloid-β pathology: a link to Alzheimer's disease? *Nat Rev Neurosci* 2010;11(5):361–70.
13. Maas AI, Stocchetti N, Bullock R. Moderate and severe traumatic brain injury in adults. *Lancet Neurol* 2008;7(8):728–41.

14. Malec JF, Brown AW, Leibson CL, Flaada JT, Mandrekar JN, Diehl NN, et al. The mayo classification system for traumatic brain injury severity. *J Neurotrauma* 2007;24(9):1417–24.
15. Cantu RC, Gean AD. Second-impact syndrome and a small subdural hematoma: an uncommon catastrophic result of repetitive head injury with a characteristic imaging appearance. *J Neurotrauma* 2010;27(9):1557–64.
16. Cernak I. Animal models of head trauma. *NeuroRx* 2005;2(3):410–22.
17. Marklund N, Hillered L. Animal modelling of traumatic brain injury in preclinical drug development: where do we go from here? *Br J Pharmacol* 2011;164(4):1207–29.
18. Marmarou A, Foda MA, van den Brink W, Campbell J, Kita H, Demetriadou K. A new model of diffuse brain injury in rats. Part I: pathophysiology and biomechanics. *J Neurosurg* 1994;80(2):291–300.
19. Foda MA, Marmarou A. A new model of diffuse brain injury in rats. Part II: morphological characterization. *J Neurosurg* 1994;80(2):301–13.
20. Engelborghs K, Verlooy J, Van Reempts J, Van Deuren B, Van de Ven M, Borgers M. Temporal changes in intracranial pressure in a modified experimental model of closed head injury. *J Neurosurg* 1998;89(5):796–806.
21. Obenaus A, Robbins M, Blanco G, Galloway NR, Snissarenko E, Gillard E, et al. Multimodal magnetic resonance imaging alterations in two rat models of mild neurotrauma. *J Neurotrauma* 2007;24(7):1147–60.
22. Dixon CE, Clifton GL, Lighthall JW, Yaghmai AA, Hayes RL. A controlled cortical impact model of traumatic brain injury in the rat. *J Neurosci Methods* 1991;39(3):253–62.
23. Zweckberger K, Erös C, Zimmermann R, Kim SW, Engel D, Plesnila N. Effect of early and delayed decompressive craniectomy on secondary brain damage after controlled cortical impact in mice. *J Neurotrauma* 2006;23(7):1083–93.
24. Zweckberger K, Plesnila N. Anatibant, a selective non-peptide bradykinin B2 receptor antagonist, reduces intracranial hypertension and histopathological damage after experimental traumatic brain injury. *Neurosci Lett* 2009;454(2):115–17.
25. Schwarzmaier SM, Gallozzi M, Plesnila N. Identification of the vascular source of vasogenic brain edema following traumatic brain injury using in vivo 2-photon microscopy in mice. *J Neurotrauma* 2015;32(13):990–1000.
26. Engel DC, Mies G, Terpolilli NA, Trabold R, Loch A, De Zeeuw CI, et al. Changes of cerebral blood flow during the secondary expansion of a cortical contusion assessed by 14C-iodoantipyrine autoradiography in mice using a non-invasive protocol. *J Neurotrauma* 2008;25(7):739–53.
27. Fukuda AM, Pop V, Spagnoli D, Ashwal S, Obenaus A, Badaut J. Delayed increase of astrocytic aquaporin 4 after juvenile traumatic brain injury: possible role in edema resolution? *Neuroscience* 2012;222:366–78.
28. Fukuda AM, Adami A, Pop V, Bellone JA, Coats JS, Hartman RE, et al. Posttraumatic reduction of edema with aquaporin-4 RNA interference improves acute and chronic functional recovery. *J Cereb Blood Flow Metab* 2013;33(10):1621–32.
29. McGinn MJ, Povlishock JT. Pathophysiology of traumatic brain injury. *Neurosurg Clin N Am* 2016;27(4):397–407.
30. Unterberg AW, Stover J, Kress B, Kiening KL. Edema and brain trauma. *Neuroscience* 2004;129(4):1021–9.
31. Winkler EA, Minter D, Yue JK, Manley GT. Cerebral edema in traumatic brain injury: pathophysiology and prospective therapeutic targets. *Neurosurg Clin N Am* 2016;27(4):473–88.
32. Michinaga S, Koyama Y. Pathogenesis of brain edema and investigation into anti-edema drugs. *Int J Mol Sci* 2015;16(5):9949–75.
33. Kamper JE, Pop V, Fukuda AM, Ajao DO, Hartman RE, Badaut J. Juvenile traumatic brain injury evolves into a chronic brain disorder: behavioral and histological changes over 6months. *Exp Neurol* 2013;250:8–19.

34. Bullock R, Statham P, Patterson J, Wyper D, Hadley D, Teasdale E. The time course of vasogenic oedema after focal human head injury—evidence from SPECT mapping of blood brain barrier defects. *Acta Neurochir Suppl (Wien)* 1990;51:286–8.
35. Bullock R, Maxwell WL, Graham DI, Teasdale GM, Adams JH. Glial swelling following human cerebral contusion: an ultrastructural study. *J Neurol Neurosurg Psychiatry* 1991;54(5):427–34.
36. Badaut J, Bix GJ. Vascular neural network phenotypic transformation after traumatic injury: potential role in long-term sequelae. *Transl Stroke Res* 2014;5(3):394–406.
37. Gazendam J, Go KG, van Zanten AK. Composition of isolated edema fluid in cold-induced brain edema. *J Neurosurg* 1979;51(1):70–7.
38. Westergren I, Nyström B, Hamberger A, Johansson BB. Amino acids in extracellular fluid in vasogenic brain edema. *Acta Neurochir Suppl (Wien)* 1994;60:124–7.
39. Walz W, Mukerji S. KCl movements during potassium-induced cytotoxic swelling of cultured astrocytes. *Exp Neurol* 1988;99(1):17–29.
40. Walz W. Mechanism of rapid K(+)-induced swelling of mouse astrocytes. *Neurosci Lett* 1992;135(2):243–6.
41. Badaut J, Ajao DO, Sorensen DW, Fukuda AM, Pellerin L. Caveolin expression changes in the neurovascular unit after juvenile traumatic brain injury: signs of blood–brain barrier healing? *Neuroscience* 2015;285:215–26.
42. Li W, Watts L, Long J, Zhou W, Shen Q, Jiang Z, et al. Spatiotemporal changes in blood–brain barrier permeability, cerebral blood flow, T2 and diffusion following mild traumatic brain injury. *Brain Res* 2016;1646:53–61.
43. Song J, Wu C, Korpos E, Zhang X, Agrawal SM, Wang Y, et al. Focal MMP-2 and MMP-9 activity at the blood–brain barrier promotes chemokine-induced leukocyte migration. *Cell Rep* 2015;10(7):1040–54.
44. Almutairi MM, Gong C, Xu YG, Chang Y, Shi H. Factors controlling permeability of the blood–brain barrier. *Cell Mol Life Sci* 2016;73(1):57–77.
45. Yeoh S, Bell ED, Monson KL. Distribution of blood–brain barrier disruption in primary blast injury. *Ann Biomed Eng* 2013;41(10):2206–14.
46. Abbott NJ, Rönnbäck L, Hansson E. Astrocyte–endothelial interactions at the blood–brain barrier. *Nat Rev Neurosci* 2006;7(1):41–53.
47. Lossinsky AS, Shivers RR. Structural pathways for macromolecular and cellular transport across the blood–brain barrier during inflammatory conditions. Review. *Histol Histopathol* 2004;19(2):535–64.
48. Povlishock JT, Becker DP, Sullivan HG, Miller JD. Vascular permeability alterations to horseradish peroxidase in experimental brain injury. *Brain Res* 1978;153(2): 223–39.
49. Knowland D, Arac A, Sekiguchi KJ, Hsu M, Lutz SE, Perrino J, et al. Stepwise recruitment of transcellular and paracellular pathways underlies blood-brain barrier breakdown in stroke. *Neuron* 2014;82(3):603–17.
50. Sohn J, Brick RM, Tuan RS. From embryonic development to human diseases: the functional role of caveolae/caveolin. *Birth Defects Res C Embryo Today* 2016;108(1):45–64.
51. Zhao YL, Song JN, Zhang M. Role of caveolin-1 in the biology of the blood-brain barrier. *Rev Neurosci* 2014;25(2):247–54.
52. van der Meer AD, Kamphuis MM, Poot AA, Feijen J, Vermes I. Lowering caveolin-1 expression in human vascular endothelial cells inhibits signal transduction in response to shear stress. *Int J Cell Biol* 2009;2009:532432.
53. Nag S, Venugopalan R, Stewart DJ. Increased caveolin-1 expression precedes decreased expression of occludin and claudin-5 during blood–brain barrier breakdown. *Acta Neuropathol* 2007;114(5):459–69.
54. Muradashvili N, Benton RL, Saatman KE, Tyagi SC, Lominadze D. Ablation of matrix metalloproteinase-9 gene decreases cerebrovascular permeability and fibrinogen deposition post traumatic brain injury in mice. *Metab Brain Dis* 2015;30(2):411–26.

55. Nag S, Manias JL, Stewart DJ. Expression of endothelial phosphorylated caveolin-1 is increased in brain injury. *Neuropathol Appl Neurobiol* 2009;35(4):417–26.
56. Minshall RD, Tiruppathi C, Vogel SM, Niles WD, Gilchrist A, Hamm HE, et al. Endothelial cell-surface gp60 activates vesicle formation and trafficking via G(i)-coupled Src kinase signaling pathway. *J Cell Biol* 2000;150(5):1057–70.
57. Wen J, Qian S, Yang Q, Deng L, Mo Y, Yu Y. Overexpression of netrin-1 increases the expression of tight junction-associated proteins, claudin-5, occludin, and ZO-1, following traumatic brain injury in rats. *Exp Ther Med* 2014;8(3):881–6.
58. Ren W, Jing G, Shen Q, Yao X, Jing Y, Lin F, et al. Occludin and connexin 43 expression contribute to the pathogenesis of traumatic brain edema. *Neural Regen Res* 2013;8(29):2703–12.
59. Zhao J, Pati S, Redell JB, Zhang M, Moore AN, Dash PK. Caffeic acid phenethyl ester protects blood–brain barrier integrity and reduces contusion volume in rodent models of traumatic brain injury. *J Neurotrauma* 2012;29(6):1209–18.
60. Yeung D, Manias JL, Stewart DJ, Nag S. Decreased junctional adhesion molecule-A expression during blood–brain barrier breakdown. *Acta Neuropathol* 2008;115(6):635–42.
61. Nitta T, Hata M, Gotoh S, Seo Y, Sasaki H, Hashimoto N, et al. Size-selective loosening of the blood–brain barrier in claudin-5-deficient mice. *J Cell Biol* 2003;161(3):653–60.
62. Abdul-Muneer PM, Pfister BJ, Haorah J, Chandra N. Role of matrix metalloproteinases in the pathogenesis of traumatic brain injury. *Mol Neurobiol* 2016;53(9):6106–23.
63. Stefano ME, Herrero MT. The multifaceted role of metalloproteinases in physiological and pathological conditions in embryonic and adult brains. *Prog Neurobiol* 2016.
64. Yuan J, Wang A, He Y, Si Z, Xu S, Zhang S, et al. Cordycepin attenuates traumatic brain injury-induced impairments of blood–brain barrier integrity in rats. *Brain Res Bull* 2016;127:171–6.
65. Abdul Muneer PM, Alikunju S, Szlachetka AM, Haorah J. The mechanisms of cerebral vascular dysfunction and neuroinflammation by MMP-mediated degradation of VEGFR-2 in alcohol ingestion. *Arterioscler Thromb Vasc Biol* 2012;32(5):1167–77.
66. Alluri H, Wilson RL, Anasooya Shaji C, Wiggins-Dohlvik K, Patel S, Liu Y, et al. Melatonin preserves blood–brain barrier integrity and permeability via matrix metalloproteinase-9 inhibition. *PLoS One* 2016;11(5):e0154427.
67. Higashida T, Kreipke CW, Rafols JA, Peng C, Schafer S, Schafer P, et al. The role of hypoxia-inducible factor-1α, aquaporin-4, and matrix metalloproteinase-9 in blood–brain barrier disruption and brain edema after traumatic brain injury. *J Neurosurg* 2011;114(1):92–101.
68. Shigemori Y, Katayama Y, Mori T, Maeda T, Kawamata T. Matrix metalloproteinase-9 is associated with blood–brain barrier opening and brain edema formation after cortical contusion in rats. *Acta Neurochir Suppl* 2006;96:130–3.
69. Michinaga S, Nagase M, Matsuyama E, Yamanaka D, Seno N, Fuka M, et al. Amelioration of cold injury-induced cortical brain edema formation by selective endothelin ETB receptor antagonists in mice. *PLoS One* 2014;9(7):e102009.
70. Michinaga S, Seno N, Fuka M, Yamamoto Y, Minami S, Kimura A, et al. Improvement of cold injury-induced mouse brain edema by endothelin ETB antagonists is accompanied by decreases in matrixmetalloproteinase 9 and vascular endothelial growth factor-A. *Eur J Neurosci* 2015;42(6):2356–70.
71. Dore-Duffy P, Wang S, Mehedi A, Katyshev V, Cleary K, Tapper A, et al. Pericyte-mediated vasoconstriction underlies TBI-induced hypoperfusion. *Neurol Res* 2011;33(2):176–86.
72. Dore-Duffy P, Owen C, Balabanov R, Murphy S, Beaumont T, Rafols JA. Pericyte migration from the vascular wall in response to traumatic brain injury. *Microvasc Res* 2000;60(1):55–69.
73. Trabold R, Erös C, Zweckberger K, Relton J, Beck H, Nussberger J, et al. The role of bradykinin B(1) and B(2) receptors for secondary brain damage after traumatic brain injury in mice. *J Cereb Blood Flow Metab* 2010;30(1):130–9.

74. Plesnila N, Schulz J, Stoffel M, Eriskat J, Pruneau D, Baethmann A. Role of bradykinin B2 receptors in the formation of vasogenic brain edema in rats. *J Neurotrauma* 2001;18(10):1049–58.
75. Donkin JJ, Cernak I, Blumbergs PC, Vink R. A substance P antagonist reduces axonal injury and improves neurologic outcome when administered up to 12 hours after traumatic brain injury. *J Neurotrauma* 2011;28(2):217–24.
76. Donkin JJ, Nimmo AJ, Cernak I, Blumbergs PC, Vink R. Substance P is associated with the development of brain edema and functional deficits after traumatic brain injury. *J Cereb Blood Flow Metab* 2009;29(8):1388–98.
77. Hladky SB, Barrand MA. Mechanisms of fluid movement into, through and out of the brain: evaluation of the evidence. *Fluids Barriers CNS* 2014;11(1):26.
78. Bogdanova A, Petrushanko IY, Hernansanz-Agustín P, Martínez-Ruiz A. "Oxygen sensing" by Na,K-ATPase: these miraculous thiols. *Front Physiol* 2016;7:314.
79. de Lores Arnaiz GR, Ordieres MG. Brain Na(+), K(+)-ATPase activity in aging and disease. *Int J Biomed Sci* 2014;10(2):85–102.
80. Lima FD, Souza MA, Furian AF, Rambo LM, Ribeiro LR, Martignoni FV, et al. Na+,K+-ATPase activity impairment after experimental traumatic brain injury: relationship to spatial learning deficits and oxidative stress. *Behav Brain Res* 2008;193(2):306–10.
81. Stokum JA, Gerzanich V, Simard JM. Molecular pathophysiology of cerebral edema. *J Cereb Blood Flow Metab* 2016;36(3):513–38.
82. Simard JM, Tsymbalyuk O, Ivanov A, Ivanova S, Bhatta S, Geng Z, et al. Endothelial sulfonylurea receptor 1-regulated NC Ca-ATP channels mediate progressive hemorrhagic necrosis following spinal cord injury. *J Clin Invest* 2007;117(8):2105–13.
83. O'Donnell ME, Lam TI, Tran L, Anderson SE. The role of the blood–brain barrier Na–K–2Cl cotransporter in stroke. *Adv Exp Med Biol* 2004;559:67–75.
84. Simard JM, Tarasov KV, Gerzanich V. Non-selective cation channels, transient receptor potential channels and ischemic stroke. *Biochim Biophys Acta* 2007;1772(8):947–57.
85. Lam TI, Wise PM, O'Donnell ME. Cerebral microvascular endothelial cell Na/H exchange: evidence for the presence of NHE1 and NHE2 isoforms and regulation by arginine vasopressin. *Am J Physiol Cell Physiol* 2009;297(2):C278–89.
86. Watanabe M, Fukuda A. Development and regulation of chloride homeostasis in the central nervous system. *Front Cell Neurosci* 2015;9:371.
87. Foroutan S, Brillault J, Forbush B, O'Donnell ME. Moderate-to-severe ischemic conditions increase activity and phosphorylation of the cerebral microvascular endothelial cell Na+-K+-Cl-cotransporter. *Am J Physiol Cell Physiol* 2005;289(6):C1492–501.
88. Brillault J, Lam TI, Rutkowsky JM, Foroutan S, O'Donnell ME. Hypoxia effects on cell volume and ion uptake of cerebral microvascular endothelial cells. *Am J Physiol Cell Physiol* 2008;294(1):C88–96.
89. Hui H, Rao W, Zhang L, Xie Z, Peng C, Su N, et al. Inhibition of Na(+)-K(+)-2Cl(−) Cotransporter-1 attenuates traumatic brain injury-induced neuronal apoptosis via regulation of Erk signaling. *Neurochem Int* 2016;94:23–31.
90. Chen M, Dong Y, Simard JM. Functional coupling between sulfonylurea receptor type 1 and a nonselective cation channel in reactive astrocytes from adult rat brain. *J Neurosci* 2003;23(24):8568–77.
91. Simard JM, Chen M. Regulation by sulfanylurea receptor type 1 of a non-selective cation channel involved in cytotoxic edema of reactive astrocytes. *J Neurosurg Anesthesiol* 2004;16(1):98–9.
92. Simard JM, Chen M, Tarasov KV, Bhatta S, Ivanova S, Melnitchenko L, et al. Newly expressed SUR1-regulated NC(Ca-ATP) channel mediates cerebral edema after ischemic stroke. *Nat Med* 2006;12(4):433–40.
93. Zweckberger K, Hackenberg K, Jung CS, Hertle DN, Kiening KL, Unterberg AW, et al. Glibenclamide reduces secondary brain damage after experimental traumatic brain injury. *Neuroscience* 2014;272:199–206.

94. de Castro Ribeiro M, Badaut J, Price M, Meins M, Bogousslavsky J, Monard D, et al. Thrombin in ischemic neuronal death. *Exp Neurol* 2006;198(1):199–203.
95. Hirt L, Ternon B, Price M, Mastour N, Brunet JF, Badaut J. Protective role of early aquaporin 4 induction against postischemic edema formation. *J Cereb Blood Flow Metab* 2009;29(2):423–33.
96. Strbian D, Durukan A, Pitkonen M, Marinkovic I, Tatlisumak E, Pedrono E, et al. The blood–brain barrier is continuously open for several weeks following transient focal cerebral ischemia. *Neuroscience* 2008;153(1):175–81.
97. Hanukoglu I. ASIC and ENaC type sodium channels: conformational states and the structures of the ion selectivity filters. *FEBS J* 2016.
98. Wu J, Xu Y, Jiang YQ, Xu J, Hu Y, Zha XM. ASIC subunit ratio and differential surface trafficking in the brain. *Mol Brain* 2016;9:4.
99. Huang C, Hu ZL, Wu WN, Yu DF, Xiong QJ, Song JR, et al. Existence and distinction of acid-evoked currents in rat astrocytes. *Glia* 2010;58(12):1415–24.
100. Patet C, Suys T, Carteron L, Oddo M. Cerebral lactate metabolism after traumatic brain injury. *Curr Neurol Neurosci Rep* 2016;16(4):31.
101. Timofeev I, Nortje J, Al-Rawi PG, Hutchinson PJ, Gupta AK. Extracellular brain pH with or without hypoxia is a marker of profound metabolic derangement and increased mortality after traumatic brain injury. *J Cereb Blood Flow Metab* 2013;33(3):422–7.
102. Chu XP, Wemmie JA, Wang WZ, Zhu XM, Saugstad JA, Price MP, et al. Subunit-dependent high-affinity zinc inhibition of acid-sensing ion channels. *J Neurosci* 2004;24(40):8678–89.
103. Yin T, Lindley TE, Albert GW, Ahmed R, Schmeiser PB, Grady MS, et al. Loss of Acid sensing ion channel-1a and bicarbonate administration attenuate the severity of traumatic brain injury. *PLoS One* 2013;8(8):e72379.
104. Zhao X, Gorin FA, Berman RF, Lyeth BG. Differential hippocampal protection when blocking intracellular sodium and calcium entry during traumatic brain injury in rats. *J Neurotrauma* 2008;25(10):1195–205.
105. Lu KT, Wu CY, Cheng NC, Wo YY, Yang JT, Yen HH, et al. Inhibition of the Na^+–K^+–$2Cl^-$-cotransporter in choroid plexus attenuates traumatic brain injury-induced brain edema and neuronal damage. *Eur J Pharmacol* 2006;548(1–3):99–105.
106. Lu KT, Cheng NC, Wu CY, Yang YL. NKCC1-mediated traumatic brain injury-induced brain edema and neuron death via Raf/MEK/MAPK cascade. *Crit Care Med* 2008;36(3):917–22.
107. Jayakumar AR, Panickar KS, Curtis KM, Tong XY, Moriyama M, Norenberg MD. Na–K–Cl cotransporter-1 in the mechanism of cell swelling in cultured astrocytes after fluid percussion injury. *J Neurochem* 2011;117(3):437–48.
108. MacAulay N, Zeuthen T. Water transport between CNS compartments: contributions of aquaporins and cotransporters. *Neuroscience* 2010;168(4):941–56.
109. Badaut J, Fukuda AM, Jullienne A, Petry KG. Aquaporin and brain diseases. *Biochim Biophys Acta* 2014;1840(5):1554–65.
110. Manley GT, Fujimura M, Ma T, Noshita N, Filiz F, Bollen AW, et al. Aquaporin-4 deletion in mice reduces brain edema after acute water intoxication and ischemic stroke. *Nat Med* 2000;6(2):159–63.
111. Bao HJ, Qiu HY, Kuai JX, Song CJ, Wang SX, Wang CQ, et al. Apelin-13 as a novel target for intervention in secondary injury after traumatic brain injury. *Neural Regen Res* 2016;11(7):1128–33.
112. Zhang M, Cui Z, Cui H, Cao Y, Zhong C, Wang Y. Astaxanthin alleviates cerebral edema by modulating NKCC1 and AQP4 expression after traumatic brain injury in mice. *BMC Neurosci* 2016;17(1):60.
113. Sturdivant NM, Smith SG, Ali SF, Wolchok JC, Balachandran K. Acetazolamide mitigates astrocyte cellular edema following mild traumatic brain injury. *Sci Rep* 2016;6:33330.

114. Kiening KL, van Landeghem FK, Schreiber S, Thomale UW, von Deimling A, Unterberg AW, et al. Decreased hemispheric aquaporin-4 is linked to evolving brain edema following controlled cortical impact injury in rats. *Neurosci Lett* 2002;324(2):105–8.
115. Liang F, Luo C, Xu G, Su F, He X, Long S, et al. Deletion of aquaporin-4 is neuroprotective during the acute stage of micro traumatic brain injury in mice. *Neurosci Lett* 2015;598:29–35.
116. Yao X, Uchida K, Papadopoulos MC, Zador Z, Manley GT, Verkman AS. Mildly reduced brain swelling and improved neurological outcome in aquaporin-4 knockout mice following controlled cortical impact brain injury. *J Neurotrauma* 2015;32(19):1458–64.
117. Xu R, Luo C. Relationship between changes of *N*-methyl-d-aspartate receptor activity and brain edema after brain injury in rats. *Chin J Traumatol* 2001;4(3):135–8.
118. Lorber A, Artru AA, Lam MA, Mueller LA, Karpas Z, Roytblat L, et al. NPS 1506: a novel NMDA receptor antagonist: neuroprotective effects in a model of closed head trauma in rats. *J Neurosurg Anesthesiol* 2000;12(4):345–55.

CHAPTER

14

Spinal Cord Edema After Spinal Cord Injury: From Pathogenesis to Management

Newton Cho, Laureen D. Hachem and Michael G. Fehlings

University of Toronto Department of Surgery, Toronto, ON, Canada

OUTLINE

Spinal cord injury (SCI) is associated with significant disability that has broad implications both for the patient and for society. Global estimates of the annual incidence of SCI range from 15 to 40 cases per million.[1] In the United States alone, more than 1 million patients currently live with an SCI and over 12,000 new cases occur every year.[2] The most common causes of traumatic SCI are motor vehicle accidents, falls, recreational

Brain Edema.
DOI: http://dx.doi.org/10.1016/B978-0-12-803196-4.00014-X

injuries, and violence.[3] The incidence of SCI follows a bimodal distribution with the highest incidence occurring in young and healthy individuals aged 15–30 years.[4] The second peak occurs in older adults over 65 years of age and is largely attributed to falls. The incidence of SCI in the older population has risen significantly in the past decade[5] and functional recovery in this age group is often limited.[6] The burden to society of SCI is immense. Annual medical costs for managing patients with SCI have been estimated to reach $4.84 billion in the United States[7] and $1.57 billion in Canada.[8] These estimates are further compounded by the additional indirect costs of lost earnings and the enormous personal loss incurred by patients and their families as a result of the injury.

At present time, there is no effective neuroregenerative strategy or widely accepted neuroprotective treatment for SCI. Research into therapies (Fig. 14.1) for SCI is ongoing, ranging from neuroprotective therapies (e.g., riluzole) to regenerative strategies (e.g., stem cells).[9] At the core of these therapies, there needs to be an understanding of the pathophysiology of SCI. One of the areas of SCI research that has garnered much attention is spinal cord edema. The exact molecular mechanisms by which edema occurs following injury has yet to be fully understood, and therapies attempting to target spinal cord edema require further investigation. The purpose of this chapter is to provide an overview of the pathophysiology of edema formation post-SCI and to focus on the major therapies that have been investigated in order to specifically address this issue.

PATHOGENESIS: MOLECULAR MECHANISMS OF SPINAL CORD EDEMA

Biphasic Nature of Spinal Cord Injury

SCI overall is a biphasic process. The primary injury to the spinal cord is the result of direct impact to the cord including hemorrhage, contusion, and necrosis.[10–13] Usually, this is secondary to distraction, fracture, or dislocation of the spine. The secondary injury response continues after the initial insult. This includes various processes that exacerbate the initial injury including neuroinflammation, excitoxicity, lipid peroxidation, apoptosis, and spinal cord edema (Table 14.1).[13]

Spinal Edema Exacerbates the Initial Injury

Spinal cord edema occurs in the acute phase following SCI and often develops about 2 to 3 days postinjury.[13] Signs of edema have been seen as soon as 5 minutes postinjury and generally resolves by 14 days after injury

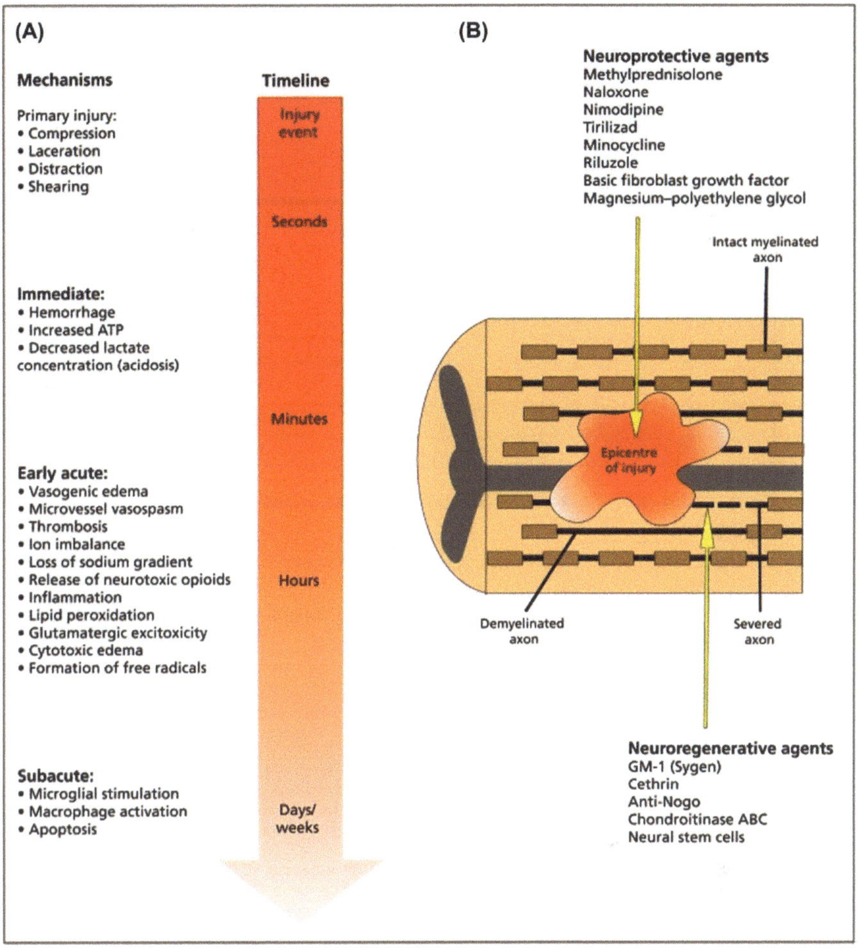

FIGURE 14.1 Various therapies attempting to address the different mechanisms of spinal cord injury have and currently are being investigated. *Source: From Wilson J, Forgione N, Fehlings M. Emerging therapies for acute traumatic spinal cord injury. CMAJ. 2013;185:485–92, with permission.*

as astrocytes form the glial scar and the blood–spinal cord barrier (BSCB) is restored.[14] This edema is not a static process: it spreads away from the initial injury site—reported as one vertebral segment every 30 hours in the first 3 days after injury in complete cervical SCI patients.[15] The degree of spread is also related to the initial force of trauma.[16]

The mechanism by which edema exacerbates injury may be the raised intrathecal pressure resulting in reduced blood flow and associated spinal cord ischemia, cell death, and worsening function.[17] Leonard et al.

TABLE 14.1 Key Phases of Spinal Cord Injury and Associated Pathology

	Time After SCI				
Phases and Key Events	**≤2h**	**≤48h**	**≤14 days**	**≤6 months**	**≥6 months**
Injury phase	Primary immediate	Early acute	Secondary subacute	Intermediate	Chronic/late
Key processes and events	Primary mechanical injury, traumatic severing of axons, gray matter hemorrhage, hemorrhagic necrosis, microglial activation, released factors: IL-1β, TNFα, IL-6, & others	Vasogenic & cytoxic edema, ROS production: lipid peroxidation, glutamate-mediated excitotoxicity, continued hemorrhage & necrosis, neutrophil invasion, peak BBB permeability, early demyelination (oligodendrocyte death), neuronal death, axonal swelling, systemic events (systemic shock, spinal shock, hypotension, hypoxia)	Macrophage infiltration, initiation of astroglial scar (reactive astrocytosis), BBB repair & resolution of edema	Continued formation of glial scar, cyst formation, lesion stabilization	Prolonged Wallerian degeneration, persistence of spared demyelinated axons, potential structural & functional plasticity of spared spinal cord tissue
Therapeutic aims	Neuroprotection	Neuroprotection, immune modulation, cell-based remyelination approaches, glial scar degradation		Glial scar degradation	Rehabilitation, neuroprostheses

From Rowland J et al. Current status of acute spinal cord injury pathophysiology and emerging therapies: promise on the horizon. Neurosurg Focus. 2008;25(5):1–17, with permission.

tested this hypothesis to determine whether increased intrathecal pressure results after a balloon compression model of SCI in rabbits, and if these increases are the result of edema.[17] In their study, they found that tissue water content significantly increased as early as 5 hours postinjury with a peak at 1 week post-SCI. Similarly, intrathecal pressure significantly increased with peak values at 3 days post-SCI but declined by 1 week after injury. This study demonstrated that edema contributes to raised intrathecal pressure post-SCI. They also found that there was a larger degree of hemorrhage contributing to lesion volume immediately after injury and lasting up to 3 days, at which point edema appeared to be playing a larger role for worsening lesion volume up to 1 week post-SCI.[17] Furthermore, the authors found that there was increased albumin immunoreactivity postinjury suggesting that there was disruption of the BSCB, and thus the resultant edema was likely vasogenic in nature.[17] The role of a vasogenic mechanism in edema formation after SCI is further borne out by a recent study by Figley et al. who described the vascular changes and BSCB disruptions that occur post-SCI.[18] In a rat clip-compression model of SCI, they found that the BSCB is disrupted as early as 1 hour post-SCI and remains open until 5 days after injury. Despite this, post-SCI edema is likely both cytotoxic and vasogenic in nature as demonstrated by the presence of both cellular swelling and increased extracellular fluid in experimental models of SCI.[19]

Spinal Edema Formation Is a Multifactorial Molecular Process

The exact mechanisms of edema formation in the spinal cord are not fully understood. Cytotoxic edema may result from cord ischemia after injury, which leads to failure of the Na^+-K^+ ATPase from depletion of adenosine triphosphate (ATP), resulting in an influx of water into cells secondary to increased intracellular ionic content.[20] Increases in the levels of arachidonic acid post-SCI also inhibit the Na^+/K^+ exchanger and similarly exacerbate cytotoxic edema.[20] Vasogenic edema results from disruption of the BSCB after injury. Disruption of endothelial cell gap junctions is seen as early as 90 seconds postinjury in experimental models of SCI.[19] Some research has suggested that endothelins play a role in this process as endothelin-1 administration into the intact spinal cord leads to BSCB disruption, and levels of endothelin-1 have been shown to increase post-SCI.[21] Other compounds implicated in increasing endothelial cell permeability and worsening vasogenic edema include prostaglandins, histamines, serotonin, and vascular endothelial growth factor (VEGF).[11,22–27]

The water channel aquaporin-4 (AQP4) has been implicated as a key molecular player in the formation of edema postinjury. The expression of AQP4 has been shown to increase post-SCI in rats and corresponds temporally and spatially with areas of increased water content in acute and chronically injured spinal cords.[28] More recent research by Saadoun and Papadopoulos proposes that spinal cord edema formation is similar to edema formation in the brain, with both cytotoxic and vasogenic mechanisms mediated by AQP4.[29] The same group demonstrated that AQP4 deletion reduces spinal cord edema at 48 hours.[30] They hypothesize that SCI results in a mixed cytotoxic–vasogenic edema picture. Cytotoxic edema is AQP4-dependent, but because vasogenic edema results from disruption of the BSCB, it is AQP4-independent.[30] Moreover, they suggest that edema is eliminated via white matter tracts, and this process is AQP4-independent.[30] Wu et al. further examined the role of AQP4 in a hemisection model of SCI in AQP4 knockout mice.[31] In contrast, they demonstrated that the absence of AQP4 resulted in worsening edema postinjury. Furthermore, there was less scar formation in the knockout mice but also prominent cyst formation. In this study, the authors also found that there was greater axonal degeneration in the absence of AQP4 with significantly reduced numbers of rubrospinal neurons. Overall, the authors concluded that AQP4 played a role in reducing secondary damage post-SCI by promoting clearance of edema and glial scar formation.[31] There is evidence to suggest that AQP4 channels play a key role in mediating edema; however, the exact molecular underpinnings of this effect along with the role of other aquaporin channels remain to be determined.

Another protein implicated in the formation of edema post-SCI is WNK1. WNK1 is a serine/threonine kinase encoded by the *WNK1* in humans.[32] It has been implicated in phosphorylation of Na^+-Cl^- cotransporter 1 (NKCC1) and K^+-Cl^- cotransporter 2 (KCC2), which are expressed in the central nervous system. Ahmed et al. hypothesized that spinal cord edema results from aberrant function of NKCC1, resulting in altered ionic homeostasis and associated osmotic homeostasis. Following a contusion SCI in rats at T9, they demonstrated increased expression of NKCC1 in the injured cords in addition to increased expression of WNK1. They propose that the edema initially results from destruction of the BSCB, causing vasogenic edema. However, after 2 weeks when the BSCB is restored, ongoing edema results from changes in ionic transport. More specifically, pharmacological inhibition or genetic deficiency of NKCC1 has been shown to decrease ischemia-induced cell swelling and cerebral edema.[32] Extending these results to the spinal cord, as WNK1 is an activator of NKCC1, the authors postulated that both proteins could play a role in the formation of spinal cord edema post-SCI. They plan to perform future experiments looking at how NKCC1 phosphorylation changes with changes in WNK1 expression.[32] The exact role of NKCC1 and its activation by WKN1 still remains to be determined but provides another pathway through which edema formation can occur.

MANAGEMENT OF SPINAL CORD EDEMA: BENCH TO BEDSIDE

General Management of Spinal Cord Injury

In general, the management of SCI involves medical and surgical interventions. Although an in-depth review of each step in the management of SCI patients is out of the scope of this chapter, any patient with a suspected SCI should be first stabilized medically, transported to the nearest neurosurgical center, undergo a comprehensive neurological examination, and receive appropriate imaging. In terms of management parameters that may help address spinal edema specifically, it is recommended that patients with SCI be managed initially in an intensive care unit.[33–35] The recent iteration of conjoint guidelines produced by the American Association of Neurological Surgeons and Congress of Neurological Surgeons Spine Section recommend that any systolic blood pressure less than 90 mmHg should be corrected immediately.[36] Furthermore, maintenance of the mean arterial blood pressure between 85 and 90 mmHg for the first 7 days post-SCI is also recommended.[36] In theory, these parameters can help to maintain spinal cord perfusion in the face of ischemia secondary to spinal cord edema. In addition, there may be a role for methylprednisolone post-SCI, although this remains a significant point of debate.[37] There is some preclinical evidence that methylprednisolone may reduce edema post-SCI.[38,39] The overall functional benefit and associated complications have resulted in significant debate around its utility that go beyond its ability to reduce edema post-SCI. Recent guidelines recommend the use of a 24 hour infusion of MPSS, within 8 hours of SCI, as an option, although it is recognized the usage of MPSS in clinical practice will vary among different centers.[40]

The findings of Leonard et al. may suggest the need for a paradigm shift in treating SCI similar to what has occurred in traumatic brain injury. More specifically, there may be a need to consider spinal cord perfusion as affected not only by the mean arterial blood pressure, but also the intrathecal pressure.[17] In this way, edema may play a more direct role as it has been shown to be associated with worsening intrathecal pressure and tissue damage post-SCI.[17] Further studies looking at the utility of measuring intrathecal pressure and its effects on patient outcomes are required to clarify this issue.

Surgical decompression of the spinal cord is an obvious intervention in order to prevent further exacerbation of injury, including spinal cord edema. Various experimental studies have confirmed that decompression of the spinal cord attenuates secondary damage.[41–43] The recent STASCIS (Surgical Timing in Acute SCI Study) trial has shed light on the potential role for early surgical decompression as providing greater benefits to patient outcomes.[44] This trial was the largest prospective multicenter study comparing early vs late surgical decompression in acute traumatic SCI. The early surgery group demonstrated a significantly greater proportion of patients recovering at least two American Spinal Injury Association Impairment Scale grades at 6-month follow-up.[44] Although the exact pathophysiological underpinnings of these results were not addressed in the study, it can be hypothesized that earlier decompression (within 24 hours) may help to prevent further secondary injury from processes like spinal cord edema formation which tends to peak around 2 to 3 days post-SCI.

Edema-Specific Therapies Under Investigation

From the point of view of therapies looking specifically at targeting edema post-SCI, there have been a wide range of different interventions that have been tested. It is important to note that none of these therapies has yet entered clinical trials. Further research into the reliability of these different methods to reduce spinal cord edema and hopefully improve neurological function post-SCI is required.

Aquaporins

As discussed previously, aquaporins have been shown to play a role in the regulation of edema in the spinal cord. Administration of the HIF-1α inhibitor 2-methoxyestradiol (2ME2) intraperitoneally in a rat model of SCI resulted in decreased expression of VEGF, AQP4, and AQP1 post-SCI.[45] Potentially, this could result in reduced edema after injury although this study did not assess functional outcomes. Mechanistically, however, this study does provide the insight that upregulation of VEGF, AQP4, and AQP1 are related to increases in HIF-1α, which is increased secondary to hypoxia. Furthermore, the study demonstrated that VEGF administered to the injured spinal cord resulted in increased permeability of the BSCB with associated upregulation of the levels of AQP4 mRNA in the injured cord.[45] The authors postulated that VEGF's direct effects on disruption of the BSCB and associated changes in the oncotic pressure of the spinal cord could encourage changes in AQP4 expression. The exact dynamics of these different signaling mediators and channels still remains to be determined, but 2ME2 may have the potential to help reduce spinal cord edema by addressing all the players in this pathway.

Another study addressed the potential utility of sulphoraphane administration in SCI.[46] Sulphoraphane is a compound occurring naturally in cruciferous vegetable and has been shown to reduce edema after traumatic brain injury with associated increases in AQP4 expression. Following from this, the authors attempted to determine its utility in SCI.[46] Using a compression model of SCI in mice, the authors then treated mice with intraperitoneal sulphaphorane (5 mg/kg) 1 hour postinjury. They found that administration of sulphaphorane resulted in increased levels of AQP4 mRNA in addition to protein expression. Furthermore, they found that there was an associated reduction in spinal cord water content in treated animals.[46] They postulate that AQP4 thus plays a role in clearance of excess water from the injured spinal cord and that sulphoraphane may play a role in enhancing this reduction of edema by increasing expression of AQP4.[46] It still remains to be determined whether this therapy has clinical relevance.

Melatonin

The administration of melatonin has also been studied to help reduce edema post-SCI. In one study, rats were subject to a clip-compression SCI at T12 followed by allocation to sham group, laminectomy with intraperitoneal saline, and laminectomy and 100 mg/kg dose of melatonin.[47] The authors found that there was significantly decreased spinal cord water content in the melatonin treatment group in addition to significantly decreased expression of AQP4.[47] They concluded that melatonin acted to reduce spinal cord edema by altering AQP4 expression. The exact mechanisms by which melatonin exerts this effect was not discussed, and functional assessment was not performed in this study. From this study alone, it is hard to know whether and how melatonin can reduce edema significantly to improve functional recovery. Another study with a similar dose of melatonin administration found similar results.[48] Again, the lack of behavioral outcome measures makes it difficult to fully ascertain the ability of melatonin to improve functional outcomes following SCI, although there appears to be evidence to suggest it helps to reduce postinjury edema.

Hormonal Therapies

Hormonal agents have also been tested and been found to have effects on spinal cord edema. One agent of interest has been growth hormone. Nyberg and Sharma looked at the role of growth hormone in SCI in the context of previous research demonstrating that its overexpression causes an increase in spinal motor neuron size and that growth hormone levels are decreased in young SCI patients.[49] They incised the right dorsal horn at the T10–11 segment in their rat model of SCI. Thirty minutes prior to injury, they applied purified growth hormone onto the surface of the spinal cord. They found that treatment with growth hormone resulted in reduced intensity of Evans blue staining and extravasation of a radiotracer. There was also significantly reduced water content in the injured cord.[49] These results suggest that growth hormone may help reduce edema post-SCI by attenuating the breakdown of the BSCB. Functional outcomes were not assessed in this study, and the exact mechanisms by which growth hormone exerted its effects are not fully understood.

Treatment of SCI with tamoxifen has also been investigated. Tamoxifen is a nonsteroidal selective estrogen receptor modulator. One study looked at the effect of intraperitoneal injection of tamoxifen (5.0mg/kg) after injury in a rat model of SCI.[50] The authors found that there were significantly reduced levels of TNF-α and IL-1β at 6hours post-SCI in the treated group. With electron microscopy, the authors also found that there was less edema in the astrocytes of perivascular glial membranes and neuropil.[50] With its potential for reducing inflammation and edema, tamoxifen may also be considered a future therapeutic agent for SCI. The authors suggest that the advantage of using tamoxifen is that although there has been research suggesting estrogen itself may have neuroprotective effects, the use of tamoxifen can potentially avoid undesired side effects from exogenous estrogen including developing breast and uterine malignancies.[50]

Other Therapies

Many other therapies have been investigated for their utility in reducing post-SCI edema; however, the evidence for these therapies still remains to be developed. Hypertonic saline is routinely used in the context of traumatic brain injury, but its role in SCI is less clear. Nout et al. performed a right-sided contusion injury at C5 in rats.[51] Following injury, the rats received either boluses of normal saline or 5% NaCl at 1.4mL/kg IV every hour starting 30minutes post-SCI for 8hours.[51] Quantification of hyperintensity on subsequent MRI of the spinal cord demonstrated significantly reduced volume in the animals treated with hypertonic saline. The authors suggested that the reduction of cord swelling should, hypothetically, improve spinal cord perfusion.[51] Further work around the strength of hypertonic saline to be used in addition to the timing of administration post-SCI likely needs to be done prior to its regular use in SCI. Other therapies, including hyperbaric oxygen, myelotomy, and the NMDA receptor antagonist MK801, all demonstrated reduction of spinal cord edema and even functional improvements post-SCI through various mechanisms including effects on matrix metalloproteinases and aquaporin channel expression.[52–54] Furthermore, a number of therapies that have shown general neuroprotective and antiinflammatory effects in SCI may also aid in reducing edema formation. Riluzole, a Na/glutamate antagonist, has been shown to confer significant neuroprotection in experimental models of SCI and is currently under clinical trial in patients.[55,56] By inhibiting the overactivation of voltage-gated sodium channels and reducing glutamate excitotoxicity in neurons following SCI, riluzole may reduce intracellular ion accumulation and subsequent water influx thus attenuating cytotoxic edema formation. Moreover, immunomodulatory therapies such as intravenous immunoglobulin G (IVIG) may hold promise in attenuating edema after SCI. Administration of IVIG following SCI in a rodent model was found to reduce the expression of MMP-9 and was associated with improved histological and functional recovery. Although not specifically examined in this study, the reduction in MMP-9 associated with IgG could potentially attenuate BSCB disruption and reduce edema formation after SCI.[57]

Overall, the management of spinal cord edema post-SCI includes medical and surgical interventions that indirectly address edema and novel therapeutics that are being investigated to specifically reduce edema. General medical management of SCI patients in an intensive care unit with ability to maintain spinal cord perfusion and to receive timely surgical decompression is important. Specific therapies aimed at various mediators of spinal cord edema, including aquaporins, have yet to enter clinical trials. Further research into the reliable efficacy of these different interventions along with their optimal timing of administration postinjury and dosing regimen is necessary (see Table 14.2 and Table 14.3).

TABLE 14.2 Pathogenesis of Edema Formation After Spinal Cord Injury

Factor	Mechanism
Disruption of blood–spinal cord barrier	• Traumatic injury to spinal cord results in disruption of blood–spinal cord barrier as early as 1 h postinjury
Alteration in ion homeostasis	• ATP depletion and arachidonic acid production after SCI impairs Na/K ATPase leading to increased intracellular ionic content and water influx • WNK1-mediated phosphorylation of Na^+-Cl^- cotransporter 1 (NKCC1) and K^+-Cl^- cotransporter 2 (KCC2), leading to ongoing edema after restoration of blood–spinal cord barrier
Alteration in aquaporin expression	• Increased expression of AQP4 allows for increased water influx into cells contributing to cytotoxic edema • Note that AQP4 may also have a counterregulatory role that aids in clearing water accumulation from the damaged spinal cord
Endothelial cell dysfunction	• Ischemia impairs endothelial cell gap and tight junctions • Endothelin-1 accumulates after SCI further impairing endothelial cell junctions
Production of inflammatory mediators	• Production of prostaglandins, histamine, serotonin, and VEGF increases after SCI and acts to increase blood vessel permeability and impair BSCB

TABLE 14.3 Summary of Therapies Aimed at Targeting Postspinal Cord Injury Edema

Therapy	Evidence
MPSS	• Reduction in post-SCI edema in multiple experimental models of SCI • Recent guidelines recommend the use of a 24 h infusion of MPSS, within 8 h of SCI, as an option, although the usage of MPSS in clinical practice will vary among different centers
Surgery	• Surgical decompression removes ongoing compression and ideally prevents exacerbation of injury • Early surgery (within 24 h) may be beneficial to ensure optimal recovery
Hypertonic saline	• Hypertonic saline (1.4 mL/kg q1 h starting 30 min after injury) led to a reduction in cord edema and hemorrhage in rodents as measured by MRI • No clinical trials to date
AQP4 targets	• Administration of HIF-1α inhibitor, 2-methoxyestradiol (2ME2), after SCI resulted in decreased expression of AQP4, AQP1, and VEGF which may aid in reducing edema formation • Sulphoraphane increased AQP4 mRNA and was associated with reduced spinal cord water content potentially due to increased clearance of excess spinal cord water
Melatonin	• Administration of melatonin into SCI-injured rodents led to decreased spinal cord water content compared to controls
Hormonal therapies	• Administration of growth hormone to the spinal cord prior to injury attenuated BSCB disruption • Tamoxifen reduced TNF-α and IL-1β 6 h post-SCI and was associated with less edema in the astrocytes of perivascular glial membranes and neuropil

MPSS, methylprednisolone sodium succinate.

CONCLUSION

Spinal cord edema plays a significant role in mediating secondary damage post-SCI within the biphasic injury paradigm of SCI. It is a complex process that tends to present relatively acutely post-SCI whose molecular underpinnings have to be fully elucidated. Similarly, treatments directed at reducing spinal cord edema have yet to be fully developed. Although many therapies are in the experimental stage, there is currently no therapy that can be administered to patients to specifically reduce edema after injury. It is likely that the treatment of SCI will require a combinatorial approach whereby interventions will not only address spinal cord edema but will also concurrently reduce injury from other secondary mechanisms including inflammation, excitotoxicity, and apoptosis. Indeed, it is likely that these mechanisms are interconnected and constantly interact and evolve as the injury occurs, and future studies will need to address this barrier moving forward.

References

1. Ackery A, Tator C, Krassioukov A. A global perspective on spinal cord injury epidemiology. *J Neurotrauma* 2004;**21**(10):1355–70.
2. [No authors listed]. One degree of separation: paralysis and SCI in the United States. *Short Hills, New Jersey, USA: Christopher and Dana Reeve Foundation*. 2009.
3. Chen Y, Tang Y, Vogel LC, Devivo MJ. Causes of spinal cord injury. *Top Spinal Cord Inj Rehabil* 2013;**19**(1):1–8.
4. Singh A, Tetreault L, Kalsi-Ryan S, Nouri A, Fehlings MG. Global prevalence and incidence of traumatic spinal cord injury. *Clin Epidemiol* 2014;**6**:309–31.
5. Jain NB, Ayers GD, Peterson EN, et al. Traumatic spinal cord injury in the United States, 1993–2012. *JAMA* 2015;**313**(22):2236–43.
6. Furlan JC, Fehlings MG. The impact of age on mortality, impairment, and disability among adults with acute traumatic spinal cord injury. *J Neurotrauma* 2009;**26**(10): 1707–17.
7. Selvarajah S, Hammond ER, Haider AH, et al. The burden of acute traumatic spinal cord injury among adults in the united states: an update. *J Neurotrauma* 2014;**31**(3):228–38.
8. Krueger H, Noonan VK, Trenaman LM, Joshi P, Rivers CS. The economic burden of traumatic spinal cord injury in Canada. *Chronic Dis Inj Can* 2013;**33**(3):113–22.
9. Wilson J, Forgione N, Fehlings M. Emerging therapies for acute traumatic spinal cord injury. *CMAJ* 2013;**185**:485–92.
10. Baptiste D, Fehlings M. Pharmacological approaches to repair the injured spinal cord. *J Neurotrauma* 2006;**23**:318–34.
11. Hagg T, Oudega M. Degenerative and spontaneous regenerative processes after spinal cord injury. *J Neurotrauma* 2006;**23**:264–80.
12. Sekhon L, Fehlings M. Epidemiology, demographics, and pathophysiology of acute spinal cord injury. *Spine (Phila Pa 1976)* 2001;**26**:S2–S12.
13. Rowland J, Hawryluk G, Kwon B, Fehlings M. Current status of acute spinal cord injury pathophysiology and emerging therapies: promise on the horizon. *Neurosurg Focus* 2008;**25**(5):1–17.
14. Yashon D, Bingham Jr. WG, Faddoul EM, Hunt WE. Edema of the spinal cord following experimental impact trauma. *J Neurosurg* 1973;**38**(6):693–7.
15. Leypold BG, Flanders AE, Burns AS. The early evolution of spinal cord lesions on MR imaging following traumatic spinal cord injury. *AJNR* 2008;**29**(5):1012–6.
16. Wagner Jr. FC, Stewart WB. Effect of trauma dose on spinal cord edema. *J Neurosurg* 1981;**54**(6):802–6.
17. Leonard A, Thornton E, Vink R. The relative contribution of edema and hemorrhage to raised intrathecal pressure after traumatic spinal cord injury. *J Neurotrauma* 2015;**32**:397–402.
18. Figley SA, Khosravi R, Legasto JM, Tseng YF, Fehlings MG. Characterization of vascular disruption and blood–spinal cord barrier permeability following traumatic spinal cord injury. *J Neurotrauma* 2014;**31**(6):541–52.
19. Goodman JH, Bingham Jr. WG, Hunt WE. Ultrastructural blood–brain barrier alterations and edema formation in acute spinal cord trauma. *J Neurosurg* 1976;**44**(4):418–24.
20. Faden AI, Chan PH, Longar S. Alterations in lipid metabolism, Na+,K+-ATPase activity, and tissue water content of spinal cord following experimental traumatic injury. *J Neurochem* 1987;**48**(6):1809–16.
21. Mautes AE, Weinzierl MR, Donovan F, Noble LJ. Vascular events after spinal cord injury: contribution to secondary pathogenesis. *Phys Ther* 2000;**80**(7):673–87.
22. Sharma HS, Olsson Y. Edema formation and cellular alterations following spinal cord injury in the rat and their modification with p-chlorophenylalanine. *Acta Neuropathol* 1990;**79**(6):604–10.

23. Sharma HS, Olsson Y, Nyberg F, Dey PK. Prostaglandins modulate alterations of microvascular permeability, blood flow, edema and serotonin levels following spinal cord injury: an experimental study in the rat. *Neuroscience* 1993;**57**(2):443–9.
24. Sharma HS, Vannemreddy P, Patnaik R, Patnaik S, Mohanty S. Histamine receptors influence blood–spinal cord barrier permeability, edema formation, and spinal cord blood flow following trauma to the rat spinal cord. *Acta Neurochir Suppl* 2006;**96**:316–21.
25. Winkler T, Sharma HS, Stalberg E, Olsson Y. Indomethacin, an inhibitor of prostaglandin synthesis attenuates alteration in spinal cord evoked potentials and edema formation after trauma to the spinal cord: an experimental study in the rat. *Neuroscience* 1993;**52**(4):1057–67.
26. Naftchi NE, Demeny M, DeCrescito V, Tomasula JJ, Flamm ES, Campbell JB. Biogenic amine concentrations in traumatized spinal cords of cats. Effect of drug therapy. *J Neurosurg* 1974;**40**(1):52–7.
27. Akiyama C, Yuguchi T, Nishio M, et al. Src family kinase inhibitor PP1 improves motor function by reducing edema after spinal cord contusion in rats. *Acta Neurochir Suppl* 2003;**86**:421–3.
28. Nesic O, Lee J, Ye Z, et al. Acute and chronic changes in aquaporin 4 expression after spinal cord injury. *Neuroscience* 2006;**143**(3):779–92.
29. Saadoun S, Papdopoulos M. Aquaporin-4 in brain and spinal cord oedema. *Neuroscience* 2010;**168**:1036–46.
30. Saadoun S, Bell B, Verkman A, Papdopoulos M. Greatly improved neurological outcome after spinal cord compression injury in AQP4-deficient mice. *Brain* 2008;**131**:1087–98.
31. Wu Q, Zhang YJ, Gao JY, et al. Aquaporin-4 mitigates retrograde degeneration of rubropsinal neurons by facilitating edema clearance and glial scar formation after spinal cord injury in mice. *Mol Neurobiol* 2014;**49**:1327–37.
32. Ahmed MM, Lee H, Clark Z, et al. Pathogenesis of spinal cord injury induced edema and neuropathic pain: expression of multiple isoforms of wnk1. *Ann Neurosci* 2014;**21**(3):97–103.
33. Levi L, Wolf A, Belzberg H. Hemodynamic parameters in patients with acute cervical cord trauma: description, intervention, and prediction of outcome. *Neurosurgery* 1993;**33**(6):1007–16. discussion 1016-1007.
34. [No authors listed] Management of acute spinal cord injuries in an intensive care unit or other monitored setting. *Neurosurgery* 2002;**50**(3 Suppl):S51–7.
35. Gschaedler R, Dollfus P, Mole JP, Mole L, Loeb JP. Reflections on the intensive care of acute cervical spinal cord injuries in a general traumatology centre. *Paraplegia* 1979;**17**(1):58–61.
36. Ryken TC, Hurlbert RJ, Hadley MN, Aarabi B, Dhall SS, Gelb DE, et al. The acute cardiopulmonary management of patients with cervical spinal cord injuries. *Neurosurgery* 2013;**72**:84–92.
37. Fehlings M, Wilson J, Cho N. Methylprednisolone for the treatment of acute spinal cord injury: counterpoint. *Neurosurgery* 2014;**61**:36–42.
38. Merola A, O'Brien MF, Castro BA, et al. Histologic characterization of acute spinal cord injury treated with intravenous methylprednisolone. *J Orthop Trauma* 2002;**16**(3):155–61.
39. Xu J, Qu ZX, Hogan EL, Perot Jr. PL. Protective effect of methylprednisolone on vascular injury in rat spinal cord injury. *J Neurotrauma* 1992;**9**(3):245–53.
40. Ismailoğlu O, Oral B, Görgülü A, Sütçü R, Demir N. Neuroprotective effects of tamoxifen on experimental spinal cord injury in rats. *J Clin Neurosci* 2010;**17**(10):1306–10. http://dx.doi.org/10.1016/j.jocn.2010.01.049. Epub 2010 Jul 13. PubMed PMID: 20630763.
41. Brodkey JS, Richards DE, Blasingame JP, Nulsen FE. Reversible spinal cord trauma in cats. Additive effects of direct pressure and ischemia. *J Neurosurg* 1972;**37**(5):591–3.
42. Carlson GD, Minato Y, Okada A, et al. Early time-dependent decompression for spinal cord injury: vascular mechanisms of recovery. *J Neurotrauma* 1997;**14**(12):951–62.

43. Delamarter RB, Sherman J, Carr JB. Pathophysiology of spinal cord injury. Recovery after immediate and delayed decompression. *J Bone Joint Surg Am* 1995;**77**(7):1042–9.
44. Fehlings MG, Vaccaro A, Wilson JR, et al. Early versus delayed decompression for traumatic cervical spinal cord injury: results of the Surgical Timing in Acute Spinal Cord Injury Study (STASCIS). *PLoS ONE* 2012;**7**(2):e32037.
45. Wang Y, Fan Z, Cao Y, Yu D, Zhang Y, Wang Y. 2-Methoxyestradiol inhibits the up-regulation of AQP4 and AQP1 expression after spinal cord injury. *Brain Res* 2010;**1370**:220–6.
46. Mao L, Wang HD, Wang XL, Qiao L, Yin HX. Sulforaphane attenuates matrix metalloproteinase-9 expression following spinal cord injury in mice. *Ann Clin Lab Sci* 2010;**40**(4):354–60.
47. Liu X, Wang Y, Yang J, et al. Anti-edema effect of melatonin on spinal cord injury in rats. *Biomed Pap Med Fac Univ Palacky Olomouc Czech Repub* 2015;**159**:220–6.
48. Li C, Chen X, Qiao S, et al. Melatonin lowers edema after spinal cord injury. *Neural Regen Res* 2014;**9**(24):2205–10.
49. Nyberg F, Sharma H. Repeated topical application of growth hormone attenuates blood–spinal cord barrier permeability and edema formation following spinal cord injury. *Amino Acids* 2002;**23**:231–9.
50. Ismailoglu O, Oral B, Gorgulu A, Sutcu R, Demir N. Neuroprotective effects of tamoxifen on experimental spinal cord injury in rats. *J Clin Neurosci* 17:1306–10.
51. Nout YS, Mihai G, Tovar CA, Schmalbrock P, Bresnahan JC, Beattie MS. Hypertonic saline attenuates cord swelling and edema in experimental spinal cord injury: a study utilizing magnetic resonance imaging. *Crit Care Med* 2009;**37**(7):2160–6.
52. Yang J, Wang G, Gao C, Shao G, Kang N. Effects of hyperbaric oxygen on MMP-2 and MMP-9expression and spinal cord edema after spinal cord injury. *Life Sci* 2013;**93**:1033–8.
53. Hu A, Li J, Sun W, et al. Myelotomy reduces spinal cord edema and inhibits aquaporin-4 and aquaporin-9 expression in rats with spinal cord injury. *Spinal Cord* 2015;**53**:98–102.
54. Li S, Tator C. Effects of MK801 on evoked potentials, spinal cord blood flow and cord edema in acute spinal cord injury in rats. *Spinal Cord* 1999;**37**:820–32.
55. Wu Y, Satkundrarajah K, Teng Y, Chow DS, Fehlings MG. Evaluation of the sodium-glutamate blocker riluzole in a preclinical model of ervical spinal cord injury. *Evid Based Spine-Care J* Aug 2010;**1**(2):71–2.
56. Grossman RG, Fehlings MG, Frankowski RF, et al. A prospective, multicenter, phase I matched-comparison group trial of safety, pharmacokinetics, and preliminary efficacy of riluzole in patients with traumatic spinal cord injury. *J Neurotrauma* 2014;**31**(3):239–55.
57. Nguyen DH, Cho N, Satkunendrarajah K, Austin JW, Wang J, Fehlings MG. Immunoglobulin G (IgG) attenuates neuroinflammation and improves neurobehavioral recovery after cervical spinal cord injury. *J Neuroinflammation* 2012;**9**:224.

CHAPTER

15

Brain Edema in Acute Liver Failure

Roger F. Butterworth

University of Montreal, Montreal, QC, Canada

OUTLINE

INTRODUCTION

The appearance of altered mental status in a patient with an acute deterioration of liver function defines the condition previously known as fulminant hepatic failure; now widely known as acute liver failure (ALF).

Currently, by far the most common cause of ALF in the United States relates to liver injury resulting from an overdose (either intentional or unintentional) of the widely used analgesic acetaminophen. Other causes

Brain Edema.
DOI: http://dx.doi.org/10.1016/B978-0-12-803196-4.00015-1

of ALF include viral hepatitis, idiosyncratic drug reactions, and ischemic liver injury. Liver transplantation provides a life-saving procedure that is currently widely available in tertiary care centers throughout Europe, North America, and other developed countries and its use has resulted in much-improved survival rates in ALF.

In addition to alterations of mental status (encephalopathy), brain edema is a common finding in patients with ALF independent of the etiology of liver failure. If severe, brain edema in ALF patients may progress resulting in intracranial hypertension which, if left untreated frequently results in brain herniation, a condition that remains one of the major causes of mortality in ALF. Increased cerebral blood flow (CBF) may also accompany the onset of brain edema in ALF.

Recent advances in understanding of pathophysiology of brain edema in ALF result from advances in knowledge relating to the neuropathology of the condition, to the selective nature of the neural cell changes involved in brain swelling related to ALF and to the importance of systemic and central inflammatory mechanisms. The present review will examine some of these issues.

Neuropathologic evaluation of the brain in both human and in experimental ALF in laboratory animals reveals significant cellular alterations that are primarily expressed by neuroglia rather than neurons. Two types of neuroglia namely astrocytes and microglia bear the brunt of brain dysfunction in ALF.

Astrocytes in brain sections from patients who died in ALF are clearly swollen as are their mitochondria (Fig. 15.1).[1] Morphological studies in brain preparations from rabbits with ALF resulting from galactosamine-induced toxic liver injury show similar changes of cortical astrocytes characterized by marked hydropic changes with vacuolation of the perinuclear cytoplasm as well as pericapillary and periaxonal foot processes.[2] The blood–brain barrier remained grossly intact in these animals.

Astrocyte swelling was again reported in cortical gray matter of rats with ALF resulting from hepatic devascularization created by end-to-side portacaval anastomosis followed 24 hours later by hepatic artery ligation.[3]

Based upon these observations, it is now widely accepted that the brain edema that characterizes ALF is primarily cytotoxic in nature and that it is the astrocyte that manifests the majority of the changes in cell volume.

It should be noted, however, that although gross alterations of blood–brain barrier (BBB) are not generally a feature of ALF, alterations of cerebrovascular endothelial cell integrity have occasionally been reported.[1,4] Moreover, a recent study in an animal model of ALF in which brain edema was precipitated by infection showed clear changes in permeability of the BBB and altered expression of tight junction proteins and allied proteins of the BBB.[5] This issue will be discussed at length in a later section of this review. These findings suggest that when ALF is accompanied by

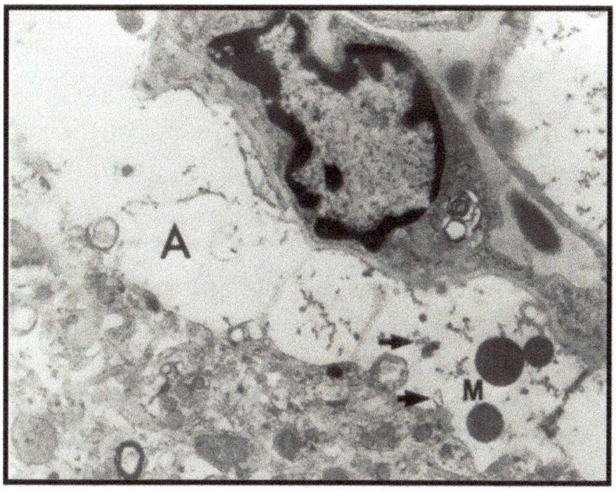

FIGURE 15.1 **Electron microscopic study of brain capillaries in brain edema from a patient who died in acute liver failure.** Note swelling and vacuolation of the perivascular astrocyte (A) and of mitochondria (MA). Endoplasmic reticulum is dilated (arrows). Original magnification: ×6000. *Source: From Kato M, Hughes RD, Keays RT, Williams R. Electron microscopic study of brain capillaries in cerebral edema from fulminant hepatic failure. Hepatology 1992:***15***;1060–6, with permission.*

significant levels of infection/inflammation, brain edema may comprise elements of both cytotoxic and vasogenic edema.

MECHANISMS OF BRAIN EDEMA IN ACUTE LIVER FAILURE

Studies relating to the nature of the mechanisms responsible for brain edema in relation to ALF are difficult to perform either clinically or using modern imaging techniques in these critically-ill patients. Consequently, advances in current understanding rely heavily on the use of animal models.[6]

In general, animal models of ALF that are known to result in brain edema and its complications are of two general types namely ALF due to toxic liver injury caused by toxins such as galactosamine, thioacetamide, or azoxymethane and ischemic liver injury resulting from hepatic devascularization. Hepatectomy has also been used as an animal model of ALF. All types primarily involve rodent species.

In the hepatectomized rat, a 5.4% increase of brain water is observed[7] in cortical gray matter at preterminal stages. Increases of brain water in the 3%–5% range were also noted in rabbits with ALF resulting from galactosamine-induced liver injury where increases in water content were particularly marked in hippocampus and cerebral cortex.[2] Brain edema was confined to gray matter structures in this latter study and increased water content of the gray matter regions of the brains of these animals paralleled the severity of encephalopathy. Qualitative morphological studies revealed striking changes in cerebral cortical astrocytes such as hydropic changes with vacuolation of the perinuclear cytoplasm and pericapillary and periaxonal foot processes.[2] The BBB remained impermeable to horseradish peroxidase indicating the absence of gross barrier changes in these animals.

At early stages of encephalopathy in rabbits with ALF resulting from galactosamine-induced toxic liver injury, brain edema is present but there are no significant increases of intracranial pressure (ICP). Pressure increases became apparent at more advanced stages of ALF.[2] In the hepatic-devascularized rat model, up to 2.5% increases of brain water were reported[3] and swelling of astrocytes was observed in cortical gray matter. However, such increases of brain water were largely prevented if the animals were allowed to develop hypothermia.

Ammonia

Complications of brain edema such as intracranial hypertension and brain herniation in patients with ALF are significantly correlated with arterial ammonia concentrations[8,9] and arterial ammonia in these patients is an independent risk factor for the development of intracranial hypertension[10] (Fig. 15.2). Brain ammonia is estimated to increase to low millimolar concentrations in ALF as a result of increases of arterial ammonia and the loss of hepatic ammonia-metabolizing enzyme systems including the urea cycle and glutamine synthetase (GS).

A wide range of actions of ammonia on metabolic and transport systems in the astrocyte have been identified, all of which have the capacity to result in impaired function of the cell and as a consequence, impaired cell–cell communication mechanisms in the brain.[11] Glial fibrillary acidic protein (GFAP) is a major component of astrocytic intermediate filaments that are implicated in the control of cell motility and morphology by providing stability to astrocyte processes.[12] Loss of expression of GFAP has been reported in autopsied brain tissue from patients who died in ALF resulting primarily from viral hepatitis.[13] Experimental ALF resulting from ischemic liver failure results in a loss of expression of the GFAP gene,[14] a finding that is selective since expression of other glial filamentous proteins such as S100beta is unchanged in the same animal model of ALF. Loss of GFAP in this model was significantly correlated with the degree of hyperammonemia and with the presence of brain edema[14] leading to the proposal that loss of GFAP in ALF had the potential to alter the control of visco-elastic properties of the cell with consequent facilitation of cell swelling and cytotoxic brain edema.

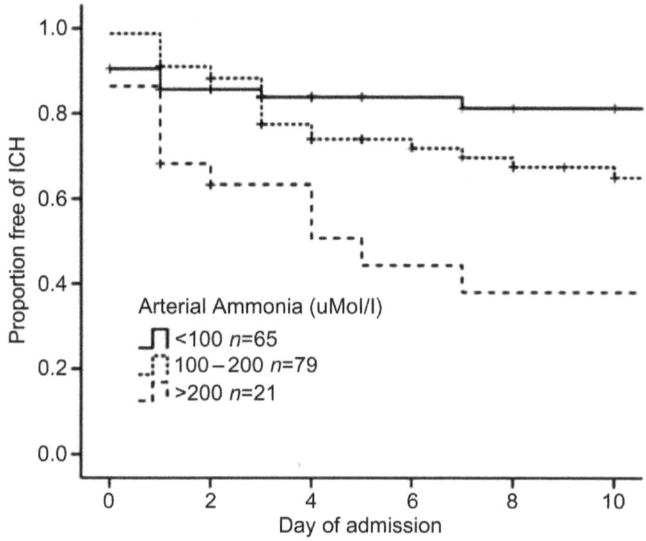

FIGURE 15.2 **Arterial ammonia is an independent risk factor for intracranial hypertension in patients with acute liver failure.** Figure shows a Kaplan–Meier plot of intracranial hypertension in 165 patients with ALF as a function of their arterial ammonia concentrations on admission. Log rank test, $p < 0.01$. *Source: From Bernal W, Hall C, Karvellas CJ, et al. Arterial ammonia and clinical risk factors for encephalopathy and intracranial hypertension in acute liver failure. Hepatology 2007:***46***;1844–52, with permission.*

Ammonia removal by brain relies almost exclusively on the formation of glutamine catalyzed by the enzyme GS that has a uniquely astrocytic localization.[15] Inhibition of GS in brain prevents glutamine accumulation and attenuates brain edema resulting from hyperammonemia[16] consistent with the notion that it is a downstream action of ammonia, rather than the ammonia molecule per se, that causes brain edema as a consequence of hyperammonemia and by extension, possibly also, in ALF. This issue is further addressed under the lactate section of this review.

Cellular Transporter Proteins

The membrane of the astrocyte is home to an extensive range of high affinity transporters for energy substrates and for amino acids many of which serve roles as excitatory or inhibitory transmitters. EAAT-2 is a cloned and well-characterized high affinity transporter for glutamate that is located on the astrocyte membrane of the mammalian forebrain. Knock-down of the EAAT-2 gene leads to increased extracellular concentrations of glutamate in brain and is accompanied by seizures and brain edema, two key features of ALF. Expression of EAAT-2 is significantly reduced in human ALF brain[13] and experimental ALF resulting from liver ischemia likewise results in a loss of expression of EAAT-2 at edema stages of encephalopathy[17] together with increased extracellular brain concentrations of glutamate.[18] This increase of extracellular glutamate resulting from ammonia-induced reduction in capacity of the astrocyte to remove glutamate released from the nerve terminal could explain at least in part the report of excess glutamatergic activity in brain in ALF and the report that reducing this excess activity using the glutamate receptor antagonist memantine significantly attenuates the cerebral complications of ALF.[19]

Synaptic concentrations of glycine in the forebrain are effectively regulated by the high affinity astrocytic transporter GLYT-1; glycine functions as an agonist of a modulator site on the N-Methyl D-Aspartate (NMDA) subclass of postsynaptic glutamate receptor. ALF in the rat with liver ischemia results in a loss of expression of GLYT-1 and a concomitant increase of extracellular brain glycine.[20] This result has the potential to result in further increases of glutamatergic neuronal activity.

Sodium-coupled neutral amino acid transporters (SNATs) have now been well characterized and two such transporters, SNAT-3 and SNAT-5, favor the release of glutamine from the astrocyte.[21] ALF in the rat results in a significant loss in expression of SNAT-5[22] leading the authors to suggest that the restriction in release of glutamine from the astrocyte offers a possible explanation for the glutamine accumulation in the astrocyte leading to brain edema in ALF.

GLUT-1 is a well-defined glucose transporter localized on the membranes of astrocytes and cerebrovascular endothelial cells that regulates glucose delivery to these cells and also functions also as a water channel.[23] GLUT-1 expression is increased in brain in parallel with brain water accumulation in experimental ALF[24] suggesting a role for the up-regulation of this glucose transporter in the pathogenesis of brain edema in ALF.

Cellular Energy Metabolism, Lactate, and Cerebral Blood Flow

Brain energy metabolism has been investigated in the context of ALF using a variety of techniques. Brain concentrations of high energy phosphates including adenosine triphosphate (ATP) and phosphocreatine have generally been found to be unaltered until very late (isoelectric EEG) stages of encephalopathy in ALF.[25] Glucose is the primary energy substrate for the adult mammalian brain and there is evidence to support the contention that glucose transport (see preceding section), glucose utilization, and glucose metabolism are seriously modified in the brain in ALF. Ammonia, in concentrations equivalent to those observed in the plasma and brain in ALF is an established inhibitor of the enzyme alpha-ketoglutarate dehydrogenase, a rate-limiting enzyme of the tricarboxylic acid cycle.[26] Inhibition of glucose oxidation in this way would be expected to result in decreased entry of pyruvate into the cycle and a consequent accumulation of lactate.

Indeed, brain lactate concentrations are increased in a range of experimental animal models of ALF[25,27–29] as well as in brain microdialysates from patients with ALF where increases of brain lactate precede surges in ICP.[30] Worsening of neurological status and the onset of brain edema in animal models of ALF is significantly correlated with increases of brain lactate synthesis[31] (Fig. 15.3). Increases of brain lactate have also been reported in dogs with ALF where the degree of intracranial hypertension was found to correlate with brain lactate concentrations[32] suggesting a key role for increases in brain lactate in the pathogenesis of brain edema in ALF and, in support of this contention, exposure of primary cortical astrocytes to pathophysiologically relevant concentrations of lactate was found to result in significant cell swelling.[33]

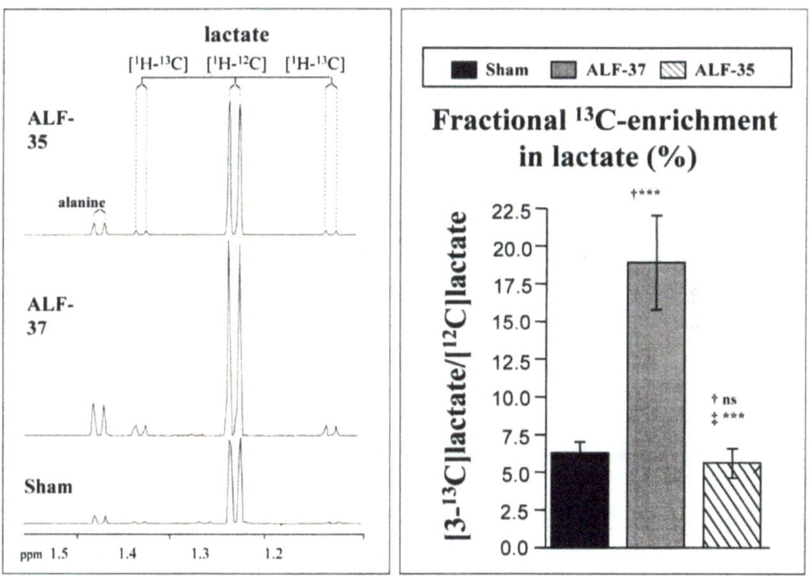

FIGURE 15.3 **Increased de novo brain lactate synthesis in an animal with ALF resulting from liver ischemia (ALF-37) compared to sham-operated control animal (Sham).** Left hand panel indicates lactate resonances (1H–13C) that appear as doublets centered at 1.25 ppm. Right hand panel shows fractional 13C-enrichment data indicative of de novo lactate synthesis rates. Note protective effects of mild hypothermia (ALF-35) leading to reduction in brain edema and decreased brain lactate synthesis. *Source: From Chatauret N, Zwingmann C, Rose C, et al. Effects of hypothermia on brain glucose metabolism in acute liver failure: a H/C nuclear magnetic resonance study. Gastroenterology 2003:***125***;815–24, with permission.*

Alterations of CBF are frequently observed in ALF where a loss of CBF autoregulation has been reported.[34] This is pertinent for decisions on clinical management since increases or decreases of mean arterial pressure may result in episodes of cerebral hyperemia or hypoperfusion, respectively. Based upon a retrospective analysis of sequential ICP and CBF measurements in 26 patients with ALF, a 5-phase sequence of events has been proposed.[35] Phase 1, an initial phase consists of low values of CBF resulting no doubt from decreased neuronal activity and brain energy requirements and normal ICP. A progressive increase of CBF follows (Phase 2) a subsequent increase of ICP (Phase 3). In final stages, intracranial hypertension leads to a decrease of CBF (Phase 4) with impaired rates of cerebral perfusion and brain death (Phase 5). The manipulation of CBF strongly influences the development of brain edema and intracranial hypertension in ALF and the proposed 5-Phase model is potentially valuable for the prognostic assessment of patients with ALF.[36]

Neuroinflammation

Systemic inflammation and infection are common in patients with ALF and, when present, are associated with more severe central nervous system complications including encephalopathy and brain edema.[37–39] Liver–brain signaling in ALF involves the activation of microglia.[40] Microglia constitute the resident macrophages of the brain with the inherent ability to respond when activated to a range of challenges including tissue damage, vascular disturbances, changes in pH, and impending cellular energy failure. Microglial activation indicative of a neuroinflammatory response was first described by Jiang et al.[41,42] in rats with ALF resulting from liver ischemia. Subsequent reports confirmed these findings in a mouse model of ALF resulting from azoxymethane-induced toxic liver injury[43,44] as well as in a patient with ALF resulting from viral hepatitis.[37]

Using the azoxymethane mouse model, it was subsequently reported that the onset of brain edema resulted in increased expression of a range of markers of a neuroinflammatory response in the brains of these mice, the response consisting of microglial activation and increased expression of genes coding for the proinflammatory cytokines tumor necrosis factor-alpha (TNF-a) and the interleukins (IL-1b and IL-6).[45] Evidence consistent with the production of these cytokines in the brains of patients with ALF was provided by arterio-venous difference studies in ALF patients where the etiology of liver failure was acetaminophen overdose.[46] Further evidence for a role of proinflammatory molecular mechanisms in ALF was provided by the reports that TNF-a or IL-1 gene deletion results in attenuation of the level of brain edema in experimental ALF[47] and that polymorphisms of the TNF-alpha gene significantly influence the clinical outcome in patients with ALF.[48]

Implications for Future Therapy in Acute Liver Failure

Current approaches to the management of the CNS complications of ALF in general and intracranial hypertension in particular rely on standard paradigms that are used in their management and treatment in other conditions. Elevation of the head of the bed, hyperventilation, administration of mannitol and barbiturates, and hemo-filtration have all been used in patients with ALF. However, none of these approaches is totally satisfactory and all have adverse effects with relapses occurring in up to 80% of patients.[49] There is a clear and urgent need for the development of novel treatment paradigms in ALF.

The use of ICP monitoring clearly confers an advantage for therapeutic management of ALF patients.[50] However, ICP monitoring has no influence on global outcomes[51] and complications of the procedure remain a concern.

Neuroprotective interventions in ALF commence with attention to general factors such as fever, arterial hypertension, and agitation, all of which have the potential to exacerbate brain edema and intracranial hypertension.

A new generation of therapeutic options based upon emerging breakthroughs in current understanding of the mechanisms responsible for the pathogenesis of brain edema in ALF is starting to emerge.

Ammonia-Lowering Strategies

Despite the clear relationship between intracranial hypertension and arterial ammonia in patients with ALF, agents aimed at the lowering of gut ammonia are of questionable value in the prevention of the CNS complications of ALF.[52,53]

N-*Acetyl Cysteine*

N-Acetyl cysteine (NAC) has been used for the prevention and treatment of brain edema and improved brain oxygen delivery in patients with ALF resulting from acetaminophen overdose.[34,54] However, beneficial effects were not confirmed in a subsequent trial.[55] Both antioxidant and central antiinflammatory actions have been proposed to explain the beneficial action of NAC in the prevention of brain edema in nonacetaminophen-related experimental animal models of ALF.[56]

Mild Hypothermia

Two or three degrees of hypothermia prevent brain edema and its complications in both experimental animal models of ALF[3,57] and in some ALF patients[58] (Fig. 15.4).

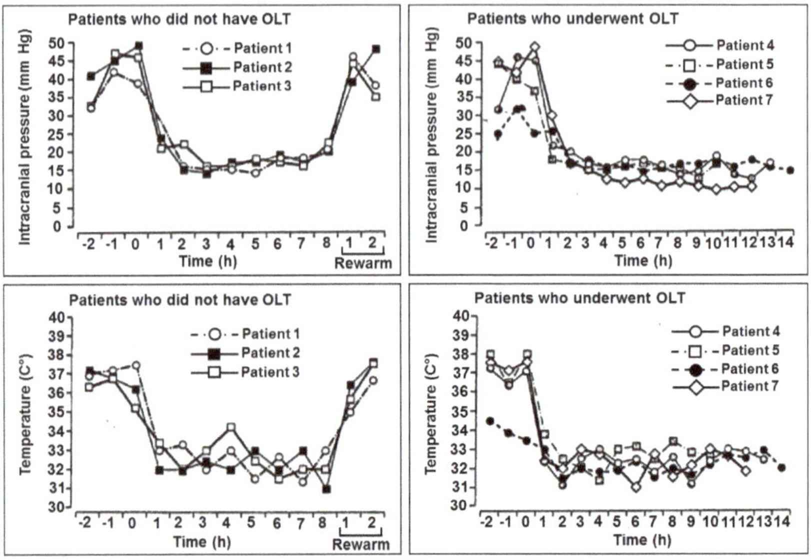

FIGURE 15.4 **Mild-to-moderate hypothermia (31–34°C) lowers intracranial pressure in patients with acute liver failure and may serve as a bridge to liver transplantation.** *Source: From Jalan R, Olde Daminck SW, Deutz NE, et al. Moderate hypothermia for uncontrolled intracranial hypertension in acute liver failure. Lancet 1999:***1***;1164–8, with permission.*

Based largely on studies in animal models, multiple mechanisms have been proposed to explain the beneficial effect of hypothermia on brain edema and its complications in ALF. Most importantly, hypothermia reduces the extent of acetaminophen-induced liver damage in mice[59] but this has not been assessed in patients to date. Mild hypothermia reduces the blood–brain transfer of ammonia,[60] and decreases the synthesis of lactate in brain (Fig. 15.3) consistent with improved brain glucose oxidation.[28] Hypothermia has also been shown to possess antiinflammatory properties in the brain in ALF consisting of decreased activation of microglia[42] and reduced brain production of proinflammatory cytokines in parallel with the reduction in brain edema.[45]

Mild hypothermia is increasingly being employed on a case-by-case basis in many tertiary ALF-treatment centers but, tragically, the lack of clear guidelines (similar to those available for the use of hypothermia in the limiting of cerebral damage in traumatic brain injury and stroke) continues to hamper its more widespread use in ALF. Randomized controlled clinical trials are urgently required in order to confirm the safety and efficacy of mild hypothermia in ALF, to critically assess its effects on global outcomes and to provide adequate guidelines for its safe use.[59]

Phenytoin

The occasional reports of clinical or subclinical seizures in ALF patients led to a report describing beneficial actions of phenytoin in reducing the incidence and severity of brain edema.[61] Unfortunately, a subsequent randomized controlled clinical trial of the agent did not confirm these findings.[62]

Propofol

A preliminary report published in 2002 suggested that propofol, a short-acting lipophylic anesthetic may help to control intracranial hypertension in ALF patients.[63]

Antiinflammatory Agents

The concept of "the inflamed brain" in ALF is rapidly gaining ground as discussed in the earlier section of this chapter. Consequently, agents with central antiinflammatory properties have been tried and shown to be effective in the prevention of brain edema in experimental models of ALF. TNF-a gene deletion in an animal model resulted in significant reduction of brain edema.[56] Furthermore, as stated above, both NAC and mild hypothermia are known to possess antiinflammatory properties in relation to their antiedema properties in experimental ALF.

Indomethacin

Indomethacin prevents the development of hyperemia and leads to reductions of CBF and ICP in both animal models and in patients with ALF.[64,65] However, potentially severe adverse effects linked to the inhibition of prostaglandin synthesis in the gut and kidneys have so far limited its clinical application.

Minocycline

Minocycline is a semi-synthetic tetracycline antibiotic with potent antiinflammatory properties that are independent of its antimicrobial action. Minocycline inhibits the neuroinflammatory response in ALF that reduces microglial activation and brain edema in treated animals with ALF due to liver ischemia.[42] Hepatotoxicity has occasionally been reported with long-term use of minocycline.

Etanercept

Etanercept is a fusion protein consisting of two ligand binding domains of the soluble TNF-alpha receptor linked to human immunoglobulin (IgG) that, when it binds to TNF-alpha, renders it inactive. Treatment of mice with ALF resulting from toxic liver injury with etanercept led to delay in the neuroinflammatory response and delayed progression of neurological symptoms.[66]

Liver-Assist Devices, Blood Depuration Strategies

Given the evidence suggesting a link between increased blood concentrations of toxins such as ammonia that are not effectively cleared by the liver in ALF and their role in the pathogenesis of brain edema, it is not surprising that several attempts have been made to assess the value of liver-assist devices for treatment of this complication. Bio-artificial liver-assist devices have been shown to be beneficial in some early studies in experimental models of ALF. One such trial used high volume plasmapheresis, a technique known to decrease arterial ammonia, increase cerebral perfusion pressure, and improve encephalopathy in ALF patients. Albumin dialysis is known to reduce ICP in patients with ALF.[67] In another trial, the use of the so-called molecular absorbent re-circulating system (MARS) aimed at clearance of albumin-bound toxins from the circulation used in an experimental animal model of ALF in pigs with liver ischemia, reduced brain edema, and ICP in support of the notion that circulating factors other than (or in addition to) ammonia are responsible for these complications.[68]

SUMMARY

Brain edema progressing to intracranial hypertension is a common occurrence in ALF. Neuropathologic studies reveal that functional and morphologic alterations of neuroglia, particularly astrocytes and microglia, are causally related to the pathogenesis of brain edema in ALF. The astrocyte is the major neural cell type uniquely responsible for removal of ammonia by the brain. Its failure to do so in ALF is considered to be a major cause of altered neural control mechanisms responsible for brain edema and its complications. Brain edema and intracranial hypertension are significantly correlated with arterial ammonia and ammonia-induced alterations in expression of genes coding for neuroglial proteins involved in cell volume regulation and in intercellular neural transport have consistently been reported in ALF.

There is an increasing body of evidence to support the notion that microglial activation indicative of a neuroinflammatory response (inflammation of the brain per se) also plays a major role in the pathogenesis of the cerebral complications of ALF. In support of this contention, studies have shown that deletion of the gene coding for the proinflammatory cytokine TNF-alpha attenuates brain edema in ALF. Moreover, polymorphisms of the TNF-alpha gene are known to significantly influence the clinical outcome in ALF.

Emerging strategies for the management and treatment of brain edema and intracranial hypertension in ALF include the use of *N*-acetyl cysteine, mild hypothermia, and centrally acting antiinflammatory agents.

Acknowledgment

Studies from the author's research unit were funded by research grants from The Canadian Institutes of Health Research (CIHR).

References

1. Kato M, Hughes RD, Keays RT, Williams R. Electron microscopic study of brain capillaries in cerebral edema from fulminant hepatic failure. *Hepatology* 1992;**15**:1060–6.
2. Traber PG, Dal Canto M, Ganger DR, et al. Electron microscopic evaluation of brain edema in rabbits with galactosamine-induced fulminant hepatic failure: ultrastructure and integrity of the blood–brain barrier. *Hepatology* 1987;**7**:1272–7.
3. Traber P, DalCanto M, Ganger D, et al. Effect of body temperature on brain edema and encephalopathy in the rat after hepatic devascularisation. *Gastroenterology* 1989;**96**:885–91.
4. Potvin M, Morrison HF, Hinchey EJ, et al. Cerebral abnormalities in hepatectomized rats with acute hepatic coma. *Lab Invest* 1984;**50**:560–4.
5. Chastre A, Belanger M, Nguyen BN, Butterworth RF. Lipopolysaccharide precipitates hepatic encephalopathy and increases blood–brain barrier permeability in mice with acute liver failure. *Liver Int* 2013 http://dx.doi.org/10.1111/liv.12252.
6. Butterworth RF. Molecular neurobiology of acute liver failure. *Semin Liver Dis* 1998;**23**(Suppl. 1):251–8.
7. Livingstone AS, Potvin M, Goresky CA, et al. Changes in the blood–brain barrier in hepatic coma after hepatectomy in the rat. *Gastroenterology* 1977;**74**:664–71.
8. Clemmesen JO, Larsen FS, Kondrup J, et al. Cerebral herniation in patients with acute liver failure is correlated with arterial ammonia concentrations. *Hepatology* 1999;**29**:648–53.
9. Tofteng F, Hauerberg J, Hansen BA. Persistent arterial hyperammonemia increases the concentration of glutamine and alanine in the brain and correlates with intracranial pressure in patients with fulminant hepatic failure. *J Cereb Blood Flow Metab* 2006;**26**:21–7.
10. Bernal W, Hall C, Karvellas CJ, et al. Arterial ammonia and clinical risk factors for encephalopathy and intracranial hypertension in acute liver failure. *Hepatology* 2007;**46**:1844–52.
11. Butterworth RF. Pathophysiology of hepatic encephalopathy: a new look at ammonia. *Metab Brain Dis* 2002;**17**:221–7.
12. Eng LF, Ghirnkar RS, Lee YL. Glial fibrillary acidic protein: GFAP-31 years (1969–2000). *Neurochem Res* 2000;**25**:1439–51.
13. Thumburu KK, Dhiman RK, Vasishta RK, et al. Expression of astrocytic genes implicated in neural excitation and brain edema is altered after acute liver failure. *J Neurochem* 2013;**128**:617–27.
14. Belanger M, Desjardins P, Chatauret N, et al. Loss of expression of glial fibrillary acidic protein in acute hyperammonemia. *Neurochem Int* 2002;**41**:155–60.
15. Martinez-Hernandez A, Bell KP, Norenberg MD. Glutamine synthetase: glial localization in brain. *Science* 1977;**195**:1356–8.
16. Takahashi H, Koehler RC, Brusilow SW, et al. Inhibition of brain glutamine accumulation prevents cerebral edema in hyperammonemic rats. *Am J Physiol* 1991;**261**:H825–9.
17. Knecht K, Michalak A, Rose C, Butterworth RF. Decreased glutamate transporter (GLT-1) expression in frontal cortex of rats with acute liver failure. *Neurosci Lett* 1997;**229**:201–3.
18. Michalak A, Rose C, Butterworth J, et al. Neuroactive amino acids and glutamate (NMDA) receptors in frontal cortex of rats with experimental acute liver failure. *Hepataology* 1996;**24**:908–13.

19. Vogels BA, Maas MA, Daalhuisen J, et al. Memantine, a non-competitive NMDA receptor antagonist improves hyperammonemia-induced encephalopathy and acute liver encephalopathy in rats. *Hepatology* 1997;**25**:820–7.
20. Zwingmann C, Desjardins P, Hazell AS, et al. Reduced expression of astrocytic glycine transporter GLYT-1 in acute liver failure. *Metab Brain Dis* 2002;**17**:263–73.
21. Chaudhry FA, Krisaj D, Larsson P, et al. Coupled and uncoupled proton movement by amino acid transport system N. *EMBO J* 2001;**20**:7041–51.
22. Desjardins P, Du T, Jiang W, et al. Pathogenesis of hepatic encephalopathy and brain edema in acute liver failure: the role of glutamine redefined. *Neurochem Int* 2012;**60**:690–6.
23. Fishbarq J, Kuang KY, Vera JC, et al. Glucose transporters serve as water channels. *Proc Natl Acad Sci USA* 1990;**87**:3244–7.
24. Belanger M, Desjardins P, Chatauret N, et al. Selectively increased expression of the astrocytic/endothelial glucose transporter protein GLUT-1 in acute liver failure. *Glia* 2006;**53**:557–62.
25. Mans AM, DeJoseph MR, Hawkins RA. Metabolic abnormalities and grade of encephalopathy in acute liver failure. *J Neurochem* 1994;**63**:1829–38.
26. Lai JCK, Cooper AJL. Alpha ketoglutarate dehydrogenase complex: kinetic properties, regional distribution and effects of inhibitors. *J Neurochem* 1986;**47**:1376–86.
27. Chatauret N, Rose C, Therrien G, et al. Mild hypothermia prevents cerebral edema and CSF lactate accumulation in acute liver failure. *Metab Brain Dis* 2001;**16**:95–102.
28. Chatauret N, Zwingmann C, Rose C, et al. Effects of hypothermia on brain glucose metabolism in acute liver failure: a H/C nuclear magnetic resonance study. *Gastroenterology* 2003;**125**:815–24.
29. Peeling J, Shoemaker L, Gauthier T, et al. Cerebral metabolic and histological effects of thioacetamide-induced liver failure. *Am J Physiol* 1993;**265**:G572–8.
30. Tofteng F, Jorgensen L, Hansen BA, et al. Cerebral microdialysis in patients with fulmionant hepatic failure. *Hepatology* 2002;**36**:1333–40.
31. Zwingmann C, Chatauret N, Liebfritz D, et al. Selective increase in brain lactate synthesis in experimental acute liver failure: results of a 1H–13C NMR study. *Hepatology* 2003;**37**:420–8.
32. Nyberg SL, Cerra FB, Gruetter R, et al. Brain lactate by magnetic resonance spectroscopy during fulminant hepatic failure in the dog. *Liver Transpl Surg* 1998;**4**:158–65.
33. Staub F, Baethmann Peters J, et al. Effects of lactic acidosis on glial cell volume and viability. *J Cereb Blood Flow Metab* 1990;**10**:866–76.
34. Wendon J, Harrison PM, Keays R, et al. Cerebral blood flow and metabolism in fulminant hepatic failure. *Hepatology* 1994;**19**:1407–13.
35. Aggarwal S, Kramer D, Yonas H, et al. Cerebral hemodynamic and metabolic changes in fulminant hepatic failure: a retrospective study. *Hepatology* 1994;**19**:80–7.
36. Vaquero J, Butterworth RF. Mild hypothermia for the treatment of acute liver failure: What are we waiting for? *Nature Clin Pract Gastroenterol Hepatol* 2007;**10**:528–9.
37. Butterworth RF. Hepatic encephalopathy: a central neuroinflammatory disorder? *Hepatology* 2011;**53**:1372–6.
38. Rolando N, Wade J, Davalos M, et al. The systemic inflammatory response syndrome in acute liver failure. *Hepatology* 2000;**32**:734–9.
39. Vaquero J, Chung C, Cahill ME, Blei AT. Pathogenesis of hepatic encephalopathy in acute liver failure. *Sem Liver Dis* 2003;**23**:259–69.
40. Bemeur C, Butterworth RF. Liver–brain proinflammatory signaling in acute liver failure: role in the pathogenesis of hepatic encephalopathy and brain edema. *Metab Brain Dis* 2013;**28**:145–50.
41. Jiang W, Qu H, Desjardins P, Butterworth RF. Unequivocal evidence for cytokine accumulation in brain in experimental acute liver failure. *Hepatology* 2006;**44**(Suppl. 1):366A.

42. Jiang W, Desjardins P, Butterworth RF. Cerebral inflammation contributes to encephalopathy and brain edema in acute liver failure: protective effect of minocycline. *J Neurochem* 2009;**109**:485–93.
43. McMillin M, Galindo C, Frampton GA, et al. Increased neuronal chemokine (CCL2/MCP1) expression is associated with hepatic encephalopathy and contributes to neurological decline. *Hepatology* 2012;**56**(Suppl 1):958A.
44. Thrane VR, THrane AS, Chanag J, et al. Real-time analysis of microglial activation and motility in hepatic and hyperammonemic encephalopathy. *Neuroscience* 2012;**220**:247–55.
45. Jiang W, Desjardins P, Butterworth RF. Direct evidence for central proinflammatory mechanisms in rats with experimental acute liver failure: protective effect of hypothermia. *J Cereb Blood Flow Metab* 2009;**29**:944–52.
46. Wright G, Shawcross D, Olde Daminck SW, Jalan R. Brain cytokine flux in acute liver failure and its relationship with intracranial hypertension. *Metab Brain Dis* 2007;**22**: 375–88.
47. Bemeur C, Vaquero J, Desjardins P, et al. *N*-Acetylcysteine attenuates cerebral complications of non-acetaminophen-induced acute liver failure in mice: antioxidant and anti-inflammatory mechanisms. *Metab Brain Dis* 2010;**25**:241–9.
48. Bernal W, Donaldson P, Underhill J, et al. Tumor necrosis factor genomic polymorphism and outcome of acetaminophen (paracetamol)-induced acute liver failure. *J Hepatol* 1998;**29**:53–9.
49. Hoofnagle JH, Carithers Jr RL, Shapiro C, et al. Fulminant hepatic failure: summary of a workshop. *Hepatology* 1995;**21**:240–52.
50. Wendon J, Larsen FS. Intracranial pressure monitoring in acute liver failure. A procedure with clear indications. *Hepatology* 2006;**44**:504–6.
51. Vaquero J, Rose C, Butterworth RF. Keeping cool in acute liver failure: rationale for the use of mild hypothermia. *J Hepatol* 2005;**43**:1067–77.
52. Acharya SK, Bhatia V, Sreenivas V, et al. Efficacy of L-ornithine L-aspartate in acute liver failure: a double-blind randomized, placebo-controlled trial. *Gastroenterology* 2009;**136**:2159–68.
53. Lee WM. Acute liver failure in the United States. *Sem Liver Dis* 2003;**23**:217–26.
54. Harrison PM, Keays R, Bray GP, et al. Improved outcome of paracetamol-induced fulminant hepatic failure by late administration of acetylcysteine. *Lancet* 1990;**335**:1572–3.
55. Walsh TS, Hopton P, Philips BJ, et al. The effect of *N*-acetyl cysteine on oxygen transport and uptake in patients with fulminant hepatic failure. *Hepatology* 1998;**27**:1332–40.
56. Bemeur C, Qu H, Desjardins P, Butterworth RF. IL-1 or TNF receptor gene deletion delays onset of encephalopathy and attenuates brain edema in experimental acute liver failure. *Neurochem Int* 2010;**56**:213–16.
57. Rose C, Michalak A, Pannunzio P, et al. Mild hypothermia delays the onset of coma and prevents brain edema and extracellular brain glutamate accumulation in rats with acute liver failure. *Hepatology* 2000;**31**:872–7.
58. Jalan R, Olde Daminck SW, Deutz NE, et al. Moderate hypothermia for uncontrolled intracranial hypertension in acute liver failure. *Lancet* 1999;**1**:1164–8.
59. Vaquero J, Belanger M, James L, et al. Mild hypothermia attenuates liver injury and improves survival in mice with acetaminophen toxicity. *Gastroenterology* 2007;**132**:3722–38.
60. Ytrebo LM, Kristiansen RG, Maehre H, et al. L-Ornithine phenylacetate attenuates increased arterial and extracellular brain ammonia and prevents intracranial hypertension in pigs with acute liver failure. *Hepatology* 2009;**50**:165–74.
61. Ellis AJ, Wendon JA, Williams R. Subclinical seizure activity and prophylactic phenytoin infusion in acute liver failure: a controlled clinical trial. *Hepatology* 2000;**32**:536–41.
62. Bhatia V, Batra Y, Acharya SK. Prophylactic phenytoin does not improve cerebral edema or survival in acute liver failure—a controlled clinical trial. *J Hepatol* 2004;**41**:89–96.
63. Wijdicks EF, Nyberg SL. Propofol to control intracranial pressure in fulminant hepatic failure. *Transplant Proc* 2002;**34**:1220–2.

64. Chung C, Gottstein J, Blei AT. Indomethacin prevents the development of experimental ammonia-induced brain edema in rats after portacaval anastomosis. *Hepatology* 2001;**34**:249–54.
65. Tofteng F, Larsen FS. The effect of indomethacin on intracranial pressure, cerebral perfusion and extracellular lactate and glutamate concentrations in patients with fulminant hepatic failure. *J Cereb Blood Flow and Metab.* 2004;**24**:798–804.
66. Chastre A, Belanger M, Beauchesne E, et al. Inflammatory cascades driven by tumor necrosis factor alpha play a major role in the progression of acute liver failure and its neurological complications. *PLos One* 2012;**7**:e49670.
67. Awad SS, Swaniker F, Magee J, et al. Results of a phase 1 trial evaluating a liver support device utilizing albumin dialysis. *Surgery* 2001;**130**:354–62.
68. Sen S, Rose C, Ytrebo LM, et al. Effect of albumin dialysis on intracranial pressure in pigs with acute liver failure: a randomized study. *Crit Care Med* 2006;**34**:158–64.

CHAPTER

16

Blood–Brain Barrier and Edema in Brain Tumors

Andreas F. Mack[1], *Hartwig Wolburg*[2] *and Petra Fallier-Becker*[2]

[1]Institute of Clinical Anatomy and Cell Analysis, University of Tübingen, Tübingen, Germany [2]Institute of Pathology and Neuropathology, University of Tübingen, Tübingen, Germany

OUTLINE

Brain Edema.
DOI: http://dx.doi.org/10.1016/B978-0-12-803196-4.00016-3

INTRODUCTION

Tumors in the central nervous system range from benign forms diagnosed sometimes accidentally, to highly malignant forms with low survival probability. Accordingly, they are graded from I to IV on a four level (WHO) classification which is based on various criteria such as clinical, histological, and prognostic aspects.[1] Research on central nervous system tumors has mainly focused on the elucidation of genetic pathways and genomic analyses.[2,3] There is a plethora of data from in vitro and in vivo studies concerning mutations in the *PTEN*, *TP53*, and *RB1* pathways. These mutations were recognized to be obligatory events in the pathogenesis of human and experimental glioblastoma.[4] The clinical signs of glioblastomas are connected to intracranial pressure, and this pressure development is in turn linked to the formation of a broadly distributed edema. Such an edema originates from disturbed water flow from the blood vessels into the parenchyma and from there possibly into the cells (cytotoxic, ionic, or vasogenic edema, for distinctions see other chapters in this volume). In order to understand how an edema is formed it is therefore necessary to gain insight into the way of function of the blood–brain barrier.

The blood–brain barrier (BBB) is one of the fences that keeps the healthy brain parenchyma separated from the environment of other body tissues and blood circulation. Other boundary gates are blood–cerebrospinal fluid barriers (BCSFB), or blood–retinal barrier for the diencephalon-derived retina.

In this chapter, we address the problem how these barrier systems are affected by the tumor growth, or vice versa, how changes in the barrier influence the tumor and edema formation. We will first outline the components of the BBB and water flow in the brain, then briefly astrocyte responses to pathological disturbances and the origins of brain tumors, to finally elucidate the events occurring in brain tumor edema, and summarize experimental approaches.

THE HEALTHY BLOOD–BRAIN BARRIER

There are three barriers separating the blood from the brain parenchyma: Firstly, all microcirculatory vessels in the brain are surrounded by astrocytic endfeet but in the strict sense, the tight junctions between microvascular endothelial cells form the BBB[5–7]; for a historical review on the emerging concepts of the BBB, see Saunders et al.[8] Secondly, the cerebrospinal fluid (CSF) filling the ventricular and subarachnoid space of the brain is not separated by a true barrier from the brain parenchyma but from the blood by specialized cells. The epithelial cells of the choroid plexus producing the bulk amount of the CSF form a barrier between the ventricles and the fenestrated capillaries running in the stroma of the choroid plexus.[9–11] The choroid plexus epithelium is continuous with "regular" ependymal cells which do not form a barrier. At the circumventricular organs like the area postrema, the subfornical organ, the median eminence of the hypothalamus, and others, neurosecretory or neurosensory cells get access to blood vessels. At these restricted locations, specialized ependymal cells called tanycytes, and astrocytes form barriers to prevent blood-born substances from diffusion through the fenestrated capillaries into the ventricle.[12–14] The third barrier separates the CSF in the subarachnoid space (outer CSF space) from the leaky blood vessels and sagittal sinus in the dura mater, and is formed by the cells of the arachnoid leptomeningeal layer.[15,16] This implies that the brain parenchyma terminated by the superficial glial limiting membrane and covered by the pia mater is not sealed off from the subarachnoid space, and brain interstitial fluid communicates with the outer CSF.

All three of these blood–brain or blood–CSF barriers show a complex network of tight junctional strands, and each of them is characterized by its own molecular composition[16–18] (see also other chapters in this volume).

Endothelial tight junctions in most organs are predominantly composed of claudin-5, yet the tight junctions of the BBB endothelial cells additionally consist of claudin-3 and claudin-12, together with other tight junction-proteins like occludin, tricellulin, and junction-associated proteins of the immunoglobulin family.[19,20] This is considered to be the basis for the barrier tightness of the brain capillary endothelium that restricts especially hydrophilic proteins to enter the brain. As a consequence, endothelial cells must be equipped with a large number of transporter proteins, ion channels, and pumps.[21]

Like all endothelial cells, those of brain capillaries are underlined by a basal lamina. Processes of astrocytes contact this basal lamina directly leaving essentially no perivascular extracellular space around capillaries.[22] At these abutting endfeet, astrocytes express a high concentration of potassium channels Kir4.1, and water channels aquaporin-4 (AQP4, Fig. 16.1A). This can be visualized by immunocytochemical staining for AQP4, and by freeze-fracture electron microscopy where AQP4 can be identified as so-called orthogonal array of particles (OAP).[23,24] Where the glial membrane recedes from the basal lamina of the endothelial cell, this accumulation of Kir4.1 and AQP4 does not occur resulting in a dramatic decrease of OAP-density in the parenchymal membrane[25] (Fig. 16.1B).

This highly polarized channel expression is a typical feature of astrocytes in healthy brain tissue, and thought to be essential for the control for potassium and water regulation in the brain.[26,27] The astrocytic polarization is the basis for spatial buffering: Potassium accumulated in the

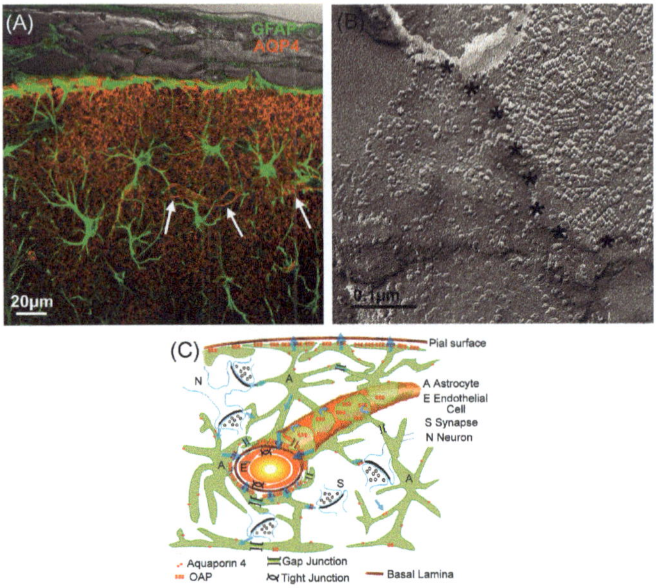

FIGURE 16.1 **Glial cells and the blood–brain barrier in the healthy brain.** (A) Cryosection through the surface layers of the rat cortex, stained with antibodies against glial fibrillary acidic protein (GFAP, green) and aquaporin-4 (AQP4, red). AQP4 reactivity is strong in the astrocytic endfeet forming the superficial glia limitans under the meninges (top of the image), and in the endfeet surrounding blood vessels (arrows). (B) Freeze-fracture replica of the cortex of the mouse brain. Right of the row of asterisks, many OAPs are aggregated in the membrane which has contact with the overlying basal lamina. Where this contact is interrupted (membrane domain left of the asterisks) the density of OAPs abruptly drops to zero. (C) Schematic drawing illustrating the AQP4 distribution in astrocytic endfeet at the brain surface and around capillaries forming the blood–brain barrier. The blue arrows suggest the water flow, the curved arrows indicate that water may spread along perivascular spaces.

interstitial space due to neuronal activity is taken up by astrocytes and redistributed at the perivascular or superficial glia limitans by the high density of potassium channels and AQP4 serving quick osmotic equilibration. This spatial buffering includes the regulation of pH.[28] The activity of Na^+/K^+-ATPase works then to restore ionic and volume conditions.[29]

The morphological observation that OAPs and/or AQP4 immunoreactivity are widely restricted to membranes covered by a basal lamina presumes a role of the extracellular matrix (ECM) in the distribution and arrangement of this water channel. Indeed, as has been suggested previously, the presence of the heparansulfate proteoglycan agrin is necessary for the complete formation of OAPs.[30] Consistent with this view, Rauch et al.[31] have highlighted the importance of agrin for the maintenance of the BBB. However, agrin is not the only component to be presumably involved in astrocyte polarization. The dystrophin–dystroglycan complexis is in contact with both the ECM and the cytoskeleton, and studies with dystrophin-null mice have shown an altered distribution of AQP4.[32] The lack of dystroglycan at the brain surface glia limitans in GFAP-Cre/Dystroglycan null mice led to a reduction of OAPs but not to a reduction of AQP4 protein.[33] This suggested that dystroglycan plays a role in AQP4 clustering, at least in superficial endfeet. In contrast, the perivascular endfeet facing the capillary basal lamina showed a reduction of AQP4 expression and not just a reduction in OAP clustering.[33] During development, AQP4 is first expressed in a nonpolarized pattern along radial glial cells although dystroglycan is already present on blood vessels.[34]

This raises a pivotal question whether endothelial cells are involved in the polarization of astrocytes, or whether glial polarization is required for the formation of endothelial tight junctions to build the barrier. Eilert-Olsen et al.[35] have shown that the BBB is not disrupted in the AQP4-deficient mouse. This somewhat perplexing finding does not change the fact that polarized astrocytes are always associated with endothelial cells interconnected by complex tight junctions in the healthy brain (Fig. 16.1C).

In addition to astrocytes, microvascular pericytes also were shown to be involved in the maintenance of the BBB.[36–38] Taken together, the BBB is part of the neurovascular unit that is formed during the development of the interaction of endothelial cells, astrocytes, neurons, and pericytes.[39]

WATER MOVEMENT IN THE BRAIN

In front and behind the capillary region of the blood brain circuitry, the vascular wall thickens, and the fused basal lamina of astrocytes and endothelial cells separate. This perivascular space is filled with fluid and is continuous with the so-called Virchow–Robin space. This in turn connects with the subarachnoid space (SAS) which contains CSF. Fluid circulation from the SAS into brain parenchyma via paravascular pathways was shown by tracers.[40] Movement of CSF produced mainly by the choroid plexus is driven by vascular pulsation and ependymal cells,[41] and drained through various passages including arachnoid villi, spinal nerves but mostly through olfactory routes.[42,43] Recently, perivascular water movement has been suggested to play a role in cerebral water balance.[44] Evidence suggests that AQP4 plays a role in this movement.[45] Moreover, this water movement might have an important function in removal of waste products from the brain parenchyma.[46,47] Since the brain parenchyma lacks lymphatic vessels and the corresponding drainage, this water movement involving astroglial endfeet was named 'the glymphatic pathway'. Thus, besides the blood circulating through vessels in the brain, CSF and interstitial fluid flow seem to play a major role in water balance in the brain (see also Beneviste et al.[48]). Aquaporins are also cotransporters which are able to translocate water along with their substrates that have significant impact on the regulation of the water homeostasis and water transport in the brain.[49]

REACTIVE ASTROCYTES

Astrocytes respond to brain injury and other pathological conditions like ischemia, tumors, and neurodegenerative diseases. There is a wide range of astrocytic responses subsumed with the term "activated" or "reactive" astrocyte and some definitions have been proposed.[50,51] This response can vary in a gradated manner depending on the severity of the insult and include increase of cell volume (hypertrophy) and number of processes, upregulation of GFAP and vimentin intermediate filaments, re-expression of nestin, and proliferation.[50,52] The biological sense of this reactivity is likely a defense or protective mechanism and thus beneficial, but also can be detrimental in pathological conditions.[53] In our context, reactive processes that involve volume changes and alterations in the expression patterns of water channels are of major interest. Prior to the discovery that OAPs are formed by AQP4, there were reports on the re-distribution and/or increase of these arrays in reactive astrocytes.[54,55] In a more recent study, reactive astrocytes after traumatic brain injury showed a re-distribution of AQP4.[56] The authors suggest that this might be a compensatory mechanism to counteract developing cytotoxic edema rather than causing it.

When astrocytes were cultured in a hypotonic environment, they transiently increased in volume and released amino acids such as glutamate.[57] Since then many studies on cultured astrocytes pursued the question of how volume changes in astrocytes are induced and regulated. For example, traumatic injury conditions caused astrocyte swelling[58] as did hypotonic challenge but less in AQP4 deficient cells.[59] Thus, AQP4 might contribute to cytotoxic edema by taking up excess water and/or by its dysregulated localization. Since the interplay with ionic drivers is not entirely clear and varies under different pathological conditions, more work is needed for clarification.[60]

BRAIN TUMORS AND THEIR CELLULAR SOURCES

Essentially all the tissues in and around the brain can form neoplasms; this includes neuroepithelial-derived tumors with gliomas, neuronal or neuronal-glial tumors, and embryonal tumors, as well as tumors of the meninges.[1] In addition, metastatic tumors from other body regions can develop in the CNS, e.g., from breast cancers. Tumors are usually named after their tissues or cells of origin. Some tumors like benign (grade I) pilocytic astrocytomas seem to be of astroglial origins while others are likely to be derived from mixed oligodendro- and astro-glial populations (oligoastrocytoma), and still others seem to be a heterogeneous mixture (anaplastic astrocytoma, glioblastoma). Since the discovery of adult stem cell niches in the mammalian brain[61,62] it has been suggested that these niches, in particular the subventricular zone, might at least provide a permissive environment for gliomagenesis.[63] Whether these stem cells, normally giving rise to new neurons and glia, are related to the small population of stem-cell like cells that are found in brain tumors[64] is not clear at this point.

These cancer stem cells are defined by their self-renewing, tumorigenic properties sustaining the re-growth of the tumor (reviewed in Ref. 65). The cancer stem cell hypothesis does not claim a stem cell as the cell of tumor origin. Important for the understanding of altered vascularization in glioblastomas is the finding that tumor endothelial cells can be recruited from cancer stem cells.[66] This suggests an enormous plasticity of tumor vessel endothelial cells.

Low-grade and benign tumors are not afflicted with edema although reactive astrocytes and loss of polarity of AQP4 can be observed in subependymomas, benign tumors located close to the ventricular zone.[67] In the following, we will focus on glioblastomas as they are the most malignant brain tumors and clearly involve changes in the BBB.

Depending on their ontogeny, glioblastomas are categorized into primary glioblastomas (more than 90%) and secondary glioblastomas (less than 10%). Secondary, anaplastic glioblastomas develop from diffuse astrocytomas (WHO grade II) to higher malignancy, and are frequently characterized by the mutation of the isocitrate dehydrogenase 1 (IDH1) gene, or less commonly of the related IDH2 gene.[68] Gliomas with IDH1R132H mutation have been discovered to express high levels of activated leukocyte cell adhesion molecule (ALCAM/CD166).[69] However, in most cases, the mechanisms of how these mutations dysregulate the function of the cells of tumor origin are not understood. Nevertheless, the mutations are used for molecular diagnostics.[68] Moreover, several potential biomarkers of glioblastoma were identified and classified for clinical prognosis without precise knowledge of their function.[70]

BBB CHANGES IN BRAIN TUMORS AND EDEMA FORMATION

General Observations

The loss of barrier function can be detected in magnetic resonance imaging (MRI) by the application of contrast medium (CM).[71] The standard

CM gadolinium does not cross the intact BBB but does so in glioblastoma indicating a leaky barrier. Pilocytic astrocytomas (WHO grade I) characterized by noninvasive growth yet microvascular proliferations, have blood vessels that appear normal, and rarely show dysfunction of the BBB. WHO grade II astrocytomas show no or little CM enhancement. WHO grade III astrocytomas enrich more CM than WHO grade II astrocytomas, but mostly less than glioblastomas.[72] Pronin et al.[73] reported that edema production is quantitatively related to the degree of breakdown of the BBB as determined by gadolinium enhancement. Chemotherapeutics normally do not cross the intact BBB yet are able to reach the main bulk of glioblastoma tissue passing through the disturbed BBB. The main problem treating the tumors is to reach single tumor cells in the infiltration zone where the BBB is not or less altered. Therefore, it is easier to get drugs into the main center of a high grade tumor than into the brain with an intact or less altered BBB (for example, hypericin 504[74]). As a result, the treatment of advanced glioblastoma with a necrotic center appears to come too late.

Tight Junctions in Glioblastoma

As pointed out above, there is a relationship between the astroglial endfoot and the endothelial barrier, but it is not clear whether the endothelial TJs evoke glial polarity or vice versa. For human glioblastomas, Wolburg et al.[75] described a loss of claudin-3 in the BBB. In addition, a thinning of the tight junction network associated with a switch from the P- to the E-face was observed in freeze fracture electron microscopy.[76] This occurred in contrast to the intact BBB where tight junctions of the BBB are associated with both the external (E-face) and the protoplasmic (P-face) leaflet of the membrane.[76] The degree of P-face-association in endothelial cell tight junctions in the brain is higher than anywhere else between endothelial cells in the body. However, after injury or any disturbance of the BBB, the tight junctions are nearly completely associated with the E-face. This ultrastructural morphological change indicates an alteration of the molecular composition of the tight junctions as well. The link between detachment of astrocytes from the vessels and the vessel basal lamina (which is extremely thickened in glioma tissue[77]) on the one hand, and down-regulation of tight junction components on the other is insufficiently known. Since proteins of the extracellular matrix were suggested to play a role in this regulation, Rascher et al.[78] investigated the correlation between the expression of endothelial tight junctions and the presence of agrin in gliomas. They reported the loss of anti-agrin immunoreactivity in gliomas coinciding with the disappearance of tight junction molecules. On a typological scale from 1 to 4, one type (1) of vessels was normal in expressing agrin and all tight junction molecules tested (occludin, claudin-1/3, claudin-5), type 2 showed loss of claudin-1, type 3 vessels in addition showed loss of claudin-5, and type 4 vessels did not show any agrin and no tight junction molecules. These latter blood vessels, however, showed strong immunoreactivity for tenascin, an extracellular matrix protein that plays a role in cell migration during brain development,[79] and has been implicated in glioma cell migration.[80]

Aquaporins and Glioblastoma

There have been several reports on the upregulation of the main astrocytic water channel AQP4 in brain tumors[58,81–83] (Fig. 16.2A). Most importantly, the polarized distribution of AQP4 described above is lost in gliomas yet also in pilocytic astrocytomas.[84]

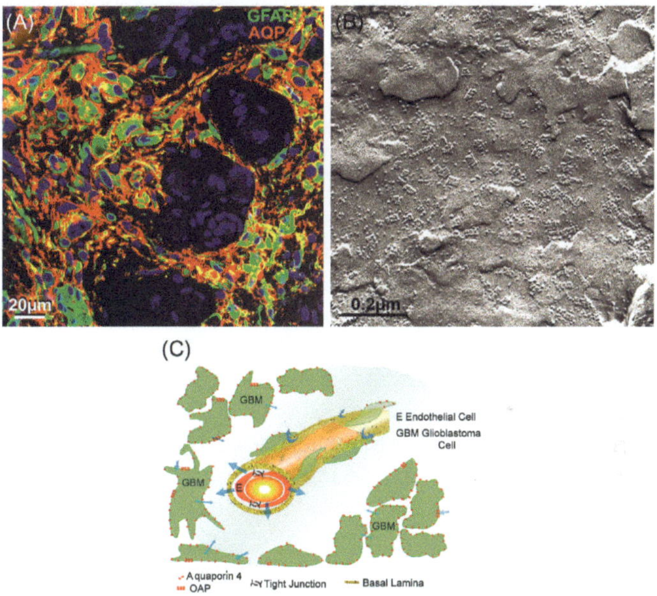

FIGURE 16.2 **Blood–brain barrier in a brain tumor.** (A) Immunohistochemical staining of AQP4 (red) and GFAP (green) in human glioblastoma. Nuclei are stained in blue. AQP4 staining is completely unpolarized. Since agrin is missing in such glioblastoma tissues, it can be hypothesized that agrin plays a role in the stabilization of AQP4 in the basal lamina-covered membrane domain. (B) Freeze-fracture replica of human glioblastoma showing OAPs spread all over the surface of the cell and not accumulated in the endfoot membrane. (C) Schematic diagram illustrating some of the changes occurring in vasogenic edema formation in a glioblastoma. The basal lamina is increased in thickness and has a composition different from the healthy brain. AQP4 is strongly expressed but in an unpolarized distribution. Tight junctions are changed and more permeable. Thus, water cannot be drained from the parenchyma and accumulates extracellularly. This draws more water from the blood circulation. The blue arrows show the suggested water flow. Some tumor cells might use the perivascular space as a migration path.

Remarkably, although the immunocytochemical reactivity is increased in glioma cells and present on the entire cell membrane, the density of OAPs is low even when in contact to the perivascular space. Therefore, AQP4 must occur as a non-OAP molecule in glioma cells (Fig. 16.2B). The functional implication of this finding is not clear. It is conceivable that the two major AQP4 isoforms M1 and M23[85] are expressed in disproportionate amounts since array formation occurs only when both isoforms are present[86] but this could not be verified so far.[82]

In addition to AQP4, two other aquaporins have been reported to be present in gliomas: AQP9, expressed by astrocytes in the healthy brain at low levels, is increased in tumors[87] but apparently only in certain cell populations.[88] AQP1 normally occurs in the apical membrane of choroid plexus epithelial cells and in nonbrain endothelial cells, and has been reported to be upregulated in gliomas.[89,90] Functional implications are not clear, and one has to be cautious about the diagnostic value of aquaporin expression patterns since low grade tumors also show altered AQP4 and AQP1 expressions[67,84] reminiscent of reactive astrocytes.

Edema Formation in Glioblastoma

As explained above, there appears to be a relationship between astrocytic polarization and the endothelial tight junction network. Reactive astrocytes show a re-distribution of AQP4 and potassium channels indicating a loss of polarity. The situation gets complicated when astrocytes proliferate uncontrolled and become malignant themselves. In glioblastomas, characterized by massive vasogenic edema formation, we were able to observe both, a loss or reduction of AQP4 in glioma cell membranes, and a strong reduction of the complexity of tight junction strands.[25,91] This coincided with a reduction of immunocytochemical signals of tight junction molecules such as claudin-3. The role of AQP4 in edema formation is elucidated by experiments with the AQP4-deficient mouse: Water intoxication causes more water uptake to the brain and astrocytes in the wild-type mice than in the knock-out.[92] This protection from cytotoxic edema by the lack of AQP4 channels is reversed in models of vasogenic edema. Intraparenchymal fluid infusion and tumor implantation led to higher intracranial pressure and lasting water accumulation in the brain,[93,94] suggesting that AQP4 is needed for clearance of excess water in the interstitial space. For a more detailed discussion of AQP4-deficient mice and water balance, see Refs. 95 and 96.

Reduced OAP/AQP4-related polarity means presence of potassium and water channels in physiologically unsuitable membrane domains. As discussed above, the channels are kept in subendothelial membrane domains by agrin and/or dystroglycan. But if these stabilizers disappear, for example by the increasing activity of matrix metallo-proteinases,[82] the OAPs migrate or form in nonendfoot membranes thereby reducing polarity.

A relationship between AQP4 expression and migration of astrocytes has been postulated[97] suggesting a facilitating role of AQP4 in the infiltration of malignant cells in glioblastoma. This may suggest that an inhibition of AQP4 would limit this infiltration strength. Indeed, a low molecular compound, 2-(nicotinamide)-1,3,4-thiadiazole, TGN-020, has recently been identified as an inhibitor of AQP4.[98] The authors showed in vivo that this AQP4 inhibitor was able to reduce ischemic cerebral edema.[99] However, there is neither any information on the influence of putative inhibitors on the AQP4 isoforms M1 and M23, nor on the link between water transport inhibition and the inhibition of migration of tumor cells.[100]

From the information presented so far we can attempt to draw a scenario of edema formation in brain tumors: Cells take up potassium from the interstitial space, and then water enters the cells osmotically. The loss of polarity in a strongly reactive astrocyte or a glioma cell has the pathophysiological consequence that this water cannot be released because most channels are localized at the wrong position. This might be the reason for the development of cytotoxic edema in astrocytes (cellular water accumulation) due to energy failure and/or disturbed volume regulation.[60] This in turn gives rise to the breakdown of the BBB with influx of water into the brain parenchyma, and as a consequence formation of vasogenic edema (extracellular water accumulation due to BBB disruption). Alternatively or in addition, new blood vessels are formed de novo in two ways: (i) By sprouting from existing vessels or (ii) from cancer (stem) cells. The new blood vessels might not have a tight endothelial junction to begin with. From these altered and leaky blood vessels, more water can enter the brain parenchyma and cannot be cleared by intact astrocytes. Interestingly, in the tumor periphery, cytotoxic edema formation has been observed. To make things worse, the newly formed perivascular edema provides new routes in which glioblastoma cells might migrate (Fig. 16.2C).

MODEL SYSTEMS TO STUDY CHANGES OF BBB IN TUMORS

As described above, BBB changes in tumors comprise a leaky BBB, tight junction alterations in endothelial cells (ECs), and unpolarized upregulation of AQP4 in tumor cells. In order to study these changes and to understand the basic mechanism, a plethora of BBB models systems have been developed. The driving force to use these models is the search for new therapies. Since currently the survival time for patients with glioblastoma is only about 15 months after resection of the tumor, radio- and chemotherapy,[101] there is an urgent need for new ways to find more effective therapeutics. For this purpose, in vitro and in vivo model systems are absolutely necessary which are briefly outlined here.

In Vitro Models

Monocultures

Monocultures of glioma cells have been used as a model for elucidating the potential of antineoplastic effects of substances.[102] For 3D cultures of glioma cells, tumor spheroid cultures were developed for anticancer therapy tests.[103,104] This 3D approach has an important advantage compared to 2D cultures concerning cell morphology and cell behavior.

In vivo, tight junctions of brain EC are mainly associated with the P-face in freeze fracture morphology (see Section "Tight Junctions in Glioblastoma"). In vitro, however, tight junctions of cultured brain ECs are always associated with the E-face regardless of growth conditions in co- or mono-culture. This is comparable to the tight junctions of ECs in tumors.[77]

Changes of endothelial cells, i.e., changes in tight junction formation at the blood-tumor barrier have been examined since the early 1980s: Endothelial cell cultures were used as BBB model measuring the permeability of sucrose under different conditions.[105] For example, lipophosphatic acid was found to modulate tight junction permeability in EC cultures from porcine brain, and possibly triggered BBB permeability in vivo under pathophysiological conditions.[106] Reactive oxygen species (ROS) caused malfunction of the BBB. Applying ROS to bovine brain EC cultures increased the permeability of ECs and decreased the electrical resistance due to disruption of tight junctions.[107] Abbott[108] reported on modulating effects of inflammatory mediators on endothelial cell tight junctions. An important aspect for these in vitro investigations is the question how to deliver drugs into the brain without destruction of the endothelial cell lining. Using primary brain microvessel endothelial cell cultures, Hülper and coworkers[109] succeeded in reversibly opening the BBB by applying short chain alkylglycerols without affecting the tight junction strand complexity.

Cocultures of Glioma and Endothelial Cells

Cocultures of glioma cells with endothelial cells were established to test for the effects on transendothelial electrical resistance, e.g., Hurst and Fritz.[110] There are several contradictory results: Some authors found that the electrical resistance of ECs to be decreased,[111,112] and others to be increased,[113–115] all using the rat glioma cell line C6 in coculture with ECs.

More recently, in a human coculture system of immortalized human brain capillary endothelial cells (hCMEC/d3) and human astrocytoma U87 cell line, transendothelial resistance decreased and permeability increased. Moreover, loss of tight junctions and barrier damage were observed in hCMEC/d3 cells when cocultured with U87 astrocytoma cells.[116]

However, applying just astroglial conditioned culture medium showed an effect on ECs. Hypoxia-induced VEGF expression by Ecs decreased suggesting also a decrease of permeability caused by astroglial conditioned medium.[117] Removal of cocultured astrocytes resulted in an increased permeability for inulin and sucrose but no changes in tight junctions molecules were detected.[118]

Explant Cultures

Explant cultures from glioma tissue are useful because at least initially, the cellular context and components of the extracellular matrix are maintained. Already in 1981, Escalona-Zapata and Diez-Nau[119] cultured human astrocytomas and glioblastomas to examine their growth pattern in vitro. They found glioblastomas to show maximal de-differentiation. More recently, Hovinga et al.[120] used a 3D organotypic explant system from surgical GBM specimen to study tumor and microenvironment interactions with the goal to find new therapeutics for brain tumor treatment.[121] Furthermore, drugs can directly be injected into explants and the results are comparable with in vivo application of therapeutic agents.[122] Explant cultures of human tumor tissue are filling the gap between in vitro experiments and animal in vivo models. Mercurio et al.[123] investigated the effect of celecoxi/fluvostatin in pilocytic astrocytoma explant cultures. The synergistic combination led to cell death of astrocytoma cells, a result that gives hope for patients with refractory pilocytic astrocytoma.

In Vivo Models to Study Changes of BBB in Tumors

To examine the blood–tumor barrier in vivo, animal models are used, implanting glioma cells into the brain thus generating tumors. The RG2-,[124] 9L rat brain tumor-,[125] and C6-cell lines[126] originate from rat gliomas. Mouse gliomas can be induced by grafting SMA cells (spontaneous murine astrocytoma) into mouse brains.[127] All these animal models have been used in preclinical astrocytoma drug development but represent the properties of human astrocytomas only poorly.[128] However, they may help to elucidate the in vivo situation at the blood–tumor barrier. Noell et al.[129] showed that AQP4 negative C6 glioma cells re-express AQP4 when implanted into mouse brain; outside the brain, C6 cells remained AQP4-negative.

Taken together, the multitude of preclinical model systems to study changes at the blood–brain barrier under tumor conditions helps to elucidate some aspects of these processes. Creating an overall picture from these puzzle pieces remains as a challenge but should help to find new therapeutics for tumor patients.

CONCLUSIONS

In this chapter, we have highlighted the barrier changes known to occur in brain tumors. This involves several structures linked by still ill-defined connections: The tight junction molecules of the brain microvessel endothelial cells representing the barrier proper, the astrocytes or tumor cells with altered channel expression, and the nonendothelial and extracellular molecules which are responsible for the regulation of the endothelial barrier. How these systems, each complicated themselves on the molecular level, interact is presently poorly understood.

We are beginning to understand how vasogenic edema develops involving dysregulation of tight junction proteins and water channels. Whether this dysregulation has a function and beneficial impact at the onset of tumor formation, or is paving the way for malignant tumor growth remains a matter of speculations. New concepts of water flow in the brain may help to understand these functional implications. We pointed out that astrocyte reactivity and aquaporin-4 dysregulation is not restricted to malignant brain tumors but occurs in other pathological conditions such as traumatic injury. There is no doubt that the molecular analysis of the BBB is of outstanding clinical relevance since insight into regulatory mechanisms are of primary significance for the development of new therapeutic strategies. Treatment of brain tumors has to consider both tumor angiogenesis and the permeability of tumor vessels. The role of the extracellular matrix for the polarity of astrocytes, and its loss in gliomas results in the incapability of the tumor cells to directed release of water out of the interstitial space. This is dependent on the brain microenvironment, and future research on this microenvironment will be pivotal in understanding the blood–brain and blood–tumor barrier.

References

1. Louis DN, Ohgaki H, Wiestler OD, Cavenee WK, Burger PC, Jouvet A, et al. The 2007 WHO classification of tumours of the central nervous system. *Acta Neuropathol* 2007;**114**(2):97–109.
2. Ohgaki H, Dessen P, Jourde B, Horstmann S, Nishikawa T, Di Patre PL, et al. Genetic pathways to glioblastoma: A population-based study. *Cancer Res* 2004;**64**(19): 6892–9.
3. Parsons DW, Jones S, Zhang X, Lin JC-H, Leary RJ, Angenendt P, et al. An integrated genomic analysis of human glioblastoma multiforme. *Science* 2008;**321**(5897):1807–12.

4. Chow Lionel ML, Endersby R, Zhu X, Rankin S, Qu C, Zhang J, et al. Cooperativity within and among Pten, p53, and Rb pathways induces high-grade astrocytoma in adult brain. *Cancer Cell* 2011;**19**(3):305–16.
5. Reese TS, Karnovsky MJ. Fine structural localization of a blood–brain barrier to exogenous peroxidase. *J Cell Biol* 1967;**34**(1):207–17.
6. Brightman MW, Reese TS. Junctions between intimately apposed cell membranes in the vertebrate brain. *J Cell Biol* 1969;**40**:648–77.
7. Abbott NJ, Patabendige AAK, Dolman DEM, Yusof SR, Begley DJ. Structure and function of the blood–brain barrier. *Neurobiol Dis* 2010;**37**(1):13–25.
8. Saunders NR, Dreifuss JJ, Dziegielewska K, Johansson P, Habgood M, Møllgård K, et al. The rights and wrongs of blood–brain barrier permeability studies: a walk through 100 years of history. *Front Neurosci* 2014:8.
9. Van Deurs B, Koehler JF. Tight junctions in the choroid plexus epithelium. A freeze-fracture study including complementary replicas. *J Cell Biol* 1979;**80**:662–73.
10. Møllgard K, Saunders NR. The development of the human blood–brain and blood–CSF barriers. *Neuropathol Appl Neurobiol* 1986;**12**(4):337–58.
11. Wolburg H, Paulus W. Choroid plexus: Biology and pathology. *Acta Neuropathol* 2010;**119**(1):75–88.
12. Krisch B, Leonhardt H, Buchheim W. Functional and structural border between the csf-milieu and blood-milieu in the circumventricular organs (organum vasculosum laminae terminalis, subfornical organ, area postrema) of the rat. *Cell Tissue Res* 1978;**195**(3):485–97.
13. Wang Q-P, Guan J-L, Pan W, Kastin A, Shioda S. A diffusion barrier between the area postrema and nucleus tractus solitarius. *Neurochem Res* 2008;**33**(10):2035–43.
14. Siso S, Jeffrey M, Gonzalez L. Sensory circumventricular organs in health and disease. *Acta Neuropathol* 2010;**120**(6):689–705.
15. Rascher G, Wolburg H. The tight junctions of the leptomeningeal blood cerebrospinal fluid barrier during development. *J Brain Res—J f Hirnforschung* 1997;**38**(4):525–40.
16. Brøchner CB, Holst CB, Møllgård K. Outer brain barriers in rat and human development. *Front Neurosci* 2015:9.
17. Kratzer I, Vasiljevic A, Rey C, Fevre-Montange M, Saunders N, Strazielle N, et al. Complexity and developmental changes in the expression pattern of claudins at the blood–CSF barrier. *Histochem Cell Biol* 2012;**138**(6):861–79.
18. Haseloff RF, Dithmer S, Winkler L, Wolburg H, Blasig IE. Transmembrane proteins of the tight junctions at the blood–brain barrier: Structural and functional aspects. *Semin Cell Dev Biol* 2015;**38**:16–25.
19. Tietz S, Engelhardt B. Brain barriers: Crosstalk between complex tight junctions and adherens junctions. *J Cell Biol* 2015;**209**(4):493–506.
20. Bauer H-C, Krizbai IA, Bauer H, Traweger A. You shall not pass—Tight junctions of the blood brain barrier. *Front Neurosci* 2014:8.
21. Hermann DM, ElAli A. The abluminal endothelial membrane in neurovascular remodeling in health and disease. *Sci Signal* 2012;**5**(236):re4.
22. Mathiisen TM, Lehre KP, Danbolt NC, Ottersen OP. The perivascular astroglial sheath provides a complete covering of the brain microvessels: An electron microscopic 3D reconstruction. *Glia* 2010;**158**:1094–103.
23. Rash JE, Yasumura T, Hudson CS, Agre P, Nielsen S. Direct immunogold labeling of aquaporin-4 in square arrays of astrocyte and ependymocyte plasma membranes in rat brain and spinal cord. *Proc Natl Acad Sci USA* 1998;**95**(20):11981–6.
24. Wolburg H, Wolburg-Buchholz K, Fallier-Becker P, Noell S, Mack AF. Structure and functions of aquaporin-4-based orthogonal arrays of particles. *Int Rev Cell Mol Biol* 2011;**287**:1–41.
25. Wolburg H, Noell S, Mack A, Wolburg-Buchholz K, Fallier-Becker P. Brain endothelial cells and the glio-vascular complex. *Cell Tissue Res* 2009;**335**(1):75–96.

26. Rash JE. Molecular disruptions of the panglial syncytium block potassium siphoning and axonal saltatory conduction: pertinence to neuromyelitis optica and other demyelinating diseases of the central nervous system. *Neuroscience* 2010;**168**(4):982–1008.
27. Mack AF, Wolburg H. A novel look at astrocytes: Aquaporins, ionic homeostasis, and the role of the microenvironment for regeneration in the CNS. *Neuroscientist* 2013;**19**(2):195–207.
28. Sykova E. Control of the extracellular ionic environment and volume. In: Kettenmann H, Ransom BR, editors. *Neuroglia.* 3rd ed. Oxford: Oxford University Press; 2013. p. 430–42.
29. Larsen BR, Assentoft M, Cotrina ML, Hua SZ, Nedergaard M, Kaila K, et al. Contributions of the Na+/K+-ATPase, NKCC1, and Kir4.1 to hippocampal K+ clearance and volume responses. *Glia* 2014;**62**(4):608–22.
30. Noell S, Fallier-Becker P, Deutsch U, Mack A, Wolburg H. Agrin defines polarized distribution of orthogonal arrays of particles in astrocytes. *Cell Tissue Res* 2009;**337**(2):185–95.
31. Rauch SM, Huen K, Miller MC, Chaudry H, Lau M, Sanes JR, et al. Changes in brain beta-amyloid deposition and aquaporin 4 levels in response to altered agrin expression in mice. *J Neuropathol Exp Neurol* 2011;**70**(12):1124–37.
32. Vajda Z, Pedersen M, Fuchtbauer EM, Wertz K, Stodkilde Jorgensen H, Sulyok E, et al. Delayed onset of brain edema and mislocalization of aquaporin-4 in dystrophin-null transgenic mice. *Proc Natl Acad Sci USA* 2002;**99**(20):13131–6.
33. Noell S, Wolburg-Buchholz K, Mack AF, Beedle AM, Satz JS, Campbell KP, et al. Evidence for a role of dystroglycan regulating the membrane architecture of astroglial endfeet. *Eur J Neurosci* 2011;**33**(12):2179–86.
34. Fallier-Becker P, Vollmer JP, Bauer H-C, Noell S, Wolburg H, Mack AF. Onset of aquaporin-4 expression in the developing mouse brain. *Int J Dev Neurosci* 2014;**36**(0):81–9.
35. Eilert-Olsen M, Haj-Yasein NN, Vindedal GF, Enger R, Gundersen GA, Hoddevik EH, et al. Deletion of aquaporin-4 changes the perivascular glial protein scaffold without disrupting the brain endothelial barrier. *Glia* 2012;**60**(3):432–40.
36. Armulik A, Genove G, Mae M, Nisancioglu MH, Wallgard E, Niaudet C, et al. Pericytes regulate the blood–brain barrier. *Nature* 2010;**468**(7323) 557-U231.
37. Daneman R, Zhou L, Kebede AA, Barres BA. Pericytes are required for blood–brain barrier integrity during embryogenesis. *Nature* 2010;**468**(7323):562–6.
38. Sweeney MD, Ayyadurai S, Zlokovic BV. Pericytes of the neurovascular unit: key functions and signaling pathways. *Nat Neurosci* 2016;**19**(6):771–83.
39. Wälchli T, Wacker A, Frei K, Regli L, Schwab Martin E, Hoerstrup Simon P, et al. Wiring the vascular network with neural cues: A CNS perspective. *Neuron* 2015;**87**(2):271–96.
40. Rennels ML, Gregory TF, Blaumanis OR, Fujimoto K, Grady PA. Evidence for a "Paravascular" fluid circulation in the mammalian central nervous system, provided by the rapid distribution of tracer protein throughout the brain from the subarachnoid space. *Brain Res* 1985;**326**(1):47–63.
41. Siyahhan B, Knobloch V, de Zelicourt D, Asgari M, Daners MS, Poulikakos D, et al. Flow induced by ependymal cilia dominates near-wall cerebrospinal fluid dynamics in the lateral ventricles. *J R Soc Interf* 2014;**11**(94).
42. Johanson CE, Duncan JA, Klinge PM, Brinker T, Stopa EG, Silverberg GD. Multiplicity of cerebrospinal fluid functions: New challenges in health and disease. *Cerebrosp Fluid Res* 2008;**5**:10.
43. Pollay M. The function and structure of the cerebrospinal fluid outflow system. *Cerebrosp Fluid Res* 2010;**7**:9.
44. Iliff JJ, Wang M, Zeppenfeld DM, Venkataraman A, Plog BA, Liao Y, et al. Cerebral arterial pulsation drives paravascular CSF–Interstitial fluid exchange in the murine brain. *J Neurosci* 2013;**33**(46):18190–9.

45. Badaut J, Ashwal S, Adami A, Tone B, Recker R, Spagnoli D, et al. Brain water mobility decreases after astrocytic aquaporin-4 inhibition using RNA interference. *J Cereb Blood Flow Metab* 2011;**31**(3):819–31.
46. Nedergaard M. Garbage truck of the brain. *Science* 2013;**340**(6140):1529–30.
47. Iliff JJ, Wang M, Liao Y, Plogg BA, Peng W, Gundersen GA, et al. A paravascular pathway facilitates CSF flow through the brain parenchyma and the clearance of interstitial solutes, including amyloid β. *Sci Transl Med* 2012;**4**(147):147ra111.
48. Benveniste H, Hof PR, Nedergaard M, Bechter K. Modern cerebrospinal fluid flow research and Heinrich Quincke's seminal 1872 article on the distribution of cinnabar in freely moving animals. *J Comp Neurol* 2015;**523**(12):1748–55.
49. MacAulay N, Zeuthen T. Water transport between CNS compartments: contributions of aquaporins and cotransporters. *Neuroscience* 2010;**168**(4):941–56.
50. Pekny M, Nilsson M. Astrocyte activation and reactive gliosis. *Glia* 2005;**50**(4):427–34.
51. Sofroniew MV. Molecular dissection of reactive astrogliosis and glial scar formation. *Trends Neurosci* 2009;**32**(12):638–47.
52. Zamanian JL, Xu L, Foo LC, Nouri N, Zhou L, Giffard RG, et al. Genomic Analysis of Reactive Astrogliosis. *J Neurosci* 2012;**32**(18):6391–410.
53. Pekny M, Pekna M. Astrocyte reactivity and reactive astrogliosis: Costs and benefits. *Physiol Rev* 2014;**94**(4):1077–98.
54. Anders JJ, Brightman MW. Assemblies of particles in cell membranes of developing, mature, and reactive astrocytes. *J Neurocytol* 1979;**8**:777–95.
55. Kästner R. Comparative-studies on the astrocytic reaction in the lesioned central-nervous-system of different vertebrates. *J Brain Res—J f Hirnforschung* 1987;**28**(2): 221–32.
56. Ren Z, Iliff JJ, Yang L, Yang J, Chen X, Chen MJ, et al. "Hit & Run" model of closed-skull traumatic brain injury (TBI) reveals complex patterns of post-traumatic AQP4 dysregulation. *J Cereb Blood Flow Metab* 2013;**33**(6):834–45.
57. Kimelberg H, Goderie S, Higman S, Pang S, Waniewski R. Swelling-induced release of glutamate, aspartate, and taurine from astrocyte cultures. *J Neurosci* 1990;**10**(5): 1583–91.
58. Jayakumar AR, Rao KVR, Panickar KS, Moriyama M, Reddy PVB, Norenberg MD. Trauma-induced cell swelling in cultured astrocytes. *J Neuropathol Exp Neurol* 2008;**67**(5):417–27.
59. Solenov E, Watanabe H, Manley GT, Verkman AS. Sevenfold-reduced osmotic water permeability in primary astrocyte cultures from AQP-4-deficient mice, measured by a fluorescence quenching method. *Am J Physiol Cell Physiol* 2004;**286**(2):C426–432.
60. Stokum JA, Kurland DB, Gerzanich V, Simard JM. Mechanisms of astrocyte-mediated cerebral edema. *Neurochem Res* 2015;**40**(2):317–28.
61. Altman J, Das GD. Autoradiographic and histological evidence of postnatal hippocampal neurogenesis in rats. *J Comp Neurol* 1965;**124**:319–36.
62. Gage FH. Mammalian neural stem cells. *Science* 2000;**287**:1433–8.
63. Siebzehnrubl FA, Reynolds BA, Vescovi A, Steindler DA, Deleyrolle LP. The origins of glioma: E Pluribus Unum? *Glia* 2011;**59**(8):1135–47.
64. Ignatova TN, Kukekov VG, Laywell ED, Suslov ON, Vrionis FD, Steindler DA. Human cortical glial tumors contain neural stem-like cells expressing astroglial and neuronal markers in vitro. *Glia* 2002;**39**(3):193–206.
65. Lathia JD, Mack SC, Mulkearns-Hubert EE, Valentim CLL, Rich JN. Cancer stem cells in glioblastoma. *Genes Dev* 2015;**29**(12):1203–17.
66. Ricci-Vitiani L, Pallini R, Biffoni M, Todaro M, Invernici G, Cenci T, et al. Tumour vascularization via endothelial differentiation of glioblastoma stem-like cells. *Nature* 2010;**468**(7325):824–8.

67. Noell S, Fallier-Becker P, Mack AF, Hoffmeister M, Beschorner R, Ritz R. Water channels aquaporin 4 and -1 expression in subependymoma depends on the localization of the tumors. *PLoS One* 2015;**10**(6):e0131367.
68. Riemenschneider M, Jeuken JM, Wesseling P, Reifenberger G. Molecular diagnostics of gliomas: state of the art. *Acta Neuropathol* 2010;**120**(5):567–84.
69. Allmendinger O, Trautmann K, Mittelbronn M, Waidelich J, Meyermann R, Tatagiba M, et al. Activated leukocyte cell adhesion molecule is expressed in neuroepithelial neoplasms and decreases with tumor malignancy, matrix metalloproteinase 2 expression, and absence of IDH1R132H mutation. *Hum Pathol* 2012;**43**(8):1289–99.
70. Sreekanthreddy P, Srinivasan H, Kumar DM, Nijaguna MB, Sridevi S, Vrinda M, et al. Identification of potential serum biomarkers of glioblastoma: Serum osteopontin levels correlate with poor prognosis. *Cancer Epidemiol Biomark Prev* 2010;**19**(6):1409–22.
71. Sage MR, Wilson AJ. The blood–brain-barrier—an important concept in neuroimaging. *Am J Neuroradiol* 1994;**15**(4):601–22.
72. Larsson HBW, Stubgaard M, Frederiksen JL, Jensen M, Henriksen O, Paulson OB. Quantitation of blood–brain-barrier defect by magnetic-resonance-imaging and gadolinium-dtpa in patients with multiple-sclerosis and brain-tumors. *Magn Reson Med* 1990;**16**(1):117–31.
73. Pronin IN, Holodny AI, Petraikin AV. MRI of high-grade glial tumors: Correlation between the degree of contrast enhancement and the volume of surrounding edema. *Neuroradiology* 1997;**39**(5):348–50.
74. Noell S, Mayer D, Strauss WSL, Tatagiba MS, Ritz R. Selective enrichment of hypericin in malignant glioma: Pioneering in vivo results. *Int J Oncol* 2011;**38**(5):1343–8.
75. Wolburg H, Wolburg-Buchholz K, Kraus J, Rascher-Eggstein G, Liebner S, Hamm S, et al. Localization of claudin-3 in tight junctions of the blood–brain barrier is selectively lost during experimental autoimmune encephalomyelitis and human glioblastoma multiforme. *Acta Neuropathol* 2003;**105**(6):586–92.
76. Kniesel U, Risau W, Wolburg H. Development of blood–brain barrier tight junctions in the rat cortex. *Dev Brain Res* 1996;**96**(1–2):229–40.
77. Liebner S, Fischmann A, Rascher G, Duffner F, Grote EH, Kalbacher H, et al. Claudin-1 and claudin-5 expression and tight junction morphology are altered in blood vessels of human glioblastoma multiforme. *Acta Neuropathol* 2000;**100**(3):323–31.
78. Rascher G, Fischmann A, Kroger S, Duffner F, Grote EH, Wolburg H. Extracellular matrix and the blood–brain barrier in glioblastoma multiforme: spatial segregation of tenascin and agrin. *Acta Neuropathol* 2002;**104**(1):85–91.
79. Faissner A, Reinhard J. The extracellular matrix compartment of neural stem and glial progenitor cells. *Glia* 2015;**63**(8):1330–49.
80. Brösicke N, Sallouh M, Prior L-M, Job A, Weberskirch R, Faissner A. Extracellular matrix glycoprotein-derived synthetic peptides differentially modulate glioma and sarcoma cell migration. *Cell Mol Neurobiol* 2015;**35**(5):741–53.
81. Warth A, Kröger S, Wolburg H. Redistribution of aquaporin-4 in human glioblastoma correlates with loss of agrin immunoreactivity from brain capillary basal laminae. *Acta Neuropathol (Berl)* 2004;**107**(4):311–18.
82. Noell S, Wolburg-Buchholz K, Mack A, Ritz R, Tatagiba M, Beschorner R, et al. Dynamics of expression patterns of AQP4, dystroglycan, agrin and matrix metalloproteinases in human glioblastoma. *Cell Tissue Res* 2012;**347**(2):429–41.
83. Ribatti D, Ranieri G, Annese T, Nico B. Aquaporins in cancer. *Biochim Biophys Acta* 2014;**1840**(5):1550–3.
84. Warth A, Simon P, Capper D, Goeppert B, Tabatabai G, Herzog H, et al. Expression pattern of the water channel aquaporin-4 in human gliomas is associated with blood-brain barrier disturbance but not with patient survival. *J Neurosci Res* 2007;**85**(6):1336–46.

85. Jung JS, Bhat RV, Preston GM, Guggino WB, Baraban JM, Agre P. Molecular characterization of an aquaporin cDNA from brain: candidate osmoreceptor and regulator of water balance. *Proc Natl Acad Sci USA* 1994;**91**(26):13052–6.
86. Furman CS, Gorelick Feldman DA, Davidson KG, Yasumura T, Neely JD, Agre P, et al. Aquaporin-4 square array assembly: opposing actions of M1 and M23 isoforms. *Proc Natl Acad Sci USA* 2003;**100**(23):13609–14.
87. Tan G, Sun SQ, Yuan DL. Expression of the water channel protein aquaporin-9 in human astrocytic tumours: Correlation with pathological grade. *J Int Med Res* 2008;**36**(4):777–82.
88. Jelen S, Ulhoi BP, Larsen A, Frokiaer J, Nielsen S, Rutzler M. AQP9 expression in glioblastoma multiforme tumors is limited to a small population of astrocytic cells and CD15(+)/CalB(+) leukocytes. *PLoS One* 2013;**8**(9).
89. Oshio K, Binder DK, Liang Y, Bollen A, Feuerstein B, Berger MS, et al. Expression of the aquaporin-1 water channel in human glial tumors. *Neurosurgery* 2005;**56**(2):375–80.
90. El Hindy N, Bankfalvi A, Herring A, Adamzik M, Lambertz N, Zhu Y, et al. Correlation of Aquaporin-1 Water channel protein expression with tumor angiogenesis in human astrocytoma. *Anticancer Res* 2013;**33**(2):609–13.
91. Wolburg H, Noell S, Wolburg-Buchholz K, Mack A, Fallier-Becker P. Agrin, Aquaporin-4, and astrocyte polarity as an important feature of the blood–brain barrier. *Neuroscientist* 2009;**15**(2):180–93.
92. Manley GT, Binder DK, Papadopoulos MC, Verkman AS. New insights into water transport and edema in the central nervous system from phenotype analysis of aquaporin-4 null mice. *Neuroscience* 2004;**129**(4):981–9.
93. Papadopoulos MC, Verkman AS. Aquaporin-4 and brain edema. *Pediatr Nephrol* 2007;**22**(6):778–84.
94. Papadopoulos MC, Saadoun S, Binder DK, Manley GT, Krishna S, Verkman AS. Molecular mechanisms of brain tumor edema. *Neuroscience* 2004;**129**(4):1009–18.
95. Nagelhus EA, Ottersen OP. Physiological roles of aquaporin-4 in brain. *Physiol Rev* 2013;**93**(4):1543–62.
96. Papadopoulos MC, Verkman AS. Aquaporin water channels in the nervous system. *Nat Rev Neurosci* 2013;**14**(4):265–77.
97. Auguste KI, Jin S, Uchida K, Yan D, Manley GT, Papadopoulos MC, et al. Greatly impaired migration of implanted aquaporin-4-deficient astroglial cells in mouse brain toward a site of injury. *FASEB J* 2007;**21**(1):108–16.
98. Huber VJ, Tsujita M, Nakada T. Identification of Aquaporin 4 inhibitors using in vitro and in silico methods. *Bioorg Med Chem* 2009;**17**(1):411–17.
99. Igarashi H, Huber V, Tsujita M, Nakada T. Pretreatment with a novel aquaporin 4 inhibitor, TGN-020, significantly reduces ischemic cerebral edema. *Neurol Sci* 2011;**32**(1):113–16.
100. Zelenina M. Regulation of brain aquaporins. *Neurochem Int* 2010;**57**(4):468–88.
101. Stupp R, Mason WP, van den Bent MJ, Weller M, Fisher B, Taphoorn MJB, et al. Radiotherapy plus concomitant and adjuvant temozolomide for glioblastoma. *N Engl J Med* 2005;**352**(10):987–96.
102. Lee GL, Hait WN. Inhibition of growth of C6-astrocytoma cells by inhibitors of calmodulin. *Life Sci* 1985;**36**(4):347–54.
103. Ghods AJ, Irvin D, Liu G, Yuan X, Abdulkadir IR, Tunici P, et al. Spheres isolated from 9L gliosarcoma rat cell line possess chemoresistant and aggressive cancer stem-like cells. *Stem Cells* 2007;**25**(7):1645–53.
104. Hirschhaeuser F, Menne H, Dittfeld C, West J, Mueller-Klieser W, Kunz-Schughart LA. Multicellular tumor spheroids: An underestimated tool is catching up again. *J Biotechnol* 2010;**148**(1):3–15.
105. Bowman PD, Ennis SR, Rarey KE, Lorris Betz A, Goldstein GW. Brain microvessel endothelial cells in tissue culture: A model for study of blood–brain barrier permeability. *Ann Neurol* 1983;**14**(4):396–402.

106. Schulze C, Smales C, Rubin LL, Staddon JM. Lysophosphatidic acid increases tight junction permeability in cultured brain endothelial cells. *J Neurochem* 1997;**68**(3):991–1000.
107. Lee H-S, Namkoong K, Kim D-H, Kim K-J, Cheong Y-H, Kim S-S, et al. Hydrogen peroxide-induced alterations of tight junction proteins in bovine brain microvascular endothelial cells. *Microvasc Res* 2004;**68**(3):231–8.
108. Abbott NJ. Inflammatory mediators and modulation of blood–brain barrier permeability. *Cell Mol Neurobiol* 2000;**20**(2):131–47.
109. Hülper P, Veszelka S, Walter FR, Wolburg H, Fallier-Becker P, Piontek J, et al. Acute effects of short-chain alkylglycerols on blood–brain barrier properties of cultured brain endothelial cells. *Br J Pharmacol* 2013;**169**(7):1561–73.
110. Hurst RD, Fritz IB. Properties of an immortalised vascular endothelial glioma cell co-culture model of the blood–brain barrier. *J Cell Physiol* 1996;**167**(1):81–8.
111. Raub TJ. Signal transduction and glial cell modulation of cultured brain microvessel endothelial cell tight junctions. *Am J Physiol Cell Physiol* 1996;**271**(2):C495–503.
112. Hurst R, Clark J. Alterations in transendothelial electrical resistance by vasoactive agonists and cyclic AMP in a blood–brain barrier model system. *Neurochem Res* 1998;**23**(2):149–54.
113. Scism J, Laska D, Horn J, Gimple J, Pratt S, Shepard R, et al. Evaluation of an in vitro coculture model for the blood–brain barrier: Comparison of human umbilical vein endothelial cells (ECV304) and rat glioma cells (C6) from two commercial sources. *In Vitro Cell Dev Biol-Anim* 1999;**35**(10):580–92.
114. Tan KH, Dobbie MS, Felix RA, Barrand MA, Hurst RD. A comparison of the induction of immortalized endothelial cell impermeability by astrocytes. *Neuroreport* 2001;**12**(7):1329–34.
115. Easton AS, Abbott NJ. Bradykinin increases permeability by calcium and 5-lipoxygenase in the ECV304/C6 cell culture model of the blood–brain barrier. *Brain Res* 2002;**953**(1–2):157–69.
116. Mendes B, Marques C, Carvalho I, Costa P, Martins S, Ferreira D, et al. Influence of glioma cells on a new co-culture in vitro blood–brain barrier model for characterization and validation of permeability. *Int J Pharm* 2015;**490**(1–2):94–101.
117. Fischer S, Wobben M, Kleinstuck J, Renz D, Schaper W. Effect of astroglial cells on hypoxia-induced permeability in PBMEC cells. *Am J Physiol Cell Physiol* 2000;**279**(4):C935–44.
118. Hamm S, Dehouck B, Kraus J, Wolburg-Buchholz K, Wolburg H, Risau W, et al. Astrocyte mediated modulation of blood–brain barrier permeability does not correlate with a loss of tight junction proteins from the cellular contacts. *Cell Tissue Res* 2004;**315**(2):157–66.
119. Escalona-Zapata J, Dieznau MD. The astrocytic nature of glioblastoma demonstrated by tissue-culture. *Acta Neuropathol* 1981;**53**(2):155–60.
120. Hovinga KE, Shimizu F, Wang R, Panagiotakos G, Van Der Heijden M, Moayedpardazi H, et al. Inhibition of Notch signaling in glioblastoma targets cancer stem cells via an endothelial cell intermediate. *Stem Cells* 2010;**28**(6):1019–29.
121. Shimizu F, Hovinga KE, Metzner M, Soulet D, Tabar V. Organotypic explant culture of glioblastoma multiforme and subsequent single-cell suspension *Current protocols in stem cell biology*. Hoboken, NJ: John Wiley & Sons, Inc.; 2007.
122. Joshi K, Demir H, Yamada R, Miyazaki T, Ray-Chaudhury A, Nakano I. Method for novel anti-cancer drug development using tumor explants of surgical specimens. *J Vis Exp* 2011(53).
123. Mercurio S, Padovani L, Colin C, Carre M, Tchoghandjian A, Scavarda D, et al. Evidence for new targets and synergistic effect of metronomic celecoxib/fluvastatin combination in pilocytic astrocytoma. *Acta Neuropathol Comm* 2013;**1**:17.

124. Aas AT, Brun A, Blennow C, Stromblad S, Salford LG. The RG2 rat glioma model. *J Neurooncol* 1995;**23**(3):175–83.
125. Weizsaecker M, Deen DF, Rosenblum ML, Hoshino T, Gutin PH, Barker M. The 9L rat-brain tumor—description and application of an animal-model. *J Neurol* 1981;**224**(3):183–92.
126. Benda P, Lightbod J, Sato G, Levine L, Sweet W. Differentiated rat glial cell strain in tissue culture. *Science* 1968;**161**(3839):370–1.
127. Serano RD, Pegram CN, Bigner DD. Tumorigenic cell-culture lines from a spontaneous vm-dk murine astrocytoma (sma). *Acta Neuropathol* 1980;**51**(1):53–64.
128. McNeill RS, Vitucci M, Wu J, Miller CR. Contemporary murine models in preclinical astrocytoma drug development. *Neuro Oncol* 2015;**17**(1):12–28.
129. Noell S, Ritz R, Wolburg-Buchholz K, Wolburg H, Fallier-Becker P. An allograft glioma model reveals the dependence of aquaporin-4 expression on the brain microenvironment. *PLoS One* 2012;**7**(5).

CHAPTER

17

Water Homeostasis Dysfunction in Epilepsy

Devin K. Binder

University of California, Riverside, CA, United States

OUTLINE

WATER AND ION HOMEOSTASIS AND EPILEPSY

Glial cells are involved in many important physiologic functions, such as sequestration and/or redistribution of K^+ during neural activity, neurotransmitter cycling, and provision of energy substrates to neurons.[1] Several recent lines of evidence strongly suggest that changes in glial cells potentially contribute to epilepsy.[2–4] First, many studies now link

Brain Edema.
DOI: http://dx.doi.org/10.1016/B978-0-12-803196-4.00017-5

glial cells to modulation of synaptic transmission.[5–9] Second, functional alterations of specific glial membrane channels and receptors have been discovered in epileptic tissue.[10–15] Third, direct stimulation of astrocytes has been shown to be sufficient for neuronal synchronization in epilepsy models[9] (although see Fiacco et al., 2007[16]). Thus, if the cellular and molecular mechanisms by which glial cells (especially astrocytes) modulate excitability are better understood, specific antiepileptic therapies based on modulation of glial receptors and channels can be contemplated.[17] It is likely that therapies directed to glial cells would have fewer deleterious side effects than current therapies targeting neurons.

Alteration of water and K^+ homeostasis could dramatically affect seizure susceptibility. First, brain tissue excitability is exquisitely sensitive to osmolarity and the size of the extracellular space (ECS).[18] Decreasing ECS volume with hypoosmolar treatment produces hyperexcitability and enhanced epileptiform activity.[19–22] At least two mechanisms could be operative: decreasing ECS volume could increase extracellular neurotransmitter and ion concentrations and magnify interactions among neurons due to increased ECS resistance.[3] Conversely, increasing ECS volume with hyperosmolar treatment powerfully attenuates epileptiform activity.[20,22–24] These experimental data parallel extensive clinical experience indicating that hypoosmolar states such as hyponatremia lower seizure threshold while hyperosmolar states elevate seizure threshold.[25–27] Second, millimolar and even submillimolar increases in extracellular K^+ concentration powerfully enhance epileptiform activity in the hippocampus.[28–31] High-K^+ reliably induces epileptiform activity in hippocampal slices from human patients with intractable temporal lobe epilepsy (TLE).[32]

AQUAPORINS

The aquaporins (AQPs) are a family of membrane proteins that function as water channels in many cell types and tissues in which fluid transport is crucial.[33–35] The AQPs are small hydrophobic integral membrane proteins (~30kDa monomer) that facilitate bi-directional water transport in response to osmotic gradients.[36] Multiple mammalian AQPs have been identified including AQP0, AQP1, AQP2, AQP4, and AQP5 which transport water only (AQPs) and AQP3, AQP7, and AQP9 which also transport glycerol (aquaglyceroporins).[35,36]

Aquaporin-4 (AQP4)[37–39] is of particular interest in neuroscience as it is expressed in brain and spinal cord by glial cells, especially at specialized membrane domains including astroglial endfeet in contact with blood vessels and astrocyte membranes that ensheathe glutamatergic synapses.[40–44] Activity-induced radial water fluxes in neocortex have been demonstrated that may represent water movement via AQP channels in response to physiological activity.[45,46] Interestingly, AQP4 is a structural component of orthogonal arrays of particles (OAPs) seen in freeze-fracture electron micrographs[47–51] (for details see Chapter 11: Role of Matrix Metalloproteinases in Brain Edema).

AQP4$^{-/-}$ Mice

AQP4$^{-/-}$ mice were originally generated by targeted gene disruption in 1997,[52] and recently a glial conditional deletion of AQP4 has been generated.[53] AQP4$^{-/-}$ mice are grossly normal phenotypically, do not manifest overt neurological abnormalities, altered blood–brain barrier properties, abnormal baseline intracranial pressure, impaired osmoregulation, or obvious brain dysmorphology.[54–56] A mild hearing impairment has been recorded and the electroretinogram is perturbed.[57,58] Mice with AQP4 deletion have normal blood chemistries and hematologies, normal expression of other brain AQPs (AQP1 and AQP9), and their only non-neural phenotype is a mild impairment in maximal urine concentrating ability.[52]

Interestingly, more detailed studies of brain tissue from AQP4$^{-/-}$ mice demonstrated reduced osmotic water permeability as measured in isolated membrane vesicles,[52] brain slices,[59] and intact brain.[60] Also, water permeability was sevenfold reduced in primary astrocyte cultures from AQP4$^{-/-}$ mice as measured by a calcein fluorescence quenching method,[61] and similar results were obtained with AQP4 RNAi knockdown experiments in wild-type astrocytes.[62] These data directly demonstrated that AQP4 provides the predominant pathway for transmembrane water movement in astrocytes.

In vivo studies of these mice demonstrated a functional role for AQP4 in brain water transport. AQP4$^{-/-}$ mice have markedly decreased accumulation of brain water (cerebral edema) following water intoxication and focal cerebral ischemia[63] and impaired clearance of brain water in models of vasogenic edema.[56] Clearance of seizure-induced edema may also be AQP4-dependent.[64] Impaired water flux into (in the case of cytotoxic edema) and out of (in the case of vasogenic edema) the brain makes sense based on the bi-directional nature of water flux across the AQP4 membrane channel. The recently generated glial conditional AQP4 knockout mouse line demonstrates a 31% reduction in brain water uptake after systemic hypoosmotic stress.[53] Similarly, mice deficient in dystrophin or α-syntrophin, in which there is mislocalization of the AQP4 protein,[65–67] show attenuated cerebral edema.[66,68]

Altered Extracellular Space in AQP4$^{-/-}$ Mice

The ECS in brain comprises ~20% of brain tissue volume, consisting of a jelly-like matrix in which neurons, glia, and blood vessels are embedded.[69] The ECS contains ions, neurotransmitters, metabolites, peptides, and extracellular matrix molecules, forming the microenvironment for all cells in the brain and mediating glia–neuron communication via diffusible messengers, metabolites, and ions.[70] The size of the ECS is likely largely controlled by astrocytic mechanisms.[71–73]

The ECS of $AQP4^{-/-}$ mice was assessed using a cortical fluorescence recovery after photobleaching (cFRAP) method to measure the diffusion of fluorescently labeled macromolecules in the cortex.[74] ECS in mouse brain was labeled by exposure of the intact dura to fluorescein-dextrans (M_r 4, 70, and 500 kDa) after craniectomy, and fluorescein-dextran diffusion was detected by fluorescence recovery after laser-induced cortical photobleaching using confocal optics. The cFRAP method was applied to brain edema, seizure initiation, and AQP4 deficiency. In contrast to the slowed diffusion produced by brain edema and seizure activity, ECS diffusion was *faster* in $AQP4^{-/-}$ mice, indicating ECS expansion in AQP4 deficiency.[74] Similar results were obtained with follow-up studies using the TMA^+ method.[75]

Seizure Phenotype of $AQP4^{-/-}$ Mice

Seizure susceptibility of $AQP4^{-/-}$ mice was initially examined using the convulsant ($GABA_A$ antagonist) pentylenetetrazol (PTZ).[55] At 40 mg/kg PTZ (i.p.), all wild-type mice exhibited seizure activity, whereas 6 out of 7 $AQP4^{-/-}$ mice did not exhibit seizure activity. At 50 mg/kg PTZ, both groups exhibited seizure activity; however, the latency to generalized (tonic-clonic) seizures was longer in $AQP4^{-/-}$ mice.[55] Since seizure propensity is exquisitely sensitive to ECS volume,[18] the expanded ECS in AQP4 deficiency is consistent with the increased seizure threshold. Thus, more intense stimuli (e.g., higher PTZ doses or a longer time after PTZ) may be required to overcome the expanded ECS of $AQP4^{-/-}$ mice in order to initiate a seizure.

In order to analyze the seizure phenotype of the $AQP4^{-/-}$ mice in greater detail, in vivo electroencephalographic (EEG) characterization with stimulation and recording was employed.[76] $AQP4^{-/-}$ mice and wild-type controls were implanted in the right dorsal hippocampus with bipolar electrodes. Following postoperative recovery, electrical stimulations were given to assess electrographic seizure threshold and duration. $AQP4^{-/-}$ mice had a higher mean electrographic seizure threshold than wild-type controls, consistent with the prior PTZ studies.[55] However, $AQP4^{-/-}$ mice were also found to have remarkably prolonged stimulation-evoked seizures compared to wild-type controls.[55]

Altered K^+ Homeostasis in $AQP4^{-/-}$ Mice

Because of the colocalization of AQP4 and inwardly rectifying K^+ (K_{ir}) channels in glial endfeet, the hypothesis arose that AQP4 was indirectly involved in K^+ reuptake.[40,77] Impaired K^+ clearance from the ECS following the intense neuronal activity accompanying the seizure would lead to prolonged depolarization of neurons and inhibit seizure termination.[10,28–30] Indeed, in addition to modulation of brain water transport, AQP4 and its known molecular partners have been hypothesized to modulate ion homeostasis.[78,79] During rapid neuronal firing, extracellular $[K^+]$ ($[K^+]_o$) increases from ~3 mM to a maximum of 10–12 mM; and K^+ released by active neurons is thought to be primarily taken up by glial cells.[69,80–82] Such K^+ reuptake into glial cells could be AQP4-dependent, as water influx coupled to K^+ influx is thought to underlie activity-induced glial cell swelling.[83,84] In support of this possibility was the known subcellular colocalization of AQP4 with the inwardly rectifying K^+ channel $K_{ir}4.1$ in the retina.[41,77,85] $K_{ir}4.1^{-/-}$ mice, like $AQP4^{-/-}$ mice,[57,58] demonstrate abnormal retinal and cochlear physiology presumably due to altered K^+ homeostasis.[86–89] $K_{ir}4.1$ is thought to contribute to K^+ reuptake and spatial K^+ buffering by glial cells,[90–93] and pharmacological or genetic inactivation of $K_{ir}4.1$ leads to impairment of extracellular K^+ regulation.[86,94–99]

To address the possibility that AQP4 deficiency was associated with a deficit in K^+ homeostasis, K^+ dynamics were examined in vivo in $AQP4^{-/-}$ mice.[76] Neither baseline $[K^+]_o$ nor the "Lux–Heinemann ceiling" level of activity-induced $[K^+]_o$ elevation (~12 mM)[80,81] were altered in AQP4 deficiency, indicating that basic K^+ homeostasis was intact. Based on K^+ measurements made from cortex with double-barreled K^+-sensitive microelectrodes, stimulation-induced rises in $[K^+]_o$ were quite different in $AQP4^{-/-}$ compared to wild-type mice; in particular, there was a markedly slower rise and decay time for poststimulus changes in $[K^+]_o$ in $AQP4^{-/-}$ mice.[76] A similar delay in K^+ kinetics was observed following cortical spreading depression in $AQP4^{-/-}$ mice using a fluorescent K^+ sensor.[100]

Slowed $[K^+]_o$ rise time is consistent with increased ECS volume fraction in $AQP4^{-/-}$ mice.[74] Slowed $[K^+]_o$ decay is possibly due to impaired K^+ reuptake into $AQP4^{-/-}$ astrocytes. Interestingly, there is no difference in expression of $K_{ir}4.1$ protein[76] or $K_{ir}4.1$ immunoreactivity[101] in $AQP4^{-/-}$ mice nor AQP4 immunoreactivity in $K_{ir}4.1^{-/-}$ mice.[101] In addition, no alterations were observed in membrane potential, barium-sensitive $K_{ir}4.1$ K^+ current, or current–voltage curves in $AQP4^{-/-}$ retinal Müller cells[102] or brain astrocytes.[103] Lack of alteration of K_{ir} channels in $AQP4^{-/-}$ mice suggests the interesting possibility that the slowed $[K^+]_o$ decay may be a secondary effect of slowed water extrusion (deswelling) following stimulation, and this has been modeled carefully[104] but not directly demonstrated.

Stroschein et al.[105] investigated the impact of AQP4 on stimulus-induced alterations of $[K^+]_o$ in hippocampal slices. Antidromic stimulation evoked smaller increases and slower recovery of $[K^+]_o$ in the stratum pyramidale of $AQP4^{-/-}$ mice, consistent with the previous in vivo studies in cortex. Interestingly, astrocyte gap junction coupling as assessed with tracer filling during patch-clamp recording demonstrated enhanced tracer coupling in $AQP4^{-/-}$ mice, and laminar profiles indicated enhanced spatial redistribution of K^+.[105] The functional consequences of alterations in gap junctional coupling are not completely clear; however the complete absence of astrocytic gap junctions impairs K^+ homeostasis.[106] To further assess the direct link between AQP4 and K^+ homeostasis, Haj-Yasein et al. examined extracellular K^+ concentration in a distinct constitutive $AQP4^{-/-}$ mouse line[107] and found a layer-specific effect of AQP4 deletion on K^+ dynamics during synaptic stimulation.[108]

Anchoring of AQP4 by the Dystrophin-Associated Protein Complex (DAPC)

AQP4 in endfeet membranes is anchored to the DAPC. For example, endfoot expression of AQP4 is dependent on α-syntrophin, a dystrophin-associated protein.[67] Accordingly, mice deficient in α-syntrophin show marked loss of AQP4 from perivascular and subpial membranes as judged by quantitative immunogold electron microscopy.[68] Similarly, dystrophin-deficient (*mdx*) mice exhibit a dramatic reduction of AQP4 in astroglial endfeet surrounding capillaries and at the glia limitans (cerebrospinal fluid–brain interface) despite no alteration in total AQP4 protein.[66] These studies clearly suggest that alterations in components of the DAPC may affect the subcellular targeting and function of AQP4. A recent study also demonstrates that dystrophin localization at the astrocytic endfoot is dependent on syntrophin.[109] Another study using the zero magnesium in vitro model of epilepsy demonstrated reduction in expression of a distinct member of the DAPC, β-dystroglycan, in cortical astrocytes following continuous seizures, which was accompanied by a decrease in AQP4 expression.[110] Astrocyte swelling is markedly impaired in response to hypoosmotic stress, oxygen-glucose deprivation or high K^+ in the cortex of α-syntrophin-negative glial fibrillary acidic protein (GFAP)/EGFP mice.[111]

What are the functional consequences to seizure susceptibility of loss of DAPC components? Amiry-Moghaddam et al. (2003) were the first to report a possible association between AQP4 and epilepsy by demonstrating an increased severity of hyperthermia-evoked seizures in α-syntrophin-deficient mice. These mice exhibited a deficit in extracellular K^+ clearance following evoked neuronal activity.[4] Dystrophin-deficient mice were found to have altered seizure susceptibility in response to various chemical convulsants; in particular, *mdx* mice showed enhanced seizure severity and a shorter latency in the development of chemical kindling produced by administration of PTZ.[112] It is interesting to note that there is an increased incidence of epilepsy in forms of human muscular dystrophy in which the dystrophin complex is affected.[113] Taken together, these data suggest that AQP4 and its molecular partners together comprise a multifunctional "unit" responsible for clearance of K^+ and/or water following neural activity and that alterations in expression of this complex can lead to alterations in seizure susceptibility.

Human Tissue Studies

The most common pathology in patients with medically intractable TLE is mesial temporal sclerosis (MTS), characterized by marked neuronal cell loss in specific hippocampal areas, gliosis, and microvascular proliferation.[114] Emerging work also demonstrates dysregulation of water and K^+ homeostasis in patients with mesial TLE. First, imaging studies demonstrate abnormal T2 prolongation by magnetic resonance imaging (MRI) in the epileptic hippocampus, possibly partially due to increased water content.[115] This is accompanied by alterations in apparent diffusion coefficient (ADC) with diffusion-weighted MRI imaging[116] (see Chapter 4: Experimental Techniques to Investigate the Formation of Brain Edema In Vivo, for details on neuroimaging techniques). Second, the expression and subcellular localization of AQP4 have been shown to be altered in sclerotic hippocampi obtained from patients with MTS. Using immunohistochemistry, RT-PCR, and gene chip analysis, Lee et al. demonstrated an overall increase in AQP4 expression in sclerotic epilepsy tissue.[117] However, using quantitative immunogold electron microscopy, the same group found that there was mislocalization of AQP4 in the human epileptic hippocampus, with reduction in perivascular membrane expression.[118] Thus, although there was an overall increase in AQP4 content by Western blot, RT-PCR, and gene chip analysis, the subcellular distribution of AQP4 in mesial TLE tissue had changed. Their hypothesis was that reduction in perivascular AQP4 expression would lead to water and K^+ dysregulation in the epileptic hippocampus, potentially contributing to hyperexcitability.

Medici et al. reported AQP4 expression in tissue from patients with focal cortical dysplasia (FCD) type IIB, normal-appearing (cryptogenic) epileptic cortex, and nonepileptic control tissue.[119] AQP4 expression and distribution in the cryptogenic cases were similar to control cases, i.e., with intact perivascular immunoreactivity. In the patients with FCD type IIB, the pattern was different, with strong AQP4 immunoreactivity around dysplastic neurons but with very weak immunoreactivity for AQP4 around blood vessels.[119]

Disruption of the DAPC has also been demonstrated in human epileptic tissue. In the study in which reduced perivascular membrane expression of AQP4 was noted,[118] these authors also studied perivascular dystrophin expression. Like AQP4 expression, perivascular dystrophin expression was also reduced in tissue from the sclerotic epileptic hippocampus.[118] The more recent study also demonstrated reduced perivascular dystrophin expression in FCD type IIB.[119] Thus, subcellular alteration in AQP4 expression may result secondarily from alterations in dystrophin and/or other members of the DAPC.

Dysfunction of astroglial K_{ir} channels has also been found in specimens from patients with TLE.[10,12] First, using ion-sensitive microelectrodes, Kivi et al. demonstrated an impairment of glial barium-sensitive K^+ uptake in the CA1 region of MTS specimens.[11] Second, using patch-clamp analysis, Hinterkeuser et al. demonstrated a dramatic reduction in astroglial K_{ir} currents in MTS tissue.[13] Reduction in K_{ir} currents would be expected to contribute to hyperexcitability; evidence for this comes from the finding of stress-induced seizures in conditional $K_{ir}4.1^{-/-}$ mice.[97]

Methodological Issues

In assessing the contribution of AQP4 and its molecular partners to epileptogenesis, it is critical to examine the methodology in the reported studies. First, studies of human tissue represent the "endpoint" of an already long-standing epileptogenic process. A sclerotic hippocampus with significant neuronal cell loss and gliosis may show many molecular changes in glial cells including the reported altered distribution of AQP4,[117,118] but what is unclear from these studies is whether AQP4 dysregulation represents a cause or consequence of epileptogenesis. Appropriate animal models of epilepsy permit dissection of the process of epileptogenesis in greater detail. In particular, in such models it is possible to examine the transition from a normal to an epileptic brain as assessed by in vivo electrophysiological recordings of spontaneous seizures following an initial epileptogenic insult. Description of molecular changes in glial cells during epileptogenesis[2] will be facilitated by studies in such in vivo models. Unlike in vivo models, studies with *ex vivo* slices enable the delineation of tissue and synaptic physiology in greater detail,[4] but at the expense of isolating only a part of the in vivo network. Thus, a fruitful combination of methodologies for future study will include in vivo electrophysiologic recording for validation of epileptogenesis, *ex vivo* slice physiology, and examination of the cellular and molecular changes in glial cells leading up to the development of spontaneous seizures.

The studies discussed above suggest novel roles for AQP4 in control of seizure susceptibility,[4,55,76] K^+ homeostasis,[4,76,100] and ECS physiology.[74] These findings together with the changes in human epileptic tissue[117,118] lead to the unifying hypothesis that AQP4 and its molecular partners may play a functional role in epilepsy.[3,120] Thus, more recent studies have examined the pattern of expression, regulation, and role(s) of AQP4 in various animal models of epileptogenesis (development of epilepsy).

Expression and Regulation of AQP4 During Epileptogenesis

Early studies demonstrated the perivascular localization of AQP4[40,41,121,122] but did not examine region-specific expression of AQP4 in the brain in detail. A recent study examined subregional expression pattern of AQP4 within the mouse hippocampus, a structure critical to epilepsy.[101] AQP4 immunohistochemistry revealed a developmentally regulated and laminar-specific pattern, with highest expression in the CA1 stratum lacunosum-moleculare (SLM) and the molecular layer (ML) of the dentate gyrus (DG). AQP4 was also ubiquitously expressed in astrocytic endfeet around blood vessels.

This study also addressed the cell type-specificity of AQP4 expression in the hippocampus. The description of novel classes of hippocampal glial cells (classical astrocytes vs NG2 cells) with different morphology and functional properties[123–126] makes it particularly important to clearly identify which "glial cell" expresses AQP4. While AQP4 is thought to be expressed predominantly by "classical" astrocytes, some have reported expression in other cell types, such as activated microglia.[127] However, a recent study found no AQP4 expression in resting microglia of the neocortex[128] and conditional AQP4 deletion driven by the human *GFAP* promoter completely removed AQP4 from brain.[53] In line with the latter studies, AQP4 was found to colocalize with the astrocyte markers GFAP and S100β in the hippocampus, and electrophysiological and postrecording RT-PCR analyses of individual cells revealed that AQP4 and $K_{ir}4.1$ were coexpressed in nearly all CA1 "classical astrocytes." In NG2 cells, AQP4 was detected at the transcript level but not with immunohistochemistry.[101]

Seizure activity is accompanied by a variety of changes in gene expression, and repeated or prolonged seizure activity leads to glial changes, ultimately including the proliferation of "reactive" astrocytes (gliosis). The human studies indicated an upregulation and altered distribution of AQP4 protein in sclerotic tissue resected from patients with mesial TLE.[117,118] However, it is unclear whether AQP4 is upregulated by seizure activity per se and/or by other processes occurring during epileptogenesis (development of epilepsy), such as cell death, inflammation, or gliosis.[129] AQP4 is known to be dramatically upregulated in reactive astrocytes following injury[130–132] but may also be regulated by physiologic stimuli.[133]

Of course, whether AQP4 dysregulation *precedes* or *follows* epileptogenesis impacts its functional role as cause or consequence. To address this question will require studies of AQP4 regulation with concurrent video-EEG monitoring during epileptogenesis to determine the exact timing of onset of spontaneous seizures relative to changes in AQP4 and its molecular partners. Three studies so far address this key issue. First, Kim et al. studied regulation of various AQPs in the rat pilocarpine model of epilepsy.[134,135] In control animals, AQP4 immunoreactivity was detected diffusely in the piriform cortex and hippocampus, with greatest expression at astrocyte endfeet. Following status epilepticus in this model the authors describe an "AQP4-deleted area" in the piriform cortex, associated with decreases in immunoreactivity for dystrophin and α-syntrophin, members of the DAPC.[135] However, EEG analysis for timing of spontaneous seizure onset was not investigated.

Second, Lee et al.[136] studied AQP4 distribution and regulation in the intrahippocampal model of epileptogenesis. In this model, stereotactic microinjection of kainic acid into the mouse dorsal hippocampus leads to limbic status epilepticus followed by a "latent" period of several days (usually about 7 days) prior to the occurrence of spontaneous recurrent seizures (epilepsy) recorded by chronic video-EEG monitoring. Lee et al. found dramatic downregulation of AQP4 in multiple hippocampal subregions/laminae during epileptogenesis in this model. Importantly, hippocampal AQP4 immunoreactivity was markedly downregulated during the latent phase prior to the occurrence of spontaneous seizures as well as during the chronic epileptic phase, with partial recovery in some hippocampal laminae. While marked reactive astrocytosis was observed, the reactive astrocytes in this model were largely AQP4-negative. Interestingly, in this study, $K_{ir}4.1$ expression on reactive astrocytes was preserved.[136]

Third, Alvestad et al.[137] characterized EM localization of AQP4 in the latent and chronic phase in the intraperitoneal kainic acid rat model. Immunogold electron microscopic analysis revealed that adluminal AQP4 expression in astrocyte endfoot membranes was decreased in KA-treated rats during the latent phase. There was an accompanying reduction in adluminal alpha-syntrophin expression.[137] These data together with those of Kim et al.[134,135] and Lee et al.[136] suggest downregulation and/or mislocalization of AQP4 during early epileptogenesis. Thus, AQP4 dysregulation may contribute a critical role to loss of ion and water homeostasis leading to excitability and decreased seizure threshold during epileptogenesis.

Further studies will need to address the coordinate regulation of AQP4 and other relevant astrocyte molecules such as $K_{ir}4.1$ in these models. Finally, distinct epilepsy models in which epileptogenesis occurs in the *absence* of detectable cell death would prove useful in determining the threshold for AQP4 regulation.[138,139]

Subcellular Alteration in AQP4 Distribution: A Common Disease Mechanism?

The hallmark of AQP4 expression in the CNS is its polarized expression at astrocyte endfeet ensheathing blood vessels.[40,41,140] Loss of such polarization is associated with pathology in mesial TLE as discussed above.[118]

During seizures, there is focal swelling in the area of the seizure focus[23,74]; and the putative effect of loss of the perivascular pool of AQP4 would be to slow water egress from astrocyte to capillary leading to local astrocyte swelling, ECS constriction, and increased excitability.[3]

Interestingly, a similar loss of AQP4 polarization has been observed in distinct models of neurological diseases. For example, in a mouse model of Alzheimer's disease, loss of AQP4 from endfoot membranes at sites of perivascular amyloid deposits was observed.[141] Such subcellular alterations could lead to perturbation of local water and K^+ homeostasis in affected brain regions, and thus contribute to cognitive decline and seizure susceptibility. Similarly, Badaut et al. found loss of polarization of AQP4 expression on astrocyte endfeet following subarachnoid hemorrhage (SAH) and in peritumoral tissue[130]; thus, posttraumatic epilepsy and tumor-associated epilepsy[142] could be associated with alterations in AQP4 expression and distribution. Further studies will require careful determination of perivascular/perisynaptic AQP4 expression ratios in a variety of conditions.

AQP4 and the Gliovascular Junction

Another area of astrocyte biology of potential importance for epilepsy is the "gliovascular junction."[78] Recent studies have shown the close relationship between astrocyte endfeet ensheathing blood vessels, the targeted expression of AQP4 and $K_{ir}4.1$ on astrocyte endfeet, and the role of astrocytes in blood–brain barrier permeability[143] and control of cerebrovascular tone.[144–148] Molecular and cellular alterations in the gliovascular junction could perturb blood flow, water, and K^+ regulation, and therefore local tissue excitability. Interestingly, recent studies have shown that transient opening of the blood–brain barrier is sufficient for focal epileptogenesis.[149,150]

Just as alterations in the perivascular pool of AQP4 on existing blood vessels could affect local water and K^+ regulation, it is also the case that many new blood vessels can be formed in epileptic tissue. MTS is characterized by neuronal cell loss in specific hippocampal areas, gliosis, synaptic reorganization, but also significant microvascular proliferation.[114] How does the regulation of AQP4 relate to angiogenesis during epileptogenesis? The increased number of microvessels in epileptic tissue may in part be due to upregulation of vascular endothelial growth factor (VEGF) following seizures.[151–153] VEGF administration itself has been shown to upregulate AQP4,[154] but it is unclear whether AQP4 distribution and/or function may be unique in "epileptic vasculature." Alterations in cerebrovascular tone in "epileptic vasculature" would be another interesting topic of investigation.

A Role for AQP4 at the Synapse?

It was clear from early immunogold studies of AQP4 localization that while the most abundant pool was localized at astrocyte endfeet, there was also a significant concentration in nonendfeet membranes, including those astrocyte membranes that ensheath glutamatergic synapses.[40,41] Given the role of AQP4 in rapid water fluxes and K^+ homeostasis, more specific roles of AQP4 in synaptic transmission, plasticity, and behavior have begun to be tested.

Skucas et al. evaluated long-term potentiation (LTP) and long-term depression (LTD) in hippocampal slices from AQP4$^{-/-}$ mice.[155] Interestingly, AQP4$^{-/-}$ mice exhibited a selective defect in LTP and LTD without a change in basal synaptic transmission or short-term plasticity. The impairment in LTP in AQP4$^{-/-}$ mice was specific for the type of LTP that depends on the neurotrophin brain-derived neurotrophic factor (BDNF), which is induced by stimulation at theta rhythm [theta-burst stimulation (TBS)-LTP], but there was no impairment in a form of LTP that is BDNF-independent, induced by high-frequency stimulation. LTD was also impaired in AQP4$^{-/-}$ mice, which was rescued by a scavenger of BDNF or a blockade of Trk receptors. The AQP4$^{-/-}$ mice also exhibited a cognitive defect in location-specific object memory but not Morris water maze or contextual fear conditioning. These results suggest that AQP4 channels in astrocytes may play an unanticipated role in neurotrophin-dependent plasticity and influence behavior.

Based on these results, downregulation of AQP4 may not only lead to increased neural excitability due to abnormalities of water and potassium homeostasis but may also lead directly to abnormalities in synaptic plasticity (both LTP and LTD). This provides a potential explanation for the way that astrocytic changes in epilepsy may contribute not only to seizures but also to cognitive deficits. Cognitive impairment is very important because patients with TLE have many alterations in cognitive function and in particular hippocampal-dependent tasks such as spatial memory.[156–159] In addition, many other forms of plasticity are operative in the epileptic brain such as potentiation of synapses, reorganization of neuronal circuitry, and alteration in postnatal neurogenesis.[160–162] More experiments are necessary to elucidate the role of AQP4 in modulation of these processes underlying epileptogenesis.

Genetic Studies of AQP4 and Epilepsy

Based on the idea that AQP4 and $K_{ir}4.1$ act in concert to regulate water and K^+ homeostasis in the brain, the hypothesis arose that variants of the *AQP4* and *KCNJ10* (*$K_{ir}4.1$*) genes may be associated with forms of epilepsy. In a study of 218 Norwegian patients with TLE and 181 controls,

Heuser et al. found several single nucleotide polymorphisms (SNPs) in the *KCNJ10* and *AQP4* genes associated with TLE.[163] In a mouse model with glial-specific deletion of *KCNJ10*, delayed stimulation-associated K^+ clearance in CA1 stratum radiatum was observed.[99]

Restoration of Water and Ion Homeostasis

The above considerations lead to a new potential therapeutic opportunity: restoration of water and K^+ homeostasis by regulation of AQP4 targeting and distribution. Whether such a therapeutic approach in epilepsy may be reasonable is an open question. Might AQP4 upregulation be therapeutically feasible? So far there is no known pharmacological method to upregulate AQP4. However, increasing understanding of the molecular partners involved (such as the DAPC) could lead to such approaches. For example, the extracellular matrix protein agrin, a heparin sulfate proteoglycan, may be a critical regulator of AQP4 membrane trafficking and polarity.[164] Astrocytes cultured with the neuronal agrin isoform A4B8 but not with the endothelial and meningeal isoform A0B0 demonstrated increases in the M23 splice variant of AQP4, increased water transport capacity, and increased membrane density of OAPs.[164] The glucocorticoid triamcinolone acetonide (TA) was shown to induce AQP4 downregulation in the normal retina but increased expression in the inflamed retina.[165] Targeting the DAPC for increased expression has the potential to be a novel therapeutic target, but a way to modulate its expression in the brain is still unknown. Current studies focus on the role of dystrophin in Duchenne muscular dystrophy and a few have found ways to upregulate the anchoring protein,[166,167] but the expression changes in brain tissue and effects on AQP4 expression have not yet been examined.

CONCLUSIONS

Compelling evidence indicates that the glial water channel AQP4 plays a fundamental role in water transport in the brain. AQP4 is expressed in astrocytes, and along with the inwardly rectifying K^+ channel $K_{ir}4.1$ is thought to be responsible for water and K^+ homeostasis during neural activity. Because osmolarity and K^+ have powerful effects on seizure susceptibility, AQP4 and its molecular partners may represent novel therapeutic targets for control of seizures. $AQP4^{-/-}$ mice have significantly prolonged seizure duration, associated with alterations in extracellular K^+ clearance and gap junctional coupling. Dysfunctional K^+ homeostasis and upregulation and altered subcellular distribution of AQP4 have been observed in human epileptic tissue. The relevance of these findings to hippocampal epileptogenesis and also to human epilepsy requires further study. Restoration of water and K^+ homeostasis in epileptic tissue constitutes a novel therapeutic concept. AQP4 occupies a pivotal position: concentrated in both astrocytic endfeet and astroglial processes investing synapses, AQP4 may be involved in glio-neuronal interaction of water and ion fluxes and also in gliovascular interactions across the blood–brain barrier and thus play a critical role both during active tissue metabolism and in pathophysiologic states.

References

1. Ransom B, Behar T, Nedergaard M. New roles for astrocytes (stars at last). *Trends Neurosci* 2003;**26**(10):520–2.
2. Binder DK, Steinhäuser C. Functional changes in astroglial cells in epilepsy. *Glia* 2006;**54**(5):358–68.
3. Wetherington J, Serrano G, Dingledine R. Astrocytes in the epileptic brain. *Neuron* 2008;**58**(2):168–78.
4. Amiry-Moghaddam M, Williamson A, Palomba M, et al. Delayed K^+ clearance associated with aquaporin-4 mislocalization: phenotypic defects in brains of alpha-syntrophin-null mice. *Proc Natl Acad Sci USA* 2003;**100**(23):13615–20.
5. Volterra A, Steinhäuser C. Glial modulation of synaptic transmission in the hippocampus. *Glia* 2004;**47**(3):249–57.
6. Volterra A, Meldolesi J. Astrocytes, from brain glue to communication elements: the revolution continues. *Nat Rev Neurosci* 2005;**6**(8):626–40.
7. Halassa MM, Haydon PG. Integrated brain circuits: astrocytic networks modulate neuronal activity and behavior. *Annu Rev Physiol* 2010;**72**:335–55.
8. Halassa MM, Fellin T, Haydon PG. The tripartite synapse: roles for gliotransmission in health and disease. *Trends Mol Med* 2007;**13**(2):54–63.
9. Tian GF, Azmi H, Takano T, et al. An astrocytic basis of epilepsy. *Nat Med* 2005;**11**:973–81.
10. Steinhäuser C, Seifert G. Glial membrane channels and receptors in epilepsy: impact for generation and spread of seizure activity. *Eur J Pharmacol* 2002;**447**(2–3):227–37.
11. Kivi A, Lehmann TN, Kovacs R, et al. Effects of barium on stimulus-induced rises of $[K^+]_o$ in human epileptic non-sclerotic and sclerotic hippocampal area CA1. *Eur J Neurosci* 2000;**12**(6):2039–48.
12. Heinemann U, Gabriel S, Jauch R, et al. Alterations of glial cell function in temporal lobe epilepsy. *Epilepsia* 2000;**41**(Suppl 6):S185–189.
13. Hinterkeuser S, Schroder W, Hager G, et al. Astrocytes in the hippocampus of patients with temporal lobe epilepsy display changes in potassium conductances. *Eur J Neurosci* 2000;**12**(6):2087–96.
14. de Lanerolle NC, Lee TS. New facets of the neuropathology and molecular profile of human temporal lobe epilepsy. *Epilepsy Behav* 2005;**7**(2):190–203.
15. Seifert G, Schilling K, Steinhäuser C. Astrocyte dysfunction in neurological disorders: a molecular perspective. *Nat Rev Neurosci* 2006;**7**(3):194–206.
16. Fiacco TA, Agulhon C, Taves SR, et al. Selective stimulation of astrocyte calcium in situ does not affect neuronal excitatory synaptic activity. *Neuron* 2007;**54**(4):611–26.
17. Seifert G, Steinhäuser C. Neuron–astrocyte signaling and epilepsy. *Exp Neurol* 2011;**244**:4–10. September 7 [Epub ahead of print].
18. Schwartzkroin PA, Baraban SC, Hochman DW. Osmolarity, ionic flux, and changes in brain excitability. *Epilepsy Res* 1998;**32**:275–85.
19. Roper SN, Obenaus A, Dudek FE. Osmolality and nonsynaptic epileptiform bursts in rat CA1 and dentate gyrus. *Ann Neurol* 1992;**31**:81–5.

20. Dudek FE, Obenhaus A, Tasker JG. Osmolality-induced changes in extracellular volume alter epileptiform bursts independent of chemical synapses in the rat: importance of non-synaptic mechanisms in hippocampal epileptogenesis. *Neurosci Lett* 1990;**120**:267–70.
21. Chebabo SR, Hester MA, Aitken PG, Somjen GG. Hypotonic exposure enhances synaptic transmission and triggers spreading depression in rat hippocampal tissue slices. *Brain Res* 1995;**695**:203–16.
22. Pan E, Stringer JL. Influence of osmolality on seizure amplitude and propagation in the rat dentate gyrus. *Neurosci Lett* 1996;**207**(1):9–12.
23. Traynelis SF, Dingledine R. Role of extracellular space in hyperosmotic suppression of potassium-induced electrographic seizures. *J Neurophysiol* 1989;**61**(5):927–38.
24. Haglund MM, Hochman DW. Furosemide and mannitol suppression of epileptic activity in the human brain. *J Neurophysiol* 2005;**94**(2):907–18.
25. Andrew RD, Fagan M, Ballyk BA, Rosen AS. Seizure susceptibility and the osmotic state. *Brain Res* 1989;**498**:175–80.
26. Carter CH. Status epilepticus treated by intravenous urea. *Epilepsia* 1962;**3**:198–200.
27. Maa EH, Kahle KT, Walcott BP, Spitz MC, Staley KJ. Diuretics and epilepsy: will the past and present meet? *Epilepsia* 2011;**52**(9):1559–69.
28. Rutecki PA, Lebeda FJ, Johnston D. Epileptiform activity induced by changes in extracellular potassium in hippocampus. *J Neurophysiol* 1985;**54**(5):1363–74.
29. Yaari Y, Konnerth A, Heinemann U. Nonsynaptic epileptogenesis in the mammalian hippocampus in vitro. II. Role of extracellular potassium. *J Neurophysiol* 1986;**56**(2):424–38.
30. Traynelis SF, Dingledine R. Potassium-induced spontaneous electrographic seizures in the rat hippocampal slice. *J Neurophysiol* 1988;**59**(1):259–76.
31. Feng Z, Durand DM. Effects of potassium concentration on firing patterns of low-calcium epileptiform activity in anesthetized rat hippocampus: inducing of persistent spike activity. *Epilepsia* 2006;**47**:727–36.
32. Gabriel S, Njunting M, Pomper JK, et al. Stimulus and potassium-induced epileptiform activity in the human dentate gyrus from patients with and without hippocampal sclerosis. *J Neurosci* 2004;**24**(46):10416–30.
33. Verkman AS. More than just water channels: unexpected cellular roles of aquaporins. *J Cell Sci* 2005;**118**(Pt 15):3225–32.
34. Amiry-Moghaddam M, Ottersen OP. The molecular basis of water transport in the brain. *Nat Rev Neurosci* 2003;**4**(12):991–1001.
35. Agre P, King LS, Yasui M, et al. Aquaporin water channels—from atomic structure to clinical medicine. *J Physiol* 2002;**542**(Pt 1):3–16.
36. Verkman AS. Physiological importance of aquaporin water channels. *Ann Med* 2002;**34**(3):192–200.
37. Hasegawa H, Ma T, Skach W, Matthay MA, Verkman AS. Molecular cloning of a mercurial-insensitive water channel expressed in selected water-transporting tissues. *J Biol Chem* 1994;**269**(8):5497–500.
38. Yang B, Ma T, Verkman AS. cDNA cloning, gene organization, and chromosomal localization of a human mercurial insensitive water channel. Evidence for distinct transcriptional units. *J Biol Chem* 1995;**270**(39):22907–13.
39. Jung JS, Bhat RV, Preston GM, Guggino WB, Baraban JM, Agre P. Molecular characterization of an aquaporin cDNA from brain: candidate osmoreceptor and regulator of water balance. *Proc Natl Acad Sci USA* 1994;**91**(26):13052–6.
40. Nielsen S, Nagelhus EA, Amiry-Moghaddam M, Bourque C, Agre P, Ottersen OP. Specialized membrane domains for water transport in glial cells: high-resolution immunogold cytochemistry of aquaporin-4 in rat brain. *J Neurosci* 1997;**17**(1):171–80.

41. Nagelhus EA, Mathiisen TM, Ottersen OP. Aquaporin-4 in the central nervous system: cellular and subcellular distribution and coexpression with $K_{ir}4.1$. *Neuroscience* 2004;**129**(4):905–13.
42. Rash JE, Yasumura T, Hudson CS, Agre P, Nielsen S. Direct immunogold labeling of aquaporin-4 in square arrays of astrocyte and ependymocyte plasma membranes in rat brain and spinal cord. *Proc Natl Acad Sci USA* 1998;**95**(20):11981–6.
43. Oshio K, Binder DK, Yang B, Schecter S, Verkman AS, Manley GT. Expression of aquaporin water channels in mouse spinal cord. *Neuroscience* 2004;**127**(3):685–93.
44. Jarius S, Paul F, Franciotta D, et al. Mechanisms of disease: aquaporin-4 antibodies in neuromyelitis optica. *Nat Clin Pract Neurol* 2008;**4**(4):202–14.
45. Niermann H, Amiry-Moghaddam M, Holthoff K, Witte OW, Ottersen OP. A novel role of vasopressin in the brain: modulation of activity-dependent water flux in the neocortex. *J Neurosci* 2001;**21**(9):3045–51.
46. Holthoff K, Witte OW. Directed spatial potassium redistribution in rat neocortex. *Glia* 2000;**29**(3):288–92.
47. Verbavatz JM, Ma T, Gobin R, Verkman AS. Absence of orthogonal arrays in kidney, brain and muscle from transgenic knockout mice lacking water channel aquaporin-4. *J Cell Sci* 1997;**110**(Pt 22):2855–60.
48. Yang B, Brown D, Verkman AS. The mercurial insensitive water channel (AQP-4) forms orthogonal arrays in stably transfected Chinese hamster ovary cells. *J Biol Chem* 1996;**271**(9):4577–80.
49. Frigeri A, Gropper MA, Umenishi F, Kawashima M, Brown D, Verkman AS. Localization of MIWC and GLIP water channel homologs in neuromuscular, epithelial and glandular tissues. *J Cell Sci* 1995;**108**(Pt 9):2993–3002.
50. Wolburg H, Berg K. Distribution of orthogonal arrays of particles in the Muller cell membrane of the mouse retina. *Glia* 1988;**1**(4):246–52.
51. Wolburg H, Wolburg-Buchholz K, Fallier-Becker P, Noell S, Mack AF. Structure and functions of aquaporin-4-based orthogonal arrays of particles. *Int Rev. Cell Mol Biol* 2011;**287**:1–41.
52. Ma T, Yang B, Gillespie A, Carlson EJ, Epstein CJ, Verkman AS. Generation and phenotype of a transgenic knockout mouse lacking the mercurial-insensitive water channel aquaporin-4. *J Clin Investig* 1997;**100**(5):957–62.
53. Haj-Yasein NN, Vindedal GF, Eilert-Olsen M, et al. Glial-conditional deletion of aquaporin-4 (AQP4) reduces blood–brain water uptake and confers barrier function on perivascular astrocyte endfeet. *Proc Natl Acad Sci USA* 2011;**108**:17815–20.
54. Saadoun S, Tait MJ, Reza A, et al. AQP4 gene deletion in mice does not alter blood–brain barrier integrity or brain morphology. *Neuroscience* 2009;**161**(3):764–72.
55. Binder DK, Oshio K, Ma T, Verkman AS, Manley GT. Increased seizure threshold in mice lacking aquaporin-4 water channels. *Neuroreport* 2004;**15**(2):259–62.
56. Papadopoulos MC, Manley GT, Krishna S, Verkman AS. Aquaporin-4 facilitates reabsorption of excess fluid in vasogenic brain edema. *FASEB J* 2004;**18**(11):1291–3.
57. Li J, Patil RV, Verkman AS. Mildly abnormal retinal function in transgenic mice without Muller cell aquaporin-4 water channels. *Invest Ophthalmol Vis Sci* 2002;**43**(2):573–9.
58. Li J, Verkman AS. Impaired hearing in mice lacking aquaporin-4 water channels. *J Biol Chem* 2001;**276**(33):31233–7.
59. Solenov EI, Vetrivel L, Oshio K, Manley GT, Verkman AS. Optical measurement of swelling and water transport in spinal cord slices from aquaporin null mice. *J Neurosci Methods* 2002;**113**(1):85–90.
60. Thiagarajah JR, Papadopoulos MC, Verkman AS. Noninvasive early detection of brain edema in mice by near-infrared light scattering. *J Neurosci Res* 2005;**80**(2):293–9.
61. Solenov E, Watanabe H, Manley GT, Verkman AS. Sevenfold-reduced osmotic water permeability in primary astrocyte cultures from AQP-4-deficient mice, measured by a fluorescence quenching method. *Am J Physiol Cell Physiol* 2004;**286**(2):C426–432.

62. Nicchia GP, Frigeri A, Liuzzi GM, Svelto M. Inhibition of aquaporin-4 expression in astrocytes by RNAi determines alteration in cell morphology, growth, and water transport and induces changes in ischemia-related genes. *FASEB J* 2003;**17**(11):1508–10.
63. Manley GT, Fujimura M, Ma T, et al. Aquaporin-4 deletion in mice reduces brain edema after acute water intoxication and ischemic stroke. *Nat Med* 2000;**6**(2):159–63.
64. Lee DJ, Amini M, Hamamura MJ, et al. Aquaporin-4-dependent edema clearance following status epilepticus. *Epilepsy Res* 2011;**98**:264–8. October 11 [Epub ahead of print].
65. Frigeri A, Nicchia GP, Nico B, et al. Aquaporin-4 deficiency in skeletal muscle and brain of dystrophic mdx mice. *FASEB J* 2001;**15**(1):90–8.
66. Vajda Z, Pedersen M, Fuchtbauer EM, et al. Delayed onset of brain edema and mislocalization of aquaporin-4 in dystrophin-null transgenic mice. *Proc Natl Acad Sci USA* 2002;**99**(20):13131–6.
67. Neely JD, Amiry-Moghaddam M, Ottersen OP, Froehner SC, Agre P, Adams ME. Syntrophin-dependent expression and localization of aquaporin-4 water channel protein. *Proc Natl Acad Sci USA* 2001;**98**(24):14108–13.
68. Amiry-Moghaddam M, Otsuka T, Hurn PD, et al. An alpha-syntrophin-dependent pool of AQP4 in astroglial end-feet confers bidirectional water flow between blood and brain. *Proc Natl Acad Sci USA* 2003;**100**(4):2106–11.
69. Sykova E. The extracellular space in the CNS: its regulation, volume and geometry in normal and pathological neuronal function. *Neuroscientist* 1997;**3**:28–41.
70. Fields RD, Stevens-Graham B. New insights into neuron-glia communication. *Science* 2002;**298**(5593):556–62.
71. Østby I, Øyehaug L, Einevoll GT, et al. Astrocytic mechanisms explaining neural-activity-induced shrinkage of extraneuronal space. *PLoS Comput Biol* 2009;**5**(1):e1000272.
72. Risher WC, Andrew RD, Kirov SA. Real-time passive volume responses of astrocytes to acute osmotic and ischemic stress in cortical slices and in vivo revealed by two-photon microscopy. *Glia* 2009;**57**(2):207–21.
73. Andrew RD, Labron MW, Boehnke SE, Carnduff L, Kirov SA. Physiological evidence that pyramidal neurons lack functional water channels. *Cereb Cortex* 2007;**17**(4):787–802.
74. Binder DK, Papadopoulos MC, Haggie PM, Verkman AS. In vivo measurement of brain extracellular space diffusion by cortical surface photobleaching. *J Neurosci* 2004;**24**(37):8049–56.
75. Yao X, Hrabetova S, Nicholson C, Manley GT. Aquaporin-4-deficient mice have increased extracellular space without tortuosity change. *J Neurosci* 2008;**28**(21):5460–4.
76. Binder DK, Yao X, Sick TJ, Verkman AS, Manley GT. Increased seizure duration and slowed potassium kinetics in mice lacking aquaporin-4 water channels. *Glia* 2006;**53**:631–6.
77. Nagelhus EA, Horio Y, Inanobe A, et al. Immunogold evidence suggests that coupling of K^+ siphoning and water transport in rat retinal Muller cells is mediated by a coenrichment of Kir4.1 and AQP4 in specific membrane domains. *Glia* 1999;**26**(1):47–54.
78. Simard M, Nedergaard M. The neurobiology of glia in the context of water and ion homeostasis. *Neuroscience* 2004;**129**(4):877–96.
79. Manley GT, Binder DK, Papadopoulos MC, Verkman AS. New insights into water transport and edema in the central nervous system from phenotype analysis of aquaporin-4 null mice. *Neuroscience* 2004;**129**(4):983–91.
80. Heinemann U, Lux HD. Ceiling of stimulus induced rises in extracellular potassium concentration in the cerebral cortex of cat. *Brain Res* 1977;**120**(2):231–49.
81. Somjen GG. Ion regulation in the brain: implications for pathophysiology. *Neuroscientist* 2002;**8**:254–67.
82. Xiong ZQ, Stringer JL. Astrocytic regulation of the recovery of extracellular potassium after seizures in vivo. *Eur J Neurosci* 1999;**11**(5):1677–84.
83. Walz W. Swelling and potassium uptake in cultured astrocytes. *Can J Physiol Pharmacol* 1987;**65**(5):1051–7.

84. Walz W. Mechanism of rapid K+-induced swelling of mouse astrocytes. *Neurosci Lett* 1992;**135**(2):243–6.
85. Connors NC, Adams ME, Froehner SC, Kofuji P. The potassium channel Kir4.1 associates with the dystrophin-glycoprotein complex via alpha-syntrophin in glia. *J Biol Chem* 2004;**279**(27):28387–92.
86. Kofuji P, Ceelen P, Zahs KR, Surbeck LW, Lester HA, Newman EA. Genetic inactivation of an inwardly rectifying potassium channel (Kir4.1 subunit) in mice: phenotypic impact in retina. *J Neurosci* 2000;**20**(15):5733–40.
87. Rozengurt N, Lopez I, Chiu CS, Kofuji P, Lester HA, Neusch C. Time course of inner ear degeneration and deafness in mice lacking the Kir4.1 potassium channel subunit. *Hear Res* 2003;**177**(1–2):71–80.
88. Marcus DC, Wu T, Wangemann P, Kofuji P. KCNJ10 (Kir4.1) potassium channel knockout abolishes endocochlear potential. *Am J Physiol Cell Physiol* 2002;**282**(2): C403–407.
89. Neusch C, Rozengurt N, Jacobs RE, Lester HA, Kofuji P. Kir4.1 potassium channel subunit is crucial for oligodendrocyte development and in vivo myelination. *J Neurosci* 2001;**21**(15):5429–38.
90. Newman EA. High potassium conductance in astrocyte endfeet. *Science* 1986;**233**(4762):453–4.
91. Newman EA. Inward-rectifying potassium channels in retinal glial (Muller) cells. *J Neurosci* 1993;**13**(8):3333–45.
92. Newman EA, Frambach DA, Odette LL. Control of extracellular potassium levels by retinal glial cell K^+ siphoning. *Science* 1984;**225**(4667):1174–5.
93. Newman EA, Karwoski CJ. Spatial buffering of light-evoked potassium increases by retinal glial (Muller) cells. *Acta Physiol Scand Suppl* 1989;**582**:51.
94. Ballanyi K, Grafe P, ten Bruggencate G. Ion activities and potassium uptake mechanisms of glial cells in guinea-pig olfactory cortex slices. *J Physiol* 1987;**382**:159–74.
95. Kofuji P, Newman EA. Potassium buffering in the central nervous system. *Neuroscience* 2004;**129**(4):1045–56.
96. Neusch C, Papadopoulos N, Muller M, et al. Lack of the Kir4.1 channel subunit abolishes K^+ buffering properties of astrocytes in the ventral respiratory group: impact on extracellular K^+ regulation. *J Neurophysiol* 2006;**95**(3):1843–52.
97. Djukic B, Casper KB, Philpot BD, Chin LS, McCarthy KD. Conditional knock-out of Kir4.1 leads to glial membrane depolarization, inhibition of potassium and glutamate uptake, and enhanced short-term synaptic potentiation. *J Neurosci* 2007;**27**(42): 11354–65.
98. Seifert G, Huttmann K, Binder DK, et al. Analysis of astroglial K^+ channel expression in the developing hippocampus reveals a predominant role of the Kir4.1 subunit. *J Neurosci* 2009;**29**(23):7474–88.
99. Haj-Yasein NN, Jensen V, Vindedal GF, et al. Evidence that compromised K^+ spatial buffering contributes to the epileptogenic effect of mutations in the human Kir4.1 gene (KCNJ10). *Glia* 2011;**59**(11):1635–42.
100. Padmawar P, Yao X, Bloch O, Manley GT, Verkman AS. K^+ waves in brain cortex visualized using a long-wavelength K^+-sensing fluorescent indicator. *Nat Methods* 2005;**2**(11):825–7.
101. Hsu MS, Seldin M, Lee DJ, Seifert G, Steinhäuser C, Binder DK. Laminar-specific and developmental expression of aquaporin-4 in the mouse hippocampus. *Neuroscience* 2011;**178**:21–32.
102. Ruiz-Ederra J, Zhang H, Verkman AS. Evidence against functional interaction between aquaporin-4 water channels and Kir4.1 potassium channels in retinal Muller cells. *J Biol Chem* 2007;**282**(30):21866–72.
103. Zhang H, Verkman AS. Aquaporin-4 independent Kir4.1 K^+ channel function in brain glial cells. *Mol Cell Neurosci* 2008;**37**(1):1–10.

104. Jin BJ, Zhang H, Binder DK, Verkman AS. Aquaporin-4-dependent K(+) and water transport modeled in brain extracellular space following neuroexcitation. *J Gen Physiol* 2013;**141**(1):119–32.
105. Strohschein S, Huttmann K, Gabriel S, Binder DK, Heinemann U, Steinhäuser C. Impact of aquaporin-4 channels on K^+ buffering and gap junction coupling in the hippocampus. *Glia* 2011;**59**(6):973–80.
106. Wallraff A, Kohling R, Heinemann U, Theis M, Willecke K, Steinhäuser C. The impact of astrocytic gap junctional coupling on potassium buffering in the hippocampus. *J Neurosci* 2006;**26**(20):5438–47.
107. Thrane AS, Rappold PM, Fujita T, et al. Critical role of aquaporin-4 (AQP4) in astrocytic Ca^{2+} signaling events elicited by cerebral edema. *Proc Natl Acad Sci USA* 2011;**108**(2):846–51.
108. Haj-Yasein NN, Bugge CE, Jensen V, et al. Deletion of aquaporin-4 increases extracellular K(+) concentration during synaptic stimulation in mouse hippocampus. *Brain Struct Funct* 2015;**220**(4):2469–74.
109. Bragg AD, Amiry-Moghaddam M, Ottersen OP, Adams ME, Froehner SC. Assembly of a perivascular astrocyte protein scaffold at the mammalian blood–brain barrier is dependent on alpha-syntrophin. *Glia* 2006;**53**(8):879–90.
110. Gondo A, Shinotsuka T, Morita A, Abe Y, Yasui M, Nuriya M. Sustained down-regulation of beta-dystroglycan and associated dysfunctions of astrocytic endfeet in epileptic cerebral cortex. *J Biol Chem* 2014;**289**(44):30279–88.
111. Anderova M, Benesova J, Mikesova M, et al. Altered astrocytic swelling in the cortex of alpha-syntrophin-negative GFAP/EGFP mice. *PLoS One* 2014;**9**(11):e113444.
112. De Sarro G, Ibbadu GF, Marra R, et al. Seizure susceptibility to various convulsant stimuli in dystrophin-deficient mdx mice. *Neurosci Res* 2004;**50**(1):37–44.
113. Tsao CY, Mendell JR. Coexisting muscular dystrophies and epilepsy in children. *J Child Neurol* 2006;**21**(2):148–50.
114. Blümcke I, Beck H, Lie AA, Wiestler OD. Molecular neuropathology of human mesial temporal lobe epilepsy. *Epilepsy Res* 1999;**36**(2–3):205–23.
115. Mitchell LA, Jackson GD, Kalnins RM, et al. Anterior temporal abnormality in temporal lobe epilepsy: a quantitative MRI and histopathologic study. *Neurology* 1999;**52**(2):327–36.
116. Hugg JW, Butterworth EJ, Kuzniecky RI. Diffusion mapping applied to mesial temporal lobe epilepsy: preliminary observations. *Neurology* 1999;**53**(1):173–6.
117. Lee TS, Eid T, Mane S, et al. Aquaporin-4 is increased in the sclerotic hippocampus in human temporal lobe epilepsy. *Acta Neuropathol (Berl)* 2004;**108**(6):493–502.
118. Eid T, Lee TS, Thomas MJ, et al. Loss of perivascular aquaporin-4 may underlie deficient water and K^+ homeostasis in the human epileptogenic hippocampus. *Proc Natl Acad Sci USA* 2005;**102**(4):1193–8.
119. Medici V, Frassoni C, Tassi L, Spreafico R, Garbelli R. Aquaporin 4 expression in control and epileptic human cerebral cortex. *Brain Res* 2011;**1367**:330–9.
120. Dudek FE, Rogawski MA. Regulation of brain water: is there a role for aquaporins in epilepsy? *Epilepsy Curr* 2005;**5**(3):104–6.
121. Badaut J, Lasbennes F, Magistretti PJ, Regli L. Aquaporins in brain: distribution, physiology, and pathophysiology. *J Cereb Blood Flow Metab* 2002;**22**(4):367–78.
122. Badaut J, Verbavatz JM, Freund-Mercier MJ, Lasbennes F. Presence of aquaporin-4 and muscarinic receptors in astrocytes and ependymal cells in rat brain: a clue to a common function? *Neurosci Lett* 2000;**292**(2):75–8.
123. Jabs R, Pivneva T, Huttmann K, et al. Synaptic transmission onto hippocampal glial cells with hGFAP promoter activity. *J Cell Sci* 2005;**118**(Pt 16):3791–803.
124. Matthias K, Kirchhoff F, Seifert G, et al. Segregated expression of AMPA-type glutamate receptors and glutamate transporters defines distinct astrocyte populations in the mouse hippocampus. *J Neurosci* 2003;**23**(5):1750–8.

125. Wallraff A, Odermatt B, Willecke K, Steinhäuser C. Distinct types of astroglial cells in the hippocampus differ in gap junction coupling. *Glia* 2004;**48**(1):36–43.
126. Nishiyama A, Yang Z, Butt A. Astrocytes and NG2-glia: what's in a name? *J Anat* 2005;**207**(6):687–93.
127. Tomas-Camardiel M, Venero JL, de Pablos RM, Rite I, Machado A, Cano J. In vivo expression of aquaporin-4 by reactive microglia. *J Neurochem* 2004;**91**(4):891–9.
128. Eilert-Olsen M, Haj-Yasein NN, Vindedal GF, et al. Deletion of aquaporin-4 changes the perivascular glial protein scaffold without disrupting the brain endothelial barrier. *Glia* 2012;**60**(3):432–40.
129. Vezzani A, Granata T. Brain inflammation in epilepsy: experimental and clinical evidence. *Epilepsia* 2005;**46**(11):1724–43.
130. Badaut J, Brunet JF, Grollimund L, et al. Aquaporin-1 and aquaporin-4 expression in human brain after subarachnoid hemorrhage and in peritumoral tissue. *Acta Neurochir Suppl* 2003;**86**:495–8.
131. Saadoun S, Papadopoulos MC, Davies DC, Krishna S, Bell BA. Aquaporin-4 expression is increased in oedematous human brain tumours. *J Neurol Neurosurg Psychiatry* 2002;**72**(2):262–5.
132. Vizuete ML, Venero JL, Vargas C, et al. Differential upregulation of aquaporin-4 mRNA expression in reactive astrocytes after brain injury: potential role in brain edema. *Neurobiol Dis* 1999;**6**(4):245–58.
133. Saito N, Ikegami H, Shimada K. Effect of water deprivation on aquaporin-4 (AQP4) mRNA expression in chickens (*Gallus domesticus*). *Brain Res Mol Brain Res* 2005;**141**(2):193–7.
134. Kim JE, Ryu HJ, Yeo SI, et al. Differential expressions of aquaporin subtypes in astroglia in the hippocampus of chronic epileptic rats. *Neuroscience* 2009;**163**(3):781–9.
135. Kim JE, Yeo SI, Ryu HJ, et al. Astroglial loss and edema formation in the rat piriform cortex and hippocampus following pilocarpine-induced status epilepticus. *J Comp Neurol* 2010;**518**(22):4612–28.
136. Lee DJ, Hsu MS, Seldin MM, Arellano JL, Binder DK. Decreased expression of the glial water channel aquaporin-4 in the intrahippocampal kainic acid model of epileptogenesis. *Exp Neurol* 2012;**235**(1):246–55.
137. Alvestad S, Hammer J, Hoddevik EH, et al. Mislocalization of AQP4 precedes chronic seizures in the kainate model of temporal lobe epilepsy. *Epilepsy Res* 2013;**105**(1–2):30–41.
138. Bender RA, Dubé C, Baram TZ. Febrile seizures and mechanisms of epileptogenesis: insights from an animal model. *Adv Exp Med Biol* 2004;**548**:213–25.
139. Dubé CM, Brewster AL, Richichi C, Zha Q, Baram TZ. Fever, febrile seizures and epilepsy. *Trends Neurosci* 2007;**30**(10):490–6.
140. Nagelhus EA, Veruki ML, Torp R, et al. Aquaporin-4 water channel protein in the rat retina and optic nerve: polarized expression in Muller cells and fibrous astrocytes. *J Neurosci* 1998;**18**(7):2506–19.
141. Yang J, Lunde LK, Nuntagij P, et al. Loss of astrocyte polarization in the Tg-ArcSwe mouse model of Alzheimer's disease. *J Alzheimers Dis* 2011;**27**:711–22.
142. Rajneesh KF, Binder DK. Tumor-associated epilepsy. *Neurosurg Focus* 2009;**27**(2):E4.
143. Abbott NJ. Astrocyte–endothelial interactions and blood–brain barrier permeability. *J Anat* 2002;**200**(6):629–38.
144. Metea MR, Newman EA. Glial cells dilate and constrict blood vessels: a mechanism of neurovascular coupling. *J Neurosci* 2006;**26**(11):2862–70.
145. Mulligan SJ, MacVicar BA. Calcium transients in astrocyte endfeet cause cerebrovascular constrictions. *Nature* 2004;**431**(7005):195–9.
146. Takano T, Tian GF, Peng W, et al. Astrocyte-mediated control of cerebral blood flow. *Nat Neurosci* 2006;**9**(2):260–7.
147. Zonta M, Angulo MC, Gobbo S, et al. Neuron-to-astrocyte signaling is central to the dynamic control of brain microcirculation. *Nat Neurosci* 2003;**6**(1):43–50.

148. Gordon GR, Mulligan SJ, MacVicar BA. Astrocyte control of the cerebrovasculature. *Glia* 2007;**55**(12):1214–21.
149. Seiffert E, Dreier JP, Ivens S, et al. Lasting blood–brain barrier disruption induces epileptic focus in the rat somatosensory cortex. *J Neurosci* 2004;**24**(36):7829–36.
150. Ivens S, Kaufer D, Flores LP, et al. TGF-beta receptor-mediated albumin uptake into astrocytes is involved in neocortical epileptogenesis. *Brain* 2007;**130**(Pt 2):535–47.
151. Croll SD, Goodman JH, Scharfman HE. Vascular endothelial growth factor (VEGF) in seizures: a double-edged sword. *Adv Exp Med Biol* 2004;**548**:57–68.
152. Nicoletti JN, Shah SK, McCloskey DP, et al. Vascular endothelial growth factor is upregulated after status epilepticus and protects against seizure-induced neuronal loss in hippocampus. *Neuroscience* 2008;**151**(1):232–41.
153. Rigau V, Morin M, Rousset MC, et al. Angiogenesis is associated with blood–brain barrier permeability in temporal lobe epilepsy. *Brain* 2007;**130**(Pt 7):1942–56.
154. Rite I, Machado A, Cano J, Venero JL. Intracerebral VEGF injection highly upregulates AQP4 mRNA and protein in the perivascular space and glia limitans externa. *Neurochem Int* 2008;**52**:897–903.
155. Skucas VA, Mathews IB, Yang J, et al. Impairment of select forms of spatial memory and neurotrophin-dependent synaptic plasticity by deletion of glial aquaporin-4. *J Neurosci* 2011;**31**(17):6392–7.
156. Amlerova J, Laczo J, Vlcek K, Javurkova A, Andel R, Marusic P. Risk factors for spatial memory impairment in patients with temporal lobe epilepsy. *Epilepsy Behav* 2013;**26**(1):57–60.
157. Chin J, Scharfman HE. Shared cognitive and behavioral impairments in epilepsy and Alzheimer's disease and potential underlying mechanisms. *Epilepsy Behav* 2013;**26**:343–51.
158. Brooks-Kayal A, Bath KG, Berg A, et al. Issues related to symptomatic and disease modifying treatments affecting cognitive and neuropsychiatric comorbidities of epilepsy. *Epilepsia* 2013;**54**(Suppl 4):44–60.
159. Bell B, Lin JJ, Seidenberg M, Hermann B. The neurobiology of cognitive disorders in temporal lobe epilepsy. *Nat Rev Neurol* 2011;**7**(3):154–64.
160. Morimoto K, Fahnestock M, Racine RJ. Kindling and status epilepticus models of epilepsy: rewiring the brain. *Prog Neurobiol* 2004;**73**(1):1–60.
161. Parent JM. The role of seizure-induced neurogenesis in epileptogenesis and brain repair. *Epilepsy Res* 2002;**50**(1–2):179–89.
162. Scharfman HE. Epilepsy as an example of neural plasticity. *Neuroscientist* 2002;**8**(2):154–73.
163. Heuser K, Nagelhus EA, Tauboll E, et al. Variants of the genes encoding AQP4 and Kir4.1 are associated with subgroups of patients with temporal lobe epilepsy. *Epilepsy Res* 2010;**88**(1):55–64.
164. Noell S, Fallier-Becker P, Beyer C, Kroger S, Mack AF, Wolburg H. Effects of agrin on the expression and distribution of the water channel protein aquaporin-4 and volume regulation in cultured astrocytes. *Eur J Neurosci* 2007;**26**(8):2109–18.
165. Zhao M, Bousquet E, Valamaneh F, et al. Differential regulations of AQP4 and Kir4.1 by triamcinolone acetonide and dexamethasone in the healthy and inflamed retina. *Investig Ophthalmol Vis Sci* 2011;**52**(9):6340–7.
166. Kayali R, Ku J, Khitrov G, Jung ME, Prikhodko O, Bertoni C. Read-through compound 13 restores dystrophin expression and improves muscle function in the mdx mouse model for Duchenne muscular dystrophy. *Hum Mol Genet* 2012;**21**(18):4007–20.
167. Malerba A, Boldrin L, Dickson G. Long-term systemic administration of unconjugated morpholino oligomers for therapeutic expression of dystrophin by exon skipping in skeletal muscle: implications for cardiac muscle integrity. *Nucleic Acid Ther* 2011;**21**(4):293–8.

CHAPTER

18

Water Homeostasis Dysfunctions and Edema in Neuroinflammatory Disease

Klaus G. Petry and Andreas Bikfalvi
Bordeaux University INSERM U1029, Pessac, France

OUTLINE

INTRODUCTION

Neuroinflammation occurs in many pathological affections of the central nervous system (CNS). Similar to other affected organs, tissue inflammation is the primary pathophysiological event in the chronic CNS disorders of multiple sclerosis (MS), neuromyelitis optica (NMO, and NMO spectrum disorder, Devic's syndrome), acute disseminated encephalomyelitis (ADEM), or acute infectious disorders (encephalitis). The CNS tissue, however, presents considerable specificities as it is protected by barriers and harbors competent cells that are reactive to inflammatory immune stimuli. We concentrate

Brain Edema.
DOI: http://dx.doi.org/10.1016/B978-0-12-803196-4.00018-7

on inflammatory neuropathological processes causing alteration of CNS water homeostasis and edema, which goes beyond the activity of immune-derived soluble and cellular compounds for CNS surveillance and supporting neuroplasticity and repair.[1] The complexity of the neuroinflammation processes includes major characteristics such as the molecular and cellular alterations of the blood–brain barrier (BBB) and the activation of microglia and astrocytes. Secondary neuroinflammatory reactions are observed upon various CNS insults such as brain ischemia, brain trauma, or eventually brain tumors (discussed in other chapters). A frequent complication to CNS insults is the development of brain edema by increased diffusion of plasma compounds, including water, from the blood circulation.

Edema formation in the brain is defined as the pathological increase of the brain water content that occurs in many brain disorders. Historically, two major types of brain edema, cellular (or cytotoxic) and vascular, have been distinguished depending upon the integrity of the BBB.[2] Despite intact BBB function, cellular edema results from a disturbance in ionic homeostasis and abnormal water uptake by injured cells, mainly by astrocytes causing their swelling and the decrease of the extracellular space. The very high heterogeneity of astrocytes, which depends on their origin and adaptation to their microenvironment, assures the multiple supportive and modulating functions for neuronal activity of the CNS (for review see Ref. 3). Although astrocytes are not electrically excitable, they are integrated into neural networks through tripartite synapses. Astrocytes regulate concentrations of protons, potassium, calcium, and magnesium in the extracellular space and participate in the modulation of the synaptic activity. They supply neurons with glutamate and glutamine (GABA precursor)[4] and contribute by glucose- and glycogen-converted lactate to support different phases of neuronal activation (according to the astrocyte-to-neuron lactate shuttle hypothesis) (for review see Refs. 5 and 6). The functional contact of astrocyte end-foot processes expressing aquaporin-4 (AQP4) water channels (see Chapter 11: Role of Matrix Metalloproteinases in Brain Edema and below) attributes to astrocytes an exposed role within the intact BBB functions including water homeostasis, blood flow regulation, and metabolic trafficking. In the context of astrocyte functions, there is further recent evidence that subtle changes in the composition of brain interstitial ions control the state-dependent patterns of neural activity as recently shown for the sleep–wake cycle.[7]

In any neuropathological insult, astrocytes may modulate their phenotype displaying distinct reactive functions. As documented for myelin degradation the reactive astrocytes are associated to molecular expression patterns that might have dual functions by expressing repair supporting[8] and inhibiting activity.[9] In cytotoxic edema, many regulatory functions of astrocytes are potentially compromised. In the context of neuroinflammation, the reactive functional expression phenotypes of astrocytes have potential impact on maintaining CNS homeostasis and neuronal activity (for review see Ref. 10).

The progressively altered BBB is reflected by an increased permeability which allows the influx of plasma proteins, ions, and water into the extracellular space within the brain parenchyma. In many brain disorders and injuries, BBB dysfunction results in the CNS as pathophysiological processes of ionic edema, vasogenic edema, and hemorrhage associated with intracerebral tissue swelling (see Chapter 12: Edema and BBB Breakdown in Stroke, and for review see Refs. 11 and 12). Although various processes and types of edema might occur simultaneously, vasogenic edema is mainly associated with neuroinflammation.[2,13]

BLOOD–BRAIN BARRIER FUNCTION IN HEALTH AND CNS INFLAMMATION

To assure the optimal function of neurons in the CNS, the BBB is an important element in controlling the molecular fluxes between blood and brain tissue by limiting the entry of blood derived circulating potentially neurotoxic compounds and pathogens into the CNS. It contributes to the control of the ionic composition and the homeostasis of fluids within the extracellular microenvironment of the CNS (for details see Chapter 3: Noninvasive Imaging Techniques for Brain Edema: From Basic Science to the Clinic). The BBB, which is lining the CNS microvessels, the capillaries, and post-capillary venules, is formed by the endothelial cells and different neighboring cell types within the "neurovascular unit" (NVU).[14] While in close contact with the astrocyte endfeet the endothelium provides the cellular link to neurons and perivascular microglial cells, the endothelial–pericyte interaction being embedded in the basal lamina is important for the vascular stability,[15] BBB development,[16] and BBB regulation.[17]

Characteristic for the endothelium is the physiological architecture and functional specialization.[18] By the formation of tight junctions between adjacent cells and very low transcellular vesicular transport activity the endothelium restricts the molecular and cellular passage between the blood circulation and the brain parenchyma. The substance influx and efflux is highly selective as it is controlled by specific transport systems, enzymes, receptors, and specific channels (for reference see Chapter 3: Noninvasive Imaging Techniques for Brain Edema: From Basic Science to the Clinic). The cells forming the BBB, notably endothelial cells, astrocytes, and pericytes, are capable to change their functional healthy status by inducing pathology associated to cellular morphology and molecular expression patterns. These cellular and molecular changes that play important roles in CNS pathogenesis might have different causes. They could primarily originate from genetic mutations of pivotal proteins in individual cell types of the NVU or present disturbed cell signaling between these cells causing their dysfunction. They could also be induced by multifactorial neurodegenerative processes causing secondary peripheral response mechanisms. In such cases, these alterations increase the permeability of the BBB for circulating blood compounds which could be converted into agents with multiple effects such as neurotoxicity, degradation of the neuronal/perivascular extracellular matrix, activation of microglia and astrocytes promoting neuroinflammation, and immune cell mediated CNS tissue degeneration (for review see Ref. 19).

The influx of plasma proteins is followed by water influx into the perivascular space and eventually into the interstitial space of the surrounding brain parenchyma presenting the characteristics for the formation of vasogenic brain edema. Besides the extravasation of circulating immune cells, among the circulating serum proteins released proinflammatory cytokines reinforce the proinflammatory reaction and BBB permeability including for water. The process is considered as the disruption of the BBB which is at the origin of the dynamically developing alteration of the water homeostasis during the neuroinflammatory process.

Disruption of the BBB is a very complex process which is still not completely understood. The extravasation of leukocytes in the CNS parenchyma while interacting with endothelial cells is probably the most extensively examined event of neuroinflammation, in particular in MS pathology, where auto-reactive leukocytes cross the BBB and initiate inflammatory cascades in the CNS that inflict damage on axons and myelin sheaths causing finally the neurological handicap. Studies in experimental rodent models of MS revealed that after an initial phase involving T cells crossing the choroid plexus into the cerebrospinal fluid, and upon stimulation of the subpial vasculature, further T cells enter the subarachnoid space and the parenchyma initiates tissue damage. Finally, disruption of the BBB allows massive infiltration of effector lymphocytes and myeloid cells.[20] The BBB disruption is characterized by an inflammatory expression pattern of endothelial cells allowing the sequential interaction with activated circulating T cells. While rolling along the endothelium and being captured, these cells express activation-dependent chemokines resulting in stable adhesion which is followed by T cell crawling on endothelial cells. Subsequently, T cells pass the endothelial cell layer by transcellular or paracellular diapedesis.[21]

Along with the expression of molecules for interactions of endothelial and immune cells, the induced neuroinflammation reactivity involves multiple cascades of molecular and cellular expressions that interact and create enhancing loops of BBB alterations. The degradation of the basal membrane mainly involving the production of matrix metalloproteinases (MPPs) is a major part in causing a compromised BBB integrity (for details see Chapter 13: Brain Edema Formation in Traumatic Brain Injury). As seen in MS pathology and experimental work, key inflammation mediators, like free radicals including reactive oxygen species,[22,23] and reactive nitrogen species (nitrite and peroxynitrite)[24–26] causing respectively oxidative and nitrosative stress. These free radicals, together with proinflammatory cytokines like TNFα and IL-1β, participate in a number of physiological and pathological processes including through the inflammatory disruption of the BBB.[27] These molecules further interact in the activation of MMPs via the regulation of the expression of proprotein convertases (PCs) which can be counterbalanced by their naturally occurring endogenous tissue inhibitors of metalloproteinases.[28] Among the PCs, furin activates MMP-14 (membrane type1 MMP) resulting in the activation of MMP-2 in particular in astrocyte end-foot. MMP-9 is mainly expressed in endothelial cells and MMP-3 in pericytes. The complexity of the induced molecular cascade of MMPs that leads to the degradation of cell-specific extracellular matrix molecules by the relevant MMPs affects their cellular functions and remodeling. In endothelial cells, MMP-2 and MMP-9 degrade agrin and affect the function of the dynamic tight junction structures in the BBB regulation, by causing a reduced expression of their transmembrane, and membrane-associated cytoplasmic tight junction proteins (see Chapter 13: Brain Edema Formation in Traumatic Brain Injury, and for review see Refs. 29 and 30).

The inflammatory interaction of the generated molecular alterations activates caveolae-mediated endothelial transcytosis. Distinct structural, biochemical, and functional properties of the caveolae convey to endothelial cells a very high plasticity in molecular dynamics as response to insults and inflammatory injury.[18,31] In the mouse ischemic stroke model, the sequential activation of the endothelial cell response involves the upregulation of endothelial transcytosis. Early events occur within 6 hours by increasing endothelial caveolae production and inducing high rate transcytosis at 24 hours, while the major remodeling and disruption of endothelial tight junctions occurs with a delay at day 2 (24–48 hours) after induction.[32] Investigations with the juvenile traumatic brain injury (jTBI) model evidenced a complex expression pattern of caveolin isoforms during acute brain injury and long-term outcome.[33] In neuroinflammation, the increased activation of caveolae and their capacity to fuse facilitates endothelial cell transport and the influx of proinflammatory chemokines and cytokines (for example, IL-6 and TNFα) into the CNS. The upregulated production of proinflammatory chemokines and the increased expression of cell adhesion proteins (for example, VCAM-1 and E-selectine) renders the BBB permissive for the amplified extravasation of activated immune competent leukocytes (effector lymphocytes and myeloid cells) from the blood circulation into the brain. The endothelial cell alterations and increased transport activity that is the early response to damaging insults might be sufficient for inflammation processes allowing the passage of immune cells. In this context, it is notable that in experimental studies of MS, T cells crossed the BBB transcellularly leaving tight junctions intact.[34]

The compromised endothelial cell function and the vasogenic edema development might form an interactive vicious circle amplifying the increase of water volume in the CNS and which therefore can lead to an increase in intracranial hydrostatic pressure, potentially leading to brain ischemia and herniation with an eventual fatal outcome. Under such condition of increased water accumulation in the extracellular spaces the function of astrocytes as regulators of the ion balance might be compromised. Furthermore, molecular gradients for cell signaling and cell structure stability might be disturbed. The astrocyte end-foot as element of the BBB function and microglia cells among the BBB neighboring cell types within the NVU need to be regarded during neuroinflammation. Under most pathological conditions of the CNS involving neuroinflammation both cell types become "reactive" presenting stereotypic morphological changes and multiple functional alterations along with secretion of cytokines and chemokines. While microglia[35] like macrophages[36] present in neuroinflammation a dual function expressing proinflammatory and immunomodulatory phenotypes, astrocyte activation part of reactive gliosis is associated with changes in the expression of many genes.[37,38] Among the neuroinflammation-induced activation of expression patterns both microglia and astrocytes express increased levels of the translocator protein 18kDa (TSPO). In vivo targeting of TSPO with radiolabeled ligands is applied in preclinical[39,40] and clinical studies of neurodegenerative diseases involving inflammatory mechanisms[41–45] to monitor microglial and astroglial reactions. This biomarker can help to determine the neuroinflammation-mediated progression of the neuropathology and neurological outcome associated with Magnetic resonance imaging (MRI) neuroimaging following the water content.

AQP4 IN NEUROINFLAMMATION

In the normal BBB, a characteristic feature is the predominant presence of AQP4 proteins in end-foot plasmalemma formed by astrocytes around the blood vessels (for details, see Chapter 11: Role of Matrix Metalloproteinases in Brain Edema). AQP4 functions as osmotic water-selective channel and their exposed polarized structure conveys them an important role in CNS water homeostasis. In neuroinflammation pathology and BBB disruption that was induced in rodent EAE models of MS, astrocytes lose their polarity and detach from the vascular structure. Observations in EAE neuropathology and other neuroinflammation models (like TBI and stab wound injury) show that AQP4 proteins are not anymore located to the perivascular astrocytic end-foot and AQP4 expression is dramatically increased throughout the plasma membrane of activated astrocytes.[46,47]

Due to their expression of receptors for proinflammatory cytokines (IL-1β, IL-6, and TNFα) astrocytes can rapidly respond to stimulation by proinflammatory cytokines[48] produced by activated microglia and infiltrating immune cells that could be further enhanced by autocrine stimulation of proinflammatory cytokines expressed by activated astrocytes.[49,50] The enhanced activation of astrocytes leads to rapid modulation of their morphological and molecular expression patterns including AQP4 expression (for review see Ref. 51). The pathology is further characterized by the alteration of tight junctions as reflected by a decrease of occludin expression.[52]

Attempts to understand the role of AQP4 in neuroinflammation, BBB disruption, and edema are mainly based on indirect experimental evidences using AQP4 knockout (k.o.) mice when compared to wild-type mice. A panel of various animal models developing vasogenic edma evidenced that their induction in mice being deficient for AQP4 expression has less pathophysiological effects. These experiments suggested the aggravating role of AQP4 in edema development and BBB disruption. In an immune-mediated response it is worth to consider that AQP4, besides in the CNS, is expressed in the cell plasma membranes of many organs including those of the immune system like lymph nodes, thymus, and spleen. The observed modulations of the immune responses in AQP k.o. animal models might thus reflect a more systemic AQP4 deficit than a specific effect on CNS inflammation and BBB alterations. Indeed, Parkinson's disease model induced in AQP4 deficient mice revealed a decreased immunoregulatory response presenting less CD4+ CD25+ regulatory T cell production, leading to increased microglial activation and a more severe dopaminergic neuronal loss.[53] AQP4 k.o. mice presented less severe neuroinflammation in EAE and in LPS-induced acute proinflammatory condition showing less proinflammatory cytokine (TNFα and IL-6) production. In comparison, AQP4 expression was causally linked to detrimental BBB disruption and neuroinflammation as seen in wild-type rodent models.[54]

Follow-up experiments with in vivo MRI-diffusion weighted imaging (DWI) monitoring of the dynamic changes of water diffusion in the brain were performed in a variety of brain injury experiments in rodents and informed in more detail about the time course of AQP4 expression during disease. In general, it was observed that AQP4 expression is upregulated over baseline very shortly (within 3 days postinjury) and persists over weeks with increase and additional expression modulations of AQP4. In the focal brain lesion model induced by injection of L-α-lysophosphatidylcholine (lysolecithine), a first phase of significant increase of AQP4 expression was observed during the first few days (1–3 days postinjection). This time frame corresponded to the build-up phase of the edema formation as defined by Apparent Diffusion

Coefficient (ADC) values that were established by in vivo MRI-DWI (for details on methods see Chapter 4: Experimental Techniques to Investigate the Formation of Brain Edema In Vivo). This phase is further characterized by BBB disruption and leakage of serum proteins into the brain parenchyma (determined by immunodetection of endogenous serum protein extravasation—IgG staining and Evans blue assay) and proinflammatory cytokine IL-1β expression. Compared to this first appearing edema buildup phase, a much more important increase of AQP4 expression was observed at the edema resolution phase (7, 14, and 20 days postinjection) defined as a return to baseline for ADC values. While IL-1β was strongly upregulated at the edema buildup phase (at day 1), strong expression of GFAP indicated activation of astrocytes during the edema resolution phase.[55] In MOG presensitized rats to develop EAE, focal EAE lesions were induced by a secondary cerebral injection of low doses of proinflammatory cytokines (TNFα + IFNγ) and were monitored with MRI. This approach validated the T2 weighted imaging (WI) signal as being related to edema formation and BBB disruption at early time points, while demyelination could be related to later time points, reflecting the histological characteristics of an EAE lesion development similar to MS pathology.[56] In a rat model of juvenile traumatic brain injury (jTBI), combining MRI monitoring approaches of ADC and T2WI allowed defining a similar pattern of edema build-up (at day 1 postinjury) and resolution phases (at day 3 postinjury). AQP4 expression was most pronounced at 3 and 7 days postinjury in astrocytes.[57] Furthermore, the increase of tight junction protein claudin-5 at 7 days post-jTBI that was persisting over 2 months reflected the long-term phenotypic changes of the BBB.[58] Further indirect evidence for the role of AQP4 in edema formation comes from experimental studies of cerebral stroke pathophysiology, where the absence of AQP4 had no or minor direct effects on edema formation, but had a beneficial effect as AQP k.o. animals presented smaller lesions with improved neuronal survival, reduced neuroinflammation, and improved behavioral outcomes at early time points and long-term development.[59,60]

Altogether, the experimental data with AQP4 deficient mice suggest variable roles for AQP4.[61] In vasogenic edema, during edema formation an osmotic water extravasation from the blood circulation across the BBB, most likely without involvement of AQP4, accumulates in extracellular spaces of the brain, while during edema resolution, it is primarily the increased AQP4 activity that is beneficial for clearance of excess water from the brain parenchyma to liquid compartments including CSF and blood vessels.[62,63] With regard to the exposed localization on astrocyte end-foot processes in the vicinity of the BBB, disturbing the correct function of AQP4 has potential functional consequences causing primarily astrocyte affection, which furthermore would also affect neuronal functions within the NVU.

In the particular situation of NMO, astrocytes become targets for pathogenic AQP4-specific antibodies. In about 70%–80% of NMO patients these auto-antibodies are found in the blood circulation which is considered as hallmark of the disease.[64] During neuroinflammatory processes the anti-AQP4 antibodies (NMO-IgG) gain access to the CNS and primarily affect optic nerves and spinal cord often leading to blindness and tetra- or paraparesis. Experimental studies revealed the presence of highly pathogenic, AQP4268–285-epitope-specific CD4+ T cells in normal Lewis rats, a commonly used EAE model. Stimulation of these T cells behind the BBB enables them to infiltrate the parenchyma. Together with NMO-IgG, these T cells initiate large astrocyte-destructive lesions which are located predominantly in spinal cord gray matter recalling similarities with NMO pathology.[65] Although the clinical development of NMO presents many pathophysiological similarities with MS, the anti-AQP4 driven astrocytoma involving auto-antigen targeting by IgG, the T cell activations by AQP4 peptides, and the innate immune system in the pathogenesis of NMO defines a distinct pathology (for review see Ref. 66).

TOOL DEVELOPMENTS FOR IN VIVO MONITORING THE DYNAMIC COMPLEXITY OF VASOGENIC EDEMA

Neuroinflammation involves BBB disruption, extravasation of substrates including immune regulatory molecules, antibodies and immune cells from blood circulation into the brain parenchyma, activation of astrocytes and microglia, and thus is characterized as a highly dynamic process. These events are time dependent, presenting sequential or simultaneous modulations of the CNS cellular environment. The multiple interactions may eventually form enhancing loops that reinforce the neuroinflammation processes, leading finally to pathogenic CNS developments. For a better understanding of the major CNS diseases involving the disturbances of the water homeostasis and edema formation during neuroinflammatory activity, several previously presented experimental approaches, like jTBI, stab wound lesions, stroke, and focal inflammatory triggering in presensitized EAE, have been validated as "workable" dynamic models.

With regard to clinical application, in vivo monitoring the dynamic of the different events is a major challenge. Promising is the development of new imaging techniques that allow a follow-up over time in vivo and non-invasively. MRI and Positron Emission Tomography (PET) scan opened new perspectives to study several events occurring in neuroinflammation. In experimental studies the generated in vivo imaging data were mainly correlated with biochemical assays and histopathological analyses to validate these novel tool developments. In translational investigations some of these approaches have already been tested and successfully applied in clinical practice. Combining the various in vivo imaging modalities in experimental studies could further help to elucidate the complex processes driving the dynamic of vasogenic edema formation in neuroinflammation and its resolution.

Determining the water content in the CNS is the key parameter in studying water accumulation and edema formation (see Chapter 5: In Vitro Models of the Blood–Brain Barrier to Better Understand the Pathophysiology of Brain Edema). The diffusion process of water molecules within the brain tissue can be determined by calculating the ADC using MRI with DWI maps of up to micrometric resolution[67,68] (see Chapter 4: Experimental Techniques to Investigate the Formation of Brain Edema In Vivo). The technique allows revealing quantitative microscopic tissue alterations associated to edema characteristics.[69]

To assess the BBB permeability and disruption in vivo, MRI using Gadolinium-chelates like diethylene-triamine-pentaacetic acid gadolinium(III) (Gadolinium-DTPA) as contrast agent is the current standard in clinical practice. After intravenous injection Gadolinium-DPTA reveals BBB disruptions by accumulation in the CNS. Another Gadolinium-based MRI contrast agent, Gado-fluorine M, has high binding affinity to plasma albumin.[70] Its accumulation in the CNS reflects the plasma protein extravasation and generates enhanced T1WI MRI signal.[71,72]

To monitor in vivo the extravasation of activated immune cells into the CNS, different cell tracking strategies have been developed. By using paramagnetic sensors of the enzyme activity myeloperoxidase which is secreted by activated macrophages and neutrophils, it allows detection of early active inflammatory CNS lesions in EAE by MRI.[73] Super-paramagnetic iron oxide (SPIO) and ultra-small SPIO (USPIO) nanoparticles accumulate in activated macrophages. After experimental evaluations in many animal studies of a variety of neurological disorders (for review see Ref. 74) the detection of infiltrating macrophages in CNS lesions that had engulfed USPIO allowed to discriminate by MRI distinct patterns of macrophage activity in CNS lesions compared to Gadolinium-DTPA detected BBB disruption in CNS inflammation of MS patients[75–77] and in brain ischemia (for review see Ref. 78).

Activation of microglia, infiltrated macrophages (as immune effector cells in the CNS parenchyma), and astrocytes are major elements in neuroinflammation activity. The upregulation of the mitochondrial translocator protein 18kDa (TSPO) system, formerly known as the peripheral benzodiazepine receptor, by these cells is one of the earliest upregulated molecules serving as indicator for their activation status. In a variety of neurological diseases in vivo exploration of this molecular target has been used to map neuroinflammation processes. A number of radiolabeled radiotracers like PK11195 have been developed to image the upregulation of TSPO expression by PET-scan in many applications of MS,[79] stroke, and other neurodegenerative disorders involving neuroinflammation (for review see Ref. 80). In experimental stroke studies, multimodal in vivo monitoring for up to 7 months using repetitive radiotracer PK11195-PET, combined with either T2WI MRI[81] or USPIO-MRI followed by histological evaluation,[82] revealed the high dynamics of short- and long-term neuroinflammation after ischemia induction. Distinguishing in vivo microglia and infiltrating macrophage activity showed the domination of phagocytes from 4 weeks after ischemia induction. Microglia activation was first seen in the early peri-infarct region, but disappeared from the lesion area over time, while at late time points microglia cells presented activity at distance in not primary affected CNS tissue of neurodegeneration. Altogether these investigations performed in various experimental animal models demonstrate the potential of multimodal imaging to monitor in vivo the different events of molecular and cellular players in neuroinflammation and the associated edema formation (for more details see Chapter 4: Experimental Techniques to Investigate the Formation of Brain Edema In Vivo).

CONCLUSION

The multicellular structure and function of the NVU involves a multitude of interactions. Upon neuroinflammatory triggering the complexity of interactions between these cells implies a cellular community effect. Stimulations by proinflammmatory cytokines induce simultaneous effects on different cell types. The activated cells change their morphological and functional phenotypes. They respond with cell movements facilitating the extravasation of serum molecules and immune competent cells together with water accumulation and edema formation eventually causing physical constraints in the CNS parenchyma. Multimodal in vivo imaging can help to monitor at least partially these sequential alterations and enlarge our understanding in defining the morphological and temporal patterns of cellular and molecular actions and interactions. With regard to the clinical importance of edema resolution, the integration of phenotypic data with expression and genomic data is necessary to understand the interactive complexity of neuroinflammation events for therapeutic interventions. This complexity is leading to a compromised water homeostasis and edema formation in the brain parenchyma as part of a protective defense mechanism to CNS injury and in case of overreaction, as contributor to pathogenic induction.

References

1. Schwartz M, Kipnis J, Rivest S, Prat A. How do immune cells support and shape the brain in health, disease, and aging? *J Neurosci* 2013;**33**(45):17587–96. http://dx.doi.org/10.1523/JNEUROSCI.3241-13.2013.
2. Betz AL, Iannotti F, Hoff JT. Brain edema: a classification based on blood–brain barrier integrity. *Cerebrovasc Brain Metab Rev* 1989;**1**(2):133–54.
3. Lundgaard I, Osório MJ, Kress B, Sanggaard S, Nedergaard M. White matter astrocytes in health and disease. *Neuroscience* 2014;**276**:161–73.
4. Simard M, Arcuino G, Takano T, Liu QS, Nedergaard M. Signaling at the gliovascular interface. Astrocytic calcium signaling is expressed at the end-foot processes enveloping the microvessels suggesting that Ca^{2+} signaling in astrocytes could modulate BBB permeability for blood substrates. *J Neurosci* 2003;**23**:9254–62.
5. Pellerin L, Magistretti PJ. Sweet sixteen for ANLS. *J Cereb Blood Flow Metab* 2012;**32**(7):1152–66. http://dx.doi.org/10.1038/jcbfm.2011.149.
6. Petit JM, Magistretti PJ. Regulation of neuron–astrocyte metabolic coupling across the sleep–wake cycle. *Neuroscience* 2016;**323**:135–56.
7. Ding F, O'Donnell J, Xu Q, Kang N, Goldman N, Nedergaard M. Changes in the composition of brain interstitial ions control the sleep-wake cycle. *Science* 2016;**352**(6285):550–5. http://dx.doi.org/10.1126/science.aad4821.
8. Franklin RJ, Ffrench-Constant C. Remyelination in the CNS: from biology to therapy. *Nat Rev Neurosci* 2008;**9**:839–55.
9. Kotter MR, Li WW, Zhao C, Franklin RJ. Myelin impairs CNS remyelination by inhibiting oligodendrocyte precursor cell differentiation. *J Neurosci* 2006;**26**:328–32.
10. Simard M, Nedergaard M. The neurobiology of glia in the context of water and ion homeostasis. *Neuroscience* 2004;**129**(4):877–96.
11. Simard JM, Kent TA, Chen M, Tarasov KV, Gerzanich V. Brain oedema in focal ischaemia: molecular pathophysiology and theoretical implications. *Lancet Neurol* 2007;**6**(3):258–68.
12. Stokum JA, Gerzanich V, Simard JM. Molecular pathophysiology of cerebral edema. *J Cereb Blood Flow Metab* 2016;**36**(3):513–38. http://dx.doi.org/10.1177/0271678X15617172.
13. Nag S, Manias JL, Stewart DJ. Pathology and new players in the pathogenesis of brain edema. *Acta Neuropathol* 2009;**118**:197–217.
14. Abbott NJ. Blood–brain barrier structure and function and the challenges for CNS drug delivery. *J Inherit Metab Dis* 2013;**36**:437–49.
15. Gerhardt H, Betsholtz C. Endothelial–pericyte interaction in angiogenesis. *Cell Tissue Res* 2003;**314**:15–23.
16. Daneman R, Zhou L, Kebede AA, Barres BA. Pericytes are required for blood–brain barrier integrity during embryogenesis. *Nature* 2010;**468**:562–6.
17. Armulik A, Genové G, Mäe M, Nisancioglu MH, Wallgard E, Niaudet C, et al. Pericytes regulate the blood–brain barrier. *Nature* 2010;**468**:557–61. http://dx.doi.org/10.1038/nature09522.
18. Luissint AC, Artus C, Glacial F, Ganeshamoorthy K, Couraud PO. Tight junctions at the blood brain barrier: physiological architecture and disease-associated dysregulation. *Fluids Barriers CNS* 2012;**9**(1):23. http://dx.doi.org/10.1186/2045-8118-9-23.
19. Zhao Z, Nelson AR, Betsholtz C, Zlokovic BV. Establishment and dysfunction of the blood–brain barrier. *Cell* 2015;**163**(5):1064–78. http://dx.doi.org/10.1016/j.cell.2015.10.067.
20. Ransohoff RM, Engelhardt B. The anatomical and cellular basis of immune surveillance in the central nervous system. *Nat Rev Immunol* 2012;**12**:623–35. http://dx.doi.org/10.1038/nri3265.
21. Engelhardt B, Ransohoff RM. Capture, crawl, cross: the T cell code to breach the blood–brain barriers. *Trends Immunol* 2012;**33**:579–89.

22. Witte ME, Mahad DJ, Lassmann H, van Horssen J. Mitochondrial dysfunction contributes to neurodegeneration in multiple sclerosis. *Trends Mol Med* 2014;**20**(3):179–87. http://dx.doi.org/10.1016/j.molmed.2013.11.007.
23. Fischer MT, Sharma R, Lim JL, Haider L, Frischer JM, Drexhage J, et al. NADPH oxidase expression in active multiple sclerosis lesions in relation to oxidative tissue damage and mitochondrial injury. *Brain* 2012;**135**(Pt 3):886–99. http://dx.doi.org/10.1093/brain/aws012.
24. Touil T, Deloire-Grassin MS, Vital C, Petry KG, Brochet B. In vivo damage of CNS myelin and axons induced by peroxynitrite. *Neuroreport* 2001;**12**:3637–44.
25. Hooper DC, Spitsin S, Kean RB, Champion JM, Dickson GM, Chaudhry I, et al. Uric acid, a natural scavenger of peroxynitrite, in experimental allergic encephalomyelitis and multiple sclerosis. *Proc Natl Acad Sci USA* 1998;**95**:675–80.
26. Boullerne AI, Rodriguez JJ, Touil T, Brochet B, Schmidt S, Abrous ND, et al. Anti-S-nitrosocysteine antibodies are a predictive marker for demyelination in experimental autoimmune encephalomyelitis: implications for multiple sclerosis. *J Neurosci* 2002;**22**:123–32.
27. Pacher P, Beckman JS, Liaudet L. Nitric oxide and peroxynitrite in health and disease. *Physiol Rev* 2007;**87**:315–424. http://dx.doi.org/10.1152/physrev.00029.2006.
28. Lapierre M, Siegfried G, Scamuffa N, Bontemps Y, Calvo F, Seidah NG, et al. Opposing function of the proprotein convertases furin and PACE4 on breast cancer cells' malignant phenotypes: role of tissue inhibitors of metalloproteinase-1. *Cancer Res* 2007;**67**:9030–4.
29. Rosenberg GA. Metalloproteinases and neurodegenerative diseases: pathophysiological and therapeutic perspectives. *Metalloproteinases Med* 2015;**2**:39–50.
30. Rivera S, Khrestchatisky M, Kaczmarek L, Rosenberg GA, Jaworski DM. Metzincin proteases and their inhibitors: foes or friends in nervous system physiology? *J Neurosci* 2010;**30**:15337–57.
31. Predescu SA, Predescu DN, Malik AB. Molecular determinants of endothelial transcytosis and their role in endothelial permeability. *Am J Physiol Lung Cell Mol Physiol* 2007;**293**(4):L823–42.
32. Knowland D, Arac A, Sekiguchi KJ, Hsu M, Lutz SE, Perrino J, et al. Stepwise recruitment of transcellular and paracellular pathways underlies blood–brain barrier breakdown in stroke. *Neuron* 2014;**82**(3):603–17. http://dx.doi.org/10.1016/j.neuron.2014.03.003.
33. Badaut J, Ajao DO, Sorensen DW, Fukuda AM, Pellerin L. Caveolin expression changes in the neurovascular unit after juvenile traumatic brain injury: signs of blood–brain barrier healing? *Neuroscience* 2015;**285**:215–26. http://dx.doi.org/10.1016/j.neuroscience.2014.10.035.
34. Wolburg H, Wolburg-Buchholz K, Engelhardt B. Diapedesis of mononuclear cells across cerebral venules during experimental autoimmune encephalomyelitis leaves tight junctions intact. *Acta Neuropathol* 2005;**109**:181–90.
35. Gertig U, Hanisch KU. Microglial diversity by responses and responders. *Front Cell Neurosci* 2014 http://dx.doi.org/10.3389/fncel.2014.00101.
36. Mikita J, Dubourdieu-Cassagno N, Deloire MS, Vekris A, Biran M, Raffard G, et al. Altered M1/M2 activation patterns of monocytes in severe relapsing experimental rat model of multiple sclerosis. Amelioration of clinical status by M2 activated monocyte administration. *Mult Scler* 2011;**17**(1):2–15. http://dx.doi.org/10.1177/1352458510379243.
37. Hanisch UK, Kettenmann H. Microglia: active sensor and versatile effector cells in the normal and pathologic brain. *Nat Neurosci* 2007;**10**:1387–94.
38. Pekny M, Wilhelmsson U, Pekna M. The dual role of astrocyte activation and reactive gliosis. *Neurosci Lett* 2014;**565**:30–8. http://dx.doi.org/10.1016/j.neulet.2013.12.071.
39. Lavisse S, Guillermier M, Hérard AS, Petit F, Delahaye M, Van Camp N, et al. Reactive astrocytes overexpress TSPO and are detected by TSPO positron emission

tomography imaging. *J Neurosci* 2012;**32**(32):10809–18. http://dx.doi.org/10.1523/JNEUROSCI.1487-12.2012.
40. Sérrière S, Tauber C, Vercouillie J, Mothes C, Pruckner C, Guilloteau D, et al. Amyloid load and translocator protein 18 kDa in APPswePS1-dE9 mice: a longitudinal study. *Neurobiol Aging* 2015;**36**(4):1639–52. http://dx.doi.org/10.1016/j.neurobiolaging.2014.11.023.
41. Owen DR, Howell OW, Tang SP, Wells LA, Bennacef I, Bergstrom M, et al. Two binding sites for [3H]PBR28 in human brain: implications for TSPO PET imaging of neuroinflammation. *J Cereb Blood Flow Metab* 2010;**30**:1608–18. http://dx.doi.org/10.1038/jcbfm.2010.63.
42. Owen DR, Guo Q, Kalk NJ, Colasanti A, Kalogiannopoulou D, Dimber R, et al. Determination of [(11)C]PBR28 binding potential in vivo: a first human TSPO blocking study. *J Cereb Blood Flow Metab* 2014;**34**(6):989–94. http://dx.doi.org/10.1038/jcbfm.2014.46.
43. Guo Q, Owen DR, Rabiner EA, Turkheimer FE, Gunn RN. A graphical method to compare the in vivo binding potential of PET radioligands in the absence of a reference region: application to [^{11}C]PBR28 and [^{18}F]PBR111 for TSPO imaging. *J Cereb Blood Flow Metab* 2014;**34**(7):1162–8. http://dx.doi.org/10.1038/jcbfm.2014.65.
44. Colasanti A, Guo Q, Muhlert N, Giannetti P, Onega M, Newbould RD, et al. In vivo assessment of brain white matter inflammation in multiple sclerosis with (18)F-PBR111 PET. *J Nucl Med* 2014;**55**(7):1112–8. http://dx.doi.org/10.2967/jnumed.113.135129.
45. Golla SS, Boellaard R, Oikonen V, Hoffmann A, van Berckel BN, Windhorst AD, et al. Quantification of [F]DPA-714 binding in the human brain: initial studies in healthy controls and Alzheimer's disease patients. *J Cereb Blood Flow Metab* 2015;**35**:766–72. http://dx.doi.org/10.1038/jcbfm.2014.261.
46. Wolburg-Buchholz K, Mack AF, Steiner E, Pfeiffer F, Engelhardt B, Wolburg H. Loss of astrocyte polarity marks blood–brain barrier impairment during experimental autoimmune encephalomyelitis. *Acta Neuropathol* 2009;**118**:219–33.
47. Badaut J, Fukuda AM, Jullienne A, Petry KG. Aquaporin and brain diseases. *Biochim Biophys Acta* 2014;**1840**(5):1554–65. http://dx.doi.org/10.1016/j.bbagen.2013.10.032.
48. Norris JG, Tang LP, Sparacio SM, Benveniste EN. Signal transduction pathways mediating astrocyte IL-6 induction by IL-1 and tumor necrosis factor-α. *J Immunol* 1994;**152**:841–50.
49. Van Wagoner NJ, Oh JW, Repovic P, Benveniste EN. Interleukin-6 (IL-6) production by astrocytes: autocrine regulation by IL-6 and the soluble IL-6 receptor. *J Neurosci* 1999;**19**(13):5236–44.
50. Erta M, Quintana A, Hidalgo J. Interleukin-6, a major cytokine in the central nervous system2012. *Int J Biol Sci* 2012;**8**(9):1254–66. http://dx.doi.org/10.7150/ijbs.4679.
51. Ikeshima-Kataoka H. Neuroimmunological implications of AQP4 in astrocytes. *Int J Mol Sci* 2016;**17**(8):1306. http://dx.doi.org/10.3390/ijms17081306.
52. Huang XN, Wang WZ, Fu J, Wang HB. The relationship between aquaporin-4 expression and blood–brain and spinal cord barrier permeability following experimental autoimmune encephalomyelitis in the rat. *Anat Rec (Hoboken)* 2011;**294**:46–54.
53. Chi Y, Fan Y, He L, Liu W, Wen X, Zhou S, et al. Novel role of aquaporin-4 in CD4+ CD25+ T regulatory cell development and severity of Parkinson's disease. *Aging Cell* 2011;**10**:368–82.
54. Li L, Zhang H, Verkman AS. Greatly attenuated experimental autoimmune encephalomyelitis in aquaporin-4 knockout mice. *BMC Neurosci* 2009;**10**:94. http://dx.doi.org/10.1186/1471-2202-10-94.
55. Tourdias T, Mori N, Dragonu I, Cassagno N, Boiziau C, Aussudre J, et al. Differential aquaporin 4 expression during edema build-up and resolution phases of brain inflammation. *J Neuroinflammation* 2011;**8**:143.
56. Tourdias T, Hiba B, Raffard G, Biran M, Nishiguchi T, Aussudre J, et al. Adapted focal experimental autoimmune encephalomyelitis to allow MRI exploration of multiple sclerosis features. *Exp Neurol* 2011;**230**:248–57.

57. Fukuda AM, Pop V, Spagnoli D, Ashwal S, Obenaus A, Badaut J. Delayed increase of astrocytic aquaporin 4 after juvenile traumatic brain injury: possible role in edema resolution? *Neuroscience* 2012;**222**:366–78.
58. Pop V, Sorensen DW, Kamper JE, Ajao DO, Murphy MP, Head E, et al. Early brain injury alters the blood–brain barrier phenotype in parallel with β-amyloid and cognitive changes in adulthood. *J Cereb Blood Flow Metab* 2013;**33**(2):205–14. http://dx.doi.org/10.1038/jcbfm.2012.154.
59. Yao X, Derugin N, Manley GT, Verkman AS. Reduced brain edema and infarct volume in aquaporin-4 deficient mice after transient focal cerebral ischemia. *Neurosci Lett* 2015;**584**:368–72. http://dx.doi.org/10.1016/j.neulet.2014.10.040.
60. Hirt L, Fukuda AM, Ambadipudi K, Rashid F, Binder D, Verkman A, et al. Improved long-term outcome after transient cerebral ischemia in aquaporin-4 knockout mice. *J Cereb Blood Flow Metab* 2017;**37**(1):277–90. Epub 2016 Jan 14 http://10.1177/0271678X15623290.
61. Verkman AS, Binder DK, Bloch O, Auguste K, Papadopoulos MC. Three distinct roles of aquaporin-4 in brain function revealed by knockout mice. *Biochim Biophys Acta* 2006;**1758**:1085–93.
62. Papadopoulos MC, Manley GT, Krishna S, Verkman AS. Aquaporin-4 facilitates reabsorption of excess fluid in vasogenic brain edema. *Faseb J* 2004;**18**:1291–3.
63. Saadoun S, Papadopoulos MC. Aquaporin-4 in brain and spinal cord oedema. *Neuroscience* 2010;**168**:1036–46. http://dx.doi.org/10.1016/j.neuroscience.2009.08.019.
64. Melamed E, Levy M, Waters PJ, Sato DK, Bennett JL, John GR, et al. Update on biomarkers in neuromyelitis optica. *Neurol Neuroimmunol Neuroinflammation* 2015;**2**(4):e134. http://dx.doi.org/10.1212/NXI.0000000000000134. eCollection 2015.
65. Zeka B, Hastermann M, Kaufmann N, Schanda K, Pende M, Misu T, et al. Highly encephalitogenic aquaporin 4-specific T cells and NMO-IgG jointly orchestrate lesion location and tissue damage in the CNS. *Acta Neuropathol Commun* 2016;**4**(1):82. http://dx.doi.org/10.1186/s40478-016-0355-y.
66. Levy M, Wildemann B, Jarius S, Orellano B, Sasidharan S, Weber MS, et al. Immunopathogenesis of neuromyelitis optica. *Adv Immunol* 2014;**121**:213–42. http://dx.doi.org/10.1016/B978-0-12-800100-4.00006-4.
67. Le Bihan D, Breton E, Lallemand D, Grenier P, Cabanis E, Laval-Jeantet M. MR imaging of intravoxel incoherent motions: application to diffusion and perfusion in neurologic disorders. *Radiology* 1986;**161**:401–7.
68. Le Bihan D. Looking into the functional architecture of the brain with diffusion MRI. *Nat Rev Neurosci* 2003;**4**:469–80.
69. Le Bihan D. The 'wet mind': water and functional neuroimaging. *Phys Med Biol* 2007;**52**:R57–90.
70. Meding J, Urich M, Licha K, Weinmann HJ. Magnetic resonance imaging of atherosclerosis by targeting extracellular matrix deposition with Gadofluorine M. *Contrast Media Mol Imaging* 2007;**2**(3):120–9.
71. Bendszus M, Ladewig G, Jestaedt L, Misselwitz B, Solymosi L, Toyka K, et al. Gadofluorine M enhancement allows more sensitive detection of inflammatory CNS lesions than T2-w imaging: a quantitative MRI study. *Brain* 2008;**131**:2341–52.
72. Stoll G, Kleinschnitz C, Meuth SG, Braeuninger S, Ip CW, Wessig C, et al. Transient widespread blood–brain barrier alterations after cerebral photothrombosis as revealed by gadofluorine M-enhanced magnetic resonance imaging. *J Cereb Blood Flow Metab* 2009;**29**:331–41.
73. Chen JW, Breckwoldt MO, Aikawa E, Chiang G, Weissleder R. Myeloperoxidase-targeted imaging of active inflammatory lesions in murine experimental autoimmune encephalomyelitis. *Brain* 2008;**131**(04):1123–33. http://dx.doi.org/10.1093/brain/awn004.
74. Petry KG, Boiziau C, Dousset V, Brochet B. Magnetic resonance imaging of human brain macrophage infiltration. *Neurotherapeutics* 2007;**4**(3):434–42.

75. Dousset V, Brochet B, Deloire MS, Lagoarde L, Barroso B, Caille JM, et al. MR imaging of relapsing multiple sclerosis patients using ultra-small-particle iron oxide and compared with gadolinium. *Am J Neuroradiol* 2006;**27**(5):1000–5.
76. Vellinga MM, Oude Engberink RD, Seewann A, Pouwels PJ, Wattjes MP, van der Pol SM, et al. Pluriformity of inflammation in multiple sclerosis shown by ultra-small iron oxide particle enhancement. *Brain* 2008;**131**(Pt 3):800–7. http://dx.doi.org/10.1093/brain/awn009.
77. Tourdias T, Roggerone S, Filippi M, Kanagaki M, Rovaris M, Miller DH, et al. Assessment of disease activity in multiple sclerosis phenotypes with combined gadolinium- and superparamagnetic iron oxide-enhanced MR imaging. *Radiology* 2012;**264**(1):225–33. http://dx.doi.org/10.1148/radiol.12111416.
78. Chauveau F, Cho TH, Berthezène Y, Nighoghossian N, Wiart M. Imaging inflammation in stroke using magnetic resonance imaging. *Int J Clin Pharmacol Ther* 2010;**48**(11):718–28.
79. Banati RB, Newcombe J, Gunn RN, Cagnin A, Turkheimer F, Heppner F, et al. The peripheral benzodiazepine binding site in the brain in multiple sclerosis: quantitative in vivo imaging of microglia as a measure of disease activity. *Brain* 2000;**3**(Pt 11):2321–37.
80. Janssen B, Vugts DJ, Funke U, Molenaar GT, Kruijer PS, van Berckel BN, et al. Imaging of neuroinflammation in Alzheimer's disease, multiple sclerosis and stroke: Recent developments in positron emission tomography. *Biochim Biophys Acta* 2016;**1862**(3):425–41. http://dx.doi.org/10.1016/j.bbadis.2015.11.011.
81. Walberer M, Jantzen SU, Backes H, Rueger MA, Keuters MH, Neumaier B, et al. In-vivo detection of inflammation and neurodegeneration in the chronic phase after permanent embolic stroke in rats. *Brain Res* 2014;**1581**:80–8. http://dx.doi.org/10.1016/j.brainres.2014.05.030.
82. Walter HL, Walberer M, Rueger MA, Backes H, Wiedermann D, Hoehn M, et al. In vivo analysis of neuroinflammation in the late chronic phase after experimental stroke. *Neuroscience* 2015;**292**:71–80. http://dx.doi.org/10.1016/j.neuroscience.2015.02.024.

CHAPTER

19

Experimental Therapies for Brain Edema and Intracranial Hypertension

Katrin Rauen[1,2] *and Nikolaus Plesnila*[2,3]

[1]University Hospital Zurich, Zurich, Switzerland [2]University of Munich Medical Center, Munich, Germany [3]Munich Cluster for Systems Neurology (Synergy), Munich, Germany

OUTLINE

Brain Edema.
DOI: http://dx.doi.org/10.1016/B978-0-12-803196-4.00019-9

KEY PHRASES

Several brain pathologies result in the final common pathway of brain edema formation and raised intracranial pressure. Brain edema formation initiates a self-perpetuating cascade of secondary events (reduced cerebral perfusion pressure, cerebral ischemia, neuronal cell death, and neuroinflammation) which finally results in intracranial hypertension and additional brain damage. Hence, therapeutic efforts to prevent secondary brain damage should start with the successful prevention of edema formation. The detailed understanding of the cellular and molecular mechanisms of brain edema formation is one of the basic requirements for the successful translation of experimental therapies targeting secondary brain damage from the bench to the bedside.

Abbreviations

AQP	aquaporin
BBB	blood–brain barrier
CNS	central nervous system
CSF	cerebrospinal fluid
CVT	cerebral venous thrombosis
DAI	diffuse axonal injury
DC	decompressive craniectomy
ECS	extracellular space
ET-1	endothelin-1
GAMES-RP	Glyburide Advantage in Malignant Edema and Stroke—Remedy Pharmaceuticals
GOS	glasgow outcome scale
GSNO	S-nitrisogluthatione
ICH	intracerebral hemorrhage
ICP	intracranial pressure
IVH	intraventricular hemorrhage
JNK	c-Jun N-terminal kinase
K_{ATP}	ATP-sensitive potassium channel
KDR/Flk-1	kinase insert domain receptor (human gene)/fetal liver kinase 1
LPS	lipopolysaccharide
MAPK	mitogen-activated protein kinase
MCAO	middle cerebral artery occlusion
mRS	modified Ranking Scale
MTH	mild therapeutic hypothermia
$NC_{Ca\text{-}ATP}$	ATP- and calcium-sensitive nonselective cation channel
NKCC	Na^+–K^+–$2Cl^-$-cotransporter
NOS	nitric oxide synthase
PGE2	prostaglandin E2
SAH	subarachnoid hemorrhage
STAIR	Stroke Therapy Academic Industry Roundtable
SUR1	sulfonylurea receptor 1
TBI	traumatic brain injury
TGFβ-1	transforming growth factor β-1

TNFα	tumor necrosis factor α
TRPM4	transient receptor potential melastatin 4 channel
VEGF	vascular endothelial growth factor
VEGFR-2	vascular endothelial growth factor receptor 2
ZO-1	zonula occludens 1

INTRODUCTION

Acute brain damage by traumatic brain injury (TBI) or stroke is the leading cause of long-term disability and mortality worldwide.[1,2] Although the detailed pathophysiology differs between TBI and stroke, both disorders result in the life-threatening condition of net water influx into brain, i.e., brain edema formation, which causes intracranial hypertension and additional (or secondary) brain damage. In order to reduce mortality and morbidity from acute brain injury it is therefore indispensable to fully understand the cellular and molecular mechanisms of brain edema formation and its consequences on intracranial pressure (ICP) and cerebral blood flow (CBF), a fact already recognized more than five decades ago.[3,4] The aim of the current chapter is to review the current concepts of brain edema formation and to present a selection of mechanisms associated with the formation of brain edema, which may have a therapeutic potential. Concepts on how to reduce brain edema by protecting the brain from damage, e.g., by hypothermia or other neuroprotective strategies, and well-established therapies of increased ICP, e.g., by hypertonic/hyperoncotic solutions, are not within the scope of the current paper and are therefore not covered.

Because the brain has a very high basal activity and no ability to store significant amounts of high energy substrates, its function and structural integrity is highly dependent on the constant delivery of oxygen and glucose through the blood stream. Experimental studies in the 1970s demonstrated that when CBF is gradually reduced below a critical threshold of 10 ml/100 g brain tissue/min, the brain first switches its metabolisms to anaerobic glycolysis, a process which produces lactate and subsequent acidification of the affected tissue.[5] Ten to fifteen seconds later, when all locally available glucose is metabolized, electrical brain activity ceases and transmembrane ion gradients start to equalize, a process which is finalized within minutes and results in a condition termed "ischemic depolarization".[6] During this process neuronal cells release neurotransmitters, particularly glutamate, and potassium into the extracellular space (ECS). The consequences of these changes are astrocytic cell swelling and tissue infarction due to the lack of high energy substrates.[5,7] Accordingly, any condition resulting in tissue acidosis or ischemic depolarization will inevitably also cause swelling of brain cells, particularly astrocytes. This process was named "cytotoxic cell swelling" or "cytotoxic edema"[4] because brain edema caused by some toxic substances like dinitrophenol resembles ischemic cell swelling observed by electron microscopy[8]; however, the term "ischemic cell swelling" may describe this process more appropriately.

Cytotoxic brain edema is defined by a fluid shift from the extra- into intracellular space while the blood–brain barrier (BBB) remains largely intact. As pointed out above, astrocytes, particularly their perivascular end feet, are the main location of cytotoxic cerebral edema. The reason for this process is that the main function of astrocytes is to maintain the extracellular homeostasis of the brain by keeping the concentration of ions and neurotransmitters in the physiological range and by the removal of other potentially harmful substances.[9,10] This occurs through the activation of highly specific transporters which take up substances that are released during ischemia into the ECS, e.g., sodium, lactate, arachnoid acid, or glutamate. Since this uptake creates an osmotic gradient, water enters the intracellular space and astrocytes swell until the ECS is used up or extracellular homeostasis is restored.[11] Water flux seems to be passive; however, it may be substantially facilitated by aquaporins (AQPs) that are highly specific and energy-independent water channels (see Chapter 11: Role of Matrix Metalloproteinases in Brain Edema).[12–14]

In addition to brain edema formation due to metabolic stress, e.g., cerebral ischemia, the brain may also swell due to disturbances of the BBB. The BBB is a functional barrier between the blood and the brain securing that serum proteins do not enter the brain and that physiological changes of plasma electrolytes do not affect brain function. In essence the BBB completely shields the brain from the blood but provides specific transport systems which actively transport all solutes necessary for proper brain function from the blood into the brain parenchyma (see Chapter 3: Noninvasive Imaging Techniques for Brain Edema: From Basic Science to the Clinic and Chapter 6: The Extracellular and Perivascular Spaces of the Brain).[15] As soon as this barrier is impaired plasma proteins may enter the brain, increase the interstitial oncotic pressure, and cause brain swelling by pulling water from the intravascular space into the brain parenchyma. Such impairments of the BBB resulting in brain edema formation, also termed vasogenic edema, may derive—among others—from (1) direct mechanical damage to blood vessels (as in TBI where vasogenic edema may be observed within seconds), (2) direct ischemic damage to blood vessels (as in ischemic stroke where edema formation may be delayed), or (3) by an indirect increase of BBB permeability by inflammatory changes induced by damage to cerebral vessels or the brain parenchyma, a process which may be delayed by many hours, i.e., the time which is necessary to activate the innate immune system. In any case, vasogenic brain edema causes an exponential raise of ICP as soon as the intracranial compliance is consumed thereby compromising CBF, impairing brain metabolism, and causing additional damage to the brain.[16]

SURGICAL TREATMENT OPTIONS—DECOMPRESSIVE CRANIECTOMY (DC)

Based on the abovementioned pathophysiological events a plethora of approaches have been suggested how to treat brain edema formation and its deleterious consequences, i.e., intracranial hypertension. The most straightforward approach for the treatment of increased ICP is certainly the opening of the skull, the space-limiting structure that is ultimately responsible for the hypertension caused by the expanding brain. This is achieved by craniectomy, a surgical procedure by which a piece of skull bone is removed in order to allow the brain to expand outside the skull. The most important practical factor when performing such a procedure is the size of the craniectomy. It needs to be large enough to create a sufficient volume effect and to prevent extensive herniation and subsequent damage of the expanding brain tissue at the rim of the craniectomy.[17] Unfortunately it is rather counterintuitive to perform a sufficiently large craniectomy because the volume effect of a craniectomy increases exponentially with its diameter.[18] Therefore, the volume effect of a craniectomy is commonly overestimated and neurosurgeons often perform craniectomies far smaller than the calculated minimal necessary diameter of 12–16 cm. Consequently, the therapeutic efficacy of DC is still under debate for many disorders associated with pathologically elevated ICP.

DC has been widely studied since its first description in modern medicine by the Swiss surgeon and Nobel laureate Theodor Kocher in 1901. It is well established experimentally and clinically that DC reduces ICP[19,20]; however, there is a considerable debate about the correct timing of DC. Pathophysiological considerations suggest that intracranial hypertension develops because of a sequence of self-amplifying events. Consequently, the earlier these pathological events are interrupted the lesser the brain damage supposed to occur. This assumption is supported by experimental and clinical data showing that DC is only beneficial when performed early, i.e., within the first few hours after the insult and that it loses its beneficial effect and may even have detrimental effects when performed too late.[21,22] Ideally, DC should be performed prophylactically in patients at risk for intracranial hypertension; however, it remains to be established if this approach is clinically feasible since it is difficult to identify such patients.

So far, the most convincing evidence for the therapeutic benefit of DC comes from the field of malignant ischemic stroke, i.e., large ischemic strokes associated with intracranial hypertension and high mortality. DC was demonstrated to decrease infarct volume and to improve survival and functional outcome after malignant experimental stroke.[23] Subsequent clinical studies performed by these investigators and others,

HAMLET, DECIMAL, and DESTINY, demonstrated that DC reduces mortality from 80% to about 20% in patients with malignant stroke.[22,24,25] Recently, these results were confirmed for patients older than 60 years, representing the majority of the cohort suffering from malignant media infarction (DESTINY II).[26] Accordingly, DC is the only experimentally evaluated therapy for ischemic stroke which was found to be effective clinically and became standard clinical practice so far.[27]

The therapeutic efficacy of DC for other disorders associated with intracranial hypertension, e.g., TBI, subarachnoid hemorrhage (SAH), or intracerebral hemorrhage (ICH), is less well established. Most contemporary experimental studies show a clear beneficial effect of DC after TBI.[28] Several authors demonstrated that early DC reduces contusion size, brain edema formation, white matter injury, and improve functional outcome following controlled cortical impact (CCI) in mice, a model for cortical contusions.[21,29] However, more recent studies challenge these findings and report even worse outcome after DC in mice.[30] The results of clinical trials are equally inconclusive[31] and complications following DC further raise concerns regarding a wider clinical use of DC.[32] The most robust clinical data available on the use of DC after TBI show that DC does not seem to be beneficial in patients with diffuse axonal injury (DAI). Decompressed patients had decreased ICP levels, less ICP-associated interventions, and shorter intensive care treatment in comparison with patients receiving conservative treatment; however, the death rate was identical between the two groups and functional outcome in patients who underwent DC was worse at 6 months after TBI.[33]

Regarding ICH two experimental studies in rodents showed beneficial effects of DC,[34,35] but overall the clinical evidence is weak.[36–38] There seem to be evidence for better survival after ICH in patients who underwent DC[39] and nonmoribund patients with ICH undergoing DC had improved neurological outcome as compared to historical controls in a recent retrospective study.[40] Nevertheless further clinical trials are needed to clarify the usage of DC after ICH.

The pathophysiology of SAH regarding ICP is quite complex and consists of two phases: an early phase characterized by a steep increase of ICP immediately after the initial bleeding and elevations of ICP associated with focal ischemia during the late phase of the disease when large artery spasms occur. The initial steep increase of ICP lasts for only a few minutes and results in global cerebral ischemia. Blunting this ICP increase by DC is detrimental since it results in more severe hemorrhage, more re-bleedings, and higher mortality as recently demonstrated experimentally.[41] The clinical evidence for the use of DC in the later phase of SAH is quite inconclusive. Some authors claim that DC might be warranted after SAH regardless of the underlying pathology of intracranial hypertension,[42] others suggest that DC is only beneficial for patients with mass effects due to hematoma formation.[43] Accordingly, also after SAH more experimental and clinical data are necessary in order to evaluate the therapeutic potential of DC.

TARGETING BLOOD–BRAIN BARRIER PERMEABILITY

In the brain endothelial cells are connected to each other by tight junctions, specialized zonulae occludens composed of proteins, e.g., occludin and zonula occludens 1 (ZO-1), which do not allow substances to freely enter or exit the brain. Transport of substances across this BBB is mediated by specific in- and export transporters located at the luminal and abluminal sides of endothelial cells and by transcytotic mechanisms.[44,45] As mentioned above, various mechanisms are able to open the BBB thereby allowing a protein-rich plasma filtrate and water to enter the brain. One of these mechanisms is physical force as occurring during TBI. Physical force results in an immediate opening of the BBB and immediate swelling of the brain within seconds after injury. Due to very rapid nature of this type of vasogenic edema, it is not possible to prevent or reduce this so-called "primary" edema. In experimental models of TBI, primary brain edema is responsible for about 50% of the brain edema observed 24 hours after injury.[21] In addition to physical force, also acute and chronic inflammatory changes triggered by tissue injury are able to open the BBB. The inflammatory cascade activated almost immediately after tissue injury is the kallikrein–kinin or contact system.

The Kallikrein–Kinin System

The kallikrein–kinin is activated by endothelial damage and contact of coagulation factor XII and plasma kallikrein with the exposed basal lamina. Kallikrein converts kininogen into bradykinin, a nonapeptide with a half-life of only 30 seconds.[46] Binding of bradykinin to bradykinin B_2 receptors dilates small arteries and capillaries, constricts veins and venules, and increases BBB permeability thereby causing brain edema formation.[47] When bradykinin is degraded, des-Arg-bradykinin is formed which binds to bradykinin B_1 receptors. While bradykinin B_2 receptors are constitutively expressed, B_1 receptors are almost absent under physiological conditions and are only expressed in the presence of cytokines, e.g., IL-1β.[48] Both B_1 and B_2 receptors mediate intracellular downstream cascades including the mitogen-activated protein kinase (MAPK) pathway.[47–49]

In the brain, bradykinin receptors are expressed on neurons, microglia, oligodendrocytes, and endothelial cells.[50] Following brain injury bradykinin has detrimental and beneficial effects depending on receptor binding. As mentioned above, bradykinin enhances vasodilatation and vascular permeability, contributing to the formation of vasogenic edema[47,51] most likely through activation of the phospholipase C pathway and production of cytokines and free radicals via B_2 receptors.[52–54] In contrast, bradykinin seems to have also beneficial effects, particularly on neuroinflammation, when it binds to B_1 receptors.[50,55] These findings suggest a crucial role of bradykinin and its receptors in the central nervous system (CNS), namely B_1 receptors most probably mediate anti-inflammatory and B_2 receptors pro-inflammatory and pro-edematous effects.[50]

These pathophysiological considerations are reflected by experimental studies using the CCI or the cryogenic lesion model for TBI. Both experimental TBI models induce a focal contusion, but the cryogenic brain lesion model produces more severe BBB leakage thereby having a stronger neuroinflammatory component. Hence, B_2-deficient mice had less brain edema, smaller infarction volume, and better functional outcome 7 days after CCI,[56] but following cryogenic brain lesion, genetic deletion of B_1 receptors reduced BBB leakage and neuroinflammation.[57] Thus, B_1 and B_2 receptor antagonists may have differential protective effects when applied at different time points after TBI.

Following ischemic stroke where the lesion develops much slower, i.e., within hours instead of minutes, the role of thrombotic and inflammatory events mediated by the kallikrein–kinin system may be of even greater importance.[58–60] Inhibition of coagulation factor XII improves neuronal survival by preventing thrombosis without increasing intracerebral hemorrhage[61] and inhibition of plasma kallikrein ameliorates stroke outcome after middle cerebral artery occlusion (MCAO) in mice by antithrombotic and anti-inflammatory mechanisms.[59] Bradykinin is produced in the brain and B_1 and B_2 receptors are upregulated on neuronal cells in the ischemic penumbra concomitant to the maximum of brain edema formation.[62] The cellular localization of bradykinin B_1 and B_2 in the ischemic brain remains to be investigated in detail; however, in vitro studies suggest that B_2 receptors may be expressed on neurons and astrocytes, while B_1 receptors are mainly found on microglia.[54] This is in line with the finding that B_1 and B_2 receptor deficiency diminishes brain edema formation, BBB leakage, and cerebral infarction as well as improves neurological function after transient MCAO[62,63], but only B_1 receptor knock-out mice have reduced cytokine production and invasion of immune cells following experimental stroke.[63]

Taking together, inhibition of bradykinin receptors is a promising therapeutic target for the treatment of brain edema formation following ischemic stroke. Further studies need to address in more detail the individual contribution of B_1 and B_2 receptors for edema formation and these effects need to be further validated in humans.

Arginine Vasopressin V_{1a} Receptors

Arginine vasopressin (AVP), also known as antidiuretic hormone, regulates systemic water homeostasis and arterial blood pressure via its V_2 and V_{1a} receptors, respectively.[64–67] AVP has various effects in the CNS and is relevant in diverse neurological disorders associated with raised ICP. AVP serum levels correlate with the severity of brain injury[68–73] and experimental studies following experimental ischemic stroke and TBI revealed that V_{1a} receptors but not V_2 receptors mediate brain edema formation and secondary brain injury.[73–77] V_{1a} receptors are upregulated mainly in the perilesional cortex up to 24 hours following experimental TBI.[78] Additionally, V_{1a} receptors were increased on perilesional cortical astrocytes and were redistributed from the cell bodies to the processes. This pattern of upregulation lasts up to 14 days postinjury and is therefore on par with the posttraumatic duration of brain edema.[73,79]

The molecular mechanisms of AVP-mediated edema formation are still not fully elucidated. AVP V_{1a} receptors seem to stimulate the Na^+/H^+ exchanger and the NKCC transporter, two ion transporters expressed on cerebro-vascular endothelial cells and astrocytic end feet.[73,80,81] Both transporters create an increase in intracellular osmolarity thereby causing water to be dragged from the extra- into the intracellular space and cell swelling. Water is transported through AQP4 water channels.[82] Thus, AVP V_{1a} receptors mediate cytotoxic brain edema formation, the modulation of transmembrane ion fluxes, and the consecutive entry of water into astrocytes through AQP4.[76,80,83]

Vascular Endothelial Growth Factor (VEGF)

The VEGF family consists of VEGF-A, VEGF-B, VEGF-C, VEGF-D and the placental growth factor PIGF has important effects on vascular permeability, angiogenesis, and neuronal repair.[84,85] Most of the important effects of VEGF are mediated through VEGF receptor 2 (VEGFR-2), also named KDR/Flk-1.[86] When VEGF is injected into the blood stream, it alters endothelial tight junctions and increases BBB permeability within 30 minutes.[87] Underlying mechanisms are synthesis and release of nitric oxide and subsequent cGMP-dependent pathways,[88–90] rapid phosphorylation of the tight junction proteins occludin and ZO-1,[91] and the subsequent transformation of continuous into fenestrated endothelium within 10 minutes.[92]

VEGF is upregulated in diverse brain pathologies such as TBI, stroke, inflammation, and brain tumors.[93–96] One of the main triggers for VEGF gene expression is hypoxia and activation of HIF-1α as demonstrated by involvement of VEGF in the pathogenesis of high-altitude illness. VEGF gene transcription is induced below a threshold of 10% atmospheric oxygen and results in vascular leakage as soon as oxygen drops below 8%. Vascular leakage can be prevented by pretreatment with an anti-VEGF antibody thereby proving the causal relationship between VEGF and vascular leakage.[90]

Following experimental TBI VEGF is expressed in a biphasic manner, namely an early release by invasive neutrophils with a maximum from 8 hours up to 24 hours post-TBI and a delayed release by astrocytes starting at 4 hours and lasting up to 4 days post-TBI. This pattern of VEGF release is associated with the biphasic opening of the BBB following experimental TBI.[97–99] More recent experiments in models of TBI and stroke, however, suggest that VEGF may not be only a permeability factor.[94,100,101] In both conditions VEGF was mainly expressed in perilesional tissue and was suggested to promote angiogenesis thereby improving tissue perfusion.[102,103] These results are supported by the REBIOS study (Research for biomarkers in ischemic stroke) which demonstrated that VEGF plasma levels were increased in stroke patients for at least 90 days after the insult.[95] These pleiotropic effects of VEGF may be mediated by differential expression of VEGF receptors. In the acute phase (days 1–3) increased NO and VEGFR-2 were found to be associated with increased vascular permeability and compromised BBB function,[101,104–106] but during the subacute phase (days 7–25) VEGFR-1 was increased and suggested to be involved in vessel stabilization and maturation.[107] These experimental results are supported by two small studies using human TBI tissue, which showed that VEGF is induced very early and triggers capillary leakage combined with vasogenic brain edema via the VEGFR-2.[108] Accordingly, VEGFR-2 seems to be more involved in angiogenesis and vascular leakage whereas VEGFR-1 might be more important for cell proliferation and differentiation. This concept was, however, challenged by a recent study using the VEGF neutralizing antibody bevacizumab following TBI. Application of bevacizumab increased lesion volumes and worsened neurological outcome while no effect on brain edema formation was observed.[94]

Taking together, the effects of VEGF on brain edema formation are still controversial. VEGF has most probably very early detrimental effects on vascular permeability and the BBB, but beneficial effects on early and late angiogenesis, revascularization, perilesional microcirculation, and neuronal repair may outweigh its detrimental effects.

Matrix Metalloproteinases (MMPs)

MMPs are zinc-dependent endopeptidases located in the extracellular matrix (for details see Chapter 13: Brain Edema Formation in Traumatic Brain Injury). When activated they are able to degrade the extracellular matrix and activate soluble signaling such as VEGF thereby promoting tissue damage, but also cell and tissue repair (see above). In the brain MMP-9 seems to play a particularly important role for BBB dysfunction and formation of vasogenic brain edema since it degrades collagen IV thereby destroying the basal lamina. In patients suffering from TBI MMP-9 was rapidly increased in pericontusional tissue and remained elevated for up to 72 hours.[109] A recent review on experimental ischemic stroke described detrimental as well as beneficial effects of MMP-9, e.g., vasogenic brain edema formation, increased hemorrhagic transformation, progressive inflammation, neuronal cell death, angiogenesis, and tissue repair.[110] Similar observations were made after ICH, a subtype of stroke with a particularly bad outcome.[111] Thus, inhibition of MMP-9 might be a promising target molecule for anti-edematous treatment; however, similar to VEGF the time-dependent detrimental and beneficial effects of MMP-9 need to be understood in much more detail and be taken into careful consideration.

Endothelin

Endothelin-1 (ET-1) is a peptide hormone synthesized in endothelial cells and released by mechanical stretch.[112] Once released ET-1 is able to bind with high affinity to two different cell surface endothelin receptors, namely ET_A and ET_B.[113] In the brain, ET_A receptors are localized on vascular smooth muscle cells and mediate vasoconstriction.[114]

ET-1 and ET-1 receptors are upregulated following ischemic stroke, hemorrhagic stroke, and TBI and seem to mediate further brain damage.[115] The mechanisms involved are suggested to be vasoconstriction thereby further reducing CBF in hypoperfused tissue,[116] opening of the BBB,[117] and enhanced postischemic inflammation.[118] These mechanistic assumptions are supported by various clinical and experimental studies. ET-1 is produced in the brain,[119] is increased in the cerebrospinal fluid (CSF), and is associated with unfavorable outcome following TBI.[120] In mice lacking ET-1 expression hypoperfusion was ameliorated following TBI.[121] Clazosentan, a specific small molecule ET_A receptor antagonist, reduced the formation of brain edema 72 hours and 7 days after experimental MCAO in rats[122] and ET_A receptor antagonism improved functional outcome following experimental TBI.[123] In experimental and clinical trials, application of ET_A receptor antagonists reduced vasospasm after SAH.[124]

Taken together, targeting ET-1 or ET_A receptors seems to have beneficial effects on CBF following cerebral ischemia, cerebral hemorrhage, and brain trauma and may therefore reduce vasogenic and cytotoxic brain edema formation.

Vasodilator-Stimulated Phosphoprotein (VASP)

VASP targets other proteins to focal cell membrane adhesions thereby stabilizing cell–cell contacts including the BBB.[125,126] In line with these findings, VASP-deficient mice develop less BBB opening, less brain edema, and less brain damage following cerebral ischemia.[127] The beneficial effect of VASP on ischemic tissue damage has been replicated in models of hepatic and cardiac ischemia suggesting that VASP may have a general function in mediating vascular permeability.[128,129] Accordingly, VASP may represent a novel target molecule for the treatment of brain edema and BBB opening, and the development of respective inhibitors may have great therapeutic potential.

REDUCING CYTOTOXIC BRAIN EDEMA

As mentioned above, the process underlying cytotoxic brain edema is swelling of astrocytes. The reason for swelling of astrocytes is the attempt of these cells to maintain the ionic and chemical homeostasis of the ECS in the brain. As soon as this homeostasis is disturbed, e.g., by high extracellular potassium, glutamate, or proton concentrations, astrocytes take up these solutes using transporters fueled by the intra-/extracellular sodium gradient. Consequently, the intracellular sodium concentration in astrocytes rises, water follows this osmotic gradient, and cell swelling occurs. Basic research into the interplay between the various transporters involved in this process (see Chapter 8: Mechanisms of Cell-Volume Regulation in the Central Nervous System and Chapter 9: Role of Aquaporins for the Formation and Resolution of Brain Edema) revealed that inhibition of individual transporters may have a potential for the treatment of brain edema.

Na^+/H^+ Antiporter

Low tissue oxygen results in a switch to anaerobic metabolism thereby resulting in the production of lactatic acid and tissue acidosis. In order to maintain intracellular pH (pH_i) cells use the Na^+/H^+ antiporter to export protons from the cytoplasm into the ECS. The price for the maintenance of pH_i is the import of sodium ions and cell swelling, a process which can be completely blocked by the removal of sodium ions from the ECS or pharmacological inhibition of the Na^+/H^+ exchanger with amiloride. Interestingly, the validity of this concept was never investigated in in vivo models of brain injury which are associated with the formation of cytotoxic brain edema, e.g., global or focal cerebral ischemia.

$Na^+/K^+/2Cl^-$ Cotransporter

The $Na^+/K^+/2Cl^-$ cotransporter (NKCC), specifically isoform 1, is expressed in astrocytes, neurons, and the choroid plexus[130,131] and is activated as a consequence of cell swelling, e.g., due to intracellular acidification.[132] The NKCC can be inhibited by the potent loop diuretic bumetanide and when applied in *in vitro* models of cytotoxic edema bumetanide significantly reduces acidosis-induced astrocytic cell swelling,[132] trauma-induced astrocyte swelling,[133] and trauma-induced AQP4 upregulation.[134] When applied in vivo bumetanide attenuates posttraumatic cytotoxic brain edema formation and neuronal cell death via the intracellular Raf/MEK/ERK cascade[135] and reduces cytotoxic brain edema[136] by modulating the perivascular AQP4 pool.[83] As suggested by the contribution of multiple cell membrane transporters to the formation of cytotoxic cell swelling, synergistic functions of NKCC and SUR1-TRPM4 (sulfonylurea receptor 1 (SUR1)-transient receptor potential melastatin 4 channel (TRPM4)) (see below) have been described. Thus, bumetanide and the SUR1/TRPM4 inhibitor glibenclamide synergistically reduce microvascular failure and secondary brain injury after cerebral ischemic and brain trauma.[137]

SUR1-TRPM4

Another ion channel involved in the formation of cytotoxic edema is the ATP- and calcium-sensitive nonselective cation channel TRPM4 (previously known as $NC_{Ca\text{-}ATP}$) which is regulated by SUR1. Cerebral ischemia induces SUR1 on neurons, astrocytes, capillaries, and microglia in experimental[138,139] and human postmortem studies.[140] Ischemia-induced loss of ATP induces depolarization and opening of the SUR1-TRPM4 channel and thus causes cytotoxic brain edema.[141,142] SUR1-TRPM4 can be inhibited by glibenclamide, a sulfonylurea type oral antidiabetic drug introduced for the treatment of diabetes mellitus type 2 in the 1960s.[143,144] Retrospective studies of patients suffering from ischemic stroke and diabetes mellitus type 2 showed a better outcome and less hemorrhagic transformation in patients receiving glibenclamide.[141] See the chapter by Simard et al. for more details (Chapter 12: Edema and BBB Breakdown in Stroke).

Similar encouraging results were also found following TBI and SAH. Continuous administration of low-dose glibenclamide for 7 days reduced brain edema formation and contusion volume[145] and glibenclamide protects from progressive secondary hemorrhage after brain contusion and improves neurobehavioral outcome following experimental TBI in rats.[143,146] Moreover, glibenclamide has beneficial cognitive long-term effects after experimental mild to moderate TBI.[143,147] Following experimental SAH glibenclamide reduces BBB disruption and vasogenic brain edema formation by decreasing ZO-1 abnormalities and reducing neuroinflammation.[143,148]

Overall, glibenclamide has multiple neuroprotective effects and reduces postischemic as well as posttraumatic brain edema formation in experimental animals and patients. Accordingly, the results of an ongoing clinical phase 2 study on the efficacy Glyburide Advantage in Malignant Edema and Stroke (GAMES-RP) are eagerly expected.[149]

CONCLUSION

Taken together primary pathomechanisms of diverse brain pathologies result in the final common pathway of brain edema formation and raised ICP. Despite significant research efforts and encouraging results emerging from the laboratory, no clinically effective therapy for brain edema could be established so far. The notable exception is decompressive craniectomy (DC); however, this approach rather addresses the deleterious consequences of brain edema formation, that is intracranial hypertension, rather than treating edema formation itself. In addition, DC has serious side effects and may not be suitable for all patients suffering from high ICP. Despite this sobering situation and upcoming therapeutic nihilism among brain edema researchers, a plethora of novel targets for the treatment of brain edema were identified experimentally in recent years. These novel targets warrant clinical evaluation and raise the hope that brain edema formation may become a treatable disorder in the near future.

References

1. Ingall T. Stroke—incidence, mortality, morbidity and risk. J Insur Med 2004;36:143–52.
2. WHO. The top 10 causes of death. http://www.who.int/mediacentre/factsheets/fs310/en/; 2015.
3. Fishman RA. Brain edema. *N Engl J Med* 1975;**293**:706–11.
4. Klatzo I. Presidental address. Neuropathological aspects of brain edema. *J Neuropathol Exp Neurol* 1967;**26**:1–14.
5. Astrup J, Siesjo BK, Symon L. Thresholds in cerebral ischemia—the ischemic penumbra. *Stroke* 1981;**12**:723–5.
6. Hossmann KA, Sakaki S, Zimmerman V. Cation activities in reversible ischemia of the cat brain. *Stroke* 1977;**8**:77–81.
7. Astrup J, Symon L, Branston NM, Lassen NA. Cortical evoked potential and extracellular K+ and H+ at critical levels of brain ischemia. *Stroke* 1977;**8**:51–7.
8. Baethmann A, Van HA. Water and electrolyte distribution in gray matter rendered edematous with a metabolic inhibitor. *J Neuropathol Exp Neurol* 1973;**32**:408–23.
9. Unterberg AW, Stover J, Kress B, Kiening KL. Edema and brain trauma. *Neuroscience* 2004;**129**:1021–9.
10. Kimelberg HK. Current concepts of brain edema. Review of laboratory investigations. *J Neurosurg* 1995;**83**:1051–9.

11. Staub F, Winkler A, Haberstok J, et al. Swelling, intracellular acidosis, and damage of glial cells. *Acta Neurochir Suppl* 1996;**66**:56–62.
12. Amiry-Moghaddam M, Ottersen OP. The molecular basis of water transport in the brain. *Nat Rev Neurosci* 2003;**4**:991–1001.
13. Nielsen S, Nagelhus EA, miry-Moghaddam M, Bourque C, Agre P, Ottersen OP. Specialized membrane domains for water transport in glial cells: high-resolution immunogold cytochemistry of aquaporin-4 in rat brain. *J Neurosci* 1997;**17**:171–80.
14. Verkman AS. Aquaporins at a glance. *J Cell Sci* 2011;**124**:2107–12.
15. Abbott NJ, Patabendige AA, Dolman DE, Yusof SR, Begley DJ. Structure and function of the blood–brain barrier. *Neurobiol Dis* 2010;**37**:13–25.
16. Mendelow AD, Teasdale GM. Pathophysiology of head injuries. *Br J Surg* 1983;**70**: 641–50.
17. Wagner S, Schnippering H, Aschoff A, Koziol JA, Schwab S, Steiner T. Suboptimum hemicraniectomy as a cause of additional cerebral lesions in patients with malignant infarction of the middle cerebral artery. *J Neurosurg* 2001;**94**:693–6.
18. Wirtz CR, Steiner T, Aschoff A, et al. Hemicraniectomy with dural augmentation in medically uncontrollable hemispheric infarction. *Neurosurg Focus* 1997;**2**:E3.
19. Timofeev I, Czosnyka M, Nortje J, et al. Effect of decompressive craniectomy on intracranial pressure and cerebrospinal compensation following traumatic brain injury. *J Neurosurg* 2008;**108**:66–73.
20. Zweckberger K, Stoffel M, Baethmann A, Plesnila N. Effect of decompression craniotomy on increase of contusion volume and functional outcome after controlled cortical impact in mice. *J Neurotrauma* 2003;**20**:1307–14.
21. Zweckberger K, Eros C, Zimmermann R, Kim SW, Engel D, Plesnila N. Effect of early and delayed decompressive craniectomy on secondary brain damage after controlled cortical impact in mice. *J Neurotrauma* 2006;**23**:1083–93.
22. Hofmeijer J, Kappelle LJ, Algra A, Amelink GJ, van GJ, van der Worp HB. Surgical decompression for space-occupying cerebral infarction (the Hemicraniectomy After Middle Cerebral Artery infarction with Life-threatening Edema Trial [HAMLET]): a multicentre, open, randomised trial. *Lancet Neurol* 2009;**8**:326–33.
23. Doerfler A, Forsting M, Reith W, et al. Decompressive craniectomy in a rat model of "malignant" cerebral hemispheric stroke: experimental support for an aggressive therapeutic approach. *J Neurosurg* 1996;**85**:853–9.
24. Vahedi K, Vicaut E, Mateo J, et al. Sequential-design, multicenter, randomized, controlled trial of early decompressive craniectomy in malignant middle cerebral artery infarction (DECIMAL Trial). *Stroke* 2007;**38**:2506–17.
25. Juttler E, Schwab S, Schmiedek P, et al. Decompressive Surgery for the Treatment of Malignant Infarction of the Middle Cerebral Artery (DESTINY): a randomized, controlled trial. *Stroke* 2007;**38**:2518–25.
26. Juttler E, Unterberg A, Woitzik J, et al. Hemicraniectomy in older patients with extensive middle-cerebral-artery stroke. *N Engl J Med* 2014;**370**:1091–100.
27. Kolias AG, Kirkpatrick PJ, Hutchinson PJ. Decompressive craniectomy: past, present and future. *Nat Rev Neurol* 2013;**9**:405–15.
28. Plesnila N. Decompression craniectomy after traumatic brain injury: recent experimental results. *Prog Brain Res* 2007;**161**:393–400.
29. Friess SH, Lapidus JB, Brody DL. Decompressive craniectomy reduces white matter injury following controlled cortical impact in mice. *J Neurotrauma* 2015.
30. Szczygielski J, Mautes AE, Muller A, et al. Decompressive craniectomy increases brain lesion volume and exacerbates functional impairment in closed head injury in mice. *J Neurotrauma* 2016;**33**:122–31.

31. Wang R, Li M, Gao WW, Guo Y, Chen J, Tian HL. Outcomes of early decompressive craniectomy versus conventional medical management after severe traumatic brain injury: a systematic review and meta-analysis. *Medicine (Baltimore)* 2015;**94**:e1733.
32. Kurland DB, Khaladj-Ghom A, Stokum JA, et al. Complications associated with decompressive craniectomy: a systematic review. *Neurocrit Care* 2015;**23**:292–304.
33. Cooper DJ, Rosenfeld JV, Murray L, et al. Decompressive craniectomy in diffuse traumatic brain injury. *N Engl J Med* 2011;**364**:1493–502.
34. Marinkovic I, Strbian D, Pedrono E, et al. Decompressive craniectomy for intracerebral hemorrhage. *Neurosurgery* 2009;**65**:780–6.
35. Omary R, Chernoguz D, Lasri V, Leker RR. Decompressive hemicraniectomy reduces mortality in an animal model of intracerebral hemorrhage. *J Mol Neurosci* 2013;**49**:157–61.
36. Morgenstern LB, Hemphill III JC, Anderson C, et al. Guidelines for the management of spontaneous intracerebral hemorrhage: a guideline for healthcare professionals from the American Heart Association/American Stroke Association. *Stroke* 2010;**41**: 2108–29.
37. Mendelow AD, Gregson BA, Fernandes HM, et al. Early surgery versus initial conservative treatment in patients with spontaneous supratentorial intracerebral haematomas in the International Surgical Trial in Intracerebral Haemorrhage (STICH): a randomised trial. *Lancet* 2005;**365**:387–97.
38. Mendelow AD, Gregson BA, Rowan EN, Murray GD, Gholkar A, Mitchell PM. Early surgery versus initial conservative treatment in patients with spontaneous supratentorial lobar intracerebral haematomas (STICH II): a randomised trial. *Lancet* 2013;**382**:397–408.
39. Ma L, Liu WG, Sheng HS, Fan J, Hu WW, Chen JS. Decompressive craniectomy in addition to hematoma evacuation improves mortality of patients with spontaneous basal ganglia hemorrhage. *J Stroke Cerebrovasc Dis* 2010;**19**:294–8.
40. Esquenazi Y, Savitz SI, El KR, McIntosh MA, Grotta JC, Tandon N. Decompressive hemicraniectomy with or without clot evacuation for large spontaneous supratentorial intracerebral hemorrhages. *Clin Neurol Neurosurg* 2015;**128**:117–22. http://dx.doi.org/10.1016/j.clineuro.2014.11.015, Epub 2014 November 27.
41. Buhler D, Azghandi S, Schuller K, Plesnila N. Effect of decompressive craniectomy on outcome following subarachnoid hemorrhage in mice. *Stroke* 2015;**46**:819–26.
42. Guresir E, Schuss P, Vatter H, Raabe A, Seifert V, Beck J. Decompressive craniectomy in subarachnoid hemorrhage. *Neurosurg Focus* 2009;**26**:E4.
43. Dorfer C, Frick A, Knosp E, Gruber A. Decompressive hemicraniectomy after aneurysmal subarachnoid hemorrhage. *World Neurosurg* 2010;**74**:465–71.
44. Mayhan WG. Regulation of blood–brain barrier permeability. *Microcirculation* 2001;**8**:89–104.
45. Cecchelli R, Berezowski V, Lundquist S, et al. Modelling of the blood–brain barrier in drug discovery and development. *Nat Rev Drug Discov* 2007;**6**:650–61.
46. Bhoola KD, Figueroa CD, Worthy K. Bioregulation of kinins: kallikreins, kininogens, and kininases. *Pharmacol Rev* 1992;**44**:1–80.
47. Dobrivojevic M, Spiranec K, Sindic A. Involvement of bradykinin in brain edema development after ischemic stroke. *Pflugers Arch* 2015;**467**:201–12.
48. Bascands JL, Pecher C, Rouaud S, et al. Evidence for existence of two distinct bradykinin receptors on rat mesangial cells. *Am J Physiol* 1993;**264**:F548–56.
49. Costa-Neto CM, Dillenburg-Pilla P, Heinrich TA, et al. Participation of kallikrein-kinin system in different pathologies. *Int Immunopharmacol* 2008;**8**:135–42.
50. Albert-Weissenberger C, Siren AL, Kleinschnitz C. Ischemic stroke and traumatic brain injury: the role of the kallikrein-kinin system. *Prog Neurobiol* 2013;**101–102**:65–82.
51. Wahl M, Whalley ET, Unterberg A, et al. Vasomotor and permeability effects of bradykinin in the cerebral microcirculation. *Immunopharmacology* 1996;**33**:257–63.

52. Noda M, Kariura Y, Amano T, et al. Kinin receptors in cultured rat microglia. *Neurochem Int* 2004;**45**:437–42.
53. Noda M, Kariura Y, Pannasch U, et al. Neuroprotective role of bradykinin because of the attenuation of pro-inflammatory cytokine release from activated microglia. *J Neurochem* 2007;**101**:397–410.
54. Noda M, Sasaki K, Ifuku M, Wada K. Multifunctional effects of bradykinin on glial cells in relation to potential anti-inflammatory effects. *Neurochem Int* 2007;**51**:185–91.
55. Abbott NJ. Inflammatory mediators and modulation of blood–brain barrier permeability. *Cell Mol Neurobiol* 2000;**20**:131–47.
56. Trabold R, Eros C, Zweckberger K, et al. The role of bradykinin B(1) and B(2) receptors for secondary brain damage after traumatic brain injury in mice. *J Cereb Blood Flow Metab* 2010;**30**:130–9.
57. Raslan F, Schwarz T, Meuth SG, et al. Inhibition of bradykinin receptor B1 protects mice from focal brain injury by reducing blood–brain barrier leakage and inflammation. *J Cereb Blood Flow Metab* 2010;**30**:1477–86.
58. Stoll G, Kleinschnitz C, Nieswandt B. Molecular mechanisms of thrombus formation in ischemic stroke: novel insights and targets for treatment. *Blood* 2008;**112**:3555–62.
59. Gob E, Reymann S, Langhauser F, et al. Blocking of plasma kallikrein ameliorates stroke by reducing thromboinflammation. *Ann Neurol* 2015.
60. Nieswandt B, Kleinschnitz C, Stoll G. Ischaemic stroke: a thrombo-inflammatory disease? *J Physiol* 2011;**589**:4115–23.
61. Kleinschnitz C, Stoll G, Bendszus M, et al. Targeting coagulation factor XII provides protection from pathological thrombosis in cerebral ischemia without interfering with hemostasis. *J Exp Med* 2006;**203**:513–18.
62. Groger M, Lebesgue D, Pruneau D, et al. Release of bradykinin and expression of kinin B2 receptors in the brain: role for cell death and brain edema formation after focal cerebral ischemia in mice. *J Cereb Blood Flow Metab* 2005;**25**:978–89.
63. Austinat M, Braeuninger S, Pesquero JB, et al. Blockade of bradykinin receptor B1 but not bradykinin receptor B2 provides protection from cerebral infarction and brain edema. *Stroke* 2009;**40**:285–93.
64. Birnbaumer M. Vasopressin receptors. *Trends Endocrinol Metab* 2000;**11**:406–10.
65. Holmes CL, Landry DW, Granton JT. Science review: vasopressin and the cardiovascular system part 1—receptor physiology. *Crit Care* 2003;**7**:427–34.
66. Maybauer MO, Maybauer DM, Enkhbaatar P, Traber DL. Physiology of the vasopressin receptors. *Best Pract Res Clin Anaesthesiol* 2008;**22**:253–63.
67. Oliver G, Schafer EA. On the physiological action of extracts of pituitary body and certain other glandular organs: preliminary communication. *J Physiol* 1895;**18**:277–9.
68. Barreca T, Gandolfo C, Corsini G, et al. Evaluation of the secretory pattern of plasma arginine vasopressin in stroke patients. *Cerebrovasc Dis* 2001;**11**:113–18.
69. Cintra EA, Maciel Jr. JA, Araujo S, et al. Vasopressin serum levels in patients with severe brain lesions and in brain-dead patients. *Arq Neuropsiquiatr* 2004;**62**:226–32.
70. Huang WD, Yang YM, Wu SD. Changes of arginine vasopressin in elderly patients with acute traumatic cerebral injury. *Chin J Traumatol* 2003;**6**:139–41.
71. Sorensen PS, Gjerris A, Hammer M. Cerebrospinal fluid vasopressin in neurological and psychiatric disorders. *J Neurol Neurosurg Psychiatry* 1985;**48**:50–7.
72. Xu M, Su W, Huang WD, Lu YQ, Xu QP, Chen ZJ. Effect of AVP on brain edema following traumatic brain injury. *Chin J Traumatol* 2007;**10**:90–3.
73. Rauen K, Trabold R, Brem C, Terpolilli NA, Plesnila N. Arginine vasopressin V1a receptor deficient mice have reduced brain edema and secondary brain damage following traumatic brain injury. *J Neurotrauma* 2013;**30**:1442–8.
74. Vakili A, Kataoka H, Plesnila N. Role of arginine vasopressin V1 and V2 receptors for brain damage after transient focal cerebral ischemia. *J Cereb Blood Flow Metab* 2005;**25**:1012–19.

75. Trabold R, Krieg S, Scholler K, Plesnila N. Role of vasopressin V(1a) and V2 receptors for the development of secondary brain damage after traumatic brain injury in mice. *J Neurotrauma* 2008;**25**:1459–65.
76. Liu X, Nakayama S, Amiry-Moghaddam M, Ottersen OP, Bhardwaj A. Arginine-vasopressin V1 but not V2 receptor antagonism modulates infarct volume, brain water content, and aquaporin-4 expression following experimental stroke. *Neurocrit Care* 2010;**12**:124–31.
77. Krieg SM, Sonanini S, Plesnila N, Trabold R. Effect of small molecule vasopressin V1a and V2 receptor antagonists on brain edema formation and secondary brain damage following traumatic brain injury in mice. *J Neurotrauma* 2015;**32**:221–7.
78. Pascale CL, Szmydynger-Chodobska J, Sarri JE, Chodobski A. Traumatic brain injury results in a concomitant increase in neocortical expression of vasopressin and its V1a receptor. *J Physiol Pharmacol* 2006;**57**(Suppl 11):161–7.
79. Szmydynger-Chodobska J, Chung I, Kozniewska E, et al. Increased expression of vasopressin v1a receptors after traumatic brain injury. *J Neurotrauma* 2004;**21**:1090–102.
80. O'Donnell ME, Duong V, Suvatne J, Foroutan S, Johnson DM. Arginine vasopressin stimulation of cerebral microvascular endothelial cell Na-K-Cl cotransporter activity is V1 receptor and [Ca] dependent. *Am J Physiol Cell Physiol* 2005;**289**:C283–92.
81. Lam TI, Wise PM, O'Donnell ME. Cerebral microvascular endothelial cell Na/H exchange: evidence for the presence of NHE1 and NHE2 isoforms and regulation by arginine vasopressin. *Am J Physiol Cell Physiol* 2009;**297**:C278–89.
82. Kleindienst A, Dunbar JG, Glisson R, Marmarou A. The role of vasopressin V1A receptors in cytotoxic brain edema formation following brain injury. *Acta Neurochir (Wien)* 2013;**155**:151–64.
83. Migliati ER, Amiry-Moghaddam M, Froehner SC, Adams ME, Ottersen OP, Bhardwaj A. Na^+–K^+–2Cl– cotransport inhibitor attenuates cerebral edema following experimental stroke via the perivascular pool of aquaporin-4. *Neurocrit Care* 2010;**13**:123–31.
84. Larpthaveesarp A, Ferriero DM, Gonzalez FF. Growth factors for the treatment of ischemic brain injury (growth factor treatment). *Brain Sci* 2015;**5**:165–77.
85. Greenberg DA, Jin K. Vascular endothelial growth factors (VEGFs) and stroke. *Cell Mol Life Sci* 2013;**70**:1753–61.
86. Merrill MJ, Oldfield EH. A reassessment of vascular endothelial growth factor in central nervous system pathology. *J Neurosurg* 2005;**103**:853–68.
87. Dobrogowska DH, Lossinsky AS, Tarnawski M, Vorbrodt AW. Increased blood–brain barrier permeability and endothelial abnormalities induced by vascular endothelial growth factor. *J Neurocytol* 1998;**27**:163–73.
88. Fischer S, Clauss M, Wiesnet M, Renz D, Schaper W, Karliczek GF. Hypoxia induces permeability in brain microvessel endothelial cells via VEGF and NO. *Am J Physiol* 1999;**276**:C812–20.
89. Mayhan WG. VEGF increases permeability of the blood–brain barrier via a nitric oxide synthase/cGMP-dependent pathway. *Am J Physiol* 1999;**276**:C1148–53.
90. Schoch HJ, Fischer S, Marti HH. Hypoxia-induced vascular endothelial growth factor expression causes vascular leakage in the brain. *Brain* 2002;**125**:2549–57.
91. Antonetti DA, Barber AJ, Hollinger LA, Wolpert EB, Gardner TW. Vascular endothelial growth factor induces rapid phosphorylation of tight junction proteins occludin and zonula occluden 1. A potential mechanism for vascular permeability in diabetic retinopathy and tumors. *J Biol Chem* 1999;**274**:23463–7.
92. Roberts WG, Palade GE. Increased microvascular permeability and endothelial fenestration induced by vascular endothelial growth factor. *J Cell Sci* 1995;**108**(Pt 6):2369–79.
93. Ferrara N, Henzel WJ. Pituitary follicular cells secrete a novel heparin-binding growth factor specific for vascular endothelial cells. *Biochem Biophys Res Commun* 1989;**161**:851–8.

94. Tado M, Mori T, Fukushima M, et al. Increased expression of vascular endothelial growth factor attenuates contusion necrosis without influencing contusion edema after traumatic brain injury in rats. *J Neurotrauma* 2014;**31**:691–8.
95. Matsuo R, Ago T, Kamouchi M, et al. Clinical significance of plasma VEGF value in ischemic stroke—research for biomarkers in ischemic stroke (REBIOS) study. *BMC Neurol* 2013;**13**:32.
96. Weis SM, Cheresh DA. Pathophysiological consequences of VEGF-induced vascular permeability. *Nature* 2005;**437**:497–504.
97. Baskaya MK, Rao AM, Dogan A, Donaldson D, Dempsey RJ. The biphasic opening of the blood–brain barrier in the cortex and hippocampus after traumatic brain injury in rats. *Neurosci Lett* 1997;**226**:33–6.
98. Chodobski A, Chung I, Kozniewska E, et al. Early neutrophilic expression of vascular endothelial growth factor after traumatic brain injury. *Neuroscience* 2003;**122**:853–67.
99. Holmin S, Mathiesen T. Biphasic edema development after experimental brain contusion in rat. *Neurosci Lett* 1995;**194**:97–100.
100. Thau-Zuchman O, Shohami E, Alexandrovich AG, Leker RR. Vascular endothelial growth factor increases neurogenesis after traumatic brain injury. *J Cereb Blood Flow Metab* 2010;**30**:1008–16.
101. Reischl S, Li L, Walkinshaw G, Flippin LA, Marti HH, Kunze R. Inhibition of HIF prolyl-4-hydroxylases by FG-4497 reduces brain tissue injury and edema formation during ischemic stroke. *PLoS One* 2014;**9**:e84767.
102. Issa R, Krupinski J, Bujny T, Kumar S, Kaluza J, Kumar P. Vascular endothelial growth factor and its receptor, KDR, in human brain tissue after ischemic stroke. *Lab Investig* 1999;**79**:417–25.
103. Krum JM, Khaibullina A. Inhibition of endogenous VEGF impedes revascularization and astroglial proliferation: roles for VEGF in brain repair. *Exp Neurol* 2003;**181**:241–57.
104. Khan M, Dhammu TS, Matsuda F, et al. Promoting endothelial function by S-nitrosoglutathione through the HIF-1alpha/VEGF pathway stimulates neurorepair and functional recovery following experimental stroke in rats. *Drug Des Dev Ther* 2015;**9**:2233–47.
105. Lee CZ, Xue Z, Hao Q, Yang GY, Young WL. Nitric oxide in vascular endothelial growth factor-induced focal angiogenesis and matrix metalloproteinase-9 activity in the mouse brain. *Stroke* 2009;**40**:2879–81.
106. Khan M, Dhammu TS, Sakakima H, et al. The inhibitory effect of S-nitrosoglutathione on blood–brain barrier disruption and peroxynitrite formation in a rat model of experimental stroke. *J Neurochem* 2012;**123**(Suppl 2):86–97.
107. Moisan A, Favre IM, Rome C, et al. Microvascular plasticity after experimental stroke: a molecular and MRI study. *Cerebrovasc Dis* 2014;**38**:344–53.
108. Suzuki R, Fukai N, Nagashijma G, et al. Very early expression of vascular endothelial growth factor in brain oedema tissue associated with brain contusion. *Acta Neurochir Suppl* 2003;**86**:277–9.
109. Guilfoyle MR, Carpenter KL, Helmy A, Pickard JD, Menon DK, Hutchinson PJ. Matrix metalloproteinase expression in contusional traumatic brain injury: a paired microdialysis study. *J Neurotrauma* 2015.
110. Lenglet S, Montecucco F, Mach F. Role of matrix metalloproteinases in animal models of ischemic stroke. *Curr Vasc Pharmacol* 2015;**13**:161–6.
111. Machida T, Takata F, Matsumoto J, et al. Brain pericytes are the most thrombin-sensitive matrix metalloproteinase-9-releasing cell type constituting the blood–brain barrier in vitro. *Neurosci Lett* 2015;**599**:109–14.
112. Ostrow LW, Sachs F. Mechanosensation and endothelin in astrocytes—hypothetical roles in CNS pathophysiology. *Brain Res Brain Res Rev* 2005;**48**:488–508.
113. Watts SW. Endothelin receptors: what's new and what do we need to know? *Am J Physiol Regul Integr Comput Physiol* 2010;**298**:R254–60.

114. Michinaga S, Nagase M, Matsuyama E, et al. Amelioration of cold injury-induced cortical brain edema formation by selective endothelin ETB receptor antagonists in mice. *PLoS One* 2014;**9**:e102009.
115. Kaundal RK, Deshpande TA, Gulati A, Sharma SS. Targeting endothelin receptors for pharmacotherapy of ischemic stroke: current scenario and future perspectives. *Drug Discov Today* 2012;**17**:793–804.
116. Legos JJ, Lenhard SC, Haimbach RE, et al. SB 234551 selective ET(A) receptor antagonism: perfusion/diffusion MRI used to define treatable stroke model, time to treatment and mechanism of protection. *Exp Neurol* 2008;**212**:53–62.
117. Matsuo Y, Mihara S, Ninomiya M, Fujimoto M. Protective effect of endothelin type A receptor antagonist on brain edema and injury after transient middle cerebral artery occlusion in rats. *Stroke* 2001;**32**:2143–8.
118. Lehmberg J, Putz C, Furst M, Beck J, Baethmann A, Uhl E. Impact of the endothelin-A receptor antagonist BQ 610 on microcirculation in global cerebral ischemia and reperfusion. *Brain Res* 2003;**961**:277–86.
119. Chatfield DA, Brahmbhatt DH, Sharp T, Perkes IE, Outrim JG, Menon DK. Juguloarterial endothelin-1 gradients after severe traumatic brain injury. *Neurocrit Care* 2011;**14**:55–60.
120. Salonia R, Empey PE, Poloyac SM, et al. Endothelin-1 is increased in cerebrospinal fluid and associated with unfavorable outcomes in children after severe traumatic brain injury. *J Neurotrauma* 2010;**27**:1819–25.
121. Petrov T. Amelioration of hypoperfusion after traumatic brain injury by in vivo endothelin-1 knockout. *Can J Physiol Pharmacol* 2009;**87**:379–86.
122. Moldes O, Sobrino T, Blanco M, et al. Neuroprotection afforded by antagonists of endothelin-1 receptors in experimental stroke. *Neuropharmacology* 2012;**63**:1279–85.
123. Reynolds CA, Schafer S, Pirooz R, et al. Differential effects of endothelin receptor A and B antagonism on behavioral outcome following traumatic brain injury. *Neurol Res* 2011;**33**:197–200.
124. Laban KG, Vergouwen MD, Dijkhuizen RM, et al. Effect of endothelin receptor antagonists on clinically relevant outcomes after experimental subarachnoid hemorrhage: a systematic review and meta-analysis. *J Cereb Blood Flow Metab* 2015.
125. Comerford KM, Lawrence DW, Synnestvedt K, Levi BP, Colgan SP. Role of vasodilator-stimulated phosphoprotein in PKA-induced changes in endothelial junctional permeability. *FASEB J* 2002;**16**:583–5.
126. Davis B, Tang J, Zhang L, et al. Role of vasodilator stimulated phosphoprotein in VEGF induced blood–brain barrier permeability in endothelial cell monolayers. *Int J Dev Neurosci* 2010;**28**:423–8.
127. Kraft P, Benz PM, Austinat M, et al. Deficiency of vasodilator-stimulated phosphoprotein (VASP) increases blood–brain-barrier damage and edema formation after ischemic stroke in mice. *PLoS One* 2010;**5**:e15106.
128. Kohler D, Straub A, Weissmuller T, et al. Phosphorylation of vasodilator-stimulated phosphoprotein prevents platelet-neutrophil complex formation and dampens myocardial ischemia–reperfusion injury. *Circulation* 2011;**123**:2579–90.
129. Kohler D, Birk P, Konig K, et al. Phosphorylation of vasodilator-stimulated phosphoprotein (VASP) dampens hepatic ischemia–reperfusion injury. *PLoS One* 2011;**6**:e29494.
130. Javaheri S. Role of NaCl cotransport in cerebrospinal fluid production: effects of loop diuretics. *J Appl Physiol (1985)* 1991;**71**:795–800.
131. Plotkin MD, Kaplan MR, Peterson LN, Gullans SR, Hebert SC, Delpire E. Expression of the Na^+–K^+–$2Cl^-$ cotransporter BSC2 in the nervous system. *Am J Physiol* 1997;**272**:C173–83.
132. Ringel F, Chang RC, Staub F, Baethmann A, Plesnila N. Contribution of anion transporters to the acidosis-induced swelling and intracellular acidification of glial cells. *J Neurochem* 2000;**75**:125–32.

133. Jayakumar AR, Panickar KS, Curtis KM, Tong XY, Moriyama M, Norenberg MD. Na–K–Cl cotransporter-1 in the mechanism of cell swelling in cultured astrocytes after fluid percussion injury. *J Neurochem* 2011;**117**:437–48.
134. Rao KV, Reddy PV, Curtis KM, Norenberg MD. Aquaporin-4 expression in cultured astrocytes after fluid percussion injury. *J Neurotrauma* 2011;**28**:371–81.
135. Lu KT, Cheng NC, Wu CY, Yang YL. NKCC1-mediated traumatic brain injury-induced brain edema and neuron death via Raf/MEK/MAPK cascade. *Crit Care Med* 2008;**36**:917–22.
136. O'Donnell ME, Tran L, Lam TI, Liu XB, Anderson SE. Bumetanide inhibition of the blood–brain barrier Na–K–Cl cotransporter reduces edema formation in the rat middle cerebral artery occlusion model of stroke. *J Cereb Blood Flow Metab* 2004;**24**:1046–56.
137. Simard JM, Kahle KT, Gerzanich V. Molecular mechanisms of microvascular failure in central nervous system injury—synergistic roles of NKCC1 and SUR1/TRPM4. *J Neurosurg* 2010;**113**:622–9.
138. Simard JM, Chen M, Tarasov KV, et al. Newly expressed SUR1-regulated NC(Ca-ATP) channel mediates cerebral edema after ischemic stroke. *Nat Med* 2006;**12**:433–40.
139. Chen M, Dong Y, Simard JM. Functional coupling between sulfonylurea receptor type 1 and a nonselective cation channel in reactive astrocytes from adult rat brain. *J Neurosci* 2003;**23**:8568–77.
140. Mehta RI, Tosun C, Ivanova S, et al. Sur1–Trpm4 cation channel expression in human cerebral infarcts. *J Neuropathol Exp Neurol* 2015;**74**:835–49.
141. Simard JM, Sheth KN, Kimberly WT, et al. Glibenclamide in cerebral ischemia and stroke. *Neurocrit Care* 2014;**20**:319–33.
142. Woo SK, Kwon MS, Ivanov A, Gerzanich V, Simard JM. The sulfonylurea receptor 1 (Sur1)-transient receptor potential melastatin 4 (Trpm4) channel. *J Biol Chem* 2013;**288**:3655–67.
143. Kurland DB, Tosun C, Pampori A, et al. Glibenclamide for the treatment of acute CNS injury. *Pharmaceuticals (Basel)* 2013;**6**:1287–303.
144. Mehnert H, Karg E. Glibenclamide (HB 419): a new oral antidiabetic of sulphonylurea type. *Ger Med Mon* 1969;**14**:373–7.
145. Zweckberger K, Hackenberg K, Jung CS, et al. Glibenclamide reduces secondary brain damage after experimental traumatic brain injury. *Neuroscience* 2014;**272**:199–206.
146. Simard JM, Kilbourne M, Tsymbalyuk O, et al. Key role of sulfonylurea receptor 1 in progressive secondary hemorrhage after brain contusion. *J Neurotrauma* 2009;**26**:2257–67.
147. Patel AD, Gerzanich V, Geng Z, Simard JM. Glibenclamide reduces hippocampal injury and preserves rapid spatial learning in a model of traumatic brain injury. *J Neuropathol Exp Neurol* 2010;**69**:1177–90.
148. Simard JM, Geng Z, Woo SK, et al. Glibenclamide reduces inflammation, vasogenic edema, and caspase-3 activation after subarachnoid hemorrhage. *J Cereb Blood Flow Metab* 2009;**29**:317–30.
149. Sheth KN, Elm JJ, Beslow LA, Sze GK, Kimberly WT. Glyburide Advantage in Malignant Edema and Stroke (GAMES-RP) trial: rationale and design. *Neurocrit Care* 2015.

SECTION V

CLINICAL FEATURES AND MANAGEMENT OF BRAIN EDEMA

CHAPTER

20

Clinical Monitoring of Brain Edema

Sandro M. Krieg and Sebastian Ille

Klinikum rechts der Isar Technische Universität München, Munich, Germany

OUTLINE

Brain Edema.
DOI: http://dx.doi.org/10.1016/B978-0-12-803196-4.00020-5

INTRODUCTION

Brain edema, independently from its origin, can cause further sequelae such as increased intracranial pressure (ICP). The basic principals in the treatment of brain edema are prevention and management of intracranial hypertension, maintenance of cerebral perfusion pressure (CPP), and optimizing of cerebral oxygenation levels thus avoiding secondary brain injury.[1] There are several techniques at hand, which are capable of not only measuring ICP on the one hand but also to predict future ICP increase or other sequelae of brain edema. These techniques allow for the measurement of cerebral physiologic and metabolic parameters related to oxygen delivery, ICP, and cerebral blood flow invasively but also noninvasively. The following chapter will give an overview on the mostly used techniques and their scientific evidence.

PATHOGENESIS

The presence of mass lesions, abscesses, or hematomas following trauma within the intracranial compartment might elevate ICP. The increase of the volume of an intracranial compartment automatically leads to the displacement of other structures, based on the fact that the overall volume of the cranial vault is constant. With this in mind the ICP is a function of the volume and the compliance of the intracranial components. This relationship is known as the Monro–Kellie doctrine.[2]

Each day the choroid plexus and further locations within the central nervous system (CNS) produce 500 mL (approximately 20 mL/h) of cerebrospinal fluid (CSF). The reabsorption of the CSF follows Pacchionini's granulations into the venous system, cortical veins, and spinal nerve roots as the main mechanisms.[3] Most clinical problems with CSF regulation are based on an impaired reabsorption. A pathologic overproduction of CSF is seldom.

The cerebral blood flow (CBF) is a second variable component within the intracranial vault, which increases in case of hypercapnia and hypoxia. In patients with traumatic brain injury (TBI) the autoregulation of CBF might be impaired. This leads to severe brain swelling, particularly in children.[4,5]

The relationship between the volumes of the intracranial components and the elevation of the ICP is defined by the compliance characteristics of the whole intracranial compartment, mathematically modeled as the volume change divided by the pressure change (dV/dP). This is a nonlinear relationship and the compensatory mechanisms of CSF displacement into the thecal sac and decrease of cerebral venous blood volume enable an increase of volume with minimal elevation of ICP.[6] However, the compensatory mechanisms are limited, especially when the increase of intracranial volumes occurs acutely. Contrarily, the compensatory mechanisms are able to balance the increase of large but slow growing mass lesions. Yet, sooner or later the compensatory mechanisms are exhausted and ICP will increase exponentially, that is quickly and suddenly. This might lead to general or characteristic clinical symptoms.

INDICATION FOR INTRACRANIAL PRESSURE MONITORING

The indication for noninvasive or invasive monitoring of ICP is difficult. Generally, the suspicion of elevated ICP is based on clinical signs, imaging, and the patient's history.[7,8] Most of our current knowledge regarding monitoring of ICP origins from the treatment of TBI.[9]

CLINICAL SIGNS OF BRAIN EDEMA

The identification of elevated ICP by clinical examination is mandatory, at least to initialize further diagnostics like computer tomography (CT) imaging. However, patients do not show specific clinical symptoms for brain edema or elevated intracranial pressure. Usually, the first symptoms start to develop when the ICP has reached a critical point. The following clinical signs and symptoms may occur in patients with elevated intracranial pressure.

Although being a symptom of highly diverse brain disorders, headache might be an initial presenting feature of cerebral edema. This global sign is probably mediated through the cranial nerve V and its pain fibers within the dura. Further symptoms are a varying state of consciousness and alertness as well as lethargy. Patients then appear disinterested and with reduced motor activity. Mostly, the latter features are accompanied with impaired orientation to person, time, place, or situation. The local effect of mass lesions, and the direct or indirect pressure to the brainstem most often cause the patient's depressed consciousness. A further sign of elevated ICP may be positive and brik extensor plantar responses.

Despite being often a clinical sign for progressed elevated ICP, bradycardia and elevated blood pressure are rarely related to brain edema. Of course, this so-called Cushing reflex only appears in patients who are continuously monitored. Another sign of the Cushing reflex might be irregular breathing, which varies from shallow to deep breathing. The combination of bradycardia, hypertension, and respiratory depression is also called the Cushing's triad. Both the underlying mechanisms of Cushing's reflex and/or Cushing's triad remain controversial. The most pragmatic explanation of this mechanism is compression of the brainstem. Another alarming clinical sign of elevated intracranial pressure are abnormal movements of the eyes and/or pupillary dilatation as well as the pupils' unresponsivenes to light. These focal symptoms of elevated ICP are most often a sign for herniation, which results in different gradients of pressure between the fossas. Herniation syndromes can be divided by their anatomical location into transcalvarial, cerebellar foramen magnum/tonsillar, upward cerebellar/transtentorial, uncal transtentorial, central transtentorial, and subfalcine. A special herniation syndrome, which is often termed as a false localizing syndrome is Kernohan's notch phenomenon. This phenomenon describes an ipsilateral hemiparesis combined with a contralateral pupillary dilatation, which is caused by transtentorial herniation and the following compression to the cerebral peduncle.[10]

However, the above-mentioned clinical signs are diagnostically less conclusive with respect to indication of surgical treatments. Moreover, the symptoms might be unreliable in different cases. Hence, on the one hand radiologic imaging might support the diagnosis. One the other hand, direct measuring of elevated ICP describes the most reliable method of diagnosing it.

NONINVASIVE TECHNIQUES

Computed Tomography

CT allows for the investigation of structural intracranial changes but it also provides surrogate parameters which may help to identify elevated ICP, such as midline shift, mass lesions, herniation, disappearance of the supratentorial sulci, or the basal cisterns. However, even patients not showing such findings on their initial CT scan may have elevated ICP.

Patients who have no pathological finding in their initial CT scan after TBI, have a chance of up to 15% of developing increased ICP while being hospitalized as shown in a study on 753 TBI patients.[11] In two other studies up to one-third of TBI patients with a normal CT scan at admission had pathologic findings during the first days after TBI.[12,13] Thus, although CT scans are one of the major modalities during diagnosis and treatment of

TBI and brain edema, further continuous monitoring of high-risk patients is crucial.

Moreover, there are even studies at hand showing different predictors of increased ICP according to CT scans, such as density measurements.[14]

Invasive ICP monitoring is indicated if clinicians have the suspicion that a patient has a high risk for developing elevated ICP, for example high acceleration trauma, pathologic findings on CT scan, as well as comatose or sedated patients.[15,16]

Besides conventional CT imaging, perfusion CT was repeatedly shown to be of significant value in the treatment and diagnosis of brain edema.[17] CT perfusion was even shown to be able to obtain information about blood–brain barrier permeability in patients and can therefore differentiate between local and global cerebral edema.[18–20]

Biomarkers

Several biomarkers, such as the levels of S-100β protein, neuron-specific enolase (NSE), and α-synuclein were evaluated for the predictive value concerning pathological CT findings as well as neurologic outcome.[21–23] At least for TBI but also for bacterial meningitis, a predictive value for a fatal outcome was already shown for α-synuclein, S-100β protein, as well as NSE.[24]

Noninvasive Monitoring of Intracranial Pressure

In the past, some techniques were developed in order to measure ICP noninvasively. However, none of them is generally accepted, and usually they are not used in clinical practice. Nevertheless, the following techniques are available for the noninvasive examination of the ICP.

Intraocular Pressure

By the use of an ultrasonic handheld optic tonometer the intraocular pressure can be measured. The hypothesis is that in the absence of glaucoma or ocular trauma, the intraocular pressure correlates with the ICP. However, although intraocular pressure and ICP correlate with each other further studies could not prove this hypothesis so far. Intraocular pressure does not allow drawing any conclusions regarding elevated ICP. Since it is not reliable enough, it does not qualify for its clinical application at the moment.[25,26]

Ocular Sonography

Another technique using the eye as approach for the measurement of ICP is ocular sonography and the determination of the optic nerve's sheath diameter. Since the optic nerve is a part of the CNS, it is surrounded by dura. The subarachnoid space between the dura and the optic nerve expands in case of elevated ICP. The diameter of the optic nerve sheath can be visualized by transocular sonography. A couple of studies showed that diameters of 5–6 mm determine the cut off between elevated and normal ICP.[27–29]

Transcranial Doppler

The transcranial Doppler (TCD) sonography is primarily used for measuring the velocity of blood flow within the proximal anterior cerebral circulation. Hence, it is particularly applied in patients suffering from acute subarachnoidal hemorrhage in order to detect vasospasm at an early stage. By characteristic changes of the waveforms due to the increased resistance, it is also possible to draw conclusions about the ICP.[30,31] Despite the TCD results correlate with the outcome after six months in trauma patients, altogether results do not warrant its sole use in clinical practice. Most importantly, TCD is not reliable enough for the diagnosis of elevated ICP in the acute phase.[32]

Tissue Resonance Analysis

A further ultrasound-based method for the noninvasive measurement of the ICP is tissue resonance analysis (TRA). The ICP is dependent on the mechanical resonance level of the brain tissue. By measuring the reflection of an ultrasound signal from the third ventricle, it is possible to obtain an echopulsogram. Based on characteristic patterns of mechanical resonance and vibration of different tissues, it is possible to determine the ICP. A study which compared ICPs measured by TRA with the results measured by invasive ICP monitoring could show a good correlation between the two methods. However, until now, the TRA is not standardly used in clinical practice.[33]

INVASIVE TECHNIQUES

The ICP can be measured at four main anatomical sites: epidural, subarachnoid, intraparenchymal, and intraventricular.[15] No matter how or where it is measured, the ICP is not a fixed measure. The cyclic variation due to the effects of cardiac action, breathing, and intracranial compliance are assigned to different characteristic waveforms. In case of normal ICP, the waveform is small, and characterized by B waves which are due to respiration and C waves, known as Traube–Hering–Mayer waves, which are related to cardiac contractions.[34,35] As opposed to this, an elevated ICP is characterized by A waves or the so-called plateau waves. These sudden elevations of the ICP of up to 50–100 mmHg indicate the decompensation of autoregulatory mechanisms.[36] With this in mind, A waves describe the necessity of urgent intervention.

Intraventricular

Intraventricular catheters are the current gold standard method of measuring and monitoring ICP. For the placement of an external ventricular drain (EVD), Kocher's point (11.5 cm from the nasion, 2.5 cm from the midline, 1 cm anterior of the coroner suture) has to be measured. If no side is preferred due to hemorrhage or an optional surgical approach, the EVD should be placed on the right side. Next, a burr hole has to be applied by a drill under sterile conditions. Here it is important that the drill is already in the same trajectory as the EVD should pass later. The trajectory is guided by the ipsilateral medial cants and the ipsilateral tragus. The trajectory targets the anterior horn of the lateral ventricle and the third ventricle. After piercing the dura, the EVD should be placed on 7 cm from skin level. After placing the EVD intraventricularly it is connected to a drainage system and a regular pressure transducer. When performed correctly, this kind of ICP monitoring is accurate and easy to handle. Moreover it enables to treat some elevated ICP by draining CSF.[37] Apart from surgical risks like hemorrhage and the difficulty of placing the catheter into small/slit ventricles, intraventricular catheters harbor the risk of infection, which can be as high as 20%.[38] This infection risk increases with the period of placement and could not be reduced by prophylactic catheter changes.[39,40]

Fig. 20.1 shows the instruments needed for the application of an EVD. It should be pointed out that it is important to work under sterile conditions since the application of an EVD always harbors the risk of infection. Fig. 20.2 shows the postoperative CT scan after the successful application of an EVD into slit ventricles. The young patient suffered from brain edema due to intoxication. Since elevated ICP could not be managed conservatively a hemicraniectomy had to be performed. Fig. 20.3 shows the trajectory for the application of an EVD on the basis of a postoperative CT scan.

Intraparenchymal

These devices are composed of a cable and a transducer at its tip. Most often the fiberoptic Camino system is used. After drilling a small borehole into the patient's skull, these probes are placed directly in brain tissue. In addition to the simplicity of placement the risk of infection or hemorrhage is much lower than with devices placed intraventricularly (<1%).[41–43] However, the intraparenchymal devices also bear disadvantages

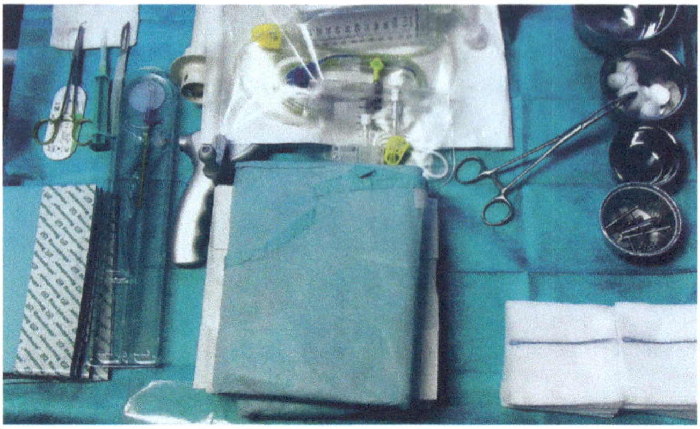

FIGURE 20.1

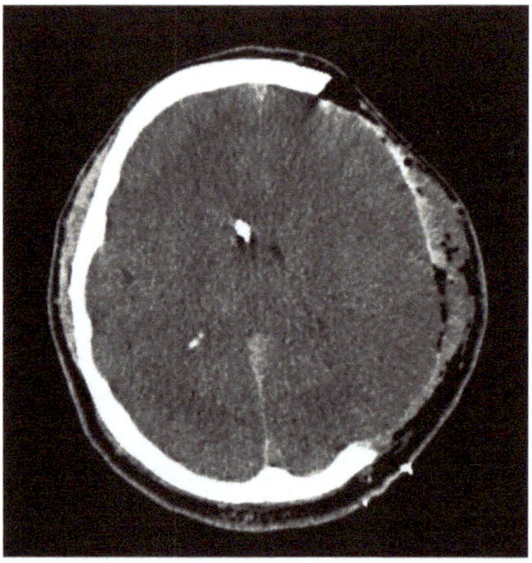

FIGURE 20.2

since it is not possible to drain CSF. Furthermore, some devices may lose their accuracy ("drift") after several days of use. Then probes need to be replaced since recalibration in the tissue is not possible. However, a recent meta-analysis regarding the reliability of intraparenchymal devices showed only a small drift in most patients (0.75 mmHg).[44]

Subarachnoid

Subarachnoid devices consist of a hollow screw and a fluid-coupled system. The hollow screw is placed through the scull near the dura. After puncturing the dura the CSF is communicating with probe and

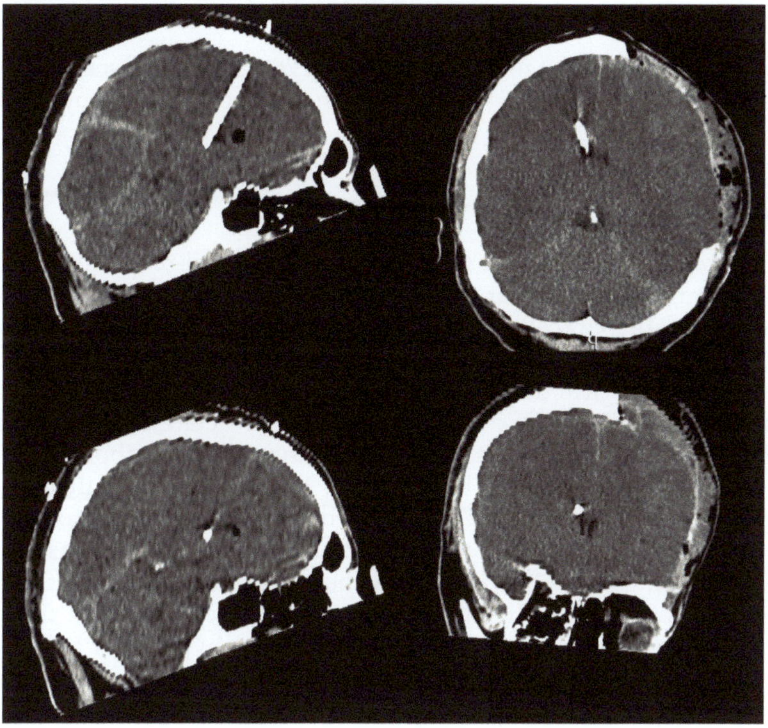

FIGURE 20.3

an associated transducer. Although the subarachnoid systems harbor a low infection risk and risk of hemorrhage, they frequently reach failure because of clogging. Hence they are unreliable as also evoidenced by their infrequent clinical use. Moreover, two studies demonstrated that subarachnoid ICP devices are less accurate than ventricular ICP devices.[15,16]

Epidural

For placing an epidural device a borehole had to be made while leaving the dura intact. The latter coincidentally describes the advantage and disadvantage of this system. On the one hand, epidural devices can be used in patients with pathologic coagulation since the intracerebral hemorrhage risk (4 vs 20 and 22% for intraparenchymal and intraventricularly placed systems) or risk of fatal hemorrhage (1 vs 5 and 4%) is lower.[45] However, epidural monitors are often imprecise, as the dura does not transmit the entire ICP to the epidural space. Hence they are also rarely used in clinical practice.[15,46,47]

Secondary Effects of Elevated Intracranial Pressure

Apart from the direct measurement of elevated ICP as described above, the secondary effects of brain edema can also be analyzed by invasive techniques.

Microdialysis

For cerebral microdialysis (MD), which allows real-time monitoring of brain chemistry, an intraparenchymal probe has to be placed that allows measurement of extracellular glucose, lactate, pyruvate, and glutamate.[48] Low glucose and high extracellular lactate has shown to be an indicator of severe neurological damage followed by a poor outcome.[49] Glutamate is part of the inflammatory cascade response. A lactate to pyruvate ratio (LPR) of more than 40 is suggestive of anaerobic metabolism, which is attributable to secondary brain injury.[48] A LPR of more than 25 has also been reported as a threshold for metabolic distress. Moreover, a LPR of more than 25 within the first 72 hours after trauma is associated with a poor outcome.[50,51] However, the outcome-modifying effect of these thresholds is unproven. In this context it has to be mentioned that LPR and lactate levels can also be abnormal without hypoxia or ischemia.[52] Especially for the targeting of interventions the complexity of interpreting the LPR has to be taken into account and further modalities such as perfusion failure (e.g., CBF) have to be considered.[53] Yet, there are studies proving clinical evidence for the implementation of MD in the clinical setup of brain edema treatment.[54,55]

Also the correct placement of the MD probe is discussed controversially. In case of focal injuries, perilesional placement is recommended. For more diffuse TBIs the probes should be placed in the right frontal lobe in order prevent additional lesions within eloquent regions.[56]

Thermal Diffusion Flowmetry

Similar to probes for measuring brain tissue oxygenation, intraparenchymal probes measuring CBF continuously can be placed in the white matter. These probes are applied by a catheter at a depth of approximately 2cm. By the application of heat and the calculation of temperature dissipation the CBF can be measured.[57] However, although data on the correlation of CBF and neuroimaging as well as clinical outcome is still sparse, there is upcoming evidence toward the clinical usability of thermal diffusion flowmetry. Current studies actually showed that the integration of thermal diffusion flowmetry to measure relative CBF in a multimodal monitoring setup can prevent delayed ischemia in patients who are at risk for secondary brain injury.[58,59]

Jugular Venous Oximetry

Oxygen saturation of blood exiting the brain can be measured via retrograde cannulating of the internal jugular vein.[60] Venous oxygen saturation (SjVO2) in the internal jugular vein is around 60%. SjVO2 values below 50% for 10 min or more were shown to correlate with decreased CPP and compromised outcome.[61]

However, the measurement of venous oxygen saturation should rather be seen as additional parameter for the monitoring of patients after TBI but not as a single diagnostic criteria. It enables clinicians to optimize the treatment after TBI but not to draw reliable conclusions regarding the ICP.

Brain Tissue Oxygen Tension

Besides measuring venous oxygen levels also brain tissue oxygen tension (PbtO2) can be measured with intraparenchymal electrodes. These are placed like fiberoptic ICP probes and are able to monitor white matter PbtO2.[62,63] PbtO2 is above 20 mmHg in the healthy brain and it could be shown that PbtO2 higher than 25 mmHg can significantly improve clinical outcome following TBI.[64] In contrast, PbtO2 lower than 15 mmHg was reported to result in poorer clinical outcome.[61]

Like the measurement of venous oxygenation saturation in the internal jugular vein, information about brain tissue oxygenation facilitate a further possibility to measure intracranial conditions. However, until now the application of the intraparenchymal electrodes is not performed standardly in clinical practice.

CONCLUSION

While a lot of monitoring modalities are at hand and used in some centers in an individual setup, in clinical practice, CT (including perfusion CT) and invasive ICP measurements are the only well established and actually widely used techniques to monitor brain edema and intracranial hypertension in unconscious patients. However, there is huge potential to further develop already existing techniques by adding clinical evidence and proving their use in patients.

References

1. Maas AI, Menon DK, Lingsma HF, Pineda JA, Sandel ME, Manley GT. Re-orientation of clinical research in traumatic brain injury: report of an international workshop on comparative effectiveness research. *J Neurotrauma* 2012;**29**(1):32–46.
2. Mokri B. The Monro–Kellie hypothesis: applications in CSF volume depletion. *Neurology* 2001;**56**(12):1746–8.

3. Sakka L, Coll G, Chazal J. Anatomy and physiology of cerebrospinal fluid. *Eur Ann Otorhinolaryngol Head Neck Dis* 2011;**128**(6):309–16.
4. Bruce DA, Alavi A, Bilaniuk L, Dolinskas C, Obrist W, Uzzell B. Diffuse cerebral swelling following head injuries in children: the syndrome of "malignant brain edema". *J Neurosurg* 1981;**54**(2):170–8.
5. Aldrich EF, Eisenberg HM, Saydjari C, et al. Diffuse brain swelling in severely head-injured children. A report from the NIH traumatic coma data bank. *J Neurosurg* 1992;**76**(3):450–4.
6. Ursino M, Lodi CA. A simple mathematical model of the interaction between intracranial pressure and cerebral hemodynamics. *J Appl Physiol (Bethesda, MD.: 1985)* 1997;**82**(4):1256–69.
7. Morgenstern LB, Hemphill 3rd JC, Anderson C, et al. Guidelines for the management of spontaneous intracerebral hemorrhage: a guideline for healthcare professionals from the American Heart Association/American Stroke Association. *Stroke* 2010;**41**(9): 2108–29.
8. Hemphill 3rd JC, Greenberg SM, Anderson CS, et al. Guidelines for the management of spontaneous intracerebral hemorrhage: a guideline for healthcare professionals from the American Heart Association/American Stroke Association. *Stroke* 2015;**46**(7):2032–60.
9. Marik PE, Varon J, Trask T. Management of head trauma. *Chest* 2002;**122**(2):699–711.
10. Binder DK, Lyon R, Manley GT. Transcranial motor evoked potential recording in a case of Kernohan's notch syndrome: case report. *Neurosurgery* 2004;**54**(4):999–1002 [discussion 1002-1003].
11. Eisenberg HM, Gary Jr. HE, Aldrich EF, et al. Initial CT findings in 753 patients with severe head injury. A report from the NIH traumatic coma data bank. *J Neurosurg* 1990;**73**(5):688–98.
12. O'Sullivan MG, Statham PF, Jones PA, et al. Role of intracranial pressure monitoring in severely head-injured patients without signs of intracranial hypertension on initial computerized tomography. *J Neurosurg* 1994;**80**(1):46–50.
13. Lobato RD, Sarabia R, Rivas JJ, et al. Normal computerized tomography scans in severe head injury. Prognostic and clinical management implications. *J Neurosurg* 1986;**65**(6):784–9.
14. Nguyen HS, Li L, Patel M, Mueller W. Density measurements with computed tomography in patients with extra-axial hematoma can quantitatively estimate a degree of brain compression. *Neuroradiol J* 2016.
15. Brain Trauma Foundation, American Association of Neurological Surgeons, and Congress of Neurological Surgeons Guidelines for the management of severe traumatic brain injury. VII. Intracranial pressure monitoring technology. *J Neurotrauma* 2007;**24**(Suppl 1):S45–54.
16. Brain Trauma Foundation, American Association of Neurological Surgeons, and Congress of Neurological Surgeons Guidelines for the management of severe traumatic brain injury. VI. Indications for intracranial pressure monitoring. *J Neurotrauma* 2007;**24**(Suppl 1):S37–44.
17. Mokin M, Ciambella CC, Masud MW, Levy EI, Snyder KV, Siddiqui AH. Whole-brain computed tomographic perfusion imaging in acute cerebral venous sinus thrombosis. *Interv Neurol* 2016;**4**(3–4):104–12.
18. Ivanidze J, Kallas ON, Gupta A, et al. Application of blood–brain barrier permeability imaging in global cerebral edema. *AJNR Am J Neuroradiol* 2016.
19. Jungner M, Siemund R, Venturoli D, Reinstrup P, Schalén W, Bentzer P. Blood–brain barrier permeability following traumatic brain injury. *Minerva Anestesiol* 2016;**82**(5): 525–33.
20. Hom J, Dankbaar JW, Soares BP, et al. Blood–brain barrier permeability assessed by perfusion CT predicts symptomatic hemorrhagic transformation and malignant edema in acute ischemic stroke. *AJNR Am J Neuroradiol* 2011;**32**(1):41–8.

21. Mondello S, Buki A, Italiano D, Jeromin A. Alpha-synuclein in CSF of patients with severe traumatic brain injury. *Neurology* 2013;**80**(18):1662–8.
22. Mercier E, Boutin A, Lauzier F, et al. Predictive value of S-100beta protein for prognosis in patients with moderate and severe traumatic brain injury: systematic review and meta-analysis. *BMJ* 2013;**346**:f1757.
23. Chabok SY, Moghadam AD, Saneei Z, Amlashi FG, Leili EK, Amiri ZM. Neuron-specific enolase and S100BB as outcome predictors in severe diffuse axonal injury. *J Trauma Acute Care Surg* 2012;**72**(6):1654–7.
24. Bartek Jr. J, Thelin EP, Ghatan PH, Glimaker M, Bellander BM. Neuron-specific enolase is correlated to compromised cerebral metabolism in patients suffering from acute bacterial meningitis; an observational cohort study. *PLoS One* 2016;**11**(3):e0152268.
25. Kirk T, Jones K, Miller S, Corbett J. Measurement of intraocular and intracranial pressure: is there a relationship? *Ann Neurol* 2011;**70**(2):323–6.
26. Yavin D, Luu J, James MT, et al. Diagnostic accuracy of intraocular pressure measurement for the detection of raised intracranial pressure: meta-analysis: a systematic review. *J Neurosurg* 2014;**121**(3):680–7.
27. Geeraerts T, Merceron S, Benhamou D, Vigue B, Duranteau J. Non-invasive assessment of intracranial pressure using ocular sonography in neurocritical care patients. *Intensive Care Med* 2008;**34**(11):2062–7.
28. Kimberly HH, Shah S, Marill K, Noble V. Correlation of optic nerve sheath diameter with direct measurement of intracranial pressure. *Acad Emerg Med* 2008;**15**(2):201–4.
29. Soldatos T, Karakitsos D, Chatzimichail K, Papathanasiou M, Gouliamos A, Karabinis A. Optic nerve sonography in the diagnostic evaluation of adult brain injury. *Crit Care* 2008;**12**(3):R67.
30. Manno EM. Transcranial Doppler ultrasonography in the neurocritical care unit. *Crit Care Clin* 1997;**13**(1):79–104.
31. Edouard AR, Vanhille E, Le Moigno S, Benhamou D, Mazoit JX. Non-invasive assessment of cerebral perfusion pressure in brain injured patients with moderate intracranial hypertension. *Br J Anaesth* 2005;**94**(2):216–21.
32. Hassler W, Steinmetz H, Gawlowski J. Transcranial Doppler ultrasonography in raised intracranial pressure and in intracranial circulatory arrest. *J Neurosurg* 1988;**68**(5):745–51.
33. Michaeli D, Rappaport ZH. Tissue resonance analysis; a novel method for noninvasive monitoring of intracranial pressure. Technical note. *J Neurosurg* 2002;**96**(6):1132–7.
34. Takizawa H, Gabra-Sanders T, Miller JD. Changes in the cerebrospinal fluid pulse wave spectrum associated with raised intracranial pressure. *Neurosurgery* 1987;**20**(3):355–61.
35. Raksin PB, Alperin N, Sivaramakrishnan A, Surapaneni S, Lichtor T. Noninvasive intracranial compliance and pressure based on dynamic magnetic resonance imaging of blood flow and cerebrospinal fluid flow: review of principles, implementation, and other noninvasive approaches. *Neurosurg Focus* 2003;**14**(4):e4.
36. Rosner MJ, Becker DP. Origin and evolution of plateau waves. Experimental observations and a theoretical model. *J Neurosurg* 1984;**60**(2):312–24.
37. You W, Feng J, Tang Q, et al. Intraventricular intracranial pressure monitoring improves the outcome of older adults with severe traumatic brain injury: an observational, prospective study. *BMC Anesthesiol* 2016;**16**(1):35.
38. Arabi Y, Memish ZA, Balkhy HH, et al. Ventriculostomy-associated infections: incidence and risk factors. *Am J Infect Control* 2005;**33**(3):137–43.
39. Mayhall CG, Archer NH, Lamb VA, et al. Ventriculostomy-related infections. A prospective epidemiologic study. *N Engl J Med* 1984;**310**(9):553–9.
40. Holloway KL, Barnes T, Choi S, et al. Ventriculostomy infections: the effect of monitoring duration and catheter exchange in 584 patients. *J Neurosurg* 1996;**85**(3):419–24.
41. Ostrup RC, Luerssen TG, Marshall LF, Zornow MH. Continuous monitoring of intracranial pressure with a miniaturized fiberoptic device. *J Neurosurg* 1987;**67**(2):206–9.

42. Bochicchio M, Latronico N, Zappa S, Beindorf A, Candiani A. Bedside burr hole for intracranial pressure monitoring performed by intensive care physicians. A 5-year experience. *Intens Care Med* 1996;**22**(10):1070–4.
43. Latronico N, Marino R, Rasulo FA, Stefini R, Schembari M, Candiani A. Bedside burr hole for intracranial pressure monitoring performed by anaesthetist-intensive care physicians. Extending the practice to the entire ICU team. *Minerva Anestesiol* 2003;**69**(3):159–64. 165-8.
44. Zacchetti L, Magnoni S, Di Corte F, Zanier ER, Stocchetti N. Accuracy of intracranial pressure monitoring: systematic review and meta-analysis. *Crit Care* 2015;**19**:420.
45. Blei AT, Olafsson S, Webster S, Levy R. Complications of intracranial pressure monitoring in fulminant hepatic failure. *Lancet* 1993;**341**(8838):157–8.
46. Adelson PD, Bratton SL, Carney NA, et al. Guidelines for the acute medical management of severe traumatic brain injury in infants, children, and adolescents. Chapter 7. Intracranial pressure monitoring technology. *Pediatr Crit Care Med* 2003;**4**(3 Suppl): S28–30.
47. Miller JD, Bobo H, Kapp JP. Inaccurate pressure readings for subarachnoid bolts. *Neurosurgery* 1986;**19**(2):253–5.
48. Goodman JC, Robertson CS. Microdialysis: is it ready for prime time? *Curr Opin Crit Care* 2009;**15**(2):110–17.
49. Sanchez JJ, Bidot CJ, O'Phelan K, et al. Neuromonitoring with microdialysis in severe traumatic brain injury patients. *Acta Neurochir Suppl* 2013;**118**:223–7.
50. Stein NR, McArthur DL, Etchepare M, Vespa PM. Early cerebral metabolic crisis after TBI influences outcome despite adequate hemodynamic resuscitation. *Neurocrit Care* 2012;**17**(1):49–57.
51. Carre E, Ogier M, Boret H, Montcriol A, Bourdon L, Jean-Jacques R. Metabolic crisis in severely head-injured patients: is ischemia just the tip of the iceberg? *Front Neurol* 2013;**4**:146.
52. Roh D, Park S. Brain multimodality monitoring: updated perspectives. *Curr Neurol Neurosci Rep* 2016;**16**(6):56.
53. Purins K, Lewen A, Hillered L, Howells T, Enblad P. Brain tissue oxygenation and cerebral metabolic patterns in focal and diffuse traumatic brain injury. *Front Neurol* 2014;**5**:64.
54. Dizdarevic K, Hamdan A, Omerhodzic I, Kominlija-Smajic E. Modified Lund concept versus cerebral perfusion pressure-targeted therapy: a randomised controlled study in patients with secondary brain ischaemia. *Clin Neurol Neurosurg* 2012;**114**(2):142–8.
55. Chen JW, Rogers SL, Gombart ZJ, Adler DE, Cecil S. Implementation of cerebral microdialysis at a community-based hospital: A 5-year retrospective analysis. *Surg Neurol Int* 2012;**3**:57.
56. Nelson DW, Thornquist B, MacCallum RM, et al. Analyses of cerebral microdialysis in patients with traumatic brain injury: relations to intracranial pressure, cerebral perfusion pressure and catheter placement. *BMC Med* 2011;**9**:21.
57. Fantini S, Sassaroli A, Tgavalekos KT, Kornbluth J. Cerebral blood flow and autoregulation: current measurement techniques and prospects for noninvasive optical methods. *Neurophotonics* 2016;**3**(3):031411.
58. Vajkoczy P, Roth H, Horn P, et al. Continuous monitoring of regional cerebral blood flow: experimental and clinical validation of a novel thermal diffusion microprobe. *J Neurosurg* 2000;**93**(2):265–74.
59. Vajkoczy P, Horn P, Thome C, Munch E, Schmiedek P. Regional cerebral blood flow monitoring in the diagnosis of delayed ischemia following aneurysmal subarachnoid hemorrhage. *J Neurosurg* 2003;**98**(6):1227–34.
60. Cruz J. The first decade of continuous monitoring of jugular bulb oxyhemoglobin saturation: management strategies and clinical outcome. *Crit Care Med* 1998;**26**(2):344–51.

61. Brain Trauma Foundation, American Association of Neurological Surgeons, Congress of Neurological Surgeons Guidelines for the management of severe traumatic brain injury. X. Brain oxygen monitoring and thresholds. *J Neurotrauma* 2007;**24**(Suppl 1):S65–70.
62. Maloney-Wilensky E, Gracias V, Itkin A, et al. Brain tissue oxygen and outcome after severe traumatic brain injury: a systematic review. *Crit Care Med* 2009;**37**(6):2057–63.
63. Stippler M, Ortiz V, Adelson PD, et al. Brain tissue oxygen monitoring after severe traumatic brain injury in children: relationship to outcome and association with other clinical parameters. *J Neurosurg Pediatr* 2012;**10**(5):383–91.
64. Stiefel MF, Spiotta A, Gracias VH, et al. Reduced mortality rate in patients with severe traumatic brain injury treated with brain tissue oxygen monitoring. *J Neurosurg* 2005;**103**(5):805–11.

CHAPTER

21

Brain Edema in Developing Brain Diseases

Raffaella Moretti[1,2,3,4], *Vibol Chhor*[1,2], *Luigi Titomanlio*[1,2,4], *Bobbi Fleiss*[1,2,5] *and Pierre Gressens*[1,2,4,5]

[1]Université Paris, Diderot, Sorbonne, Paris Cité, Paris, France
[2]PremUP, Paris, France [3]Università degli studi di Udine, Udine, Italy
[4]APHP, Robert Debré Hospital, Paris, France
[5]St. Thomas' Hospital, London, United Kingdom

OUTLINE

Brain Edema.
DOI: http://dx.doi.org/10.1016/B978-0-12-803196-4.00021-7

INTRODUCTION

Brain edema was initially defined as an increase in net brain water content leading to an increase in tissue volume.[1] Extensive research has led to a more accurate definition, including pathophysiological characteristics such as molecular, cellular, structural, and functional changes in the blood–brain barrier (BBB), microcirculation, cell volume regulation, and cell death. With even small changes in cellular and extracellular volume, cerebral edema can compromise regional or global cerebral blood flow (CBF) and metabolism, or result in the compression of vital structures, given the fixed volume of the rigid skull and the secondary neuronal injury caused by swelling.[2]

Historically, two forms of cerebral edema have been described: "vasogenic," caused by impaired BBB function, and "cytotoxic," caused by sustained intracellular water accumulation independent of BBB leakage.[3] A new concept of "ionic/cellular" edema has since been introduced to characterize the entry of water into nonischemic tissues as opposed to cytotoxic edema, which is typical of dying cells (see Fig. 21.2, and pathophysiological descriptions below).[4] Other kinds of edema have also been described, such as "osmotic" edema secondary to osmotic imbalances between blood and brain tissue, and "hydrocephalic-interstitial" edema, related to an obstruction of cerebrospinal fluid (CSF) outflow.[5]

Edema is better understood in the adult field, but various forms of brain edema associated with different pathological conditions can be found in pediatric patients. In this chapter, we will characterize brain edema associated with various pediatric disorders, with a special focus on animal models and pathological findings, summarized in Table 21.1.

Neonatal Encephalopathy (NE)

A factor considered to be a cause of NE is perinatal and intrauterine asphyxia. NE associated with perinatal asphyxia and intrauterine asphyxia are causes of neonatal brain injury and are associated with substantial morbidity and mortality[6] including developmental and motor delays, cognitive impairment, epilepsy, and cerebral palsy.[7–9]

The classic model used to study NE associated with hypoxia–ischemia was created by Rice and Vannucci in 1981, and consists of the permanent occlusion of a carotid artery leading to systemic hypotension.[10–13] Within the first hours following carotid artery ligation, there is a regional increase of CBF followed by a decrease due to regional vasoconstriction or to a diminution of cardiac output. These effects lead to excitotoxicity, generation of free radicals, cell death, and edema.[14] This acute phase is followed within a few hours/days by a secondary phase characterized by a neuroinflammatory response, mitochondrial permeabilization, reperfusion, the

TABLE 21.1 Animal Models of Pediatric Brain Edema and Principal Findings

Author and Year	Animals Used and Age	Type of Insult	Results	Ref.
Nico, B. 2001	Chicks, 14-day embryos, 20-day embryos, and postnatal day (P) 2 chicks	–	Increased AQP4 levels and astrocyte end-foot membrane localization correlates with decreased BBB permeability (→ maturation)	1
Wen, H. 1999	Human brain	–	Increased AQP4 protein from the second week of development	2
Gomori, E. 2006	Human brain	–	Increase in AQP1 and 4 with gestational age	3
Hsu, M.S. 2011	Mouse brain	–	Developmentally regulated and layer-specific pattern of AQP4 immunolabeling	4
Dobbing, J. 1981	Young and adult rats	–	Increased water content in young compared to adult rats	5
Bauer, R. Fritz, H. 2004, referred to Dobbing 1981	Piglets [newborn, P14, 7 weeks (w), 14w], and rats (newborn, P8, P14, 6w)	Evaluation of age dependency of regional brain water content expressed by wet weight/dry weight measurements	Increase in water content of brain tissue in newborns compared to adults, with marked interspecies differences	6
Manley, G.T. 2000	AQP4 knock-out mice	Acute water intoxication (only cytotoxic edema) and pMCAo	Reduction of brain edema and improved neurologic outcome (possible compensation by other aquaporins?)	7

(Continued)

TABLE 21.1 Animal Models of Pediatric Brain Edema and Principal Findings (Continued)

Author and Year	Animals Used and Age	Type of Insult	Results	Ref.
Badaut, J. 2006	P10 spontaneous hypertensive rats	tMCAO (90 min)	MRI at 24 h: Global decrease in cellular swelling (ADC) and increase in vasogenic edema (T2) at 1 day after stroke, in the cortex and in the striatum but not in the border (perilesional) region AQP4 staining at 24 h: higher in the ipsilateral regions, with greater enhancement in the perilesional region MRI at 72 h: T2 signals augmented compared to 24 h, particularly in the striatum AQP4 staining at 72 h: lower in the border region. MRI and AQP4 staining at 28 days: return to normal of all indices except ADC in the perilesional area	8
Derugin, N. 2000	P7 rats	tMCAO	Decrease in ADC values immediately after tMCAO. Area of histological discoloration with triphenyltetrazolium (TTC) at 24 h corresponds closely with areas of decreasing ADC.	9
Adelson, P.D. 2011	Children	Retrospective study of children with TBI	Younger ages, later or early low mean CBF and lower vasoreactivity to CO_2 correlates with poorer outcome	10
Prins, M.L. 1996	P17, P28, and adult rats	Fluid percussion brain injury (FPI)	Greater mortality in younger rats with increased injury severity and more pronounced hypotension after TBI	11
Armsted, W.M. 1999	Newborn piglets (P1–3) and juvenile (3–4 w) pigs	FPI	CBF, pial artery diameter and cerebral oxygenation reduction greater in newborn vs juvenile pigs. Decreased response to nitric oxide (NO), cGMP- and cAMP-dependent stimuli following FPI in piglets	12

Armstead, W.M.	P1 and P4 piglets	FPI	Endothelin-1 significantly more upregulated after FPI in newborn piglets than in young pigs, corresponding to a greater reduction in regional CBF in neonates, suggesting that FPI effects on cerebral autoregulation during hypotension could be mediated by endothelin-1	13
Fukuda, a.m. 2012	P14 rats	CCI (controlled cortical impact)	MRI: global decrease in water mobility (ADC) and increase in edema (T2) at 1 day after trauma that normalize at 3 days	14
			AQP4 staining: increased in astrocyte end-feet at 3 days and 7 days as edema resolves. AQP1 does not correlate with edema changes	
Badaut, J. 2015	P14 rats	CCI	Cav-1 and Cav-2 augmented at 1 day and 7 days. Cld-5 and P-gp reduced at 3 days (concomitant with increased IgG staining), back to normal at 7 days (when no more IgG staining). eNOS augmented at 1 day, and decreased at 3 days	15
Pop, V. 2013	P14 rats	CCI	At 60 d: no change in IgG, Cld-5 augmented→no BBB disruption	16
			P-gp decreased→BBB changes	
			Accumulation of endogenous Amyloid β in several brain regions + behavioral changes on water maze test→early events cause long-term modifications	
Ajao, D.O. 2012	P14 rats	CCI	MRI evidence of brain edema at 3, 30 and 60 dpi	17
Timaru-Kast, R. 2012	2-month-old (young) and 21-month-old (aged) C57Bl/6N mice	CCI calculation of BWC (wet/dry brain weight)	Old animals have higher mortality rates compared to young ones. Increased ipsilateral Brain Water Content (BWC) (% of ipsiBWC-contraBWC) at 24 h is higher in aged mice compared to young ones	18

(Continued)

TABLE 21.1 Animal Models of Pediatric Brain Edema and Principal Findings (Continued)

Author and Year	Animals Used and Age	Type of Insult	Results	Ref.
Scott, R.C. 2002	Children	35 children with status epilepticus +/− history of prolonged febrile convulsions studied by MRI within 5 days of acute episode	T2 relaxation time augmented in children with PFC, returns to normal at day 2, and hippocampal volume augmented (hippocampal edema) at day 5. Nonfebrile status epilepticus patients have no hippocampal edema	19
Figueroa, R.E. 2005	Human pediatric patients	Diabetic ketoacidosis (DKA)	MRI-DWI and T2 relaxation time (T2R) at 6–12h after initial treatment of DKA and 96h after correction of DKA. T2 relaxation time increased during treatment in both white and gray matter, in comparison to absolute T2R values 96h after correction of DKA ($P = 0.034$). ADC values elevated instead, implying a significant water diffusion across cell membranes, correlated with the elevated white and gray matter T2R	20

1. Nico B, Frigeri A, Nicchia GP, Quondamatteo F, Herken R, Errede M, et al. Role of aquaporin-4 water channel in the development and integrity of the blood – brain barrier. *J Cell Sci* 2001;114(Pt 7):1297–307.

2. Wen H, Nagelhus EA, Amiry-Moghaddam M, Agre P, Ottersen OP, Nielsen S. Ontogeny of water transport in rat brain: postnatal expression of the aquaporin-4 water channel. *Eur J Neurosci* March 1999;11(3):935–45.

3. Gomori E, Pal J, Abraham H, Vajda Z, Sulyok E, Seress L, et al. Fetal development of membrane water channel proteins aquaporin-1 and aquaporin-4 in the human brain. *Int J Dev Neurosci: Official J Int Soc Dev Neurosci* August 2006;24(5):295–305.

4. Hsu MS, Seldin M, Lee DJ, Seifert G, Steinhauser C, Binder DK. Laminar-specific and developmental expression of aquaporin-4 in the mouse hippocampus. *Neuroscience* March 2011;178:21–32.

5. Dobbing J. *The later development of the brain and its vulnerability* 1981.

6. Bauer R, Fritz H. Pathophysiology of traumatic injury in the developing brain: an introduction and short update. *Exp Toxicol Pathol: Official J Ges Toxikologische Pathol* October 2004;56(1–2):65–73.

7. Manley GT, Fujimura M, Ma T, Noshita N, Filiz F, Bollen AW, et al. Aquaporin-4 deletion in mice reduces brain edema after acute water intoxication and ischemic stroke. *Nat Med* 2000;6(2):159–63.

8. Badaut J, Ashwal S, Tone B, Regli L, Tian HR, Obenaus A. Temporal and regional evolution of aquaporin-4 expression and magnetic resonance imaging in a rat pup model of neonatal stroke. *Pediatr Res* September 2007;62(3):248–54.

9. Derugin N, Wendland M, Muramatsu K, Roberts TP, Gregory G, Ferriero DM, et al. Evolution of brain injury after transient middle cerebral artery occlusion in neonatal rats. *Stroke; J Cereb Circ* 2000;31(7):1752–61.
10. Adelson PD, Srinivas R, Chang Y, Bell M, Kochanek PM. Cerebrovascular response in children following severe traumatic brain injury. *Pediatr Crit Care Med: J Soc Crit Care Med World Fed Pediatr Intensive Crit Care Soc* September 2011;27(9):1465–76.
11. Prins ML, Lee SM, Cheng CL, Becker DP, Hovda DA. Fluid percussion brain injury in the developing and adult rat: a comparative study of mortality, morphology, intracranial pressure and mean arterial blood pressure. *Brain Res Dev Brain Res* 1996;95(2):272–82.
12. Armstead WM. Cerebral hemodynamics after traumatic brain injury of immature brain. *Exp Toxicol Pathol: Official J Ges Toxikologische Pathol* 1999;51(2):137–42.
13. Armstead WM. Role of endothelin-1 in age-dependent cerebrovascular hypotensive responses after brain injury. *Am J Physiol* November 1999;277(5 Pt 2):H1884–94.
14. Fukuda AM, Pop V, Spagnoli D, Ashwal S, Obenaus A, Badaut J. Delayed increase of astrocytic aquaporin 4 after juvenile traumatic brain injury: possible role in edema resolution? *Neuroscience* October 2012;222:366–78.
15. Badaut J, Ajao DO, Sorensen DW, Fukuda AM, Pellerin L. Caveolin expression changes in the neurovascular unit after juvenile traumatic brain injury: signs of blood – brain barrier healing? *Neuroscience* January 2015;285:215–26.
16. Pop V, Sorensen DW, Kamper JE, Ajao DO, Murphy MP, Head E, et al. Early brain injury alters the blood – brain barrier phenotype in parallel with beta-amyloid and cognitive changes in adulthood. *J Cereb Blood Flow Metab: Official J Int Soc Cereb Blood Flow Metab* February 2013;33(2):205–14.
17. Ajao DO, Pop V, Kamper JE, Adami A, Rudobeck E, Huang L, et al. Traumatic brain injury in young rats leads to progressive behavioral deficits coincident with altered tissue properties in adulthood. *J Neurotrauma* July 2012;29(11):2060–74.
18. Timaru-Kast R, Luh C, Gotthardt P, Huang C, Schafer MK, Engelhard K, et al. Influence of age on brain edema formation, secondary brain damage and inflammatory response after brain trauma in mice. *PloS One* 2012;7(8):e43829.
19. Scott RC, Gadian DG, King MD, Chong WK, Cox TC, Neville BG, et al. Magnetic resonance imaging findings within 5 days of status epilepticus in childhood. *Brain: J Neurol* September 2002;125(Pt 9):1951–9.
20. Figueroa RE, Hoffman WH, Momin Z, Pancholy A, Passmore GG, Allison J. Study of subclinical cerebral edema in diabetic ketoacidosis by magnetic resonance imaging T2 relaxometry and apparent diffusion coefficient maps. *Endocr Res* 2005;31(4):345–55.

loss of CBF autoregulation, and more free-radical production.[15,16] Recent hypotheses also describe a tertiary phase, which occurs weeks–months–years after the injury and contributes to a persistent inflammatory state and possibly a reduction of brain plasticity.[17]

Neuropathological data suggest that brain edema is not the primary injurious process in either the term newborn suffering from NE related to hypoxic–ischemic encephalopathy (HIE) or in premature infants (born before 37 weeks gestation) with a known sentinel asphyxic event. It may appear within the first days of life in term infants and is related to the severity of tissue necrosis, but is rarely severe enough to cause transtentorial or transmarginal herniation.[18]

Stroke

Neonatal arterial ischemic stroke occurs in 0.25–0.4 per 1000 full-term births, and in approximately 7 per 1000 preterm births.[19] Many of the pathophysiological mechanisms that lead to brain injury are common to NE and neonatal stroke. However, the final patterns of injury can differ significantly in the case of neonatal stroke, depending on the extent and vulnerability of the areas irrigated by the occluded artery as well as the duration of the occlusion[19] (see Fig. 21.1). Brain edema is a serious complication of stroke. Early edema formation can significantly contribute to infarct formation. Swelling is commonly observed within the first week, reaching a peak at 24–72 hours following the ischemic event,[20] and it represents a key aggravating factor. Brain volume is limited by the rigidity of the skull, so that even minimum increases in volume lead to an increase in intracranial pressure (ICP) and compression of neural tissue

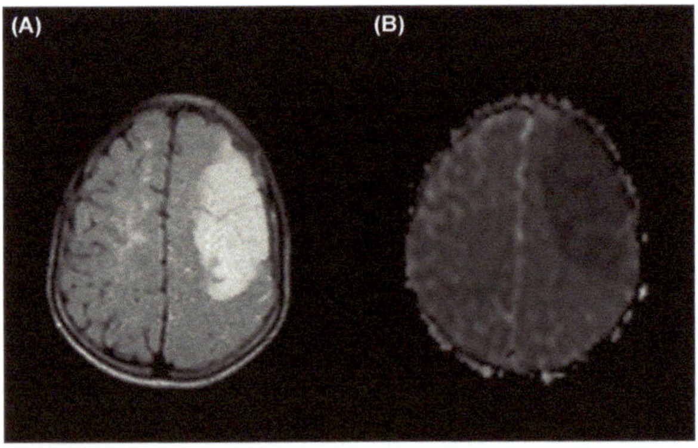

FIGURE 21.1 **MRI images (axial plane) of an 11-year-old child following an acute ischemic occlusion of the medial cerebral artery (MCA).** (A) T_2-weighted imaging (T_2WI) FLAIR acquisition. (B) Apparent diffusion coefficient (ADC) acquisition.

and vasculature. Edema therefore has a crucial impact on morbidity and mortality in stroke,[21] even if in neonates skull is less rigid, and elevation in ICP is less important.

Germinal Matrix Hemorrhage (GMH)

GMH is defined as the rupture of immature blood vessels in the subependymal brain tissue during the first week of life.[22] GMH is the most common neurological disorder of the preterm infant with an incidence of 3.5 per 1000 live births,[23] but GMH is associated with trauma, asphyxia, and clotting disorders. A model of GMH based on the intracerebral injection of collagenase (a hemorrhage-inducing agent) was created by Lekic et al. in order to generate a rupture of the ganglionic eminence into the lateral ventricles.[24] In this model, levels of free radicals, inflammation, blood cells, other blood components, and gliosis are higher in the forebrain of collagenase-injected rat pups compared to controls, and correspond to an early developmental delay followed by juvenile cognitive and motor disabilities.

In patients with severe GMH, edema is a secondary to tissue damage and inflammation caused by blood components and altered vascular permeability. The underlying causes of GMH are assumed to involve vascular instability leading to damage to the immature vasculature and the congestion of the vessels due to their shape. However, the specific underlying causes have not been confirmed and animal models chiefly rely on physical blockage of the vessels of the germinal matrix (such as collagenase, as above). However, it is speculated that GMH may occur secondary to localized edema and disruption of this vulnerable vascular-enriched area but the underlying pathological mechanism for the initiation of the edema is not explained.

Traumatic Brain Injury (TBI)

TBI affects 1.7 million people annually in the United States alone.[25] Motor impairment is the most common morbidity of TBI, but cognitive and behavioral deficits are also common. TBI is associated with long-term chronic neuroinflammation and increased risk of further neurodegeneration.[26–28] TBI causes more severe cognitive and behavioral deficits in children than comparable injuries in matured brains.[25,29,30] This is of particular importance when we consider that children under the age of four years are injured with TBI more frequently than any other age group,[31,32] and that injured infants under the age of 12 months are more likely to require intensive care support and are at high risk for a poor outcome.[33–35] A similar developmental sensitivity to TBI can be seen in rodent models, with the most severe lesions in postnatal day (P) 7 mice.[36] TBI is characterized by a primary injury resulting from an external mechanical impact on the brain, followed by various secondary events, including decreased CBF associated with hypometabolism, increased ICP, and BBB disruption leading to edema formation and brain swelling. Other mechanisms such as inflammation, oxidative stress, apoptosis, and excitotoxicity.[37]

Astrocytes and microglial cells seem to be affected by posttraumatic cellular edema (edema not associated with BBB disruption) within 2–3 hours after TBI.[38] Vasogenic edema displays a biphasic pattern of evolution, with a primary phase of BBB opening for 4–6 hours after the injury and a secondary phase for 2–3 days later.[39] TBI in children is more frequently associated with severe brain swelling than in adults.[32,40] Furthermore, children with diffuse brain swelling after TBI have a threefold higher mortality rate than those without.[41] Among the factors that could be responsible for this aggravated impact in pediatric brains are a reduced capacity for microvascular autoregulation,[42] increased postinjury blood flow, the different developmental and mechanical properties of the brain and skull,[32,43] exacerbated diffusion of excitotoxic neurotransmitters, increased water content, and an increased inflammatory response.[44]

Infectious Processes

Infections caused by viruses are ubiquitous during childhood and adolescence.[45] Viral infections that progress in severity to include edema typically present as cytotoxic edema in the early phase and vasogenic edema at later stages.[46] Magnetic resonance imaging (MRI) is an important tool to evaluate the type of edema (see later), and thus to differentiate between early and late stages of viral encephalitis. Interestingly, there is an excellent correlation between the apparent diffusion coefficient (ADC) in diffusion weighted imaging (DWI), which corresponds to cytotoxic edema (the first phase of viral infection), and the severity of the encephalitis.[46]

Bacterial meningitis is estimated to cause 125,000 child deaths each year world-wide,[47] with rates especially high in developing countries. Studies in rodents have revealed that cerebral edema observed in sufferers of meningitis is caused by BBB breakdown.[48–50] In fact, in neonatal wistar rats subjected to pneumococcal meningitis or meningitis due to *Streptococcus agalactiae* infection, pathogens readily cross the intact BBB and stimulate resident immune cells, which rapidly produce (within 6 hours) several pro-inflammatory cytokines such as IL-1ß, TNFα, IL-10 and IL-6, chemokines, and cytokine-induced neutrophil chemoattractant-1 (CINC1). These molecules recruit polymorphonuclear leukocytes (PMNs) and activate matrix metalloproteinases that contribute to BBB breakdown 12 hours after the insult.[50] Activated resident microglia and recruited PMNs induce oxidative stress through the production of reactive oxygen species (ROS) and reduction of endogenous antioxidant levels.[49,50]

Epilepsy

Epilepsy is a common neurological disorder that occurs more frequently in children than in adults. As with other processes discussed above, seizure-induced effects in the immature brain are different from those in the adult brain due to differences in neuronal vulnerability, synaptic reorganization, and regenerative processes.[51] Status epilepticus leads to vasogenic edema and this is considered to be detrimental in adults as well as in children. Specifically, in a study examining 35 children presenting with acute status epilepticus with or without a history of prolonged febrile convulsions (PFCs), there was an association between acute hippocampal edema and PFC, as suggested by increased hippocampal volume and relaxation time on T2-weighted MRI.[52]

Another interesting epileptic syndrome first described by Gastaut et al.[53] is hemiconvulsion–hemiplegia–epilepsy (HHE) syndrome. This disorder is an uncommon consequence of prolonged focal febrile seizures in infancy and early childhood. It is clinically characterized by prolonged clonic seizures with a unilateral predominance, occurring in children with fever, usually before the age of four years, and followed by the appearance of hemiplegia. Neuroimaging shows the acute occurrence of unilateral brain swelling, followed by unilateral hemispheric atrophy, and the subsequent appearance of epilepsy. During the acute phase, edema can be substantial, resulting in life-threatening events such as temporal lobe herniation, and reduced ADC values, consistent with the cytotoxic nature of the edema.[54] Auvin et al. have suggested that the combination of prolonged seizure activity, hyperthermia and BBB damage (due to virus-related inflammation) with genetic factors, the progressive maturation of the corpus callosum or preexisting epileptogenic foci are the possible mechanism in HHE syndrome.[55] Among genetic factors, some patients have a heterozygous S218L mutation in the *CACNA1A* gene that encodes the main subunit (1A) of the neuronal P/Q type voltage-gated ion channel.[56] A mutation in this gene has been described in familial hemiplegic migraine, where some reports indicate the presence of localized brain edema.[57] These data strongly suggest the shrinkage of the extracellular space and the accumulation of intracellular water.[55]

Diabetic Ketoacidosis (DKA)

DKA occurs in 25%–40% of newly diagnosed diabetes cases. Cerebral edema is the most significant complication in children with DKA, excluding iatrogenic brain edema due to rapid sodium correction. In a study of pediatric patients with severe DKA, T2 relaxation time was significantly increased within the first 6–12 hours of corrective treatment in both the white and gray matter, in comparison to values for 96 hours after correction. ADC values were elevated, implying a significant amount of water diffusion across the cell membrane. That was correlated with an elevated white and gray matter T2 relaxation time,[58] showing that vasogenic edema is more prevalent secondary to BBB disruption. In another study examining two cases of fatal brain edema associated with DKA, there were typical signs of BBB disruption: the absence of tight-junction proteins such as occludin, claudin-5 (Cld-5), ZO-1, and JAM-1 in parenchymal blood vessels, and albumin extravasation. Concomitantly, the neuroinflammatory markers CCL2, NF-kappa B, and nitrotyrosine were localized in areas perivascular to the disrupted BBB and present in the adjacent brain parenchyma, suggesting a role for neuroinflammation in BBB disruption.[59]

Hereditary Encephalopathy/Leukodystrophies

The leukodystrophies are a heterogeneous group of untreatable genetic diseases characterized by defects in central nervous system (CNS) myelin formation or maintenance. Most leukodystrophies appear during the first year of life or later in childhood and represent an important cause of progressive mental and motor disability.[60] They are traditionally divided into two main groups: (1) hypomyelinating leukodystrophies, where the amount of myelin is low due to defects in the production or quality of myelin and (2) demyelinating leukodystrophies, where myelin degenerates because of an impaired enzyme or because of astrocyte dysfunction, causing progressive cysts or myelin vacuolation and degeneration.[61] In the second category, edema is a consequence of this vacuolization, and can be clinically manifested as macrocephaly. For example, Alexander disease which is due to a coding mutation in the glial fibrillary acidic protein (GFAP) gene or megalencephalic leukoencephalopathy with subcortical cysts (MLC). In a transgenic mouse line (Tg73.7) overexpressing human GFAP, water content is significantly augmented compared to controls.[62] In megalencephalic leukoencephalopathy, brain MRI reveals swelling of the white matter with highly increased water content and subcortical cysts.[63] In most patients, the disease is caused by a mutation of the *MLC1* gene, encoding a plasma membrane protein found at astrocyte–astrocyte junctions and the brain–CSF barrier.[64] Its mutation disrupts barrier permeability. *Mlc1* mutation in mice is associated with decreased levels of the adhesion protein GlialCAM and the chloride channel ClC-2.[65,66] In addition, ClC-2 deficiency is found in other leukoencephalopathies, affecting brain ion homeostasis and characterized by intramyelinic edema.[67,68]

CNS Tumors

In a recent epidemiological study that carried out a 10-year retrospective evaluation of the characteristics of CNS tumors in children under two years of age, astrocytoma was the most common tumor type (31.3%), followed by ependymoma (28.1%), and medulloblastoma (18.8%).[69] Microvessels of brain tumors lose their BBB properties and allow fluid to enter the brain parenchyma.[70,71] In addition to BBB disruption, the deregulation of water entry should be considered; tumor astrocytes present with a massive up-regulation of AQP4. In gliomas, AQP4 is strongly upregulated and distributed throughout the cell surface. In human astrocytomas, AQP4 overexpression in reactive astrocytes surrounding the tumor is associated with vasogenic edema.[72]

Tools for Functional Edema Studies: MRI

In clinical practice, DWI and T2-weighted imaging are routinely used for diagnosis/prognosis in patients with stroke and other brain diseases. The ADC, the quantitative DWI parameter, is thought to reflect water mobility within brain tissues. Under conditions of reduced extracellular space from cell swelling, as seen in anoxic or ionic edema, the mobility of water molecules is restricted, resulting in a lower ADC. In the case of vasogenic edema, a high signal on DWI is observed in the parenchyma, but is still lower than that in the CSF.[73] The T2 value is representative of water content and increased T2 values correspond to water accumulation (vasogenic edema, for more details see Chapter 4: Experimental Techniques to Investigate the Formation of Brain Edema In Vivo).[73] Aquaporins (AQPs) facilitate water diffusion through the plasma membrane and they can contribute to altered ADC values in the normal brain. Preclinical studies in adult rodents show that there is a correlation between AQP4 expression and ADC values. In a model of hydrocephalus, periventricular AQP4 are related to the severity of hydrocephalus, which is probably a protective response. This study also suggested that ADC informed about cell sizes and interstitial bulk water, but might indirectly reflect quantitative water channel expression. This was confirmed in a study wherein knocking down AQP4 expression induced a significant decrease in ADC values without histological modifications.[74] ADC histograms of the entire intracranial content change with age, reflecting changes in water content: ADC peaks at 0–2 years, then decreases exponentially up to two years, and then, with a lower rate from 2 to 20 years.[75] To our knowledge, no studies have directly evaluated ADC changes across ages and AQP4 levels, but a correlation may be probable.

Another increasingly utilized technique is diffusion tensor imaging (DTI), which is an advanced MRI acquisition technique that can be used to investigate tissue microstructure and fiber architecture of cerebral white matter. DTI provides valuable information about axon orientation and the integrity and coherence of white matter tracts. The most common method to evaluate the degree of anisotropy (i.e., the directional nature of the signal) and the direction of nerve fibers is to evaluate the average coefficient of water diffusion and fractional anisotropy (FA). FA increases during myelination and brain maturation, and is decreased in various diseases.[76,77]

PATHOPHYSIOLOGY

Historically, brain edema has been divided into two main forms: "vasogenic" edema, due to BBB disruption resulting in extracellular water accumulation, and "cytotoxic/cellular" edema, due to sustained intracellular water accumulation.[5] Recent advances in BBB function and molecular mechanisms of edema formation suggest more complex pathological processes[4] (Fig. 21.2).

Cytotoxic/anoxic edema: Anoxic edema is the passage of liquid into the intracellular space, and causes a reduction of the fluid content of the extracellular space, and this occurs within minutes after cessation of blood flow (Fig. 21.2a). Cytotoxic/anoxic edema is not characterized by swelling of necrotic tissues and is not dependent on the recruitment of fluids or ions from the intravascular space. Cytotoxic/anoxic edema is a result of energy depletion (such as caused by oxygen and glucose deprivation), followed by a slow rise in extracellular k^+, that induces a disruption of the normal cellular ionic gradients due to a failure of the Na^+/k^+-ATPase. The potassium channel $K_{ir}4.1$ is often in close proximity with aquaporin channels and may facilitate water input through AQP4 on astrocytic end-feet.[4] Although cytotoxic/anoxic edema is not characterized by tissue swelling, this process of water and ion movement out of the extracellular space induces an ion and water gradient that generates a transcapillary effect. This can lead to the formation of ionic and vasogenic edema (see below).

Ionic/cellular edema: In accordance with the basic principle established by Starling, edema formation requires two features: the driving force that "pushes" substance to the brain and the "permeability pore," which allows a transcapillary passage of these molecules from the intravascular to the extracellular space.[80] In this type of edema the driving force is generated by cytotoxic/anoxic edema, but new fluid entry into the CNS is required. The majority of the theories suggest that this fluid originates from change of BBB permeability properties. This change is induced by ischemia, and it is clearly distinct from vasogenic edema (see below) because it involves abnormal Na^+ transport when there is normal exclusion of protein by the BBB.[3] Ischemia leads to the cessation of primary active transport via Na^+/k^+-adenosine triphosphatase (ATPase). As a result, the secondary active transport functions (mediated by constitutive or induced cotransporters) or passive transport (mediated by ionic channels) try to maintain cellular processes. By doing so, neurons and neuroglia accumulate osmotically active solutes intracellularly that cause cellular swelling within 30 minutes after reperfusion.[4] AQPs, and in particular AQP4 (the most abundant water channel in the brain), are heavily implicated in brain edema formation, but in many theories the first driver seems to be Na^+. Na^+ generates the electrical gradient for Cl^- and the osmotic gradient for water. During the replacement of Na^+, Cl^-, and water in the extracellular space that was depleted by formation of anoxic edema Na^+ accumulated exceeds the amount of k^+ lost. This is because k^+ ions that should have been extruded to the cells by the nonselective cation channels remain largely bound to negatively charged intracellular proteins and other macromolecules[81] giving a net inflow of Na^+ into edematous brain.[78] Intracellular accumulation of Na^+ in this "ionic edema" phase occurs via a multitude of transporters, including ion channels that can both constitutive channels activated by a hypoxic insult, or ischemia-induced transporters. The constitutive channels expressed by neurons and neuroglia are tetrodotoxin-sensitive Na^+ channels, $Na^+/k^+/Cl^{2-}$ cotransporter (NKCC1), or *N*-methyl-D-aspartate receptor channels. Examples of nonselective-induced cation channels are sulfonylurea receptor-1 (SUR1)-regulated $NC_{Ca\text{-}ATP}$ channel (see also Chapter 12: Edema and BBB Breakdown in Stroke). The intracellular Na^+ cannot then be extruded, due to a reduction in ATP levels due to ischemia. During ionic/cellular edema, the integrity of the BBB is maintained, or is only transiently leaking (Fig. 21.2b).

Ionic/cellular edema is also proposed to be developed via another mechanism, involving glial cells and in particular astrocytes (part of the neurovascular units, NVUs) constitute a second "barrier." This glial barrier is called the glymphatic pathway and this is proposed to create a paravascular highway for fluid exchange. This paravascular compartment is continuous along penetrating arterioles to capillaries and along veins. The glymphatic pathway hypothesis suggests that fluid is entering the brain parenchyma as CSF via the low-resistance para-arterial space. CSF is produced by cells of the choroid plexus that have higher NKCC1 expression levels compared to endothelial cells.[82] This NKCC1 can be induced by ischemic injury.[83] Para-arterial CSF flux increases during periods of high ICP when pericytes constriction and microvascular collapse could force the paravascular space to expand, decreasing resistance to influx of fluid. As opposed to the BBB the perivascular astrocyte end-foot barrier provides no osmotic resistance to diffusion due to AQP4 expression. As

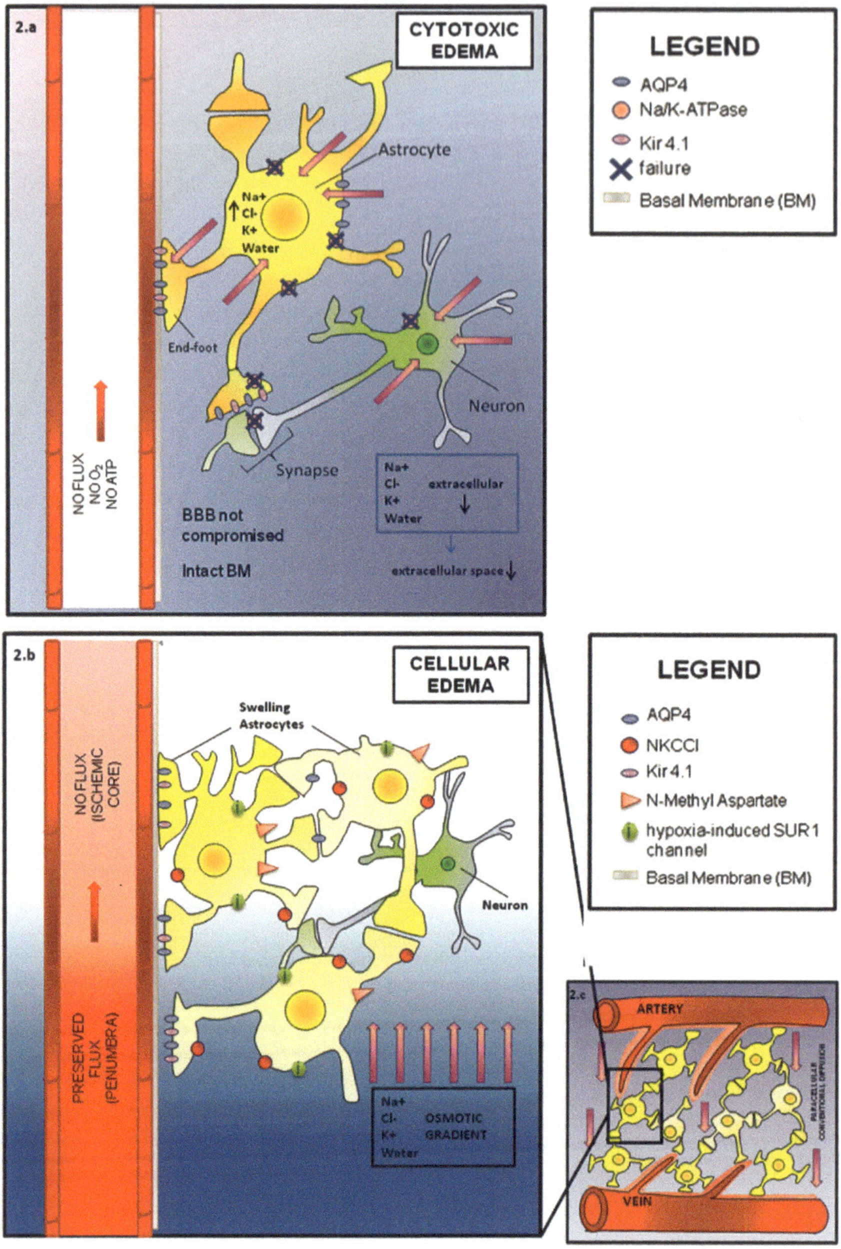

FIGURE 21.2 **Schematic representation of brain edema formation including different types of edema.** *2A. Cytotoxic/anoxic edema:* A loss of perfusion leads to oxygen and glucose deprivation and a slow rise in extracellular K^+ and failure of the Na^+/K^+-ATPase, leading disrupted cellular ion gradients. The K^+ channel $K_{ir}4.1$ is often located close to aquaporin (AQP) channels and may facilitate the entrance of water via AQP4 on astrocytic end-feet.[4] *2B Ionic/cellular edema:* Ischemia leads to the cessation of primary active transport via the Na^+/K^+-ATPase leading to the intracellular accumulation of osmotically active solutes and cell swelling. Na^+ accumulation generates an electrical gradient for Cl^- and an osmotic gradient for water leading to a replenishing influx of Na^+, Cl^-, and water into the extracellular space (depleted by the formation of anoxic edema), and a net inflow of Na^+ into the edematous brain.[78] The intracellular accumulation of Na^+ during ionic edema is made possible by various transporters, such as constitutive ion channels (i.e., tetrodotoxin-sensitive Na^+ channels,

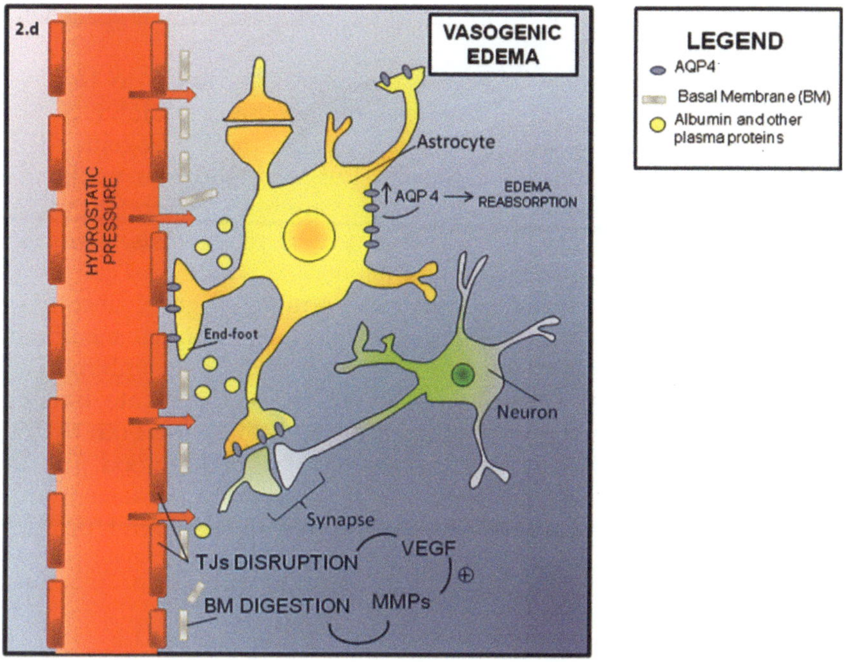

◀ the $Na^+/K^+/Cl^-$ cotransporter (NKCC1), or *N*-methyl-D-aspartate receptors) or ischemia-induced channels (i.e., sulfonylurea receptor-1 (SUR1)-regulated $NC_{Ca\text{-}ATP}$ channel). At this stage, the integrity of the BBB is maintained, or there are transient leaks after reperfusion due to excess pressure and shear stress. *2C. Glymphatic paravascular fluid flow:* Astrocytes, which cover the 97% of the vasculature with their end-feet, constitute a second barrier, called the "glymphatic" pathway, creating a paravascular "highway" to fluid exchange. In this case, edema arises from the brain parenchyma in the same way as the cerebrospinal fluid (CSF), i.e., via the low-resistance para-arterial space. The para-arterial CSF flux increases with intracranial hypertension, and pericyte constriction and microvascular collapse could force the paravascular space to expand, decreasing resistance to the influx of fluid. The perivascular astrocytic end-foot barrier provides no osmotic resistance to diffusion thanks to AQP4 expression, and the osmotic gradient could thus drive CSF into the extracellular space.[79] *2D. Vasogenic edema:* Vasogenic edema is characterized by the breakdown of BBB, with the leakage of plasma proteins into the extracellular space. Changes in BBB permeability are due to reverse pinocytosis, the disruption of Ca^{2+} signaling, actin-polymerization-dependent endothelial cell rounding-up or retraction with the formation of inter-endothelial gaps, the uncoupling of tight junctions, and enzymatic degradation of the basement membrane. Once the integrity of the BBB is lost, both osmotic and hydrostatic pressure gradients contribute to edema formation. A second peak of AQP4 expression has been noted at this stage, probably related to water re-absorption.[4]

such, the osmotic gradient could drive CSF into extracellular space without restrictions[79] (Fig. 21.2c).

Vasogenic edema: Vasogenic edema develops at >3 hours to 14 days[79] after injury and in humans is characterized by the breakdown of BBB, with leakage of plasma proteins into extracellular space. Vasogenic edema involves the extravasculation of proteins but not blood cells via the disrupted BBB. Several mechanisms have been proposed to account for these changes in BBB permeability, including reverse pinocytosis, disruption of Ca^+ signaling, actin-polymerization-depending endothelial cell rounding or retraction with formation of inter-endothelial gaps, uncoupling of tight junctions, and enzymatic degradation of basement membrane. A lot of ischemia-induced factors or enzymes, such as thrombin, VEGF, and MMPs, are founded to promote these mechanisms into ischemic experimental models. Once the integrity of the BBB is lost, both osmotic and hydrostatic pressure gradients, related to systemic blood pressure and ICP, contribute to edema formation. In addition to a transient MCA occlusion model, a second peak of AQP4 expression was found, probably linked to water re-absorption[4] (Fig. 21.2d).

The Molecular Protagonist of Edema Formation in the Developing Brain

Water channels: Little is known about AQPs in the developing brain. In a study conducted in 14- and 20-day-old chick embryos and 2-day-old chicks, Nico et al. demonstrated that AQP4 expression in astrocytic end-feet occurs in parallel with the maturation and functioning of the NVU, suggesting that there is a close relationship between water transport regulation and brain development.[84] As brain water content is higher in juvenile rats than in adults,[85] the implication of AQPs can be supposed. This is supported by evidence that AQP expression in humans and rodents increases during development.[86–88] In a murine model of neonatal stroke, unlike in adults, MRI data show a global decrease in cellular swelling and an increase in vasogenic edema 1 day after stroke in the cortex and striatum but not in border (perilesional) regions.[89] AQP4 labeling at this time point is higher in the ipsilateral regions, with greater enhancement in astrocytic end-feet of the perilesional region. These data suggest that the expression of AQP4 is increased during the period of active edema formation in the perilesional region, and that its induction in astrocytic end-feet could participate in tissue preservation in the immature brain.[89] In 17-day-old rats, compared to adult rats, after controlled cortical impact (CCI), there is an acute bilateral increase in brain water content as shown by T2-weighted MRI, and decreased ADC values. AQP4 labeling in astrocytic end-feet is decreased at 1 day, but increases at 3 and 7 days as the edema resolves.[44] Overall, MRI and immunohistochemical data suggest that edema formation in the immature brain may have anoxic and vasogenic components at different time points.

The Neurovascular Unit (NVU)

The BBB is a physical barrier essential for the maintenance of a regulated intracerebral microenvironment. Its characteristics limit paracellular diffusion but allow a controlled endocytosis of larger molecules and the intake of smaller nutrients such as glucose, insulin, and iron. Endothelial cells interact closely with other CNS cells such as neurons, pericytes, and astrocytes, through adherens junctions, influx and efflux transporters, metabolic enzymes, extracellular matrix, astrocytic end-feet, etc. All these structures together constitute the NVU, and are believed to be essential for the regulation of BBB permeability.[90]

Considered immature for years, studies during the last few decades have demonstrated that tight junctions at the BBB are already effective at the earliest stages of development.[91] Indeed, the high protein concentration of the embryonic CSF appears to be a consequence of transcellular protein transfer across the epithelial cells of the choroid plexus rather than a consequence of tight-junction immaturity.[92] Tight junctions are impermeable even to small molecules, and selective protein transfer is necessary for ventricular expansion and CNS development.[93,94] On the other hand, differences in BBB permeability during development could be an important contributing factor to more severe and extensive edema formation in many pediatric neurological disorders.

Caveolins are structural proteins involved in the formation of caveolae, which are known to be involved in transcytosis, endocytosis, and exocytosis in endothelial cells. Caveolin-1 (Cav-1) is found in endothelial cells, in neuronal membranes, and in reactive astrocytes, and is implicated in the regulated endocytosis of receptors and ligands, the inhibition of eNOS and the stabilization of Cld-5 and P-gp, a tight-junction protein, and an ATP-binding protein of the BBB, respectively. After juvenile TBI, Cav-1, and Cav-2 show a biphasic behavior with increased levels at 1 and 7 days after TBI and a slight decrease at 3 days, compared to sham animals.[95] Cld-5 and P-gp labeling, however, decreases at 3 days while IgG labeling (a method to evaluate BBB disruption) is increased. Conversely, Cld-5 and P-gp significantly increase at 7 days with a concomitant decrease in IgG labeling to normal levels. The activation of endothelial nitric oxide synthase (eNOS) via its phosphorylation increases only at 1 day,[95] suggesting that following juvenile TBI, molecular changes involving caveolins in endothelial cells might participate in BBB repair independently of eNOS activation. The mechanisms by which they act are still unknown, but interestingly, changes in Cav-1 levels in astrocyte cultures correspond to changes in the expression of the glutamate transporter GLT1, and the application of transforming growth factor 1 (TGF-1) induces a decrease in Cav-1 and an increase in GLT1, which is associated with reduced glutamate reuptake and increased excitotoxicity.[95] Excitotoxicity and brain inflammation contribute to ionic edema, and BBB disruption is the main mechanism of the vasogenic edema formation. Studies that correlate caveolin changes with edema formation are needed to further explore this possible link.

Concerning transporters, a recent study evaluated monocarboxylate transporters (MCTs), which are expressed on the membranes of neurons, endothelial cells, astrocytes, and microglia. The MCTs are proton-linked membrane carriers involved in the transport of monocarboxylates such as lactate, pyruvate, and ketone bodies.[96] MCTs are highly expressed on the BBB and their expression is decreased after P15 in rodents, and levels in the brain decline with maturity in humans. Concordant with this statement, a ketogenic diet, which increases levels of MCTs, leads to preserved cortical tissue and behavioral improvements one-week post-CCI only in young rats. These effects could be the consequence of a global reduction in ROS and anoxic edema, induced by increased ketone metabolism.[97,98]

Another domain of research concerns intracellular signaling pathways, in particular the c-Jun N-terminal kinase (JNK) signaling pathway. Mitogen-activated protein (MAP) kinases are important molecules that play a critical role in the regulation of several cellular processes such as growth, proliferation, differentiation, and cell survival/apoptosis.[99] The JNKs represent a subfamily of the MAP kinase group, and are also known as stress-activated protein kinases because they are activated by a wide variety of external stressors. There are three main JNK genes, *Jnk1, 2,* and *3*, that after transcription lead to 10 different JNK proteins. The JNK3 isoform is the most prevalent in the brain.[100] The JNK proteins are activated by a typical three-step phosphorylation cascade: JNK is activated by the MAPK kinases MKK7 and MKK4, which themselves are activated by MKK kinases (MEKKs or MAP3Ks). The scaffold protein JNK-interaction protein (JIP) facilitates the activation pathway by binding all elements. Once activated, the JNKs phosphorylate c-Jun, the dominant component of transcriptional factor Ap-1, inducing the transcription of numerous target genes involved in proliferation, cell survival, cell death, DNA repair, metabolism, and inflammation.[101] This pathway is immediately activated following both hypoxic–ischemic insult and TBI.[102] High levels of JNK are acutely induced in animal models of fluid percussion injury (FPI),[102] and seem to impair the stress-induced vasodilatation mediated by NMDA activation and thus contribute to poorer outcomes.[103] c-JNK activation starts 3 hours after ischemia in the ischemic core and peaks at 6 hours both in the core and in the penumbra. I.p. administration of the JNK inhibitor D-JNKI1 in young rats using the Renolleau model of permanent cerebral artery occlusion (pMCAo) and transient common carotid artery (tCCA) occlusion reduces lesion size by 77%, and caspase-3 activation by 40% in the core.[102] Concerning brain edema, effects of inhibition of JNK pathway have been evaluated in a mice model of ICH induced by intrastriatal collagenase injection. Inhibition of JNK pathway improves neurological outcome and decreases lesion volume early after ICH, and causes a decreased hemispheric swelling.

THERAPEUTIC TARGETS

Symptomatic Treatment and Current Treatment Strategies

Despite differences between pediatric and adult edema, treatments for pediatric edema are still derived from adult therapeutics, without taking into consideration the structural and functional differences of the developmental stages.[104] Compared to adult, pediatric brains have reduced capacity for microvascular autoregulation,[42] increased postinjury blood flow, different developmental and mechanical properties of the brain and skull,[32,43] exacerbated diffusion of excitotoxic neurotransmitters, increased water content, and an increased inflammatory response.[44]

Indeed, there is strong preclinical evidence that the developing brain displays age-dependent parameters for CBF and metabolism, development-specific biomarkers, increased incidence of posttraumatic seizures,[105] different sensitivity to medications (e.g., increased apoptosis triggered by anesthetics and anticonvulsants), and different mechanisms of neuroplasticity during recovery. In addition, differences between the respective contributions of inflammation and excitotoxicity in the developing brain suggest that the pediatric brain may benefit from therapeutic interventions that are less useful in adults, such as decompressive craniectomy or hypothermia.[43]

Osmotic therapy: The 2012 guidelines for the management of raised ICP in severe brain injury[106] suggest the use of mannitol or osmotic agents to create an osmotic gradient in well-perfused cerebral structures. However, there are concerns that it may cross the injured BBB and cause an exacerbation of cerebral edema. Hypertonic saline is also a possible therapeutic tool that decreases ICP and improves outcome in pediatric TBI patients.[106] Additional benefits include the restoration of normal cellular resting membrane potential and cell volume, stimulation of atrial natriuretic peptide release, inhibition of inflammation, and enhancement of cardiac output. Potential concerns with the use of hypertonic saline include a rebound increase in ICP, central pontine myelinolysis, subarachnoid hemorrhage, and hyperchloremic acidosis.[107]

Hypothermia: Independently of the type of injury, brain metabolism is reduced by about 5%–7% per degree (°C) of reduction in the core temperature,[108] and this reduced metabolic rate causes vasoconstriction, resulting in reduced cerebral blood volume, and a more favorable balance between oxygen and glucose supply and demand. It is standard clinical practice to apply either selective head cooling or whole body hypothermia in infants diagnosed with HIE and this reduces brain swelling, improves outcomes, and reduces mortality.[109–111] These effects on tissue swelling were demonstrated across animal models in the preclinical testing of hypothermia.[112] A meta-analysis of TBI cases conducted in 2014 did not show any evidence of improvement with hypothermia,[113] but studies are needed in the pediatric population.

Decompressive craniectomy: In this procedure, part of the skull is removed to allow the brain to expand, and thus to reduce ICP. At the stage of ionic edema, hydrostatic pressure and by extension tissue pressure is unimportant for edema formation. In contrast, in the vasogenic phase, tissue pressure is a critical determinant of edema formation. Thus, decompressive craniectomy can be safe if performed early, when the BBB is intact, because it may help avoid hypoxia by promoting reperfusion. In contrast, if decompressive therapy is carried out later, it can decrease tissue pressure and drive the formation of vasogenic edema, leading to deleterious effects.[3] Specific characteristics lead to the observations that children may benefit even more from decompressive craniectomy compared with adult patients especially with regard to long-term clinical outcome.[114] Specifically, (1) myelination is not complete before the age of 18–24 months, making the brain of an infant more pliable and (2) the skull is still smooth and the cranial sutures are open up to the age of 1.5 years allowing expansion during high ICP. Thus, brain swelling may occur even without significant elevation of the ICP. In contrast, in older children, the brain is more voluminous and has less compensatory reserve to tolerate increased ICP.

Novel Treatments for the Developing Brain

In the past few decades, several series of experiments have demonstrated that injured tissues release substances that enhance both vasogenic and cytotoxic/cellular brain edema, in particular glutamate, H^+ ions, Ca^{2+} ions, arachidonic acid and its metabolites, oxygen free radicals, histamine, and kinins.[5] Unfortunately, there are no reports concerning the clinical effects after the administration of steroids, glutamate receptor antagonists, Ca^{2+} antagonists, superoxide dismutase, or bradykinin receptor inhibitors.[115] In addition, most of them have never been tried in the pediatric population or can actually increase the opposite response. For instance, while NMDA receptors' inhibition seems to be antiapoptotic and reduces the excitotoxic response, it also substantially exacerbates neuronal death during development.[43,116] Studies using neonatal animal models to study

TABLE 21.2 Studies Regarding Therapies for Pediatric Brain Edema

Author and Year	Animals Used and Age	Type of Insult	Results	Ref.
Fernandez-Lopez, 2007	P7 rats	Rice-Vannucci HI model (unilateral common carotid ligation + systemic 8% O_2 hypoxia) + cannabinoid receptor inhibitor WIN	Reduced area of necrosis and ADC at 7 days	1
Vannucci 1999	P7 rats	Rice-Vannucci HI model (unilateral common carotid ligation + systemic 8% O_2 hypoxia +hypothermia of 3°C or 6°C)	Tissue swelling 26.9%, 5.3%, and 0.3% greater than controls at 37°C, 34°C, and 31°C, respectively	2
Mazarati, a. 2009	P11, P14, Pp30 wistar rats	Kindling procedure (electrical stimulation of ventral hippocampus) + bumetanide (0.5 mg/kg) injection	At P11, bumetanide increases the baseline hippocampal after-discharge threshold and shortens the after-discharge duration At P14 and P30; bumetanide induces no significant antiepileptic effects (the presence of more NKCC1 transporters in P11)	3
Dubey, M. 2015	Mlc1-null mice P21→adult heterozygote E18→adult	–	MLC1 expression augmented from P21 in heterozygotes and wild-type mice (WT). MLC1-null mice presented a significant increase in brain wet weight (brain edema) compared to controls from P21.	4

(Continued)

TABLE 21.2 Studies Regarding Therapies for Pediatric Brain Edema (Continued)

Author and Year	Animals Used and Age	Type of Insult	Results	Ref.
Repici, M. 2007	P14 wistar rats	pMCAo+ tCCA occlusion (Renolleau model) +/− JNK inhibitor (D-JNKI1) i.p. injection	77% reduction in lesion size in D-JNKI1 animals. 40% reduction in caspase-3 mediated neuronal death in the core	5
Borsello, T. 2003	P14 wistar rats	pMCAO+ tCCA occlusion (Renolleau model) +/− JNK inhibitor (D-JNKI1) i.p. injection	78% and 49% reduction in lesion size, compared to control, at 3h and 6h JNK inhibitor i.p. administration, respectively	6
NiJober, C.H. 2010	P7 rats	Rice-Vannucci model: CCA occlusion + 120min hypoxia +/− short-lived inhibitor of JNK (TAT-JBD) 1mg/kg i.p. injection	Treatment at 0 and 3h after HI: 30% reduction in loss of ipsilateral MAP2 (neuronal marker); concomitant 57% reduction in HI-induced caspase-3 activation, functional motor and cognitive improvement. Treatment at 6h does not reduce MAP2 loss	7
Ginet, V. 2009	P7 rats	Rice-Vannucci model: CCA occlusion + 120min hypoxia +/− JNK inhibitor (D-JNKI1) 0.3mg/kg, repeated i.p. injections (30min, 3h, 5h, 8h, 12h, 20h after HI)	26%–29% reduction of caspase-3 in cortex and striatum and 35% reduction in the thalamus in D-JNKI1 treated animals, but no reduction in lesion size	8
Ditelberg, J.S. 1996	P7 transgenic (Tg) mice expressing three times normal amounts of copper/zinc-superoxide dismutase (CuZn/SOD)	Right CCA ligation followed by 90min of hypoxia	Lesion volume and histological scoring in cerebral cortex and striatum 5 days after injury: Tg mice have higher lesion volumes and worse histologic score	9
Armstead, W.M. 2013	P1–P5 piglets	FPI+/− Dopamine (DA) i.v. injections	DA increases cerebral perfusion pressure and prevents sex-dependent reduction in pial artery diameter after FPI (which is greater in male). DA blocks endothelin-1 and ERK MAPK up-regulation equally in both sexes (even if FPI induces greater up-regulation in males)	10

Pohl, D. 1999	P7 rat	Weight-drop model of TBI + NMDA	Reduced excitotoxicity but enhanced neuronal death (in developing brain)	11
Robertson, C.L. 2015	P17–21 male and female rats	CCI+/− progesterone i.p at 1 h, s.c. at 6 h	In males: TBI reduces mitochondrial respiratory control ratio (RCR; a ratio of the active state to the resting state), RCR is preserved with progesterone treatment. In females: reduced tissue loss at 7 days after CCI	12
Fukuda, A. 2013	P17 rats	TBI + siAQP4 (inhibitor)	SiAQP4 reduces AQP4 expression by 30%, decreases T2 (edema) and ADC (water mobility) at 1 day and 3 days, reduces BBB disruption (decreased IgG extravasation, augmented Cld5 immunoreactivity), reduces GFAP labeling (astrogliosis) and Iba-1 labeling (microglial activation), and increases NeuN-positive cells (neuroprotection) associated with neurobehavioral improvement both at 3 days and 60 days	13
Prins, M.L. 2005	P17, P35, P45, P65 rats	CCI + ketogenic diet	58% and 39% reduction in cortical contusion in P35 and P45 rats, no effects in P17 or P65 rats	14

(*Continued*)

TABLE 21.2 Studies Regarding Therapies for Pediatric Brain Edema (Continued)

Author and Year	Animals Used and Age	Type of Insult	Results	Ref.
Appelberg, k. s. 2009	P35 and P45 rats	CCI+/− ketogenic diet	58% and 39% reduction in cortical contusion volume in P35 rats with ketogenic diet	15

1. Fernandez-Lopez D, Pazos MR, Tolon RM, Moro MA, Romero J, Lizasoain I, et al. The cannabinoid agonist WIN55212 reduces brain damage in an in vivo model of hypoxic–ischemic encephalopathy in newborn rats. *Pediatr Res* September 2007;62(3):255–60.
2. Vannucci RC, Connor JR, Mauger DT, Palmer C, Smith MB, Towfighi J, et al. Rat model of perinatal hypoxic – ischemic brain damage. *J Neurosci Res* 1999;55(2):158–63.
3. Mazarati A, Shin D, Sankar R. Bumetanide inhibits rapid kindling in neonatal rats. *Epilepsia* September 2009;50(9):2117–22.
4. Dubey M, Bugiani M, Ridder MC, Postma NL, Brouwers E, Polder E, et al. Mice with megalencephalic leukoencephalopathy with cysts: a developmental angle. *Ann Neurol* 2015;77(1):114–31.
5. Repici M, Centeno C, Tomasi S, Forloni G, Bonny C, Vercelli A, et al. Time-course of c-Jun N-terminal kinase activation after cerebral ischemia and effect of D-JNKI1 on c-Jun and caspase-3 activation. *Neuroscience* November 2007;150(1):40–9.
6. Borsello T, Clarke PG, Hirt L, Vercelli A, Repici M, Schorderet DF, et al. A peptide inhibitor of c-Jun N-terminal kinase protects against excitotoxicity and cerebral ischemia. *Nat Med* September 2003;9(9):1180–6.
7. Nijboer CH, van der Kooij MA, van Bel F, Ohl F, Heijnen CJ, Kavelaars A. Inhibition of the JNK/AP-1 pathway reduces neuronal death and improves behavioral outcome after neonatal hypoxic—ischemic brain injury. *Brain Behav Immun* 2010;24(5):812–21.
8. Ginet V, Puyal J, Magnin G, Clarke PG, Truttmann AC. Limited role of the c-Jun N-terminal kinase pathway in a neonatal rat model of cerebral hypoxia—ischemia. *J Neurochem* 2009;108(3):552–62.
9. Ditelberg JS, Sheldon RA, Epstein CJ, Ferriero DM. Brain injury after perinatal hypoxia—ischemia is exacerbated in copper/zinc superoxide dismutase transgenic mice. *Pediatr Res* 1996;39(2):204–8.
10. Armstead WM, Riley J, Vavilala MS. Dopamine prevents impairment of autoregulation after traumatic brain injury in the newborn pig through inhibition of up-regulation of endothelin-1 and extracellular signal-regulated kinase mitogen-activated protein kinase. *Pediatr Crit Care Med: J Soc Crit Care Med World Fed Pediatr Intensive Crit Care Soc* February 2013;14(2):e103–11.
11. Pohl D, Bittigau P, Ishimaru MJ, Stadthaus D, Hubner C, Olney JW, et al. *N*-Methyl-ᴅ-aspartate antagonists and apoptotic cell death triggered by head trauma in developing rat brain. *Proc Natl Acad Sci USA* 1999;96(5):2508–13.
12. Robertson CL, Saraswati M. Progesterone protects mitochondrial function in a rat model of pediatric traumatic brain injury. *J Bioenerg Biomembr* April 2015;47(1–2):43–51.
13. Fukuda AM, Adami A, Pop V, Bellone JA, Coats JS, Hartman RE, et al. Posttraumatic reduction of edema with aquaporin-4 RNA interference improves acute and chronic functional recovery. *J Cereb Blood Flow Metab: Official J Int Soc Cereb Blood Flow Metab* October 2013;33(10):1621–32.
14. Prins ML, Fujima LS, Hovda DA. Age-dependent reduction of cortical contusion volume by ketones after traumatic brain injury. *J Neurosci Res* November 2005;82(3):413–20.
15. Appelberg KS, Hovda DA, Prins ML. The effects of a ketogenic diet on behavioral outcome after controlled cortical impact injury in the juvenile and adult rat. *J Neurotrauma* 2009;26(4):497–506.

possible neuroprotective therapies for brain edema are summarized in Table 21.2.

As the BBB is predominantly involved in vasogenic edema and to a lesser extent in ionic edema (see Fig. 21.2), it may be an important target. Strategies that have been tried in preclinical studies are:

1. *Antiinflammatory and antioxidant molecules*: After brain injury, an excess of glutamate leads to augmented intracellular levels of Ca^{2+},[38] which can induce the transcription of several enzymes that generate free radicals, such as NOS, xanthine oxidase, or the phospholipase A2-Cox2 pathway. Such changes in intracellular Ca^{2+} concentrations also alter mitochondrial function, leading to a leakage of superoxide ions.[117] In addition, activated microglia, and later, systemic neutrophils release a variety of cytotoxic mediators and other edema-promoting molecules.[118] Experimental studies have focused on drugs with a known capacity for reducing oxidative stress (such as melatonin for which our lab group has considerable expertise[25,119–124] and SOD[117,125,126]), modulating the neuroinflammatory response (allopurinol,[127] minocycline, and cannabinoids[19,127,128]), or inhibiting glutamate excitotoxicity (magnesium sulfate, topiramate, and xenon) (see Table 21.2).
2. *Inducible factors and intracellular pathways that lead to tight-junction protein and basal membrane disruption*: The most studied are the hypoxia-inducible factor (HIF)-1 and JNK pathways. The role of HIF-1 in edema formation has been well studied in adult models of stroke and subarachnoid hemorrhage.[129–131] These studies show that HIF-1 is significantly upregulated after the insult, accompanied by a decrease in tight-junction proteins and laminin; the inhibition of HIF-1 by i.v. injection of 2-methoxyestradiol leads to the repression of brain edema as well as MMP-9 and AQP4 levels, which probably constitute the molecular pathways underlying HIF-1-mediated BBB disruption and edema formation.[131] Several types of JNK inhibitors have also been tested in adult rat models of subarachnoid hemorrhage and ischemic injury. JNK inhibition is significantly neuroprotective, by reducing both lesion size and behavioral consequences.[132] In the developing brain, the i.p. injection of the short-lived JNK-inhibitor TAT-JBD reduces neuronal damage in P7 rats if administrated at 0–3 hours after hypoxia–ischemia.[133] The JNK inhibitor D-JNKI1 could be considered an interesting therapeutic tool that easily crosses the BBB and accumulates in the ischemic region.[102] An interesting translational aspect is that these drugs seem to be neuroprotective even if administrated 6 hours after the injury.[102] Effects of inhibition of JNK pathway have been evaluated in a mice model of ICH induced by intrastriatal collagenase injection.

Inhibition of JNK pathway improves neurological outcome and decreases lesion volume early after ICH, and causes a decreased hemispheric swelling. This reduction was correlated with increased AQP4 expression, confirming a possible link between JNK pathway and induction of AQP4 expression. This reduction was correlated with increased AQP4 expression, confirming a possible link between JNK pathway and induction of AQP4 expression.[134]

3. *MCT induction and ketogenic diet*: The ketogenic diet induces MCT expression in the BBB augmenting the detoxifying properties of the BBB. Rats at P35 and P45 placed on a ketogenic diet for 7 days immediately after CCI show a 58% and 39% reduction in cortical contusion volume, respectively, compared to injured age-matched control animals on a standard diet.[135] However, because a metabolic shift effects neuronal tolerance to injury and the microglial responses specifically target studies are needed to assess the role of ketone mediated BBB integrity.
4. *Pleiotropic agents*: Progesterone is considered to be a pleiotropic agent following various types of CNS injury. It reduces necrotic damage and cell loss as well as edema formation, restores cognitive performance,[136] has antioxidant properties, regulates the expression of trophic factors such as brain-derived neurotrophic factor, increases the expression of antiapoptotic molecules such as Bcl-2 and Bcl-XL, and decreases the expression of pro-apoptotic molecules such as Bax, Bad, and caspase-3.[137] Regarding cerebral edema, there are several possibilities as to how progesterone is involved: it seems to limit the pro-inflammatory cytokines IL-1β and TGFβ as well as NOS-2 production, and to reduce BBB permeability to sodium, either through a direct effect on brain capillary Na–K-ATPase or through a generalized effect on BBB permeability that could account for the reduction in edema formation and lesion volume observed following progesterone treatment.[136,138] In a study conducted in immature P17 rats subjected to CCI, mitochondrial damage (a process typically triggered by glutamate-induced intracellular Ca^{2+} influx) was significantly reduced in young male animals.[139]
5. *AQP4 inhibition*: As previously mentioned, AQP4 plays a central role in edema formation in several cerebrovascular diseases, even if its beneficial or deleterious function is still unclear and depends on the pathologic model.[140] Pups treated with siRNA to AQP4 display a 30% decrease in AQP4 levels at the site of injection, and show acute (at 3 days) improvements in motor function and long-lasting (at 60 days) improvements in spatial memory. These improvements are associated with decreased edema formation, decreased BBB disruption, and reduced microglial activation, as well as reduced astrogliosis and cell death.[141]

6. *The sodium–potassium–chloride cotransporter NKCC1*: NKCC1 is found to be upregulated in cortical neurons and in whole brain lysates in an adult rat model of cerebral ischemia,[142,143] probably due to elevated levels of potassium and glutamate and/or cytokines.[2] Preliminary studies indicate that bumetanide, a relatively specific NKCC1 inhibitor at low doses that is commonly used as antidiuretic, may be effective in the treatment of edema due to several causes such as TBI,[119] hemorrhagic stroke,[120] ischemic stroke,[121] and tumors.[122] There are no direct studies that evaluate the effects of the inhibition of NKCC1 on edema formation. However, several studies have demonstrated that the intracellular level of chloride in immature neonatal neurons, compared with mature adult neurons, is about 20–40 mM higher due to the robust activity of NKCC1. The best-known effect of this overexpression is the increased tendency to epileptic seizures induced by the binding of GABA to ligand-gated GABA(A) receptor-associated chloride channels, which triggers a chloride efflux and depolarization. A study conducted in P11, P14, and P30 rats shows that the use of the NKCC1 blocker bumetanide induces a reduction in seizure activity only in P11 rats, confirming developmental differences in NKCC1 function.[123]
7. *The SUR1/TRPM4 Channel*: SUR1 is a regulatory subunit that associates with pore-forming subunits to form hetero-octameric K_{ATP} channels. In addition, it regulates the activity of a nonselective cation channel, SUR1/TRPM4. Its opening causes cell depolarization and blebbing, typical of anoxic edema. Glibenclamide, a sulfonylurea inhibitor, directly inhibits SUR1, causing a reduction in edema formation after ischemia by more than 50% in a wistar rat adult model of stroke,[2,124] and preliminary data from a phase II clinical trial suggest that glibenclamide significantly reduces cerebral edema and lowers the rate of hemorrhagic conversion following ischemic stroke in humans[144] (see Chapter 12: Edema and BBB Breakdown in Stroke). Its action via the inhibition of the sulfonylurea receptor (Sur1)-Trpm4 channel and in some cases via brain K_{ATP} channels, glibenclamide has shown to be beneficial in several clinically relevant models of ischemic and hemorrhagic stroke, TBI, NE of prematurity, and metastatic brain tumor[145] (for more details see Chapter 12: Edema and BBB Breakdown in Stroke). There are currently no studies that evaluate possible differences between adults and neonates, but in a moderate HI in P10 rat pups glibenclamide 0.01 mg/kg administered immediately after the injury and again 24 hours later improved long-term neurological parameters, suggesting a possible neuroprotective role.[146]
8. *CBF regulation*: Finally, another strategy to consider is the early improvement of blood flow to reduce cerebral ischemia, or a

later reduction in cerebral perfusion pressure to reduce vasogenic edema formation. In a piglet model of TBI, as in the clinical setting, autoregulation is impaired to a greater extent in newborn compared to juvenile pigs.[147] Endothelin is released following cerebral insults such as TBI, and has been found in several studies to contribute to age-related hemodynamic impairment in pigs.[147] In neonates, these effects have been studied by administering an i.v. injection of dopamine (DA) or an endothelin-1 (ET-1) inhibitor (BQ-123) 30 minutes after a FPI in newborn (P1–P5) pigs. DA prevents the up-regulation of HIF-1 after the insult, thus limiting the increase in intracranial perfusion pressure mediated by ET-1,[148] and BQ-123 blunts both postinjury ET-1-mediated vasoconstriction and neuronal loss in the parietal cortex.[43,149]

CONCLUSION

Brain edema during development, as well as in adulthood, is far from being fully understood, and our knowledge of new pathophysiological concepts and the different responses of the developing and adult brain to therapeutic agents are imperfect. More studies are needed to explore this field, both from a pathophysiological point of view by analyzing pathological changes in glial cell-mediated water movement or modifications to BBB permeability, and from the point of view of therapeutics, by testing molecules previously used in the adult brain as well as new molecules or strategies focused on the developing brain.

References

1. Pappius HM. *Fundamental aspects of brain oedema. Part I.* New York: North Holland Publishing Company; 1974.
2. Walcott BP, Kahle KT, Simard JM. Novel treatment targets for cerebral edema. *Neurotherapeutics: J Am Soc Exp NeuroTher* 2012;**9**(1):65–72.
3. Simard JM, Kent TA, Chen M, Tarasov KV, Gerzanich V. Brain oedema in focal ischaemia: molecular pathophysiology and theoretical implications. *Lancet Neurol* 2007;**6**(3):258–68.
4. Badaut J, Ashwal S, Obenaus A. Aquaporins in cerebrovascular disease: A target for treatment of brain edema? *Cerebrovasc Dis* 2011;**31**(6):521–31.
5. Unterberg AW, Stover J, Kress B, Kiening KL. Edema and brain trauma. *Neuroscience* 2004;**129**(4):1021–9.
6. Low JA. Determining the contribution of asphyxia to brain damage in the neonate. *J Obstet Gynaecol Res* August 2004;**30**(4):276–86.
7. Low JA, Robertson DM, Simpson LL. Temporal relationships of neuropathologic conditions caused by perinatal asphyxia. *Am J Obstet Gynecol* March 1989;**160**(3):608–14.
8. Glass HC, Glidden D, Jeremy RJ, Barkovich AJ, Ferriero DM, Miller SP. Clinical neonatal seizures are independently associated with outcome in infants at risk for hypoxic–ischemic brain injury. *J Pediatr* September 2009;**155**(3):318–23.

9. Baburamani AA, Ek CJ, Walker DW, Castillo-Melendez M. Vulnerability of the developing brain to hypoxic–ischemic damage: Contribution of the cerebral vasculature to injury and repair? *Front Physiol* 2012;**3**:424.
10. Rice 3rd JE, Vannucci RC, Brierley JB. The influence of immaturity on hypoxic–ischemic brain damage in the rat. *Ann Neurol* February 1981;**9**(2):131–41.
11. Ohshima M, Tsuji M, Taguchi A, Kasahara Y, Ikeda T. Cerebral blood flow during reperfusion predicts later brain damage in a mouse and a rat model of neonatal hypoxic–ischemic encephalopathy. *Exp Neurol* January 2012;**233**(1):481–9.
12. Derugin N, Ferriero DM, Vexler ZS. Neonatal reversible focal cerebral ischemia: a new model. *Neurosci Res* December 1998;**32**(4):349–53.
13. Charriaut-Marlangue C, Bonnin P, Leger PL, Renolleau S. Brief update on hemodynamic responses in animal models of neonatal stroke and hypoxia–ischemia. *Exp Neurol* October 2013;**248**:316–20.
14. Ferrari DC, Nesic O, Perez-Polo JR. Perspectives on neonatal hypoxia/ischemia-induced edema formation. *Neurochem Res* December 2010;**35**(12):1957–65.
15. McLean C, Ferriero D. Mechanisms of hypoxic–ischemic injury in the term infant. *Semin Perinatol* December 2004;**28**(6):425–32.
16. Hamrick SE, Ferriero DM. The injury response in the term newborn brain: Can we neuroprotect? *Curr Opin Neurol* April 2003;**16**(2):147–54.
17. Fleiss B, Gressens P. Tertiary mechanisms of brain damage: A new hope for treatment of cerebral palsy? *Lancet Neurol* June 2012;**11**(6):556–66.
18. Volpe J. *Neurology of the newborn*, 5th edition Philadelphia, PA: Saunders; May 2008.
19. Fernandez-Lopez D, Lizasoain I, Moro MA, Martinez-Orgado J. Cannabinoids: well-suited candidates for the treatment of perinatal brain injury. *Brain Sci* 2013;**3**(3):1043–59.
20. Marmarou A. A review of progress in understanding the pathophysiology and treatment of brain edema. *Neurosurg Focus.* 2007;**22**(5):E1.
21. Klatzo I. Brain oedema following brain ischaemia and the influence of therapy. *Br J Anaesth* January 1985;**57**(1):18–22.
22. Ballabh P. Intraventricular hemorrhage in premature infants: mechanism of disease. *Pediatr Res* January 2010;**67**(1):1–8.
23. Heron M, Sutton PD, Xu J, Ventura SJ, Strobino DM, Guyer B. Annual summary of vital statistics: 2007. *Pediatrics* January 2010;**125**(1):4–15.
24. Lekic T, Manaenko A, Rolland W, et al. Rodent neonatal germinal matrix hemorrhage mimics the human brain injury, neurological consequences, and post-hemorrhagic hydrocephalus. *Exp Neurol* July 2012;**236**(1):69–78.
25. Roozenbeek B, Maas AI, Menon DK. Changing patterns in the epidemiology of traumatic brain injury. *Nat Rev Neurol* April 2013;**9**(4):231–6.
26. Pop V, Sorensen DW, Kamper JE, et al. Early brain injury alters the blood–brain barrier phenotype in parallel with beta-amyloid and cognitive changes in adulthood. *J Cereb Blood Flow Metab: Official J Int Soc Cereb Blood Flow Metab* February 2013;**33**(2):205–14.
27. Ajao DO, Pop V, Kamper JE, et al. Traumatic brain injury in young rats leads to progressive behavioral deficits coincident with altered tissue properties in adulthood. *J Neurotrauma* July 2012;**29**(11):2060–74.
28. Kamper JE, Pop V, Fukuda AM, Ajao DO, Hartman RE, Badaut J. Juvenile traumatic brain injury evolves into a chronic brain disorder: behavioral and histological changes over 6months. *Exp Neurol* December 2013;**250**:8–19.
29. Rivara FP, Vavilala MS, Durbin D, et al. Persistence of disability 24 to 36 months after pediatric traumatic brain injury: a cohort study. *J Neurotrauma* October 2012;**29**(15):2499–504.
30. Ponsford J, Willmott C, Rothwell A, et al. Impact of early intervention on outcome after mild traumatic brain injury in children. *Pediatrics* December 2001;**108**(6):1297–303.
31. Koepsell TD, Rivara FP, Vavilala MS, et al. Incidence and descriptive epidemiologic features of traumatic brain injury in King County, Washington. *Pediatrics* November 2011;**128**(5):946–54.

32. Bauer R, Fritz H. Pathophysiology of traumatic injury in the developing brain: an introduction and short update. *Exp Toxicol Pathol: Official J Ges Toxikologische Pathol* October 2004;**56**(1-2):65–73.
33. Keenan HT, Runyan DK, Marshall SW, Nocera MA, Merten DF, Sinal SH. A population-based study of inflicted traumatic brain injury in young children. *JAMA*. August 2003;**290**(5):621–6.
34. Adelson PD, Srinivas R, Chang Y, Bell M, Kochanek PM. Cerebrovascular response in children following severe traumatic brain injury. *Child's Nervous Syst: ChNS: Official J Int Soc Pediatr Neurosurg* September 2011;**27**(9):1465–76.
35. Giza CC, Mink RB, Madikians A. Pediatric traumatic brain injury: not just little adults. *Curr Opin Crit Care* 2007;**13**(2):143–52.
36. Bittigau P, Sifringer M, Pohl D, et al. Apoptotic neurodegeneration following trauma is markedly enhanced in the immature brain. *Ann Neurol* June 1999;**45**(6):724–35.
37. Pop V, Badaut J. A neurovascular perspective for long-term changes after brain trauma. *Transl Stroke Res* December 2011;**2**(4):533–45.
38. Chodobski A, Zink BJ, Szmydynger-Chodobska J. Blood–brain barrier pathophysiology in traumatic brain injury. *Transl Stroke Res* December 2011;**2**(4):492–516.
39. Baskaya MK, Rao AM, Dogan A, Donaldson D, Dempsey RJ. The biphasic opening of the blood–brain barrier in the cortex and hippocampus after traumatic brain injury in rats. *Neurosci Lett* April 1997;**226**(1):33–6.
40. Lang DA, Teasdale GM, Macpherson P, Lawrence A. Diffuse brain swelling after head injury: More often malignant in adults than children? *J Neurosurg* April 1994;**80**(4):675–80.
41. Aldrich EF, Eisenberg HM, Saydjari C, et al. Diffuse brain swelling in severely head-injured children. A report from the NIH Traumatic Coma Data Bank. *J Neurosurg* March 1992;**76**(3):450–4.
42. Mehta A, Kochanek PM, Tyler-Kabara E, et al. Relationship of intracranial pressure and cerebral perfusion pressure with outcome in young children after severe traumatic brain injury. *Dev Neurosci* 2010;**32**(5-6):413–9.
43. Kochanek PM. Pediatric traumatic brain injury: quo vadis? *Dev Neurosci* 2006;**28**(4–5):244–55.
44. Fukuda AM, Pop V, Spagnoli D, Ashwal S, Obenaus A, Badaut J. Delayed increase of astrocytic aquaporin 4 after juvenile traumatic brain injury: Possible role in edema resolution? *Neuroscience* October 2012;**222**:366–78.
45. Alter SJ, Bennett JS, Koranyi K, Kreppel A, Simon R. Common childhood viral infections. *Curr Probl Pediatr Adolesc Health Care* February 2015;**45**(2):21–53.
46. Kumar G, Kalita J, Misra UK. Raised intracranial pressure in acute viral encephalitis. *Clin Neurol Neurosurg* June 2009;**111**(5):399–406.
47. Murray CJ, Lopez AD. Global and regional cause-of-death patterns in 1990. *Bull World Health Organ* 1994;**72**(3):447–80.
48. Quagliarello VJ, Long WJ, Scheld WM. Morphologic alterations of the blood–brain barrier with experimental meningitis in the rat. Temporal sequence and role of encapsulation. *J Clin Investig* April 1986;**77**(4):1084–95.
49. Barichello T, Fagundes GD, Generoso JS, et al. Brain–blood barrier breakdown and pro-inflammatory mediators in neonate rats submitted meningitis by *Streptococcus pneumoniae*. *Brain Res* August 2012;**1471**:162–8.
50. Barichello T, Lemos JC, Generoso JS, et al. Oxidative stress, cytokine/chemokine and disruption of blood–brain barrier in neonate rats after meningitis by *Streptococcus agalactiae*. *Neurochem Res* October 2011;**36**(10):1922–30.
51. Holopainen IE. Seizures in the developing brain: cellular and molecular mechanisms of neuronal damage, neurogenesis and cellular reorganization. *Neurochem Int* May 2008;**52**(6):935–47.

52. Scott RC, Gadian DG, King MD, et al. Magnetic resonance imaging findings within 5 days of status epilepticus in childhood. *Brain: J Neurol* September 2002;**125**(Pt 9):1951–9.
53. Gastaut H, Poirier F, Payan H, Salamon G, Toga M, Vigouroux M. H.H.E. syndrome; hemiconvulsions, hemiplegia, epilepsy. *Epilepsia* June 1960;**1**:418–47.
54. Toldo I, Calderone M, Boniver C, Dravet C, Guerrini R, Laverda AM. Hemiconvulsion–hemiplegia–epilepsy syndrome: early magnetic resonance imaging findings and neuroradiological follow-up. *Brain Dev* March 2007;**29**(2):109–11.
55. Auvin S, Bellavoine V, Merdariu D, et al. Hemiconvulsion–hemiplegia–epilepsy syndrome: current understandings. *Eur J Paediatr Neurol: EJPN: Official J Eur Paediatr Neurol Soc* September 2012;**16**(5):413–21.
56. Ophoff RA, Terwindt GM, Vergouwe MN, et al. Familial hemiplegic migraine and episodic ataxia type-2 are caused by mutations in the Ca2+ channel gene CACNL1A4. *Cell*. November 1996;**87**(3):543–52.
57. Vecchia D, Tottene A, van den Maagdenberg AM, Pietrobon D. Abnormal cortical synaptic transmission in CaV2.1 knockin mice with the S218L missense mutation which causes a severe familial hemiplegic migraine syndrome in humans. *Front Cell Neurosci* 2015;**9**:8.
58. Figueroa RE, Hoffman WH, Momin Z, Pancholy A, Passmore GG, Allison J. Study of subclinical cerebral edema in diabetic ketoacidosis by magnetic resonance imaging T2 relaxometry and apparent diffusion coefficient maps. *Endocr Res* 2005;**31**(4):345–55.
59. Hoffman WH, Stamatovic SM, Andjelkovic AV. Inflammatory mediators and blood brain barrier disruption in fatal brain edema of diabetic ketoacidosis. *Brain Res* February 2009;**1254**:138–48.
60. Kohlschutter A, Eichler F. Childhood leukodystrophies: a clinical perspective. *Expert Rev Neurother* October 2011;**11**(10):1485–96.
61. Boespflug-Tanguy O, Labauge P, Fogli A, Vaurs-Barriere C. Genes involved in leukodystrophies: a glance at glial functions. *Curr Neurol Neurosci Rep* May 2008;**8**(3): 217–29.
62. Meisingset TW, Risa O, Brenner M, Messing A, Sonnewald U. Alteration of glial–neuronal metabolic interactions in a mouse model of Alexander disease. *Glia* August 2010;**58**(10):1228–34.
63. van der Voorn JP, Pouwels PJ, Hart AA, et al. Childhood white matter disorders: quantitative MR imaging and spectroscopy. *Radiology* November 2006;**241**(2):510–7.
64. Duarri A, Lopez de Heredia M, Capdevila-Nortes X, et al. Knockdown of MLC1 in primary astrocytes causes cell vacuolation: a MLC disease cell model. *Neurobiol Dis* July 2011;**43**(1):228–38.
65. Ridder MC, Boor I, Lodder JC, et al. Megalencephalic leucoencephalopathy with cysts: defect in chloride currents and cell volume regulation. *Brain: J Neurol* November 2011;**134**(Pt 11):3342–54.
66. van der Knaap MS, Boor I, Estevez R. Megalencephalic leukoencephalopathy with subcortical cysts: chronic white matter oedema due to a defect in brain ion and water homoeostasis. *Lancet Neurol* November 2012;**11**(11):973–85.
67. Depienne C, Bugiani M, Dupuits C, et al. Brain white matter oedema due to ClC-2 chloride channel deficiency: an observational analytical study. *Lancet Neurol* July 2013;**12**(7):659–68.
68. Jentsch TJ. From mice to man: chloride transport in leukoencephalopathy. *Lancet Neurol* July 2013;**12**(7):626–8.
69. Yu J. Epidemiology of brain tumors in children aged two and under: a 10-year single-institute study. *Oncol Lett* 2015;**9**(4):1651–6.
70. Liebner S, Fischmann A, Rascher G, et al. Claudin-1 and claudin-5 expression and tight junction morphology are altered in blood vessels of human glioblastoma multiforme. *Acta Neuropathol* September 2000;**100**(3):323–31.

71. Papadopoulos MC, Saadoun S, Woodrow CJ, et al. Occludin expression in microvessels of neoplastic and non-neoplastic human brain. *Neuropathol Appl Neurobiol* October 2001;**27**(5):384–95.
72. Nico B, Ribatti D. Role of aquaporins in cell migration and edema formation in human brain tumors. *Exp Cell Res.* October 2011;**317**(17):2391–6.
73. Obenaus A, Ashwal S. Magnetic resonance imaging in cerebral ischemia: focus on neonates. *Neuropharmacology* September 2008;**55**(3):271–80.
74. Badaut J, Ashwal S, Adami A, et al. Brain water mobility decreases after astrocytic aquaporin-4 inhibition using RNA interference. *J Cereb Blood Flow Metab: Official J Int Soc Cereb Blood Flow Metab* March 2011;**31**(3):819–31.
75. Tasker RC, Acerini CL. Cerebral edema in children with diabetic ketoacidosis: Vasogenic rather than cellular? *Pediatr Diabetes* June 2014;**15**(4):261–70.
76. D'Arceuil H, Liu C, Levitt P, Thompson B, Kosofsky B, de Crespigny A. Three-dimensional high-resolution diffusion tensor imaging and tractography of the developing rabbit brain. *Dev Neurosci* 2008;**30**(4):262–75.
77. Vieira Karuta SC, Raskin S, de Carvalho Neto A, Gasparetto EL, Doring T, Teive HA. Diffusion tensor imaging and tract-based spatial statistics analysis in Friedreich's ataxia patients. *Parkinsonism Relat Disord* 21(5), May 2015, 504–508. http://dx.doi.org/10.1016/j.parkreldis.2015.02.021. Epub 2015 Mar 11.
78. Young W, Rappaport ZH, Chalif DJ, Flamm ES. Regional brain sodium, potassium, and water changes in the rat middle cerebral artery occlusion model of ischemia. *Stroke; A J Cerebr Circ* July–August 1987;**18**(4):751–9.
79. Thrane AS, Rangroo Thrane V, Nedergaard M. Drowning stars: reassessing the role of astrocytes in brain edema. *Trends Neurosci* November 2014;**37**(11):620–8.
80. Starling EH. On the absorption of fluids from the connective tissue spaces. *J Physiol* May 1896;**19**(4):312–26.
81. Young W. *Ionic and water shifts in injured central nervous tissues.* New York: Oxford University Press; 1994.
82. Daneman R, Zhou L, Agalliu D, Cahoy JD, Kaushal A, Barres BA. The mouse blood–brain barrier transcriptome: a new resource for understanding the development and function of brain endothelial cells. *PloS One* 2010;**5**(10):e13741.
83. Lu KT, Wu CY, Cheng NC, et al. Inhibition of the Na+ –K+ –2Cl-cotransporter in choroid plexus attenuates traumatic brain injury-induced brain edema and neuronal damage. *Eur J Pharmacol* October 2006;**548**(1-3):99–105.
84. Nico B, Frigeri A, Nicchia GP, et al. Role of aquaporin-4 water channel in the development and integrity of the blood–brain barrier. *J Cell Sci* 2001;**114**(Pt 7):1297–307.
85. Dobbing J. The later growth of the brain and its vulnerability. *Pediatrics* **53**(1), Januay 1974, 2–6.
86. Wen H, Nagelhus EA, Amiry-Moghaddam M, Agre P, Ottersen OP, Nielsen S. Ontogeny of water transport in rat brain: postnatal expression of the aquaporin-4 water channel. *Eur J Neurosci* March 1999;**11**(3):935–45.
87. Gomori E, Pal J, Abraham H, et al. Fetal development of membrane water channel proteins aquaporin-1 and aquaporin-4 in the human brain. *Int J Dev Neurosci: Official J Int Soc Dev Neurosci* August 2006;**24**(5):295–305.
88. Hsu MS, Seldin M, Lee DJ, Seifert G, Steinhauser C, Binder DK. Laminar-specific and developmental expression of aquaporin-4 in the mouse hippocampus. *Neuroscience* March 2011;**178**:21–32.
89. Badaut J, Ashwal S, Tone B, Regli L, Tian HR, Obenaus A. Temporal and regional evolution of aquaporin-4 expression and magnetic resonance imaging in a rat pup model of neonatal stroke. *Pediatr Res* September 2007;**62**(3):248–54.
90. Moretti R, Pansiot J, Bettati D, et al. Blood–brain barrier dysfunction in disorders of the developing brain. *Front Neurosci* 2015;**9**:40.

91. Saunders NR, Ek CJ, Habgood MD, Dziegielewska KM. Barriers in the brain: A renaissance? *Trends Neurosci* June 2008;**31**(6):279–86.
92. Liddelow SA, Dziegielewska KM, Ek CJ, et al. Mechanisms that determine the internal environment of the developing brain: a transcriptomic, functional and ultrastructural approach. *PloS One* 2013;**8**(7):e65629.
93. Johansson PA, Dziegielewska KM, Liddelow SA, Saunders NR. The blood–CSF barrier explained: when development is not immaturity. *BioEssays: News Rev Mol, Cell Dev Biol* March 2008;**30**(3):237–48.
94. Ek CJ, Dziegielewska KM, Stolp H, Saunders NR. Functional effectiveness of the blood–brain barrier to small water-soluble molecules in developing and adult opossum (*Monodelphis domestica*). *J Comp Neurol* May 2006;**496**(1):13–26.
95. Badaut J, Ajao DO, Sorensen DW, Fukuda AM, Pellerin L. Caveolin expression changes in the neurovascular unit after juvenile traumatic brain injury: signs of blood–brain barrier healing? *Neuroscience* January 2015;**285**:215–26.
96. Pierre K, Pellerin L. Monocarboxylate transporters in the central nervous system: distribution, regulation and function. *J Neurochem* July 2005;**94**(1):1–14.
97. Sullivan PG, Rippy NA, Dorenbos K, Concepcion RC, Agarwal AK, Rho JM. The ketogenic diet increases mitochondrial uncoupling protein levels and activity. *Ann Neurol* April 2004;**55**(4):576–80.
98. Ziegler DR, Ribeiro LC, Hagenn M, et al. Ketogenic diet increases glutathione peroxidase activity in rat hippocampus. *Neurochem Res* December 2003;**28**(12):1793–7.
99. Laher I, Zhang JH. Protein kinase C and cerebral vasospasm. *J Cereb Blood Flow Metab: Official J Int Soc Cereb Blood Flow Metab* 2001;**21**(8):887–906.
100. Badaut J, Lasbennes F, Magistretti PJ, Regli L. Aquaporins in brain: distribution, physiology, and pathophysiology. *J Cereb Blood Flow Metab: Official J Int Soc Cereb Blood Flow Metab* April 2002;**22**(4):367–78.
101. Bogoyevitch MA, Kobe B. Uses for JNK: the many and varied substrates of the c-Jun N-terminal kinases. *Microbiol Mol Biol Rev* December 2006;**70**(4):1061–95.
102. Repici M, Centeno C, Tomasi S, et al. Time-course of c-Jun N-terminal kinase activation after cerebral ischemia and effect of D-JNKI1 on c-Jun and caspase-3 activation. *Neuroscience* November 2007;**150**(1):40–9.
103. Armstead WM, Kiessling JW, Riley J, Cines DB, Higazi AA. tPA contributes to impaired NMDA cerebrovasodilation after traumatic brain injury through activation of JNK MAPK. *Neurol Res* September 2011;**33**(7):726–33.
104. Bratton SL, Chestnut RM, Ghajar J, et al. Guidelines for the management of severe traumatic brain injury. VIII. Intracranial pressure thresholds. *J Neurotrauma* 2007;**24**(Suppl 1):S55–8.
105. Nichols J, Perez R, Wu C, Adelson PD, Anderson T. Traumatic brain injury induces rapid enhancement of cortical excitability in juvenile rats. *CNS Neurosci Ther* 2015;**21**(2):193–203.
106. Kochanek PM, Carney N, Adelson PD, et al. Guidelines for the acute medical management of severe traumatic brain injury in infants, children, and adolescents—second edition. *Pediatr Crit Care Med: J Soc Crit Care Med World Fed Pediatr Intensive Crit Care Soc* January 2012;**13**(Suppl 1):S1–82.
107. Kukreti V, Mohseni-Bod H, Drake J. Management of raised intracranial pressure in children with traumatic brain injury. *J Pediatr Neurosci* September–December 2014;**9**(3):207–15.
108. Finkelstein RA, Alam HB. Induced hypothermia for trauma: current research and practice. *J Intensive Care Med* July 2010;**25**(4):205–26.
109. Rutherford M, Ramenghi LA, Edwards AD, et al. Assessment of brain tissue injury after moderate hypothermia in neonates with hypoxic–ischaemic encephalopathy: a nested substudy of a randomised controlled trial. *Lancet Neurol* January 2010;**9**(1):39–45.

110. Jacobs SE, Berg M, Hunt R, Tarnow-Mordi WO, Inder TE, Davis PG. Cooling for newborns with hypoxic ischaemic encephalopathy. *Cochrane Database Syst Rev* 2013(1):CD003311.
111. Simbruner G, Mittal RA, Rohlmann F, Muche R, neo.n EnTP Systemic hypothermia after neonatal encephalopathy: outcomes of neo.nEURO.network RCT. *Pediatrics* October 2010;**126**(4):e771–778.
112. Choi HA, Badjatia N, Mayer SA. Hypothermia for acute brain injury—mechanisms and practical aspects. *Nat Rev Neurol* April 2012;**8**(4):214–22.
113. Sandestig A, Romner B, Grande PO. Therapeutic hypothermia in children and adults with severe traumatic brain injury. *Ther Hypothermia Temp Manag* March 2014;**4**(1):10–20.
114. Mhanna MJ, Mallah WE, Verrees M, Shah R, Super DM. Outcome of children with severe traumatic brain injury who are treated with decompressive craniectomy. *J Neurosurg Pediatr* 2015:1–7.
115. Bullock MR, Lyeth BG, Muizelaar JP. Current status of neuroprotection trials for traumatic brain injury: lessons from animal models and clinical studies. *Neurosurgery* 1999;**45**(2):207–17. Discussion 217–220.
116. Pohl D, Bittigau P, Ishimaru MJ, et al. *N*-Methyl-D-aspartate antagonists and apoptotic cell death triggered by head trauma in developing rat brain. *Proc Natl Acad Sci USA* 1999;**96**(5):2508–13.
117. Potts MB, Koh SE, Whetstone WD, et al. Traumatic injury to the immature brain: inflammation, oxidative injury, and iron-mediated damage as potential therapeutic targets. *NeuroRx: J Am Soc Exp NeuroTher* April 2006;**3**(2):143–53.
118. del Zoppo GJ. Stroke and neurovascular protection. *N Engl J Med* 2006;**354**(6):553–5.
119. Lu KT, Wu CY, Yen HH, Peng JH, Wang CL, Yang YL. Bumetanide administration attenuated traumatic brain injury through IL-1 overexpression. *Neurol Res* June 2007;**29**(4):404–9.
120. Simard JM, Kahle KT, Gerzanich V. Molecular mechanisms of microvascular failure in central nervous system injury—synergistic roles of NKCC1 and SUR1/TRPM4. *J Neurosurg* September 2010;**113**(3):622–9.
121. Chen H, Luo J, Kintner DB, Shull GE, Sun D. Na(+)-dependent chloride transporter (NKCC1)-null mice exhibit less gray and white matter damage after focal cerebral ischemia. *J Cereb Blood Flow Metab: Official J Int Soc Cereb Blood Flow Metab* January 2005;**25**(1):54–66.
122. Conti L, Palma E, Roseti C, et al. Anomalous levels of Cl-transporters cause a decrease of GABAergic inhibition in human peritumoral epileptic cortex. *Epilepsia* September 2011;**52**(9):1635–44.
123. Mazarati A, Shin D, Sankar R. Bumetanide inhibits rapid kindling in neonatal rats. *Epilepsia* September 2009;**50**(9):2117–22.
124. Simard JM, Chen M, Tarasov KV, et al. Newly expressed SUR1-regulated NC(Ca-ATP) channel mediates cerebral edema after ischemic stroke. *Nat Med* April 2006;**12**(4):433–40.
125. Mikawa S, Kinouchi H, Kamii H, et al. Attenuation of acute and chronic damage following traumatic brain injury in copper, zinc-superoxide dismutase transgenic mice. *J Neurosurg* November 1996;**85**(5):885–91.
126. Ditelberg JS, Sheldon RA, Epstein CJ, Ferriero DM. Brain injury after perinatal hypoxia-ischemia is exacerbated in copper/zinc superoxide dismutase transgenic mice. *Pediatr Res* February 1996;**39**(2):204–8.
127. Vannucci RC, Connor JR, Mauger DT, et al. Rat model of perinatal hypoxic–ischemic brain damage. *J Neurosci Res* 1999;**55**(2):158–63.
128. Fernandez-Lopez D, Pazos MR, Tolon RM, et al. The cannabinoid agonist WIN55212 reduces brain damage in an in vivo model of hypoxic–ischemic encephalopathy in newborn rats. *Pediatr Res* September 2007;**62**(3):255–60.
129. Higashida T, Kreipke CW, Rafols JA, et al. The role of hypoxia-inducible factor-1alpha, aquaporin-4, and matrix metalloproteinase-9 in blood–brain barrier disruption and brain edema after traumatic brain injury. *J Neurosurg* January 2011;**114**(1):92–101.

130. Higashida T, Peng C, Li J, et al. Hypoxia-inducible factor-1alpha contributes to brain edema after stroke by regulating aquaporins and glycerol distribution in brain. *Curr Neurovascular Res* 2011;**8**(1):44–51.
131. Wang Z, Meng CJ, Shen XM, et al. Potential contribution of hypoxia-inducible factor-1alpha, aquaporin-4, and matrix metalloproteinase-9 to blood–brain barrier disruption and brain edema after experimental subarachnoid hemorrhage. *J Mol Neurosci* September 2012;**48**(1):273–80.
132. Borsello T, Clarke PG, Hirt L, et al. A peptide inhibitor of c-Jun N-terminal kinase protects against excitotoxicity and cerebral ischemia. *Nat Med* September 2003;**9**(9):1180–6.
133. Nijboer CH, Heijnen CJ, Groenendaal F, van Bel F, Kavelaars A. Alternate pathways preserve tumor necrosis factor-alpha production after nuclear factor-kappaB inhibition in neonatal cerebral hypoxia–ischemia. *Stroke* October 2009;**40**(10):3362–8.
134. Michel-Monigadon D, Bonny C, Hirt L. c-Jun N-terminal kinase pathway inhibition in intracerebral hemorrhage. *Cerebrovasc Dis* 2010;**29**(6):564–70.
135. Prins ML, Fujima LS, Hovda DA. Age-dependent reduction of cortical contusion volume by ketones after traumatic brain injury. *J Neurosci Res* November 2005;**82**(3):413–20.
136. Sayeed I, Stein DG. Progesterone as a neuroprotective factor in traumatic and ischemic brain injury. *Prog Brain Res* 2009;**175**:219–37.
137. Gibson CL, Gray LJ, Bath PM, Murphy SP. Progesterone for the treatment of experimental brain injury; a systematic review. *Brain* February 2008;**131**(Pt 2):318–28.
138. Gibson CL, Constantin D, Prior MJ, Bath PM, Murphy SP. Progesterone suppresses the inflammatory response and nitric oxide synthase-2 expression following cerebral ischemia. *Exp Neurol* 2005;**193**(2):522–30.
139. Robertson CL, Saraswati M. Progesterone protects mitochondrial function in a rat model of pediatric traumatic brain injury. *J Bioenerg Biomembr* April 2015;**47**(1-2):43–51.
140. Badaut J, Fukuda AM, Jullienne A, Petry KG. Aquaporin and brain diseases. *Biochim Biophys Acta* May 2014;**1840**(5):1554–65.
141. Fukuda AM, Adami A, Pop V, et al. Posttraumatic reduction of edema with aquaporin-4 RNA interference improves acute and chronic functional recovery. *J Cereb Blood Flow Metab: Official J Int Soc Cereb Blood Flow Metab* October 2013;**33**(10):1621–32.
142. Yan Y, Dempsey RJ, Flemmer A, Forbush B, Sun D. Inhibition of Na(+)–K(+)–Cl(-) cotransporter during focal cerebral ischemia decreases edema and neuronal damage. *Brain Res* January 2003;**961**(1):22–31.
143. Yan Y, Dempsey RJ, Sun D. Na+–K+–Cl-cotransporter in rat focal cerebral ischemia. *J Cereb Blood Flow Metab: Official J Int Soc Cereb Blood Flow Metab* June 2001;**21**(6):711–21.
144. Khanna A, Walcott BP, Kahle KT, Simard JM. Effect of glibenclamide on the prevention of secondary brain injury following ischemic stroke in humans. *Neurosurg Focus* January 2014;**36**(1):E11.
145. Kurland DB, Tosun C, Pampori A, et al. Glibenclamide for the treatment of acute CNS injury. *Pharmaceuticals (Basel)* 2013;**6**(10):1287–303.
146. Zhou Y, Fathali N, Lekic T, Tang J, Zhang JH. Glibenclamide improves neurological function in neonatal hypoxia–ischemia in rats. *Brain Res* 2009;**1270**:131–9.
147. Armstead WM. Role of endothelin-1 in age-dependent cerebrovascular hypotensive responses after brain injury. *Am J Physiol* November 1999;**277**(5 Pt 2):H1884–1894.
148. Armstead WM, Riley J, Vavilala MS. Dopamine prevents impairment of autoregulation after traumatic brain injury in the newborn pig through inhibition of up-regulation of endothelin-1 and extracellular signal-regulated kinase mitogen-activated protein kinase. *Pediatr Crit Care Med: J Soc Crit Care Med World Fed Pediatr Intensive Crit Care Soc* February 2013;**14**(2):e103–11.
149. Armstead WM, Raghupathi R. Endothelin and the neurovascular unit in pediatric traumatic brain injury. *Neurol Res* March 2011;**33**(2):127–32.

CHAPTER

22

Cerebral Edema in Cerebrovascular Diseases

Stéphane Olindo and Igor Sibon

CNRS-University of Bordeaux, Bordeaux, France

OUTLINE

Brain Edema.
DOI: http://dx.doi.org/10.1016/B978-0-12-803196-4.00022-9

INTRODUCTION

Stroke is the most frequent cause of acquired disability in adults, the second cause of major cognitive impairment, and the third cause of death in many developed countries.[1] The worldwide prevalence of stroke gradually increases due to population aging.[2] Ischemic stroke (IS) accounts for approximately 85% of cases while hemorrhagic strokes represent 15% (including intracerebral hemorrhage (ICH) 10% and subarachnoid hemorrhage (SAH) 5%) and cerebral venous thrombosis (CVT) less than 1% of all cerebrovascular diseases.[1] To date, admission to intensive care stroke units, acute control of blood pressure (BP), and recanalization of the occluded vessel using intravenous tissue plasminogen activator (tPA) or mechanical thrombectomy for IS represent the main effective acute stroke treatments.[3] Despite these strategies the impact of stroke often remains devastating and new approaches have to be developed in order to reduce the consequences of the acute vascular insult.

Targeting poststroke cerebral edema might represent one promising therapeutic goal in a near future. Indeed, cerebral edema is commonly observed within the first week of IS, reaching a peak between 24 and 72 hours[4] and within the second week following ICH.[5] The intensity and severity of cerebral edema strongly varies depending on the extent of the stroke lesion. Brain swelling is a feared complication because it impairs the functioning of capillaries surrounding the ischemic core and reduces collateral leptomeningeal blood flow therefore compromising the cerebral perfusion that in turn leads to secondary neuronal injuries.[6] Given the confined space within the fixed-volume of the rigid skull, very large cerebral edema can also result in increased ICP and can cause brain herniation leading to a direct compression of the nonischemic brain. These complications are mainly observed in cases of very large ischemic (i.e., malignant middle cerebral artery infarction (MMCAI))[7] or hemorrhagic lesions[5] or when the lesion is located in a so-called "closed" area such as the cerebellum in the posterior fossa.[8] These conditions are rare (i.e., MMCAI accounts for 10%–15% of all MCA infarctions)[7,9–11] but need to be anticipated in order to prevent death and poor functional prognosis through dedicated treatments based on medical and surgical procedures.[12]

In this review we will describe the predictors, treatments, and functional consequences of cerebral edema associated with acute cerebrovascular lesions.

PATHOPHYSIOLOGY OF CEREBROVASCULAR EDEMA IN ISCHEMIC STROKE

The mechanisms of brain swelling in stroke are initially different in ischemic and hemorrhagic stroke but share one common endpoint represented by the increased permeability of the blood–brain barrier (BBB). Regarding IS, the edema build-up phase can be divided into three major types: anoxic, ionic, and vasogenic edema.[13] Anoxic edema is characterized as swelling of the astrocytes and neuronal dendrites that occur within minutes after oxygen and glucose deprivation. This depletion induces major changes in the cellular ionic gradients due to the failure of efficient energy-dependent cotransporters which leads to a massive entry of ions into cells, followed by water, which induces cellular swelling in astrocytes first, and then in neuronal dendrites. The astroglial swelling becomes apparent within 30 minutes of onset of focal ischemia[14] and will initiate a variety of secondary effects, which exacerbate brain damage. Indeed, swelling of perivascular astrocytes and astrocyte end-feet may compress brain vessels and limit vascular circulation.[15] Vascular congestion from this mechanism contributes to the incomplete filling of brain vessels during reperfusion following focal ischemia, known as the *no-reflow* phenomenon.[16] Swelling of astrocytes can also result in opening of volume-regulated ion channels which are permeable to glutamate and other excitatory amino acids whereas release of the latter can induce or exacerbate excitotoxic cell death.[17] Prominent swelling of astrocytes can also severely reduce the extracellular space volume which contributes to a concentration of extracellular glutamate and K^+. Anoxic edema is quickly followed by an ionic edema secondary to the alteration of the ionic gradients of endothelial cells, including transcapillary flux of Na^+ with tissue swelling. Suffering of endothelial cells therefore results in early transient leakage of the BBB. This results in further entry of water through endothelial cells leading to brain swelling associated with further increased BBB permeability.[18] The increased BBB permeability mainly results from disturbances of endothelial tight junctions and extracellular matrix degradation[19] and leads to the leak of plasma proteins such as albumin from the blood into the brain.[20] This event initiates a delayed vasogenic edema that enhances the water content of the tissue.[21] Besides the "vascular" origin of the increased water content of the brain, water accumulation could also result from a decrease of water exchange between brain parenchyma and the ventricular system as well as an impaired functioning of the recently coined "glymphatic system" that constitute a paravascular route between glial and vascular cells.[22] The resulting increased ICP can cause secondary brain ischemia through decreased cerebral perfusion and blood flow, brain tissue hypoxia, and metabolic crisis.[23] The brain swelling can also cause direct compression of the initially noninjured brain and of the intracranial arteries leading to infarction, especially in the anterior and posterior cerebral arteries' territories. Tissue shifts can also stretch and tear cerebral vessels, causing ICH such as Duret's hemorrhage of the brainstem.[24]

CLINICAL ASPECTS

Cerebral edema symptoms can result from three different mechanisms: lesion extension, brainstem and arousal systems compression, and ICP increase.

Symptoms Related to the Extent of the Lesion

The most common findings observed in patients with MMCAI are hemiplegia, severe aphasia or dysarthria, neglect, gaze preference, and visual field defect.[7] The initial National Institutes of Health Stroke Scale score (NIHSS), which is commonly used in acute stroke to evaluate the severity and range from 0 (no symptoms) to a maximum of 48, is often >20 with dominant hemispheric infarction and >15 with nondominant hemispheric infarction.

Because cerebral edema usually peaks within the first days of stroke onset, it usually leads to an early neurological deterioration observed in about 30% of all stroke patients within the first 3 days.[25] Recently, Seners et al. reported that malignant edema accounts from 14% to 27% of all early neurological deterioration causes at 24 hours.[26] In this period the main differential diagnoses to be evoked are collateral failure with expansion of the core of the infarct, symptomatic ICH, and stroke recurrence.[27]

Symptoms Related to Compression by Brain Herniation

The most specific sign of significant cerebral swelling after stroke is a decline in the level of consciousness and arousal related to brain or cerebellar edema shifting the thalamus and brainstem, where major components of the ascending arousal system are situated.[28] The Glasgow Coma Scale (GCS) is often considered as the gold standard of neurologic assessment for consciousness. This score is based on motor responsiveness, verbal performance, and eye opening to appropriate stimuli. Individual elements as well as the sum of the score are important. The final score ranges from 3 to 15. Generally, brain injury is classified as severe (GCS <9), moderate (GCS 9–12), and minor (GCS ≥ 13). Unfortunately this score is difficult to use in stroke patients with severe aphasia. In this setting, no reliable clinical feature has been validated to measure the level of consciousness. In recent studies evaluating decompressive craniectomy, only item 1a of the NIHSS has been used to assess this symptom.[29] In addition, brainstem shift related to brain herniation can lead to ipsilateral mydriasis, ophthalmoparesis, breathing irregularities including neurogenic hyperventilation, ataxic respiratory patterns or periodic breathing, and cardiac dysrhythmias.[30] A Babinski sign (extension of the big toe following stimulation of the outside of the sole of the foot) contralateral to the hemiparesis can also be observed as a result of brainstem notching against the tentorium.[30]

Symptoms Related to Intracranial Hypertension

Headache, nausea, and vomiting are major clinical symptoms of increased ICP and are often present in patients with cerebrovascular edema.[10]

Altogether these symptoms help clinicians to evoke the diagnosis of malignant edema associated with IS or ICH but none is specific and can formally drive clinicians in their treatment strategies. Additionally, most of these symptoms are observed in severe brain-swelling condition and no relevant clinical information are available to detect early mild-to-moderate edema.[31]

BRAIN IMAGING DIAGNOSIS

Because clinical symptoms are not specific for cerebral edema the diagnosis is mainly based on brain CT-scan or MRI. On CT-scan, cerebral edema appears as a hypodensity. While hypodensity surrounding ICH hyperdensity can be easily detected (Fig. 22.1A) its identification is far more difficult in the context of IS where cerebral edema cannot be distinguished from the core of the lesion (Fig. 22.1B). For more than two decades MRI has provided a clear improvement in the early identification of ischemic lesions but also in acute edema using diffusion-weighted imaging (DWI), T2-weighted imaging (T2WI), and more recently diffusion tensor

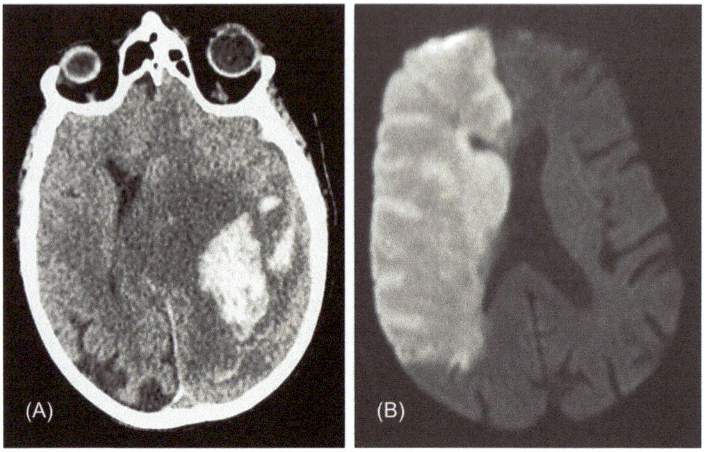

FIGURE 22.1 Large left parietal intracranial hemorrhage with perihematomal edema characterized by hyperdensity with surrounding hypodensity and severe mass effect on lateral ventricle and midline shift on CT-scan (A). Large right middle cerebral artery infarction with mass effect and no clear differentiation between ischemia and edema on standard diffusion weighted imaging (B).

imaging (DTI). Cerebral edema shows different MRI "signature" according to sequence type and edema cytotoxic or vasogenic mechanism.[32]

Although cerebral edema volume can be easily measured on DWI sequences in acute stroke, measure can be challenging when increased signal intensity is moderate such as in chronic cerebrovascular lesions. In this context, the ratio of cortical surface area to the volume of brain hemispheres has been used as a proxy of underlying white matter volume allowing the specific measure of edema in this compartment.[33] However, validity of this method in acute stroke patients still needs to be evaluated.

Brain imaging can also identify the structural consequences of the edema on the surrounding brain structures. Brain swelling causes successively ipsilateral sulcal effacement, ventricular system compression, and then midline structures such as the septum pellucidum and the pineal gland shift. The degree of midline shift is sometimes used as a biomarker for radiographic deterioration.[34] The shift can block the foramen of Monro or the third ventricle, leading to entrapment and dilatation of the contralateral lateral ventricle and obstructive hydrocephalus. Compression and compromise of the anterior or posterior cerebral arteries may sometimes be seen, leading to infarctions in the corresponding vascular territories.[35] In cerebellar lesions, effacement of the fourth ventricle is a key marker, followed by basal cistern compression, brainstem displacement, hydrocephalus, downward tonsillar herniation, and upward transtentorial herniation.[36]

PREDICTORS OF CEREBRAL EDEMA

As cerebral edema can impair functional prognosis and can be life threatening, it is mandatory to identify patients at risk as early as possible in order to offer them the opportunity to start medical and surgical procedures before irreversible brain damage occurs. For this purpose, biomarkers are typically used as determinants of prognosis or in the assessment of the risk of developing disease. They are broadly categorized into clinical evaluation, serum markers, continuous neuromonitoring techniques, or neuroimaging modalities. Among the many candidates that have been evaluated for the prediction of cerebral edema in stroke, only few biomarkers have potential application.[37]

Clinical Predictors

Several studies have tried to identify clinical determinants of malignant edema, providing heterogeneous results. Among the different factors, age seems to be the most reliable predictor, older patients being less likely to suffer the consequences of cerebral edema because of increased intracranial compliance secondary to relative atrophy. Most of the clinical studies[38] have identified the age of 60 as a potential cutoff for surgical management but one more recent study has also demonstrated the potential benefit of this type of treatment in some patients older than 60[39] suggesting that atrophy is probably a most reliable biomarker than age itself. Other clinical factors associated with edema after large stroke include early nausea and vomiting, female sex, and congestive heart failure.[10]

Unfortunately, these clinical factors alone are not sufficient to identify patients at risk of cerebral edema.[7,10,40]

Influence of the Genetic Background

Although some patients develop little edema, without midline shift, other patients with a comparable stroke size proceed to life-threatening brain swelling.[7] The cause of these differences has hardly been investigated so far. The influence of genetic variations, mainly in the AQP4 and ApoE genes, has been investigated in few studies. Klefner et al. demonstrated a possible dominant protective role of an AQP4 mutation (G allele of SNP rs9951307) after MCA occlusion.[41] Future works will be necessary to get more insight in this inter-individual vulnerability.

Influence of Stroke Mechanism

Some studies have looked for an association between IS mechanism and risk of malignant edema. Controversial results have been obtained;

no association being reported in most of the studies while some have described a higher risk among patients with carotid artery dissection or cardio-embolic stroke.[34] In these settings, the acute occlusion of a large artery could lead to large infarct size due to the absence of progressive development of collateral supplies as it is observed in progressive carotid stenosis related to atherosclerosis.

Should We Monitor Intracranial Pressure?

The raise in ICP is supposed to precede the incidence of neurological deterioration. No randomized clinical trial of ICP monitoring was performed in patients with stroke and controversial results were obtained from observational studies. Although raised ICP often correlates with clinical deterioration and computed tomography findings[42] one study reported that pupillary abnormalities and signs of severe brainstem compression were sometimes present despite normal ICP in patients with MCA.[43] Therefore, to date, routine ICP monitoring is not recommended in the acute stroke phase of patients at risk of cerebral and cerebellar malignant edema and treatment decisions should not be based on this parameter.[44] However, it may be helpful in comatose patients, especially during heavy sedation and ventilator care because in these situations neurological examinations are not sensitive enough to detect secondary brain injury, and repeated neuroimaging evaluations are not feasible.

Biological Biomarkers

Serum biomarkers of the different pathophysiological mechanisms involved in the stroke process could potentially be helpful in predicting patients at risk of malignant edema, whether by the detection of large clot using coagulation markers, by the identification of the impaired integrity of the BBB, or by evaluation of the extent of brain tissue damage.[45] Among these biomarkers, leukocytosis and ultrasensitive CRP serum level measured at baseline have been associated with an increased risk of malignant edema.[10] Measurements of plasma concentrations of neuronal or glial damage and inflammation using various biomarkers such as N-terminal pro-brain natriuretic peptide, S100B, glutamate, glycine, gamma-aminobutyric acid, interleukin-6 (IL-6), interleukin-10, tumor necrosis factor-alpha provided inconsistent results.[46] Among them S100-B serum levels >1.03 μg/L at 24 hours were associated with large infarction but not specifically to cerebral edema.[47] After 48 hours of stroke onset N-terminal pro-brain natriuretic peptide was found to be a marker of edema, suggesting that the time of evaluation of the biomarker should also been taken into account in the interpretation of the results.[48] Because the loss of integrity of the endothelial basal lamina of the BBB is believed to be one of the primary causes of edema after focal cerebral ischemia circulating markers that relate to the BBB integrity have been studied extensively. Many studies have focused on the potential role of matrix metalloproteinase, a group of proteolytic zinc-dependent enzymes that degrade the endothelial basal lamina. Increased matrix metalloproteinase-9 expression has been shown to play an essential role in producing edema and hemorrhagic transformation (HT) during stroke in experimental models of focal ischemia. One study reported that a concentration ≥140 ng/mL has a 64% sensitivity and a 88% specificity for predicting edema.[49] Another component of the basal lamina, the serum cellular-fibronectin, was shown to be a more accurate predictor of MCA than matrix metalloproteinase-9 in patients treated with tPA. A concentration >16.6 μg/mL at admission predicted the development of MMCAI with a sensitivity of 90% and a specificity of 100%. However, the relevance of this marker in clinical practice has not yet been confirmed.[46]

Despite encouraging results, the validity of serum biological measurements in the context of acute cerebrovascular disorders is still a matter a debate. Discrepancies between studies might be due to methodological factors but also to the incomplete rupture of the BBB impeding a linear association between plasma measurements and intracerebral lesions. To avoid this limitation, some studies have evaluated the potential role of measuring biochemical markers using a microdialysis probe inserted into the brain tissue for monitoring the course of these markers in patients with IS. Associated with increased ICP, a rise in the lactate-to-pyruvate ratio and hypoxanthine, indicating a change from aerobic to anaerobic metabolism and increased breakdown of ATP was reported. However, this technique is complex, not widely available, invasive, provides only focal information, and does not predict fatal outcome early enough for the successful implementation of invasive therapies which strongly limit its utility.[11,50] This method could be of interest in patients with early comatose state for which an intracranial probe has already been placed to monitor the ICP.

Electrophysiology

Few studies have investigated the potential usefulness of electrophysiological studies such as evoked potential or electroencephalography in the early prediction of malignant cerebral edema. A single study reported that diffuse slowing and increased delta activity documented on a continuous electroencephalography monitoring in the first 24 hours may identify patients who are likely to have a neurological deterioration related to brain swelling.[51] Again the limited access to continuous electrophysiological monitoring in clinical practice does not allow using this method in routine.

Brain Imaging

Stroke Volume and Severity of Ischemia

Several studies have showed that CT-scan has an acceptable sensitivity in identifying patients at risk of brain swelling that could be candidates for hemicraniectomy.[52] A frank hypodensity and involvement of one-third or more of the MCA territory including basal ganglia on head CT within the first 6 hours findings has been repeatedly identified to be the best predictor of malignant edema.[40,53,54] CT delineation of the infarct core in acute IS is a difficult issue that could account for its low specificity in identifying patients at risk of malignant edema.[52] Conversely, lesion volume measured by early DWI demonstrated excellent inter-rater agreement[35] and was identified as a major determinant of malignant infarct prediction with a high positive and negative predictive value.[55–59] Within 6 hours of stroke onset, the optimal DWI cutoff is largely in agreement between studies, with values of 80–90 mL predicting a rapid fulminant course.[60] When MRI is obtained within 14 hours after symptom onset, a DWI volume of >145 mL was found to be predictive of clinical deterioration.[57] This cutoff is currently used in clinical practice and international recommendations to decide which patient could benefit from decompressive craniectomy. Besides the lesion size, the severity of cytotoxic injury measured on ADC maps has also been associated with the probability of brain swelling.[31] The size and severity of the lesion can also be assessed using positron emission tomography (PET) of C-flumazenil, a marker of neuronal integrity. Heiss et al. demonstrated the predictive value of this method in identifying irreversible ischemic damage and the extent of the final infarct.[61,62] A cutoff value of 95 mL of neuronal damage has been observed in patients with malignant course which is lower than values measured on DWI at the same time-point probably because flumazenil-PET predominantly marks damage in cortical areas.[55]

Because the temporal course of cerebral edema formation can be highly different from one patient to another, the optimal imaging parameter to individualize those at risk of malignant edema might be the evaluation of the percent change in hemisphere volume, relative to the baseline hemisphere size. In rat MCA stroke model, hemisphere growth measured by MRI was reported to be strongly correlated to the absolute increase in brain water content ($r = 0.95$).[63] In humans, this marker accurately predicts early neurological deterioration and latter functional prognosis with an AUC ≥ 0.80.[35] Unfortunately, the currently low availability of MRI makes it difficult to broadly apply this technique in clinical practice.

Imaging Markers of BBB Permeability

Increased permeability of the BBB is critical in the pathophysiology of cerebral edema; therefore its evaluation might be of interest in identifying patients at risk of such complication. An intact BBB prevents large molecules such as gadolinium diethylenetriamine penta-acetic acid (Gd-DTPA) entering the brain tissue. Thus, leakage of gadolinium in the brain tissue reflects BBB disruption and can be used as a surrogate marker for BBB injury. Gadolinium enhancement in sulci over the infarct area can be visualized on MRI-FLAIR images several hours after injection. This parameter is called "hyperintense acute reperfusion marker (HARM)" and is observed in one-third of IS patients. Those with this sign have a higher risk of HT and worse clinical outcome.[64,65] Cho et al. also reported this pattern in about 18% of patients following carotid revascularization and described a neurological deterioration in half of them.[66] Another quantitative method studying the functional integrity of the BBB in vivo is dynamic contrast-enhanced MRI (DCE-MRI) that consists of a dynamic image acquisition during which an intravenous bolus of gadolinium-based contrast agent is performed. Images of the resultant changes in signal intensity provide parametric maps of specific microvascular biomarker. It has been proposed as a potential biomarker to identify patients with an increased risk for HT or severe edema.[67] Kassner et al. reported significantly different permeability scores between stroke patients who proceeded to hemorrhage and those who did not with and without thrombolytic therapy[68] making this sequence potentially useful in identifying patients suitable for thrombolytic treatments. However, large variations in the type of DCE-MRI sequence, the imaging protocols and the contrast agents used, and methods for data analysis have been reported across studies (for review, see Ref. 69). Consequently, estimated K-Trans values varied widely which limit the current use of this technique in current practice.

Perfusion Abnormalities

Together with the evaluation of the core of the lesion, CT and MRI also allow the evaluation of brain tissue at risk of infarction by using perfusion imaging. Results of trials that used this method to identify patients at risk of malignant edema have provided inconsistent results[70] and no recommendation can be made to manage patients based on these data. Analysis of MRI-perfusion sequences can also provide information about BBB permeability.[71] Using this method Leigh et al. demonstrated that BBB permeability is strongly associated with the increased risk of intracerebral HT, the ultimate step of BBB leakage following intravenous rt-PA.[72] Similar results were obtained using CT-perfusion, where an estimate of BBB permeability can be obtained by measuring the rate of iodinated contrast extravasation that can be used to calculate permeability-surface area product.[67] However, these results need to be replicated.

Pre-stroke Brain Morphological Structure

The space-occupying effect of ischemic edema not only depends on lesion size, but also on the intracranial volume reserve that is determined by congenital structural parameters, such as the tentorial opening, and acquired factors such as brain atrophy. A narrow tentorial opening and a high degree of contiguity between the brainstem and the tentorial edge may increase the risk of symptomatic cerebral edema, whereas a wide tentorial opening and a large perimesencephalic space could prevent its consequences.[73] The potential impact of preexisting brain atrophy on the course cerebral swelling in space-occupying infarction remains unclear. Moderate-to-severe brain atrophy might protect patients from a malignant course by giving the swollen brain more space to expand.[74,75] The methodology of evaluation of cerebral atrophy is often a matter of debate. Evaluation of cerebrospinal fluid (CSF) volume turned out to be one of the easiest and most reliable methods. A CT-based study demonstrated that its measurement improved the prediction of a malignant course compared to the prediction using only the cerebral blood volume (CBV) lesion volume.[76] Similarly, Minnerup et al. demonstrated that the ratio of the ischemic lesion volume to intracranial volume reserve, measured as CSF volume, was a strong predictor of a MMCAI.[75] However, these results were not confirmed by a recent study suggesting that additional studies are mandatory.[77] Besides the inconsistency of the results, this method is limited by the fact that volumetric analysis to determine the grade of cerebral atrophy is time consuming and often not clinically applicable. To overcome this limitation, brain atrophy can also be assessed by determining the intercaudate distance, which represents a linear brain atrophy marker that can be easily applied even in an emergency setting.[78] In the setting of acute stroke, the hemi-intercaudate distance of the intact hemisphere is measured as the shortest length between the medial border of the caudate nucleus and the midline account for local brain swelling. The best cutoff value to predict the development of early MMCAI was a hemi-intercaudate distance $<9.4\,mm$. A combination of this measure with the acute DWI lesion volume was shown to improve the positive predictive value of early MMCAI from 0.70 to 0.93.[77]

Vascular Imaging of Cerebral Arteries

The site of arterial occlusion is a well-known predictor of the size of the infarct which is strongly linked to the severity of the brain edema. A "T occlusion" of the distal internal carotid artery has been frequently reported to be a risk factor for malignant edema because it will lead to involvement of the anterior and middle cerebral arteries and reduce the potential of efficient collateral supply. Similarly, an incomplete circle of Willis, which leads to involvement of multiple vascular territories (e.g., the MCA and either the anterior cerebral artery or posterior cerebral artery), is also predictive of worse outcome.[34] Similarly, the presence of a dense MCA sign within the first 2 days, indicating persisting proximal occlusion, was associated with brain swelling, neurological deterioration, and early mortality.[54]

FUNCTIONAL PROGNOSIS

Ischemic Strokes

Brain edema constitutes one of the most lethal complications of IS during the acute phase.[79] Inconsistent data are currently available on the association between age and outcome in patients with severe ischemic brain edema. Susceptibility to develop edema in IS may decrease with increasing age.[80] It is also suggested that relative brain atrophy in older patients results in less consequences of cerebral edema.[81] Conversely, younger patients with decreased intracranial compliance may be at high risk for edema brain compression.[34] Additionally, magnitude and localization of edema-associated stroke impact the functional outcomes.

Hemispheric Infarction

MMCAI with significant edema usually results from proximal MCA, internal carotid artery, or both occlusions. Life threatening space-occupying brain swelling associated with large infarction usually manifests itself between the second and fifth day after stroke onset. The short-term prognosis is poor with case fatality rates around 80%.[81,82]

Outcomes in such large MCA swelling infarctions have been more recently studied in three therapeutic trials using decompression (hemi-) craniectomy (DHC).[29,39,56] In patients <60 years old, the pooled analysis confirmed that medical conservative management led to high fatality rate (71%) whereas DHC significantly reduced mortality (22%). The one-year functional outcomes assessed by the modified Rankin Scale (mRS) showed death or condition requiring assistance from others (mRS, 4–6) in 78% for medical and 57% for surgical treated patients. Only 21% did not require a helper (mRS, 0–3) in the medical group whereas this rate was twice as high (43%) in the surgical groups. Eventually, DHC constitutes an efficient lifesaving therapy, but only 14% of survivors are able to take care of their own affairs without assistance (mRS ≤2). There is still criticism that DHC is a worthwhile surgical option since it often averts death to severe dependent condition. However, the vast majority of patients and caregivers are satisfied with surgery outcome and would provide consent again for the procedure.[83]

In elderly patients, the 1-year mortality rates ranges from 69.6% to 76% in DHC patients and from 16.7% to 43% in medically treated patients.[84] However, overall stroke prognosis is age-dependent and only 6% of old patients had a favorable outcome whereas 43% of younger patients experienced this condition.[39]

Cerebellar Infarction

Mass effect occurs in 10%–25% of all cerebellar infarction cases[85] and in 40% of massive infarction.[86] Swelling associated with cerebellar infarction may result in direct brainstem compression and/or acute hydrocephalus secondary to of the fourth ventricle obstruction. Since the mid-20th century it has been apparent that patients who deteriorate after brain swelling from cerebellar infarction benefit from surgery.[87] Thus no relevant data on outcomes of patients who clinically deteriorates are available. Prognosis after surgery is closely associated with concomitant brainstem infarction and functional outcomes of isolated swelling cerebellar infarction requiring surgery are quite good since 74% experiences no disability 3 months after stroke onset.[86]

Hemorrhagic Transformation of Ischemic Stroke

HT, which refers to brain hemorrhage related to ischemia, is a frequent spontaneous complication of IS that could be considered as the ultimate consequence of BBB rupture. Massive brain infarction, embolic stroke, clinical severity, and rt-PA treatment are typical predictors of HT. Clinical severity of HT depends on its type. Parenchymal hematoma type 2, defined as a dense hematoma occupying >30% of the IS volume associated with edema and exerting a significant mass effect, appears as the only type of HT leading to early neurological deterioration and higher mortality[88] (Fig. 22.2). Conversely, although HT results of BBB disruption with vasogenic edema, no serious effect on the outcome is usually noted.

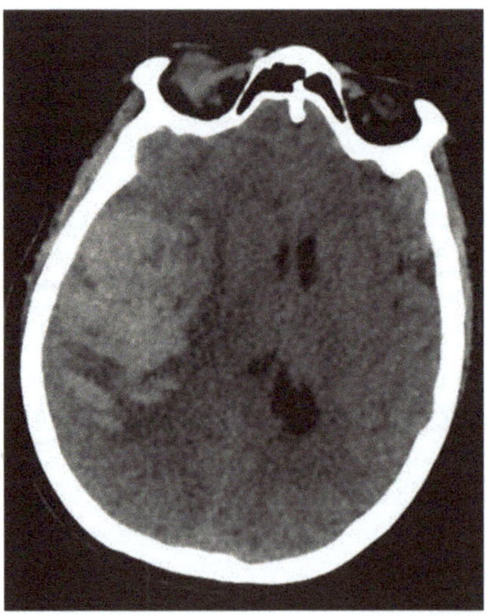

FIGURE 22.2 Hemorrhagic transformation of a right middle cerebral infarction characterized by a large parenchymal hematoma type II on a CT-scan.

Moreover, mild-to-moderate HT is often a sign of vascular recanalization and may be associated with better prognosis.[89]

TREATMENTS

Skilled physicians should perform early identification of stroke patients who may experience swelling. Ideally, medical care should be given in a stroke unit since this improves outcome.[79]

Prevention of Brain Edema Development After a Stroke

Mechanical Ventilation

A decline in consciousness and an inability to maintain a patent airway, leading to inadequate ventilation, are the most common reason for endotracheal intubation and mechanical ventilation. There is no evidence of benefit or harm with prophylactic hyperventilation in this population. Patients experiencing discomfort from the endotracheal tube may benefit from low doses of short-acting anesthetics.

Hemodynamic Support

Hemodynamic support by fluid management includes the use of isotonic saline and the avoidance of hypo-osmolar fluids. No-dextrose fluids are preferred. There are insufficient data to recommend mannitol or hypertonic saline in patients with mild CT swelling signs, but practices could vary.

Blood Pressure Control

Because of the large variation in practice and the few data from randomized, controlled trials, specific BP recommendations are difficult to propose. In IS, extreme hypertension increases the risk of HT[90] and consequently the risk of associated swelling. No specific BP targets are established but it seems reasonable to use continuous intravenous BP-lowering drugs for the treatment of systolic BP >220 mmHg or diastolic pressure >105 mmHg.

Glucose Management

Hyperglycemia is associated with increased edema in patients with cerebral ischemia and with an increased risk of HT.[91,92] Hyperglycemia should be avoided and glucose levels between 140 and 180 mg/dL are recommended while tight glycemic control (glucose <110 mg/dL) is not indicated and hypoglycemia should be avoided.

Temperature Management

Until now, no study has prospectively evaluated the impact of fever control on the outcome of stroke.[93] Normothermia is preferred, but therapeutic hypothermia has not been sufficiently studied prospectively. Treating temperatures over 37.5°C is usually recommended.[60] Early fever after acute stroke warrants complete assessment for an infectious or a drug-induced cause.

MEDICAL TREATMENT IN A DETERIORATED PATIENT

Several urgent measures are needed to treat a patient with ischemic brain edema who is deteriorating. However, management should focus on reducing the space-occupying effects of brain swelling and medical management should constitute a bridge therapy in case of surgical indication.

Elevated Head Position

Several authors recommend elevation of the head of the bed to 30°.[43,94] It is supposed to decrease ICP via increasing the venous outflow, reducing venous hydrostatic pressure and volume at the cranial level.

Osmotherapy

Osmotherapy acts mostly through an osmotic gradient, drawing water from interstitial and intracellular spaces of the brain into arteries. In addition, it may improve microvascular cerebral blood flow (CBF) by hemodilution and may lead to vasoconstriction with a subsequent reduction of CBV and ICP.[95]

Mannitol

Mannitol, a sugar alcohol with osmotic properties, is one of the most widely used medical treatment agents for increased ICP. It is given at various doses as multiple boluses upon increases of ICP. However, various observational studies and clinical trials have been unable to provide evidence of efficiency on clinical outcomes so far.[96] Generally, the treatment consists of repeated boluses of 0.25–0.5 mg/kg every 4–6 hours. A range between 310 and 320 mOsm/L has been commonly recommended as target serum osmolality.[97] Treatment is usually administered during 3 days followed by a progressive decrease in the bolus administration. Mannitol is also suggested to be effective as a therapeutic bridge to DHC.

Hypertonic Saline

At present, there is no clear information on the optimum form of application of hypertonic saline. Multiple concentrations (3%–23%) have been used in continuous and bolus infusion and the results are controversial.[98] Efficiency comparison between mannitol and hypertonic saline does not show significant difference.[99]

Other Osmotic Agents

The lack of evidence of benefit in long-term survival does not support the routine or selective use of glycerol treatment in patients with acute stroke.[100]

Some authors considered that tromethamine administration by continuous infusion at 3mmol/h or by bolus of 1mmol/kg body weight is an option for treating brain edema secondary to large MCA infarction.[97,101]

Furosemide cannot be recommended regarding the risk of substantial volume depletion and the lack of evidence of benefit on brain edema.[102,103]

Steroids

A systematic review of randomized stroke trials concluded that steroids lack any benefit and may have even adverse effects when applied after cerebral ischemia. Therefore the use of steroids is not recommended for the treatment of brain swelling after stroke.[104]

Hypothermia

Hypothermia likely exerts multiple and synergistic neuroprotective effects including brain edema reduction.[105] It has been used for various degrees such as 33–36°C and for several durations such as 12–48hours.[106] Prospective randomized studies are currently ongoing to further assess therapeutic hypothermia in patients with IS.

Barbiturates

Barbiturate administration in large MCA infarction seems to be associated with only temporarily ICP decrease,[101] whereas significant decline in cerebral perfusion pressure has been described.[107] Thus, barbiturates could offer limited benefits that may be counterbalanced by severe adverse effects, such as hypotension and critical decline in cerebral perfusion pressure.

Hyperventilation

The ICP-lowering capacity of hyperventilation results from vasoconstriction with a subsequent decrease of CBV and CBF. Data on the use of

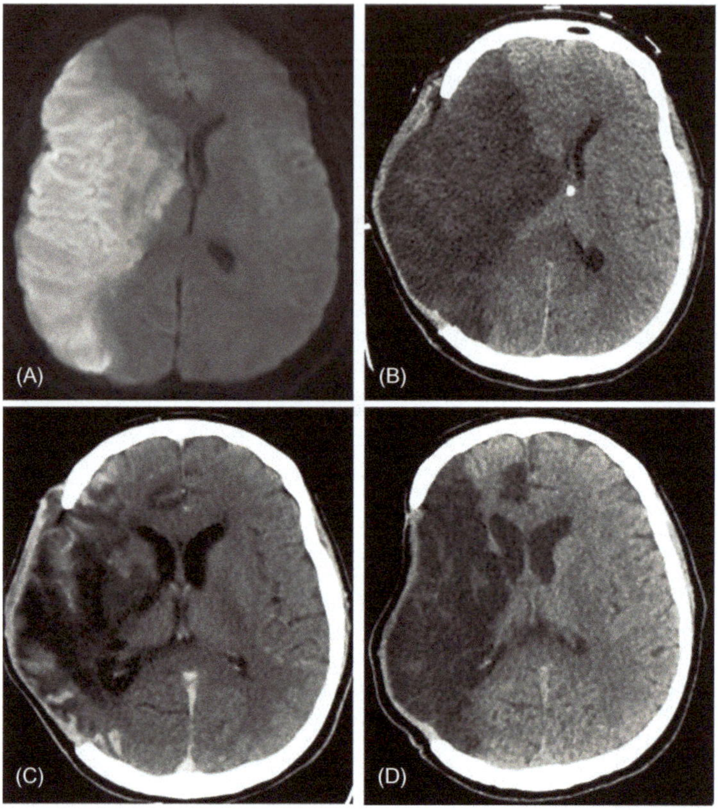

FIGURE 22.3 **MRI and CT-scan performed in a 56-year-old man who experienced a sudden onset left hemiplegia (A–D).** MRI-DWI sequence performed 8h of the onset that showed a DWI infarct volume of 186 cm^3 consistent with a right malignant middle cerebral artery infarction (A). CT-scan performed at 2 days after a decompressive hemicraniectomy that showed a large infarction with brain swelling mass on lateral ventricle and brain protrusion through the large bone defect (B). CT-scan performed at 1 month showing the regression of the midline shift and mass effect (C). CT-scan performed at 6 months showing a right hemispheric brain atrophy (D).

hyperventilation in stroke patients are derived from uncontrolled clinical studies from the pre-CT era.[108,109] CBF decrease to ischemic level and deleterious effects on cerebral metabolism constitute a major drawback of hyperventilation.[110] Thus, regarding the possible cerebral deleterious consequences and the lack of evidence of any beneficial effect of hyperventilation, this technique cannot be recommended for stroke patients.

Neurosurgical Options in a Deteriorated Patient

Malignant MCA Infarction

DHC is the only proven therapy in MMCAI and is highly recommended (Fig. 22.3). The conditions requiring surgical decompression associate: (1) young age patients ≤60 years; (2) large hemispheric infarction defined as brain CT signs involving >50% of the MCA territory or DWI infarct volume on MRI >145 cm^3; and (3) clinical deterioration within 48 hours after the onset defined as a decreased level of consciousness (score ≥1 on item 1a of the NIHSS or a gradual decrease in consciousness to a score ≤13 on the GCS for right-sided lesions or ≤9 for the left-sided lesions). Additionally, MMCAI seems to be early predicted, within 6 hours after the onset, by a lesion volume >82 mL on DWI sequences.[58] The efficacy of DHC in patients >60 years of age and the effect over 48 hours are uncertain but should be strongly considered.[84]

Brain Swelling Following Cerebellar Infarction

In the case of extensive mass effect, surgical treatment is widely accepted and strongly recommended in current stroke guidelines[44] (Fig. 22.4). Surgical therapy includes ventriculostomy and suboccipital decompressive craniectomy performed by removal of the skull and expansion of the dura to alleviate the volume constraints of the cranial bone. The procedure may also include the resection of necrotic cerebellar tissue to create extra space within the posterior fossa and reduce development of cytotoxic edema. Thus, alert and clinically stable patients with swelling cerebellar stroke should be monitored closely using neurological scales and repeated neuroimaging. In patients whose condition deteriorates, neuroimaging should be performed immediately to rule out extensive brainstem infarction and to determine the optimal surgical strategy.[111]

CONCLUSION

Cerebral edema is part of the natural history of acute cerebrovascular diseases. To date, most studies have focused on the malignant edema which is responsible for a significant alteration of the functional and vital prognosis of patients. Many biomarkers were evaluated, but their sensitivity is still unsatisfactory and the identification of new markers whether serum or imaging is still needed. Moreover future studies will focus on evaluating the predictive value of the combination of different markers.[112] Similarly, therapeutic strategies have been developed primarily to deal with malignant edema. However, the influence of minor-to-moderate

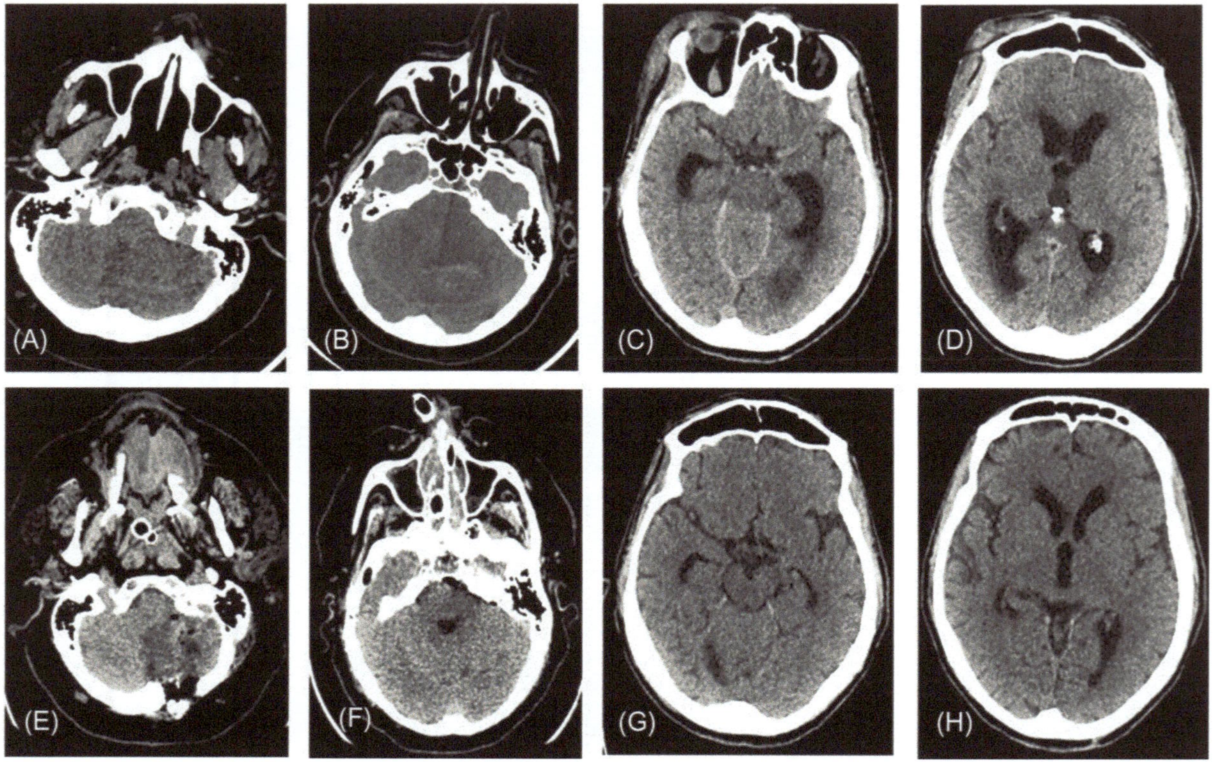

FIGURE 22.4 **CT-scan performed 16h after a sudden vertigo with left cerebellar signs in a 61-year-old man (A–D).** Left cerebellar infarction with edema leading to a compression of the fourth ventricle (A,B). Dilatation of the third and lateral ventricles secondary to no communicant hydrocephalus (C,D). CT-scan performed at 2 days after a suboccipital craniectomy and necrotic tissue resection (E–H). Regression of the compression of the fourth ventricle (F) and consequently regression of the dilatation of the third and lateral ventricles (G,H).

edema could be an important therapeutic goal in the acute phase in addition to pharmacological or mechanical recanalization. The development of new molecules having the objective of limiting the permeability of the BBB[18] could be a major therapeutic advance in the treatment of cerebral vascular pathology in the near future.

References

1. Feigin VL, et al. Global and regional burden of stroke during 1990–2010: findings from the Global Burden of Disease Study 2010. *Lancet Lond Engl* 2014;**383**:245–54.
2. Bejot Y, et al. Epidemiology of stroke in Europe: geographic and environmental differences. *J Neurol Sci* 2007;**262**:85–8.
3. Fisher M, Saver JL. Future directions of acute ischaemic stroke therapy. *Lancet Neurol* 2015;**14**:758–67.
4. Marmarou A. A review of progress in understanding the pathophysiology and treatment of brain edema. *Neurosurg Focus* 2007;**22**:E1.
5. Balami JS, Buchan AM. Complications of intracerebral haemorrhage. *Lancet Neurol* 2012;**11**:101–18.
6. Beard DJ, et al. Intracranial pressure elevation reduces flow through collateral vessels and the penetrating arterioles they supply. A possible explanation for "collateral failure" and infarct expansion after ischemic stroke. *J Cereb Blood Flow Metab Off J Int Soc Cereb Blood Flow Metab* 2015;**35**:861–72.
7. Information NC for B, Pike USNL of M 8600 R, MD, B & USA, 20894. 'Malignant' middle cerebral artery territory infarction: clinical course and prognostic signs. [Arch Neurol. 1996] – PubMed – NCBI. Available from: http://www.ncbi.nlm.nih.gov/pubmed [accessed 26.09.15].
8. Yang J, et al. Prognostic significance of perihematomal edema in acute intracerebral hemorrhage: pooled analysis from the intensive blood pressure reduction in acute cerebral hemorrhage trial studies. *Stroke J Cereb Circ* 2015;**46**:1009–13.
9. Moulin DE, Lo R, Chiang J, Barnett HJ. Prognosis in middle cerebral artery occlusion. *Stroke J Cereb Circ* 1985;**16**:282–4.
10. Kasner SE, et al. Predictors of fatal brain edema in massive hemispheric ischemic stroke. *Stroke J Cereb Circ* 2001;**32**:2117–23.
11. Dohmen C, et al. Prediction of malignant course in MCA infarction by PET and microdialysis. *Stroke J Cereb Circ* 2003;**34**:2152–8.
12. Zha AM, Sari M, Torbey MT. Recommendations for management of large hemispheric infarction. *Curr Opin Crit Care* 2015;**21**:91–8.
13. Simard JM, Kent TA, Chen M, Tarasov KV, Gerzanich V. Brain oedema in focal ischaemia: molecular pathophysiology and theoretical implications. *Lancet Neurol* 2007;**6**:258–68.
14. Pantoni L, Garcia JH, Gutierrez JA. Cerebral white matter is highly vulnerable to ischemia. *Stroke J Cereb Circ* 1996;**27**:1641–6. Discussion 1647.
15. Garcia JH, Liu KF, Yoshida Y, Chen S, Lian J. Brain microvessels: factors altering their patency after the occlusion of a middle cerebral artery (Wistar rat). *Am J Pathol* 1994;**145**:728–40.
16. Ames A, Wright RL, Kowada M, Thurston JM, Majno G. Cerebral ischemia. II. The no-reflow phenomenon. *Am J Pathol* 1968;**52**:437–53.
17. Kimelberg HK. Volume activated anion channel and astrocytic cellular edema in traumatic brain injury and stroke. *Adv Exp Med Biol* 2004;**559**:157–67.
18. Michinaga S, Koyama Y. Pathogenesis of brain edema and investigation into anti-edema drugs. *Int J Mol Sci* 2015;**16**:9949–75.

19. Wang CX, Shuaib A. Critical role of microvasculature basal lamina in ischemic brain injury. *Prog Neurobiol* 2007;**83**:140–8.
20. Yang Y, Rosenberg GA. Blood–brain barrier breakdown in acute and chronic cerebrovascular disease. *Stroke J Cereb Circ* 2011;**42**:3323–8.
21. Vella J, Zammit C, Di Giovanni G, Muscat R, Valentino M. The central role of aquaporins in the pathophysiology of ischemic stroke. *Front Cell Neurosci* 2015;**9**:108.
22. Iliff JJ, Nedergaard M. Is there a cerebral lymphatic system? *Stroke J Cereb Circ* 2013;**44**:S93–95.
23. Ayata C, Ropper AH. Ischaemic brain oedema. *J Clin Neurosci Off J Neurosurg Soc Australas* 2002;**9**:113–24.
24. Alexander E, Kushner J, Six EG. Brainstem hemorrhages and increased intracranial pressure: from Duret to computed tomography. *Surg Neurol* 1982;**17**:107–10.
25. Serena J, Rodríguez-Yáñez M, Castellanos M. Deterioration in acute ischemic stroke as the target for neuroprotection. *Cerebrovasc Dis Basel Switz* 2006;**21**(Suppl 2):80–8.
26. Seners P, Turc G, Oppenheim C, Baron J-C. Incidence, causes and predictors of neurological deterioration occurring within 24h following acute ischaemic stroke: a systematic review with pathophysiological implications. *J Neurol Neurosurg Psychiatry* 2015;**86**:87–94.
27. Seners P, et al. Unexplained early neurological deterioration after intravenous thrombolysis: incidence, predictors, and associated factors. *Stroke J Cereb Circ* 2014;**45**:2004–9.
28. Frank JI. Large hemispheric infarction, deterioration, and intracranial pressure. *Neurology* 1995;**45**:1286–90.
29. Vahedi K, et al. Early decompressive surgery in malignant infarction of the middle cerebral artery: a pooled analysis of three randomised controlled trials. *Lancet Neurol* 2007;**6**:215–22.
30. Ropper AH, Shafran B. Brain edema after stroke. Clinical syndrome and intracranial pressure. *Arch Neurol* 1984;**41**:26–9.
31. Battey TWK, et al. Brain edema predicts outcome after nonlacunar ischemic stroke. *Stroke J Cereb Circ* 2014;**45**:3643–8.
32. Albers GW. Diffusion-weighted MRI for evaluation of acute stroke. *Neurology* 1998;**51**:S47–9.
33. De Guio F, et al. White matter edema at the early stage of cerebral autosomal-dominant arteriopathy with subcortical infarcts and leukoencephalopathy. *Stroke J Cereb Circ* 2015;**46**:258–61.
34. Jaramillo A, Góngora-Rivera F, Labreuche J, Hauw J-J, Amarenco P. Predictors for malignant middle cerebral artery infarctions: a postmortem analysis. *Neurology* 2006;**66**:815–20.
35. Yoo AJ, et al. Validating imaging biomarkers of cerebral edema in patients with severe ischemic stroke. *J Stroke Cerebrovasc Dis Off J Natl Stroke Assoc* 2013;**22**:742–9.
36. Koh MG, Phan TG, Atkinson JL, Wijdicks EF. Neuroimaging in deteriorating patients with cerebellar infarcts and mass effect. *Stroke J Cereb Circ* 2000;**31**:2062–7.
37. Kernagis DN, Laskowitz DT. Evolving role of biomarkers in acute cerebrovascular disease. *Ann Neurol* 2012;**71**:289–303.
38. Neugebauer H, Jüttler E. Hemicraniectomy for malignant middle cerebral artery infarction: current status and future directions. *Int J Stroke Off J Int Stroke Soc* 2014;**9**:460–7.
39. Jüttler E, et al. Hemicraniectomy in older patients with extensive middle-cerebral-artery stroke. *N Engl J Med* 2014;**370**:1091–100.
40. Krieger DW, Demchuk AM, Kasner SE, Jauss M, Hantson L. Early clinical and radiological predictors of fatal brain swelling in ischemic stroke. *Stroke J Cereb Circ* 1999;**30**:287–92.
41. Kleffner I, et al. The role of aquaporin-4 polymorphisms in the development of brain edema after middle cerebral artery occlusion. *Stroke J Cereb Circ* 2008;**39**:1333–5.
42. Schwab S, Aschoff A, Spranger M, Albert F, Hacke W. The value of intracranial pressure monitoring in acute hemispheric stroke. *Neurology* 1996;**47**:393–8.

43. Poca MA, et al. Monitoring intracranial pressure in patients with malignant middle cerebral artery infarction: Is it useful? *J Neurosurg* 2010;**112**:648–57.
44. Wijdicks EFM, et al. Recommendations for the management of cerebral and cerebellar infarction with swelling: a statement for healthcare professionals from the American Heart Association/American Stroke Association. *Stroke J Cereb Circ* 2014;**45**:1222–38.
45. Miao Y, Liao JK. Potential serum biomarkers in the pathophysiological processes of stroke. *Expert Rev Neurother* 2014;**14**:173–85.
46. Castellanos M, et al. Plasma cellular-fibronectin concentration predicts hemorrhagic transformation after thrombolytic therapy in acute ischemic stroke. *Stroke J Cereb Circ* 2004;**35**:1671–6.
47. Missler U, Wiesmann M, Friedrich C, Kaps M. S-100 protein and neuron-specific enolase concentrations in blood as indicators of infarction volume and prognosis in acute ischemic stroke. *Stroke J Cereb Circ* 1997;**28**:1956–60.
48. Modrego PJ, Boned B, Berlanga JJ, Serrano M. Plasmatic B-type natriuretic peptide and C-reactive protein in hyperacute stroke as markers of CT-evidence of brain edema. *Int J Med Sci* 2008;**5**:18–23.
49. Castellanos M, et al. Plasma metalloproteinase-9 concentration predicts hemorrhagic transformation in acute ischemic stroke. *Stroke J Cereb Circ* 2003;**34**:40–6.
50. Bosche B, et al. Extracellular concentrations of non-transmitter amino acids in peri-infarct tissue of patients predict malignant middle cerebral artery infarction. *Stroke J Cereb Circ* 2003;**34**:2908–13.
51. Burghaus L, et al. Early electroencephalography in acute ischemic stroke: Prediction of a malignant course? *Clin Neurol Neurosurg* 2007;**109**:45–9.
52. von Kummer R, et al. Sensitivity and prognostic value of early CT in occlusion of the middle cerebral artery trunk. *AJNR Am J Neuroradiol* 1994;**15**:9–15. Discussion 16–18.
53. Kucinski T, et al. The predictive value of early CT and angiography for fatal hemispheric swelling in acute stroke. *AJNR Am J Neuroradiol* 1998;**19**:839–46.
54. Manno EM, Nichols DA, Fulgham JR, Wijdicks EFM. Computed tomographic determinants of neurologic deterioration in patients with large middle cerebral artery infarctions. *Mayo Clin Proc* 2003;**78**:156–60.
55. Dohmen C, et al. Prediction of malignant course in MCA infarction by PET and microdialysis. *Stroke J Cereb Circ* 2003;**34**:2152–8.
56. Hofmeijer J, et al. Surgical decompression for space-occupying cerebral infarction (the Hemicraniectomy After Middle Cerebral Artery infarction with Life-threatening Edema Trial [HAMLET]): a multicentre, open, randomised trial. *Lancet Neurol* 2009;**8**:326–33.
57. Oppenheim C, et al. Prediction of malignant middle cerebral artery infarction by diffusion-weighted imaging. *Stroke J Cereb Circ* 2000;**31**:2175–81.
58. Thomalla G, et al. Prediction of malignant middle cerebral artery infarction by magnetic resonance imaging within 6 hours of symptom onset: a prospective multicenter observational study. *Ann Neurol* 2010;**68**:435–45.
59. Thomalla GJ, et al. Prediction of malignant middle cerebral artery infarction by early perfusion- and diffusion-weighted magnetic resonance imaging. *Stroke J Cereb Circ* 2003;**34**:1892–9.
60. Arenillas JF, et al. Prediction of early neurological deterioration using diffusion- and perfusion-weighted imaging in hyperacute middle cerebral artery ischemic stroke. *Stroke J Cereb Circ* 2002;**33**:2197–203.
61. Heiss WD, et al. Early [(11)C]Flumazenil/H(2)O positron emission tomography predicts irreversible ischemic cortical damage in stroke patients receiving acute thrombolytic therapy. *Stroke J Cereb Circ* 2000;**31**:366–9.
62. Heiss WD, Kracht LW, Thiel A, Grond M, Pawlik G. Penumbral probability thresholds of cortical flumazenil binding and blood flow predicting tissue outcome in patients with cerebral ischaemia. *Brain J Neurol* 2001;**124**:20–9.

63. Gerriets T, et al. Noninvasive quantification of brain edema and the space-occupying effect in rat stroke models using magnetic resonance imaging. *Stroke J Cereb Circ* 2004;**35**:566–71.
64. Henning EC, Latour LL, Hallenbeck JM, Warach S. Reperfusion-associated hemorrhagic transformation in SHR rats: evidence of symptomatic parenchymal hematoma. *Stroke J Cereb Circ* 2008;**39**:3405–10.
65. Ostwaldt A-C, et al. Early time course of FLAIR signal intensity differs between acute ischemic stroke patients with and without hyperintense acute reperfusion marker. *Cerebrovasc Dis Basel Switz* 2014;**37**:141–6.
66. Cho A-H, et al. Reperfusion injury on magnetic resonance imaging after carotid revascularization. *Stroke J Cereb Circ* 2014;**45**:602–4.
67. Dankbaar JW, et al. Dynamic perfusion-CT assessment of early changes in blood brain barrier permeability of acute ischaemic stroke patients. *J Neuroradiol* 2011;**38**:161–6.
68. Kassner A, Mandell DM, Mikulis DJ. Measuring permeability in acute ischemic stroke. *Neuroimaging Clin N Am* 2011;**21**:315–25, x–xi.
69. Heye AK, Culling RD, Valdés Hernández MDC, Thrippleton MJ, Wardlaw JM. Assessment of blood–brain barrier disruption using dynamic contrast-enhanced MRI. A systematic review. *NeuroImage Clin* 2014;**6**:262–74.
70. Mlynash M, et al. Refining the definition of the malignant profile: insights from the DEFUSE-EPITHET pooled data set. *Stroke J Cereb Circ* 2011;**42**:1270–5.
71. Leigh R, Jen SS, Varma DD, Hillis AE, Barker PB. Arrival time correction for dynamic susceptibility contrast MR permeability imaging in stroke patients. *PloS One* 2012;**7**:e52656.
72. Leigh R, et al. Pretreatment blood–brain barrier damage and post-treatment intracranial hemorrhage in patients receiving intravenous tissue-type plasminogen activator. *Stroke J Cereb Circ* 2014;**45**:2030–5.
73. Adler DE, Milhorat TH. The tentorial notch: anatomical variation, morphometric analysis, and classification in 100 human autopsy cases. *J Neurosurg* 2002;**96**:1103–12.
74. Lee SH, et al. The effect of brain atrophy on outcome after a large cerebral infarction. *J Neurol Neurosurg Psychiatry* 2010;**81**:1316–21.
75. Minnerup J, et al. Prediction of malignant middle cerebral artery infarction using computed tomography-based intracranial volume reserve measurements. *Stroke J Cereb Circ* 2011;**42**:3403–9.
76. Dohmen C, Galldiks N, Bosche B, Kracht L, Graf R. The severity of ischemia determines and predicts malignant brain edema in patients with large middle cerebral artery infarction. *Cerebrovasc Dis Basel Switz* 2012;**33**:1–7.
77. Beck C, et al. A simple brain atrophy measure improves the prediction of malignant middle cerebral artery infarction by acute DWI lesion volume. *J Neurol* 2014;**261**:1097–103.
78. Butzkueven H, et al. Validation of linear cerebral atrophy markers in multiple sclerosis. *J Clin Neurosci Off J Neurosurg Soc Australas* 2008;**15**:130–7.
79. Stroke Unit Trialists' Collaboration. Organised inpatient (stroke unit) care for stroke. Cochrane Database Syst Rev 2013: CD000197. http://dx.doi.org/10.1002/14651858.CD000197.pub3.
80. Liu F, Yuan R, Benashski SE, McCullough LD. Changes in experimental stroke outcome across the life span. *J Cereb Blood Flow Metab Off J Int Soc Cereb Blood Flow Metab* 2009;**29**:792–802.
81. Hacke W, et al. Malignant' middle cerebral artery territory infarction: clinical course and prognostic signs. *Arch Neurol* 1996;**53**:309–15.
82. Berrouschot J, Rössler A, Köster J, Schneider D. Mechanical ventilation in patients with hemispheric ischemic stroke. *Crit Care Med* 2000;**28**:2956–61.
83. Rahme R, Zuccarello M, Kleindorfer D, Adeoye OM, Ringer AJ. Decompressive hemicraniectomy for malignant middle cerebral artery territory infarction: Is life worth living? *J Neurosurg* 2012;**117**:749–54.

84. Zhao J, et al. Decompressive hemicraniectomy in malignant middle cerebral artery infarct: a randomized controlled trial enrolling patients up to 80 years old. *Neurocrit Care* 2012;**17**:161–71.
85. Kase CS, et al. Cerebellar infarction. Clinical and anatomic observations in 66 cases. *Stroke J Cereb Circ* 1993;**24**:76–83.
86. Jauss M, Krieger D, Hornig C, Schramm J, Busse O. Surgical and medical management of patients with massive cerebellar infarctions: results of the German-Austrian Cerebellar Infarction Study. *J Neurol* 1999;**246**:257–64.
87. Fairburn B, Oliver LC. Cerebellar softening: a surgical emergency. *Br Med J* 1956;**1**:1335–6.
88. D'Amelio M, et al. Early and late mortality of spontaneous hemorrhagic transformation of ischemic stroke. *J Stroke Cerebrovasc Dis Off J Natl Stroke Assoc* 2014;**23**:649–54.
89. Kablau M, et al. Predictors and early outcome of hemorrhagic transformation after acute ischemic stroke. *Cerebrovasc Dis Basel Switz* 2011;**32**:334–41.
90. de Courten-Myers GM, et al. Hemorrhagic infarct conversion in experimental stroke. *Ann Emerg Med* 1992;**21**:120–6.
91. Gilmore RM, Stead LG. The role of hyperglycemia in acute ischemic stroke. *Neurocrit Care* 2006;**5**:153–8.
92. Pulsinelli WA, Waldman S, Rawlinson D, Plum F. Moderate hyperglycemia augments ischemic brain damage: a neuropathologic study in the rat. *Neurology* 1982;**32**:1239–46.
93. Badjatia N. Fever control in the neuro-ICU: Why, who, and when? *Curr Opin Crit Care* 2009;**15**:79–82.
94. Bardutzky J, Schwab S. Antiedema therapy in ischemic stroke. *Stroke J Cereb Circ* 2007;**38**:3084–94.
95. Diringer MN, Zazulia AR. Osmotic therapy: fact and fiction. *Neurocrit Care* 2004;**1**:219–33.
96. Bereczki D, Fekete I, Prado GF, Liu M. Mannitol for acute stroke. *Cochrane Database Syst Rev* 2007:CD001153. http://dx.doi.org/10.1002/14651858.CD001153.pub2.
97. Schwab S. Therapy of severe ischemic stroke: breaking the conventional thinking. *Cerebrovasc Dis Basel Switz* 2005;**20**(Suppl 2):169–78.
98. Ogden AT, Mayer SA, Connolly ES. Hyperosmolar agents in neurosurgical practice: the evolving role of hypertonic saline. *Neurosurgery* 2005;**57**:207–15. Discussion 207–215.
99. Diringer MN, Scalfani MT, Zazulia AR, Videen TO, Dhar R. Cerebral hemodynamic and metabolic effects of equi-osmolar doses mannitol and 23.4% saline in patients with edema following large ischemic stroke. *Neurocrit Care* 2011;**14**:11–17.
100. Righetti E, et al. Glycerol for acute stroke. *Cochrane Database Syst Rev* 2004:CD000096. http://dx.doi.org/10.1002/14651858.CD000096.pub2.
101. Steiner T, et al. Multimodal online monitoring in middle cerebral artery territory stroke. *Stroke J Cereb Circ* 2001;**32**:2500–6.
102. Paczynski RP, et al. Effects of fluid management on edema volume and midline shift in a rat model of ischemic stroke. *Stroke J Cereb Circ* 2000;**31**:1702–8.
103. Toung TJK, Hurn PD, Traystman RJ, Bhardwaj A. Global brain water increases after experimental focal cerebral ischemia: effect of hypertonic saline. *Crit Care Med* 2002;**30**:644–9.
104. Qizilbash N, Lewington SL, Lopez-Arrieta JM. Corticosteroids for acute ischaemic stroke. *Cochrane Database Syst Rev* 2002:CD000064. http://dx.doi.org/10.1002/14651858.CD000064.
105. Georgiadis D, Schwab S. Hypothermia in acute stroke. *Curr Treat Options Neurol* 2005;**7**:119–27.
106. Linares G, Mayer SA. Hypothermia for the treatment of ischemic and hemorrhagic stroke. *Crit Care Med* 2009;**37**:S243–249.
107. Schwab S, Spranger M, Schwarz S, Hacke W. Barbiturate coma in severe hemispheric stroke: Useful or obsolete? *Neurology* 1997;**48**:1608–13.

108. Christensen MS, Paulson OB. Prolonged artificial hyperventilation in severe cerebral apoplexy. Clinical results and cerebrospinal fluid findings in a controlled study. *Eur Neurol* 1972;**8**:137–41.
109. Simard D, Paulson OB. Artifical hyperventilation in stroke. *Trans Am Neurol Assoc* 1973;**98**:309–10.
110. Marion DW, et al. Effect of hyperventilation on extracellular concentrations of glutamate, lactate, pyruvate, and local cerebral blood flow in patients with severe traumatic brain injury. *Crit Care Med* 2002;**30**:2619–25.
111. Neugebauer H, Witsch J, Zweckberger K, Jüttler E. Space-occupying cerebellar infarction: complications, treatment, and outcome. *Neurosurg Focus* 2013;**34**:E8.
112, Chen R, Deng Z, Song Z. The prediction of malignant middle cerebral artery infarction: a predicting approach using random forest. *J Stroke Cerebrovasc Dis Off J Natl Stroke Assoc* 2015;**24**:958–64.

Further Reading

Aksoy D, et al. Magnetic resonance imaging profile of blood–brain barrier injury in patients with acute intracerebral hemorrhage. *J Am Heart Assoc* 2013;**2**:e000161.

Kassner A, Roberts T, Taylor K, Silver F, Mikulis D. Prediction of hemorrhage in acute ischemic stroke using permeability MR imaging. *Am J Neuroradiol* 2005;**26**:2213–17.

CHAPTER

23

Traumatic Brain Injury and Edema Treatment

Klaus Zweckberger
University of Heidelberg, Heidelberg, Germany

OUTLINE

Brain Edema.
DOI: http://dx.doi.org/10.1016/B978-0-12-803196-4.00023-0

INTRODUCTION

Traumatic brain injury (TBI) is an important public health concern since it is one of the leading causes of death and disability worldwide.[1,2] TBI is called a "silent epidemic" because consequences of the trauma are not always immediately visible.[3] The term "silent" furthermore reflects the common underestimation and unawareness of the impact of TBI by the society. This is particularly reflected by the fact that a clear definition of "TBI" has been lacking until recently. TBI can now be defined as "an alteration in brain function, or other evidence of brain pathology, caused by an external force."[4]

EPIDEMIOLOGY

Estimating epidemiological data of TBI cases in Europe, Tagliaferri et al. have conducted a systematic review in 2006[5]: 23 studies between 1980 and 2003 were analyzed. They calculated an aggregated (i.e., fatal plus hospitalized) incidence rate of 235 cases per 100,000 inhabitants per year, an average mortality rate of 15 per 100,000 people per year, and a case fatality rate of 2.7%.[5] In the last decade, new insights into the epidemiology have emerged with an increasing incidence in the elderly population and a decreasing trend of mortality.[6,7] A current review by the group of Peeters et al. stated a stable incidence of TBI cases in Europe, but could not find a decreasing trend of TBI cases.[8] Referring to the 28 epidemiological studies published between 1990 and 2014, they recently described an incidence of 262 per 100,000 per year. Falls and road traffic accidents were the two most frequent causes of TBI. In most studies furthermore a peak TBI incidence was seen in older age groups.[8]

TBI has a remarkable socioeconomic impact, although, given the number of untreated TBIs, it is difficult to calculate real economic costs. Langlois Orman[9] extrapolated data from Finkelstein et al.[10] estimating the economic burden of TBI in the USA for 2009 to be 221 billion US$ (14.6 billion US$ for medical costs, 69.2 billion US$ for loss of work years, and 137 billion US$ for the value of lost quality of life). Estimates from the US Center of Disease Control focusing more on medical costs and loss of work years indicated costs of 56 billion US$.[11,12]

PATHOPHYSIOLOGY

Due to the variety of different trauma mechanisms, TBI is a very heterogeneous disease. The principal pathophysiological concept, however, encompasses two phases of TBI—the primary and secondary injuries.

The term "primary injury" describes the damage initially caused by the impact itself. It results in brain tissue damage and rupture of vessels thus leading to the development of contusions, diffuse axonal injury, and hemorrhages. The primary injury is therefore not treatable, but only preventable by using seat belts and airbags or wearing helmets.

The secondary injury develops over hours and days, and includes uncontrolled neurotransmitter release (excitotoxicity), free radical generation, calcium-mediated damage, mitochondrial dysfunction, inflammatory responses, edema, intracranial hypertension, and ischemia. Extra- and intracranial pathologies thereby support the development of secondary injury. Extracranial reasons are anemia, hypotension and respiratory insufficiency leading to hypoxia and hypercapnia, as well as fever. Intracranial triggers are brain edema, hemorrhages, inflammatory reactions, metabolic impairments and infections. Continuous swelling of the brain results in an elevation of intracranial pressure (ICP) and decrease in cerebral perfusion pressure (CPP) and therefore hypoperfusion and ischemia, which further propagates edema formation. One of the main aims of medical treatment after TBI is therefore to control edema development.

BRAIN EDEMA

The development of brain edema is a common and characteristic feature of patients with traumatic brain injury (TBI). Historically, Pappius defined "brain edema" as an increase in net brain water content resulting in increased intracranial tissue volume.[13] In the last decades, intensive research was performed leading to better understanding of pathophysiological mechanisms and a redefined characterization of brain edema. This involves molecular and cellular changes, as well as structural and functional alterations. Since the limitation of the intracranial space by the skull, expansion of brain edema and thus an increase in intracranial volume, results in an elevation of ICP. Therefore the development of brain edema is one of the most challenging and complex conditions physicians are faced with when treating patients with TBI on intensive care units.

Classification

In the 1960s, the neuropathologist Igor Klatzo (1967) introduced two major different concepts of brain edema, cytotoxic and vasogenic edema, following cold lesion injury experiments.[14] According to the described concepts, water accumulates within the extra- or intracellular space due to breakdown of the blood–brain barrier (BBB) or loss of energy-dependent functions within the cell. In 1975, Fishman added the term "interstitial edema" that was observed in patients with hydrocephalus, an entity not particularly frequent in patients with acute TBI.[15] A fourth term is "osmotic edema" caused by an accumulation of osmotically active substances (e.g., sodium) in the intracellular space thereby resulting in increased water influx into the cell and cell swelling.[16]

Vasogenic Brain Edema

The integrity of the BBB is of fundamental importance for the maintenance of cerebral hemostasis and brain volume. Reese and Karnovsky described the endothelial cell layer to be the anatomical correlate of the BBB, which is functionally supported by adjacent astrocytic end-feet.[17] Endothelial cells of the BBB differ in many ways from endothelial cells of other organs since they are sealed together by tight junctions preventing uncontrolled passage of proteins and other small molecules, e.g., glucose and amino acids, across the wall of cerebral vessels. TBI causes both structural and functional impairment of the BBB. This results in an increased permeability of cerebral vessels and extravasation of a protein-rich exudate from the blood into the extracellular space of the brain.[18]

Cytotoxic Brain Edema

Cytotoxic brain edema is characterized by an accumulation of intracellular water in all cells of the brain parenchyma, but mainly within astrocytes. Generally, cells swell because of an accumulation of osmotically active solutes within the intracellular space. Reasons for such an accumulation are the increase in the Na^+ permeability of the cell membrane and the continuous uptake of osmotically active solutes due to the efforts of astrocytes to maintain extracellular homeostasis.

A model to describe these pathophysiological mechanisms is the so-called "pump-leak-equilibrium disturbance."[19] Under physiological conditions, the influx of osmotically active solutes is tightly regulated by a plethora of transporters located on astrocytic end-feet. They control the intracellular accumulation of osmotically active substances and thus the consecutive influx of water and changes in cell volume. Under pathophysiological conditions and energy depletion, however, the increased uptake of Na^+ cannot be prevented, the "pump-leak-equilibrium" is shifted, and cell swelling occurs. Concomitant energy depletion ensuing ischemia and structural and functional mitochondrial impairment propagates Na^+/K^+-ATPase failure and may finally aggravate cytotoxic edema.[19] Furthermore, the excitatory neurotransmitter amino acid glutamate disturbs the electrochemical equilibrium after TBI.[20,21] Activation of ionotropic and metabotropic glutamate receptors as well as glial and neural glutamate transporters is associated with an extensive influx of Na^+ into the intracellular space. Despite the cotransport and antiport of mainly Cl^-, these processes are resulting in a continuous shift of water from the extracellular into the intracellular space. Neural and glial efforts to maintain or reestablish ionic equilibrium fail and result in loss of cell volume control and finally cell swelling, a process that may further be compromised by Ca^{2+}-mediated activation of intracellular second messenger cascades resulting in oxygen-free radicals and arachidonic acid metabolites.[22,23]

As mentioned above the main cell type responsible for cytotoxic edema is astrocytes. One reason is that astrocytes outnumber neurons and can swell five times their normal size. The main reason, however, is that the physiological function of astrocytes is the maintenance of extracellular homeostasis in the brain. Solutes released from neurons into the extracellular space during TBI at high concentrations are taken up by astrocytes along the extra-intracellular sodium gradient. This results in high intracellular sodium concentrations, influx of water, and cell swelling. Since this process occurs only in astrocytes, it becomes clear why glial swelling represents the main component of cytototxic brain edema (see Chapter 20, Clinical Monitoring of Brain Edema).

For many decades vasogenic edema was thought to be mainly responsible for the development of posttraumatic brain edema. Nowadays, we know that cytotoxic edema also plays a key role for the emergence of posttraumatic brain[19] edema.

TRAUMATIC BRAIN EDEMA

Following TBI, ischemic conditions of the brain tissue are frequently observed. As long as cerebral blood flow (CBF) does not drop below a critical threshold, cerebral water content, and electrolytes remain stable, even if electrophysiological disturbances are already evident. If, however, CBF drops below a critical level of $10\,mL/100\,g\,min^{-1}$, ischemia occurs, and water and electrolyte homeostasis are disturbed. During this "low flow status" cytotoxic edema develops. In the case of irreversible tissue damage, BBB breakdown causes extravasation of serum proteins and thus, vasogenic edema. TBI encompasses a variety of intracranial lesions, including contusions, hematomas, subarachnoid hemorrhages, and diffuse axonal injuries. Type, extent, and speed of the development of posttraumatic brain edema are therefore dependent on the expansion of the primary lesion and associated conditions, such as arterial hypotension or hypoxia.[19,24]

Focal Traumatic Brain Edema

A hemorrhagic core with necrotic tissue that is poorly or no longer perfused characterizes a traumatic contusion. Perfusion of the surrounding perifocal tissue is significantly diminished and brain edema expands from the core into initially uninjured tissue adjacent to the lesion within 2–3 days.[25–28] Experimental and clinical data strongly suggest that the main type of edema present under these conditions is cytotoxic. The main findings are that TBI patients suffering from increased ICP show no evidence of BBB leakage within the first 24 hours after injury and that in patients with focal injuries reduced water diffusion, a marker of cytotoxic brain edema was observed in peri-contusional tissue by magnetic resonance imaging.[29,30] Hence, it was suggested that vasogenic edema occurs in the necrotic core of the contusion and that cytotoxic edema is mainly present in the rim of the contusion, which is affected by secondary injury.

Diffuse/Global Traumatic Brain Edema

In addition to focal brain edema, brain swelling may also occur globally after TBI. This condition is in most cases associated with diffuse brain injury following acceleration/deceleration injuries. Initially it was believed that global brain swelling was caused by trauma-induced vasodilatation leading to increased intracranial blood volume, also termed "vascular engorgement." The main reason for this assumption was that studies in TBI patients could not detect cerebral ischemia in the subacute stage after trauma.[31,32] More recently, however, diligently designed measurements in humans demonstrated that the major contributor for the development of global/diffuse posttraumatic brain swelling and thus increasing ICP is not a vascular engorgement, but critically decreased CBF levels.[33] Therefore, it is mainly believed that swelling and ICP increase after diffuse brain injury is related to cytotoxic brain edema.

TREATMENT OF TRAUMATIC BRAIN EDEMA

Despite extensive experimental and clinical research, current treatment of brain edema is still largely symptomatic focusing on stabilizing and reducing increased ICP.

The most effective treatment strategy is to administer osmotic substances in order to create an osmotic gradient dehydrating well-perfused cerebral tissue. Other modalities, such as hyperventilation, hypothermia, or barbiturates, are used to decrease CBF due to vasoconstriction or decreased metabolism. In case of therapy-resistant increased ICP, decompression craniectomy is discussed to have beneficial potential to control ICP and to improve outcome.

Up to date, however, there is no drug or other treatment available that prevents the development of brain edema or interrupts secondary brain injury development. This also holds true for steroids, which were postulated to seal the endothelial lining, thus attenuating vasogenic edema formation. The underestimated prevalence of cytotoxic edema formation may explain the limited efficacy of steroids to treat traumatic brain edema. Since the publication of the "CRASH" trial showing, compared to controls, severe side-effects of steroids leading to worse clinical outcome with an increased mortality rate within 14 days after trauma, steroids are no longer accepted as a treatment option for TBI, but are obsolete and recommendations were removed from all guidelines.[34,35]

Despite several promising experimental series addressing mediators that are responsible for the development of both vasogenic or cytotoxic edema, such as glutamate, H^+-, K^+-, or Ca^{2+} ions, arachidonic acids, oxygen-free radicals, histamine, or kinins, the administration of a variety of different antagonists to patients did not provide any convincing beneficial effects.[36,37] Among those, shaping glutamate-mediated excitotoxicity was considered as most promising since noncompetitive (e.g., aptiganel, dextrorphan) and competitive inhibitors (e.g., selfotel), as well as allosteric modulators of postsynaptic ionotropic glutamate receptor channel complexes (polyamine, glycine, zinc, magnesium) were proven beneficial in animal models of TBI and stroke. The majority of these substances, however, showed undesirable or adverse neuropsychological effects and therefore, did not find access in clinical use.

In order to reduce the vasogenic component of brain edema, specific bradykinin receptor antagonists, metalloprotease inhibitors or L-histidines, which interfere with oxygen-free radicals were tested in experimental settings and showed promising effects. However, none of them showed any beneficial effects in randomized clinical trials and therefore are not used in clinical practice.[38–40]

The list of substances targeting the cytotoxic component of posttraumatic edema is rather long and encompasses a variety of mechanisms and different drugs such as cannabinoid receptor modulators, bradykinin antagonists, K^+-channel opener, citicoline, or cyclosporine.[41–46] The promising cannabinoid receptor agonist KN-38-7271 was tested in a clinical randomized placebo-controlled phase II trial.[47] Despite the fact that the acute mortality within 1 month was significantly reduced, and the substance was well tolerated, these effects vanished after longer observation periods.[47]

It has been assumed that cytotoxic edema is not exclusively caused by ischemia since brain edema also develops despite sufficient CBF, but could be caused by direct mechanical injury inducing mitochondrial impairment.[33] Mitochondrial dysfunction induces energetic impairment and thus excessive Ca^{2+} overload. In conjunction with an increasing opening of mitochondrial permeability transition pores resulting in loss of the transmembrane electrogenic gradient, oxidative phosphorylation is uncoupled and mitochondrial swelling becomes inevitable.[19] Ensuing ATP deficits result in a failure of ion pumps and a breakdown of membrane stability, which results in uncontrolled cell swelling.

Furthermore, cerebral Aquaporin 4 (AQP4), a bidirectional transmembrane water channel mainly localized in astrocytic end-feet may play a crucial role for the aggravation and resolution of cytotoxic brain edema. Physiologically, AQP4 works together with the inward rectifying K^+ channel Kir4.1. AQP4 clears water and Kir4.1 K^+ via K^+- siphoning from extracellular compartment of the brain.[48] Under pathophysiological conditions, AQP4 and Kir4.1 may dissociate and AQP4 may move water bi-directionally across cell membranes along osmotic gradients leading to cell swelling.[49] Therefore AQP4 might be a promising therapeutic target for cerebral edema that should be tested in further clinical settings.[50]

CLINICAL MANAGEMENT OF BRAIN EDEMA AND INCREASED INTRACRANIAL PRESSURE

Cerebral Perfusion Pressure

Following severe TBI, patients are at risk to develop secondary cerebral ischemic damage mainly due to raised ICP or arterial hypotension resulting in insufficient CPP. This compromises CBF and causes impaired brain tissue oxygenation resulting in propagating brain edema and functional or structural tissue damage that consequently negatively affect outcome. Maintaining sufficient CPP, e.g., by increasing mean arterial blood pressure is therefore a widely accepted treatment strategy.[35]

Under physiological conditions, cerebral oxygenation involves three main factors, namely CBF, arterial content of oxygen, and cerebral metabolic rate of oxygen consumption. Normal the brain autoregulates blood flow in order to maintain a constant flow regardless of changes in blood pressure.[51] Following trauma, however, autoregulation is often impaired by hypoxia, hypercapnia or iatrogenically by volatile anesthetics. Historically, CPP >50 mmHg was accepted for trauma patients. Since, however, it was recognized that CPP levels higher than 50 mmHg also minimize ICP by reducing intracranial blood volume and cerebral edema via autoregulatory vasoconstriction, Rosner et al. recommended CPP levels >70 mmHg for the prevention of cerebral edema.[52] In trauma patients, however, autoregulation is frequently impaired and therefore, high CPP values may result in increased vasogenic edema and ICP. Therefore, there is a controversy regarding the optimal CPP following TBI. CPP-targeted treatment aims for sufficient tissue perfusion, preservation of the ischemic penumbra and reduction of secondary insults. Alternatively, a group from Lund in Sweden focused on the reduction of ICP by decreasing intracranial blood volume, an approach that later became known as the "Lund concept."[53,54] However, this concept is only applied in a few centers.

Studies using brain tissue microdialysis, jugular venous oxymetry and brain tissue oxygenation saturation (ptiO2) have demonstrated that the injured brain may show signs of ischemia if CPP drops below 50 mmHg while raising CPP above 60 mmHg may avoid cerebral oxygen desaturation.[55–57] Referring to the promising results published by Rosner et al.[52] elevating CPP to levels >70 mmHg, subsequently randomized clinical trials were performed. Maintaining CPP >70 mmHg showed five times greater risk of acute respiratory distress syndrome without any outcome benefits.[58,59] In accordance with the current literature, the CPP to target therefore may lie within 50–70 mmHg.

Recently, Steiner et al. developed an on-line method to estimate the CPP at which autoregulation is most robust and created the term "optimal CPP."[60] The more closely the mean CPP at which individual patients were maintained approximated the CPP at which their autoregulation was optimal, the more likely that patients could achieve a favorable neurological outcome.[60] Therefore current research suggests that not fixed CCP values should be used for the therapy of posttraumatic brain swelling, but rather individually calculated CPP values adapted to the current status of the patient's autoregulation, oxygenation, and cerebral perfusion.

Osmotic Substances

Elevation of ICP >20 mmHg can result in impaired cerebral perfusion and, hence, poor neurological outcome. Increasing posttraumatic brain edema is one of the main reasons for increasing ICP. There are various methods controlling elevated ICP; however, one of the main pharmacological interventions is hyperosmolar therapy. Despite a lack of high-quality clinical trials, Mannitol and hypertonic saline (HTS) are commonly used for this purpose. Both agents are lowering ICP predominately by creating an osmotic gradient mobilizing and shifting fluid from the extravascular brain tissue to the intravascular compartment. While Mannitol was previously thought to reduce ICP simply by brain dehydration, both substances furthermore work at least partly by reducing blood viscosity, leading to improved microcirculation and thus brain perfusion.[35] Furthermore for both treatment modalities neuroprotective effects on the cellular level were described based on the prevention of oxidative stress and secondary insults from inflammation and the release of inflammatory cytokines.[61,62]

Historically, 20% Mannitol intravenously administered over 20 minutes was used as the reference osmotic substance to lower ICP and is a level II recommendation in both the Brain Trauma Foundation and European Brain injury Consortium guidelines.[35,63] Since the BBB is impermeable to Mannitol, the drug is in principle restricted to the vascular compartment of the brain.[64] After trauma, however, the BBB is damaged and there are concerns that Mannitol may permeate into extravascular tissue thereby causing a rebound phenomenon with secondary edema development and thus ICP elevation. For clinical use, Mannitol should be withheld if serum osmolality exceeds 320 mOsm since with increasing serum osmolality the osmotic gradient between blood and brain tissue diminishes and the efficiency of the therapy is reduced, while the risk for kidney failure is still increasing. Although published data for this threshold are weak, Mannitol should only be used under close control of serum osmolarity.[64]

More recently, however, interest in HTS with varying concentrations (3%–23.4%) has increased. Until now, however, the optimal concentration of HTS and timing are still a matter of research and not clarified. A current review of seven trials performed on 191 patients compared the effects of Mannitol and HTS.[61] Conceding limitations referring to the quality of the enrolled studies (different initial GCS scores, small number of randomized trials), there were no clinical relevant differences in mean ICP reduction, mortality, and neurological outcome between Mannitol and HTS.[61] Accordingly, no specific guideline recommendations supporting any agent or treatment regime are given right now.[35]

Hyperventilation

Patients with severe TBI require airway protection and ventilation in order to protect them from the risk of pulmonary aspiration or compromised airway function. Normal ventilation with partial pressure of carbon dioxide in arterial blood ($PaCO_2$) between 35 and 45 mmHg is currently the goal for patients with severe TBI without any signs of herniation.[35] Under physiological conditions, $PaCO_2$ is the most powerful determinant of CBF. Low $PaCO_2$ causes vasoconstriction and lowers ICP, but also results in low CBF down to ischemic levels, while high $PaCO_2$ levels cause vasodilatation and result in cerebral hyperemia and high ICP. In patients with severe TBI and consecutive brain edema, low $PaCO_2$ induced by hyperventilation may therefore offer a potential treatment option. Hypocapnia causes vasoconstriction and therefore may reduce blood volume and edema and furthermore restores the ability to pressure autoregulation that frequently is impaired for up to two weeks after TBI.[65,66] On the other hand, it is well described in several studies using $ptiO_2$-monitoring and microdialysis that forced hyperventilation ($PaCO_2$ < 25–30 mmHg) might create critical events with hypoxic or ischemic tissue oxygen saturation (<10 mmHg) or metabolic derangements.[55,66–68] While older studies suggested that hyperemia was common, in fact, more recent studies suggest that cerebral ischemia occurs more frequently after trauma nowadays.[69–72] Therefore guideline recommendations have changed in recent years. While in the prior third Brain Trauma Foundation (BTF) guideline, hyperventilation was recommended 24 hours after TBI as a measure to reduce elevated ICP, current guidelines do not give any recommendations on this topic anymore, referring to the weak evidence in the literature.[35] Currently only a level IIb recommendation is provided that does not recommend prolonged and prophylactic hyperventilation with $PaCO_2$ of 25 mmHg or less.[35] Therefore hyperventilation can only be considered as an individual treatment modality in order to control brain swelling and elevated ICP.

Hypothermia

Fay firstly described the use of therapeutic hypothermia (TH) for TBI already in 1945.[73] Since that time, major efforts have been made to successfully establish hypothermia for the clinical use. Currently TH, characterized by a controlled lowering of core body temperature below 36°C, is used as a treatment modality for neonatal hypoxic ischemic encephalopathy and after cardiac arrest.[74,75] There is also emerging evidence that it may be beneficial in the management of ischemic stroke.[76] In the last 20 years, approximately 18 randomized controlled trials have been performed to assess the efficiency of TH in TBI patients, however, showing conflicting results. Referring to the duration of cooling, target temperature, and treatment regimes, trials can be divided into two major categories: a first one, in which TH is maintained only for a short period of time (24–48 hours), and a second one with TH that continued for a longer duration (3–7 days). The efficiency of short-term hypothermia was evaluated in two high-quality trials, the NABIS H I and II studies. Both trials, however, failed to prove any beneficial effects on short- or long-term outcome for TBI patients.[76,77] It was suggested that the period of 48 hours might be too short to provide efficacy, especially considering the fact that brain edema gradually develops within 3–5 days following trauma. The rewarming phase exactly matches this time window and might have caused a rebound phenomenon with increased brain edema and elevated ICP. Therefore, a "long-term mild hypothermia" trial (34–35°C) was initiated.[78] Furthermore the results of the Eurotherm study were recently published assessing outcome after TBI and hypothermia (32–35°C).[79] It was planned to enroll 600 patients with severe TBI and increased ICP > 20 mmHg into this randomized controlled clinical multicenter trial. After 387 patients, however, randomization was stopped for safety reasons from the steering committee. Although ICP control was better in the treatment group, outcome and mortality showed worse results. Six months after injury the distribution of GOS-E scores as the primary outcome parameter was shifted in an unfavorable direction in the hypothermia group and favorable outcome (GOS-E score 5–8) could be achieved significantly less frequent (25.7% vs 36.5%, $P = 0.03$). Furthermore, risk for death and the report of severe adverse events favored the control group. Therefore, HT did not result in better outcome compared with standard treatment alone.[79] Based on these findings, current guidelines do not recommend early (within 2.5 hours) and short-term (48 hours) prophylactic hypothermia for the treatment TBI patients.[35]

Barbiturates

Analgesics and sedation are beneficial for the management of patients with severe TBI for a variety of reasons, e.g., prevention of increased intraabdominal pressure, coughing or movements of monitoring devices and probes, aiming to improve outcome.[69]

Beyond that, barbiturates have a long history for the treatment of TBI, in particular for the control of elevated ICP. Already several decades ago, it was shown that high-dose barbiturates could reduce ICP.[80–82] This was explained by the reduction of cerebral metabolisms by barbiturates.[83–85] In a small number of patients, Thorat et al. could demonstrate that barbiturate coma shows favorable changes in ICP, pressure reactivity index, and $ptiO_2$.[86] In a randomized controlled trial the use of pentobarbital was tested vs thiopental favoring thiopental being more effective to control ICP, although these results have to be interpreted with caution due to the small number of investigated patients.[87] Since barbiturates also lower blood pressure, a large Cochrane database analysis assessed the effects of barbiturates on mortality, disability, and ICP control.[83] Data from seven trials involving 341 patients were included, but there was no evidence that barbiturate therapy improves outcome after TBI. On the other hand, barbiturates might cause severe side-effects, including hypotension, decreased cardiac output, as well as increased intrapulmonary shunting, which may lead to hypoxia and therefore might be responsible for the poor outcome. Referring to the available literature and the current guideline recommendations, administration of barbiturates to induce burst suppression measured by EEG as prophylaxis against elevated ICP is not recommended. High-dose barbiturates are only recommended to control ICP refractory to maximal standard medical or surgical treatment in patients, who are hemodynamically stable.[35] In any case, barbiturates are a second line treatment choice, and should not be considered as a prophylactic option to reduce edema or control ICP.

Decompressive Surgery

In case of therapy-refractory ICP elevation, decompressive surgery might represent an option to reduce ICP. The idea is not new since already in 1901, Theodor Kocher, a Swiss surgeon, already stated, "if there is no CSF pressure, but brain pressure exists, the pressure relief must be achieved by opening the skull." By opening the skull and the dura, decompressive surgery offers the compressed and swollen brain more space to enlarge, and therefore reduces ICP and improves CBF. Despite the simple principle of this technique, the efficiency and beneficial effect of decompressive surgery on outcome was under discussion for many years.

In an experimental study on mice using a controlled cortical impact trauma model, it was shown that early decompression avoids ICP increase and prevents secondary brain edema formation resulting in diminished contusion volume 24 hours after trauma. Decompression surgery also positively influenced functional outcome seven days after TBI.[88] Furthermore, it could also be demonstrated that not only preventive but also decompression surgery performed up to 3 hours after trauma was protective.[89]

In addition to experimental data, several retrospective clinical single center trials assessing the effect of decompression craniectomy in patients with severe TBI have been published. These studies demonstrated that decompression surgery might also reduce ICP and brain edema. For the first time, Polin et al. described on 35 patients the technique of a bifrontal craniectomy for patients with diffuse brain swelling and could achieve good or moderate outcome in 37% of patients, nevertheless, afflicted with a mortality rate of 23%.[90] This study was finally the foundation of the later performed DECRA trial. Guerra et al. focused especially on younger patients under the age of 50 years. In this special cohort, they could achieve good recovery with the ability to reintegrate into professional and personal life in 58% of patients.[91] The time point when decompression surgery should be performed has always been a critical point of discussion. Münch et al. grouped their data in a very early decompression cohort (<4.5 hours after trauma) and in a late decompression cohort (>56.2 hours after trauma) and could impressively demonstrate that early decompression was associated with better neurological outcome.[92] Schneider et al. confirmed these results. In average, patients were decompressed 82 hours after trauma and with high ICP (>40 mmHg) values. Not surprisingly, 71% of patients resulted in poor neurological outcome or death.[93] Despite promising results from several retrospective series, no prospective data have been available for many years and therefore, the clinical use of decompression surgery for the prevention of brain edema formation and intracranial hypertension remained a controversially discussed issue.

The first prospective, randomized controlled multicenter trial assessing the effect of decompression craniectomy on functional outcome of patients with severe TBI was published by Cooper et al.[94] Between 2002 and 2010, 155 patients were recruited and a bifrontal decompression craniectomy was performed if ICP remained above 20 mmHg for at least 15 minutes despite intensive conservative treatment. Following surgery and compared to conservatively treated controls, ICP was significantly lower and patients spend less time on intensive care wards. Neurological outcome, however, was worse and 70% of decompressed patients (vs 51% of conservatively treated ones) had a poor neurological outcome with an extended GOS of less than four points after six months. Furthermore, decompressed patients remained significantly more likely in a vegetative status compared to controls (12% vs 2%), whereas there was no difference in mortality (18% vs 19%).[94] Design, inclusion and exclusion criteria of this study caused severe criticism and a broad discussion aiming to explain these results. Most critique was addressing the surgical technique having performed a bifrontal craniectomy with only small frontal dural incisions. It is well describe that small bone flaps or dural openings may cause brain tissue herniation and CBF impairment leading to worse pathophysiological conditions and finally poor outcome. Furthermore, all patients with initial GCS of three points or the majority of patients with dilated pupils predicting poor outcome were randomized into the surgical arm. Therefore the results of the RescueICP trial, the European analog to the DECRA study, which were published in 2016, were urgently expected.[95] In this prospective, randomized controlled multicenter trial, 408 patients with severe TBI were enrolled between 2004 and 2014. In case, ICP exceeded 25 mmHg for 1–12 hours despite conservative treatment, patients were randomized in a control arm (including barbiturates) or a surgical arm (without barbiturates) where patients received either a bifrontal craniectomy or a hemicraniectomy. Following craniectomy, ICP was significantly reduced within the acute stage. Furthermore, mortality could significantly and impressively be reduced at discharge (22.7% vs 48.5%) as well as at 6 months (26.9% vs 48.9%) and 12 months (30.4% vs 52%) after trauma. Dichotomizing extended GOS into favorable and nonfavorable outcome, results demonstrated significant more patients achieving favorable outcome 6 and 12 months after trauma. On the other hand, data also showed that significant more patients stayed in a vegetative status (6.2% vs 1.7%), although the total amount is relatively small compared to other studies, e.g., on malignant stroke.[95]

Accurately evaluating the available literature, decompression craniectomy still may offer a treatment option for patients with therapy-refractory ICP elevation, but remains a second line treatment and cannot be recommended for prophylactic purposes. Further studies need to address the question, which patients might benefit from decompression craniectomy and when it should be performed.

SUMMARY

Posttraumatic brain edema is the leading pathophysiological condition causing ICP elevation and poor neurological outcome following severe TBI.

Therefore progressive conservative and surgical treatment is essential to preserve cerebral perfusion and oxygenation tissue saturation in order to increase the chance for survival and favorable neurological outcome. Especially since TBI is a very heterogeneous disease, treatment must be adopted to each individual conditions aiming for optimal personalized therapy.

References

1. Cole TB. Global road safety crisis remedy sought: 1.2 million killed, 50 million injured annually. *JAMA* 2004;**291**:2531–2.
2. Ghajar J. Traumatic brain injury. *Lancet* 2000;**356**:923–9.
3. Koskinen S, Alaranta H. Traumatic brain injury in Finland 1991–2005: a nationwide register study of hospitalized and fatal TBI. *Brain Inj* 2008;**22**:205–14.
4. Menon DK, Schwab K, Wright DW, et al. Demographics and clinical assessment working group of the international and interagency initiative towards common data elements for research on traumatic brain injury and psychological health. Position statement: Definition of traumatic brain injury. *Arch Phys Med Rehabil* 2010;**91**:1637–40.
5. Tagliaferri F, Compagnone C, Korsic M, et al. A systematic review of brain injury epidemiology in Europe. *Acta Neurochir* 2006;**148**:255–68.
6. Gerber LM, Chiu YL, Carney N, et al. Marked reduction in mortality in patients with severe traumatic brain injury. *J Neurosurg* 2013;**119**:1583–90.
7. Lu J, Marmarou A, Choi S, et al. Mortality from traumatic brain injury. *Acta Neurochir* 2005;**95**:281–5.
8. Peeters W, van den Brande R, Polinder S, et al. Epidemiology of traumatic brain injury in Europe. *Acta Neurochir* 2015;**157**:1683–96.
9. Langlois Orman JA. Epidemiology. In: Silver JM, McAllister TW, Yudofsky SC, editors. *Textbook of traumatic brain injury*. 2nd ed. Washington: American Psychiatric Publishing; 2011.
10. Finkelstein E, Corso PS, Miller TR. *The incidence of economic burdon on injuries in the United States*. Oxford: Oxford University Press; 2006.
11. Leo P, McCrea M. Epidemiology in translational research in traumatic brain injury. In: Laskowitz D, Grant G, editors. *NCBI bookshelf*; 2016, p. 1–12.
12. The reports to Congress on mild traumatic brain injury in the United States: steps to prevent a serious public health problem. Atlanta, GA; 2003.
13. Pappius HM. Fundamental aspects of brain edema. Vinken P.J., Bruyn GW, editors. *Handbook of clinical neurology. Part I: tumors of the brain and skull*, Vol. 16. New York: North Holland Publishing Company; 1974, p. 167–85.
14. Klatzo I. Presidential address: neuropathological aspects of brain edema. *J Neuropathol Exp Neurol* 1967;**26**:1–14.
15. Fishman RA. Brain edema. *New Engl J Med.* 1975;**293**:706–11.
16. Katayama Y, Kawamata T. Edema fluid accumulation within necrotic brain tissue as a cause of the mass effect of cerebral contusion in head trauma patients. *Acta Neurochir Suppl* 2003;**86**:323–7.
17. Reese TS, Karnovsky MJ. Fine structural localization of a blood-brain barrier to exogenous peroxidase. *J Cell Biol* 1967;**34**:207–17.
18. Betz AL, Iannotti F, Hoff JT. Brain edema: a classification based on blood-brain barrier integrety. *Cerebrovas Brain Metab Rev* 1989(1):133–54.
19. Unterberg AW, Stover J, Kress B, Kiening KL. Edema and brain trauma. *Neuroscience* 2004;**129**:2021–9.
20. Schneider GH, Baethmann A, Kempski O. Mechanisms of glial swelling induced by glutamate. *Can J Physiol Pharmacol Suppl* 1992;**70**:334–43.

21. Staub F, Peters J, Kempski O, et al. Swelling of glial cells in lactacidosis and by glutamate: significance of Cl^- transport. *Brain Res* 1993;**610**:69–74.
22. Gegelashvili G, Schousboe A. High affinity glutamate transporters: regulation of expression and activity. *Mol Pharmacol* 1997;**52**:6–15.
23. Amara SG, Fontana AC. Excitatory amino acid transporters: keeping up with glutamate. *Neurochem Int* 2002;**41**:313–18.
24. Hossmann KA, Schuier FJ. Experimental brain infarct in cats: I. Pathophysiological observations. *Stroke* 1980;**11**:583–92.
25. Bullock R, Statham P, Patterson J, et al. The time course of vasogenic edema after focal human head injury – evidence from SPECT mapping of blood brain barrier defects. *Acta Neurochir Suppl* 1990;**51**:286–8.
26. Bareyre F, Wahl F, McIntosh TK, et al. Time course of cerebral edema after traumatic brain injury in rats: effects of riluzole and mannitol. *J Neurotrauma* 1997;**14**:839–49.
27. Stoop R, Thomale UW, Pauser S, et al. Magnetic resonance imaging studies cluster algorithm for characterization of brain edema after controlled cortical impact injury (CCII). *Acta Neurochir* 1998;**71**:303–5.
28. Unterberg AW, Stroop R, Thomale UW, et al. Characterization of brain edema following controlled cortical impact injury in rats. *Acta Neurochir Suppl* 1997;**70**:106–8.
29. Kawamata T, Katayama Y, Aoyama N, et al. Heterogeneous mechanisms of early edema formation in cerebral contusion: diffusion MRI and ADC mapping study. *Acta Neurochir Suppl* 2000;**76**:9–12.
30. Maeda T, Katayama Y, Kawamata T, et al. Ultra-early study of edema formation in cerebral contusion using diffusion MRI and ACD mapping. *Acta Neurochir Suppl* 2003;**89**:329–31.
31. Bouma GJ, Muizelaar JP, Stringer WA, et al. Ultra-early evaluation of regional cerebral blood flow in severely head-injured patients using xenon-enhanced computerized tomography. *J Neurosurg* 1992;**77**:360–8.
32. Diringer MN, Yundt K, Videen TO, et al. No reduction in cerebral metabolisms as a result of early moderate hyperventilation following severe traumatic brain injury. *J Neurosurg* 2000;**92**:7–13.
33. Marmarou A, Fatouros PP, Barzo P, et al. Contribution of edema and cerebral blood volume to traumatic brain swelling in head-injured patients. *J Neurosurg* 2000;**93**:183–93.
34. Roberts I, Yates D, Sandercock P, et al. Effect of intravenous corticosteroids on death within 14 days in 10.008 adults with clinically significant head injury (MRC CRASH trial): randomized placebo-controlled trial. *Lancet* 2004;**9442**:1321–8.
35. Brain trauma foundation. Guidelines for the management of severe traumatic brain injury, 4th ed.; 2016.
36. Unterberg AW, Sarrafzadeh AS. Brain edema. In: Crockard A, Hayward R, Hoff JT, editors. *Neurosurgery: the scientific basis of clinical practice*. Oxford: Blackwell Science. 2000, p. 410–30.
37. Bullock R, Lyeth BG, Muizelaar JP. Current status of neuroprotection trials for traumatic brain injury: lessons from animal models and clinical studies. *Neurosurgery* 1999;**45**:207–20.
38. Plesnila N, Schulz J, Stoffel M, et al. Role of bradykinin B2 receptors in the formation of vasogenic brain edema in rats. *J Neurotrauma* 2001;**18**:1049–58.
39. Kawai N, Kawanishi M, Okada M, et al. Treatment of cold injury-induced brain edema with nonspecific matrix metalloproteinase inhibitor MMI270 in rats. *J Neurotrauma* 2003;**20**:649–57.
40. Ikeda Y, Mochizuki Y, Nakamura Y, et al. L-histidine but not D-histidine attenuates brain edema following cryogenic injury in rats. *Acta Neurochir Suppl* 2000;**76**:195–7.
41. Panikashvili D, Simeonidou C, Ben-Shabat S, et al. An endogenous cannabinoid (2-AG) is neuroprotecitve after brain injury. *Nature* 2001;**413**:527–31.

42. Stover JF, Beyer TF, Unterberg AW. Riluzole reduces brain swelling and contusion volume in rats following controlled cortical impact injury. *J Neurotrauma* 2000;**17**:1171–8.
43. Cheney JA, Weisser JD, Bareyre FM, et al. The maxi-K channel opener BMS-204352 attenuates regional cerebral edema and neurologic motor impairment after experimental brain injury. *Cereb Blood Flow Metab* 2001;**21**:396–403.
44. Baskaya MK, Dogan A, Rao AM, et al. Neuroprotective effects of citicoline on brain edema and blood-brain-barrier break-down after traumatic brain injury. *J Neurosurg* 2000;**92**:448–52.
45. Fukui S, Signoretti S, Dunbar JG, et al. The effect of cyclosporine A on brain edema formation following experimental cortical contusion. *Acta Neurochir Suppl* 2003;**86**:301–3.
46. Shohami E, Cohen-Yeshurun A, Magid L, et al. Endocannabinoids and traumatic brain injury. *Br J Pharmacol* 2011;**163**:1402–10.
47. Firsching R, Piek J, Skalej M, et al. Early survival of comatose patients after severe traumatic brain injury with the dual cannabinoid CB1/CB2 receptor agonist KN38-7271: a randomized, double-blinded, placebo-controlled phase II trial. *J Neurol Surg A Cent Eur Neurosurg* 2012;**73**:204–16.
48. Nagelhus EA, Horio Y, Inanobe A, et al. Immunogold evidence suggests that coupling of K^+ siphoning and water transport in rat retinal Muller cells ins mediated by a coenrichment of Kir4.1 and AQP4 in specific membrane domains. *Glia* 1999;**26**:47–54.
49. Sasoun S, Papadopouos MC, Krishna S. Water transport becomes uncoupled from K^+ siphoning in brain contusion, bacterial meningitis, and brain tumors: immunohistochemical case review. *J Clin Pathol* 2003;**56**:972–5.
50. Tang G, Yang GY. Aquaporin-4: a potential therapeutic target for cerebral edema. *Int J Mol Sci* 2016;**17**:E1413.
51. Prabhakar H, Sandhu K, Bhagat H, et al. Current concepts of optimal cerebral perfusion pressure in traumatic brain injury. *J Anaesthesiol Clin Pharmacol* 2014;**30**:318–27.
52. Rosner MJ, Rosner SD, Johnson AH. Cerebral perfusion pressure: Management protocol and clinical results. *J Neurosurg* 1995;**83**:949–62.
53. Asgeirsson B, Grande PO, Nordstrom CH. A new therapy of post-trauma brain edema based on haemodynamic principles for brain volume regulation. *Intensive Care Med* 1994;**20**:260–7.
54. Nordstrom CH. Physiological and biochemical principles underlying volume-targeted therapy – the "Lund concept". *Neurocrit Care* 2005;**2**:83–95.
55. Kiening KL, Härtl R, Unterberg AW, et al. Brain tissue pO_2-monitoring in comatose patients: implications for therapy. *Neurol Res* 1997;**19**:233–40.
56. Robertson CS, Valadka AB, Hannay HJ, et al. Prevention of secondary ischemic insults after severe head injury. *Crit Care Med* 1999;**27**:2086–95.
57. Nordstrom CH. Assessment of critical thresholds for cerebral perfusion pressure by performing bedside monitoring of cerebral energy metabolism. *Neurosurg Focus* 2003;**15**:E5.
58. Contant CF, Valadka AB, Gopinath SP, et al. Adult respiratory distress syndrome: a complication of induced hypertension after severe head injury. *J Neurosurg* 2001;**95**:560–8.
59. Clifton GL, Miller EF, Choi SC, et al. Fluid threshold and outcome from severe brain injury. *Crit Care Med* 2002;**30**:739–45.
60. Steiner LA, Czosnyka M, Piechnik SK, et al. Continuous monitoring of cerebrovascular pressure reactivity allows determination of optimal cerebral perfusion pressure in patients with traumatic brain injury. *Crit Care Med* 2002;**30**:733–8.
61. Burgess S, Pharm D, Abu-Laban RB, et al. A systematic review of randomized controlled trials comparing hypertonic sodium solutions and mannitol for traumatic brain injury: implications for emergency department management. *Ann Pharmacol* 2016;**50**:291–300.
62. Mojtahedzadeh M, Ahmadi A, Mahmoodpoor A, et al. Hypertonic saline reduces the oxidative stress responses in traumatic brain injury patients. *J Res Med Sci* 2014;**19**:867–74.

63. Maas AIR, Dearden M, Teasdale GM, et al. EBIC-guidelines for management of severe head injury in adults. *Acta Neurochir* 1997;**139**:286–94.
64. Diringer MN, Zazulia AR. Osmotic therapy. *Neurocrit Care* 2004;**1**:219–34.
65. Sarrafzadeh AS, Kiening KL, Callsen TA, et al. Metabolic changes during impending and manifest cerebral hypoxia in traumatic brain injury. *Br J Neurosurg* 2003;**17**:340–6.
66. Lee JH, Kelly DF, Oertel M, et al. Carbon dioxide reactivity, pressure autoregulation, and metabolic suppression reactivity after head injury: a transcranial Doppler study. *J Neurosurg* 2001;**95**:222–32.
67. Schneider GH, Sarrafzadeh AS, Kiening KL, et al. Influence of hyperventilation on brain tissue-PO_2, PCO_2, and pH in patients with intracranial hypertension. *Acta Neurochir Suppl* 1998;**71**:62–5.
68. Kiening KL, Schoening WN, Stover JF, et al. Continuous monitoring of intracranial compliance after severe head injury: relation to data quality, intracranial pressure and brain tissue PO2. *Br J Neurosurg* 2003;**17**:311–18.
69. Liu S, Wan X, Wang S, et al. Posttraumatic cerebral infarction in severe traumatic brain injury: characteristics, risk factors and potential mechanisms. *Acta Neurochir* 2015;**157**:1697–704.
70. Tawil I, Stein DM, Mirvis SE, et al. Posttraumatic cerebral infarction: incidence, outcome an risk factors. *J Trauma* 2008;**64**:849–53.
71. Carrera E, Schmidt JM, Fernandez L, et al. Spontaneous hyperventilation and brain tissue hypoxia in patients with severe brain injury. *J Neurol Neurosurg Psychiatry* 2010;**81**:793–7.
72. Stein NR, McArthur DL, Etchepare M, et al. Early cerebral metabolic crisis after TBI influences outcome despite adequate hemodynamic resuscitation. *Neurocrit Care* 2012;**17**:49–57.
73. Fay T. Observation of generalized refrigeration incase of severe cerebral trauma. *Assoc Res Nerv Ment Dis Proc* 1943;**24**:611–19.
74. Arrich J, Holzer M, Herkner H, et al. Hypothermia for neuroprotection in adults after cardiopulmonary resuscitation. *Cochrane Database Syst Rev* 2009(4):CD004128.
75. Van der Wrop HB, Macleod MR, Kollmar R. European Stroke Research Network for H. Therapeutic hypothermia for acute ischemic stroke: ready to start large randomized trials? *J Cereb Blood Flow Metab* 2010;**30**:1079–93.
76. Clifton GL, Miller ER, Choi SC, et al. Lack of effect of induction of hypothermia after acute brain injury. *N Engl J Med* 2001;**344**:556–63.
77. Clifton GL, Valadka A, Zygun D, et al. Very early hypothermia induction in patients with sever brain injury (the National Acute Brain Injury Study: Hypothermia II): a randomized trial. *Lancet Neurol* 2011;**10**:131–9.
78. Lei J, Gao G, Mao Q, et al. Rationale, methodology, and implementation of a nationwide multicenter randomized controlled trial of long-term mild hypothermia for sever traumatic brain injury (the LTH-1-trial). *Contemporary Clinical Trials* 2015;**40**:9–14.
79. Andrews PJD, Sinclair HL, Rodriguez A, et al. Hypothermia for intracranial hypertension after traumatic brain injury. *N Engl J Med* 2015;**373**:2403–12.
80. Marshall LF, Smith RW, Shapiro HM. The outcome with aggressive treatment in severe head injuries: Part II: Acute and chronic barbiturate administration in the management of head injury. *J Neurosurg* 1979;**50**:26–30.
81. Rea GL, Rockswold GL. Barbiturate therapy in uncontrolled intracranial hypertension. *Neurosurgery* 1983;**12**:401–4.
82. Eisenberg HM, Frankowski RF, Contant CF. High dose barbiturate control of elevated intracranial pressure in patients with severe head injury. *J Neurosurg* 1988;**69**:15–23.
83. Roberts I, Sydenham E. Barbiturates for acute traumatic brain injury. *Cochrane Database Syst Rev* 2012;**12**:DC0000333.

84. Roberts DJ, Hall RI, Kramer AH, et al. Sedation for critically ill adults with severe traumatic brain injury: a systematic review of randomized controlled trials. *Crit Care Med* 2011;**39**:2743–51.
85. McKinley BA, Parmley CL. Effects of injury and therapy on brain parenchyma PO2, CO2, pH and ICP following severe closed head injury. *Acta Neurochir Suppl* 1988;**71**:177–82.
86. Thorat JD, Wang EC, Lee KK, et al. Barbiturate therapy for patients with refractory intracranial hypertension following severe traumatic brain injury: its effects on tissue oxygenation, brain temperature and autoregulation. *J Clin Neurosci* 2008;**15**:143–8.
87. Perez-Barcena J, Liompart-Pou JA, Homar J, et al. Pentobarbital versus thiopental in the treatment of refractory intracranial hypertension in patients with traumatic brain injury: a randomized controlled trial. *Crit Care* 2008;**12**:R112.
88. Zweckberger K, Stoffel M, Baethmann A, et al. Effect of decompression craniotomy on increase of contusion volume and functional outcome after controlled cortical impact in mice. *J Neurotrauma* 2003;**20**:1307–14.
89. Zweckberger K, Erös C, Zimmermann R, et al. Effect of early and delayed decompressive craniectomy on secondary brain damage after controlled cortical impact in mice. *J Neurotrauma* 2006;**23**:1083–93.
90. Polin RS, Shaffrey ME, Bogaev CA, et al. Decompressive bifrontal craniectomy in the treatment of severe refractory posttraumatic cerebral edema. *Neurosurgery* 1997;**41**:84–92.
91. Guerra WK, Piek J, Gaab MR. Surgical decompression for traumatic brain swelling: indications and results. *J Neurosurg* 1999;**90**:187–96.
92. Münch E, Horn P, Schürer L, et al. Management of severe traumatic brain injury by decompressive craniectomy. *Neurosurgery* 2000;**47**:315–22.
93. Schneider GH, Brandt T, Lanksch WR, et al. Decompressive craniectomy following traumatic brain injury: ICP, CPP and neurological outcome. *Acta Neurochir Suppl* 2002;**81**:77–9.
94. Cooper DJ, Rosenfeld JV, Murray. Decompressive craniectomy in diffuse traumatic brain injury. *N Engl J Med* 2011;**364**:1493–502.
95. Hutchinson PJ, Kolias AG, Timofeev IS, et al. Trial of Decompressive Craniectomy for Traumatic Intracranial Hypertension. *N Engl J Med* 2016;**375**:1119–30.

Further Reading

Rangel-Castilla L, Lara LR, Gopinath S, et al. Cerebral hemodynamic effects of acute hyperoxia and hyperventilation after severe traumatic brain injury. *J Neurotrauma* 2010;**27**:1853–63.

Roberts DJ, Hall RI, Kramer AH, et al. Sedation for critically ill adults with severe traumatic brain injury: a systematic review of randomized controlled trials. *Crit Care Med* 2011(39):2743–51.

CHAPTER

24

Treatment of Edema Formation in Oncology

Oliver Schnell[1,2] *and Jörg-Christian Tonn*[1]

[1]Klinikum der Ludwig-Maximilians-Universität München, Campus Großhadern, Munich, Germany [2]University of Freiburg, Freiburg, Germany

OUTLINE

Brain Edema.
DOI: http://dx.doi.org/10.1016/B978-0-12-803196-4.00024-2

INTRODUCTION

The formation of brain edema is a hallmark of primary central nervous system (CNS) tumors and will rapidly lead to neurological symptoms. Therefore, treatment should be started as early as possible in order to avoid further clinical deterioration or permanent neurological deficits. However, despite increasing knowledge about the pathophysiology of the formation of peritumoral brain edema its treatment is still limited. It has to be kept in mind that there are different primary and secondary CNS tumors and the formation of cerebral edema can result from different pathophysiologies. Brain edema can be both (i) directly tumor-associated, and also (ii) treatment-induced. Whereas surgical procedures may not only lead to ischemia or infection-associated edema, radiation-induced complications like radionecrosis, but also (local) chemotherapy (BCNU wafer)-associated forms of cerebral edema are frequently seen. Since these treatment-associated brain edemas are due to different forms of cerebral edema (vasogenic or cytotoxic), it is obvious that treatment is not always consistent in these cases. Moreover, there are some anatomo-functional issues, which have to be taken into account when dealing with brain edema in clinical oncology.

Primary CNS Tumors

Brain tumors include different malignant (33%) and nonmalignant (66%) neoplasms with an average annual age-adjusted incidence in adults of 21–22 cases per 100,000 population and a slightly higher rate in females as compared to males (1.2:1).[1] They can be further divided into primary brain tumors deriving from glial or precursor cells, which are therefore often summarized as gliomas and have an incidence rate of 6–7/100,000 per year.[1] Glioblastoma multiforme being the most common subtype (15.7% of all tumors and 45.6% of malignant tumors) is also the most malignant primary brain tumor and despite extensive treatment with surgery, chemoradiation and often multiple lines of chemotherapy its prognosis remains dismal with a median overall survival of approximately 14–16 months.[2–6] Secondary brain tumors are brain metastases originating from different cancer types outside the CNS[7] that are almost twice as common as gliomas (approximately 11/100,000/year) and their prognosis depends mainly on the underlying primary cancer.[8,9] Importantly, metastases to the brain should no longer be considered the limiting turning point of the disease in terms of overall prognosis, since patients almost always succumb to the originating systemic disease and not due to metastatic disease to the brain.[10] Additionally, there are other malignant and nonmalignant primary brain and CNS tumors of which nonmalignant meningiomas are the most common ones representing 35.6% of all and 53.7% of nonmalignant tumors.[1]

Tumor-Associated Brain Edema

A common feature of malignant and nonmalignant primary CNS tumors is their space occupying mass effect which may lead to clinical symptoms either by direct compression or indirectly due to the formation of brain edema. Depending on the biological behavior and growth pattern of the tumor, history of symptoms may be short and comprises headache, seizures or functional neurological deficits like hemiparesis, speech problems and personality changes depending on the location of the lesion. It is noteworthy that in gliomas and brain metastases the extent of brain edema correlates with malignancy (Fig. 24.1) but benign meningiomas can cause vasogenic edema due to local breakdown of the blood–brain-barrier (BBB) within or surrounding the tumor as well (Fig. 24.2).[11,12]

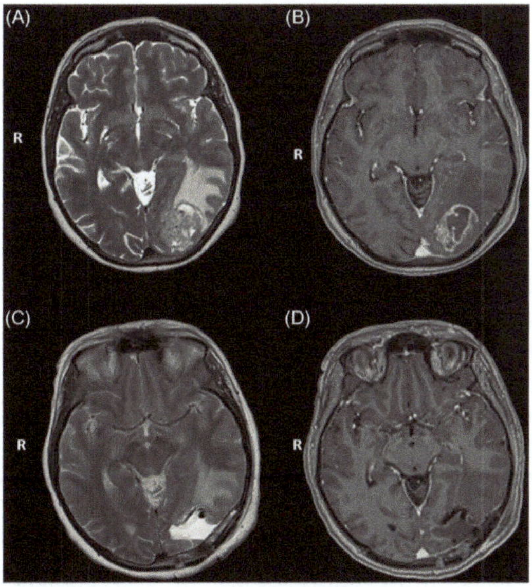

FIGURE 24.1 **Tumor-associated brain edema in a patient with a primary malignant brain tumor.** A 61-year-old woman presented with headache and disturbance of her visual field (right-sided homonymous hemianopsia). Axial T2-wighted magnetic resonance imaging (MRI) shows hyperintense signal alteration in the left temporo-occipital region including the left-sided optic radiation suggestive for peritumoral brain edema (A). T1w-MRI with contrast-enhancement (CE) of the same axial plane strengthens this assumption by demonstrating excessive CE through a disrupted BBB in the center of the tumor-associated brain edema (B). After tumor surgery peritumoral brain edema will stay visible (C), despite complete resection of the contrast-enhancing tumor parts as can be seen on T1w-CE MRI (D).

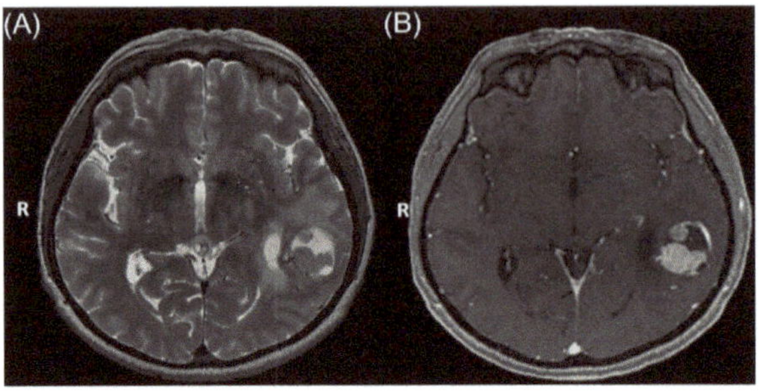

FIGURE 24.2 **Peritumoral brain edema in a patient with a benign meningioma.** MRI was performed due to screening in a 58-year-old patient with breast carcinoma showing a intracerebral, partly cystic temporo-basal lesion with surrounding brain edema (A) and contrast-enhancement within the solid part of the suggested tumor (B). Tumor resection was performed and histopathological examination revealed a benign meningioma.

One factor, which is produced by different primary CNS tumors as well as meningiomas is the vascular endothelial growth factor (VEGF), originally described as vascular permeability factor.[13–15] VEGF does not only increase permeability of normal cerebral microvasculature surrounding the tumor by opening of the tight junctions between endothelial cells, enhances blood flow and induces vasodilatation but also induces tumor angiogenesis with vascular abnormalities, leading to additional BBB disturbance and leakage of intravascular fluid into the extracellular space of the brain.[16–18] Other factors like arachnoid acid metabolites (e.g., leukotriene C4, prostaglandin E2), nitric oxide (NO) and especially various members of the aquaporin family may also be involved in the formation of peritumoral brain edema that have been discussed in more detail in the previous chapters. Noteworthy, brain edema will not be differentiated pathophysiologically due to its anoxic, ionic or vasogenic origin. For simplification and according to the clinical relevance, the terms vasogenic and cytotoxic brain edema will be used throughout this chapter.

Imaging of Brain Tumors and Tumor-Associated Edema

In order to get an impression about the extent of an existing brain edema and its underlying cause, brain imaging plays an important role in neurooncology. Computed tomography (CT) is broadly available and helps to identify brain edema distribution by its perilesional hypodensity. Additionally, it allows an easy and fast illustration of brain shift or imminent compression of critical structures within the skull. A more detailed image of the brain edema and the underlying cause will be generated by magnetic resonance imaging (MRI), which is not only the most essential diagnostic tool in order to verify suspicion of a brain tumor but also depicts location and expansion of the tumor and the peritumoral edema with high resolution (Figs. 24.1 and 24.2). Whereas T1-weighted images demonstrate permeability of the BBB by extravasation of a contrast agent into the brain parenchyma around the disrupted vascular structures, T1-weighted images with or without contrast enhancement but especially T2-weighted MRI or fluid attenuated inversion recovery (FLAIR)-weighted MR will display brain edema distribution by signal alteration (hyperintense in T2/FLAIR).[19] Both methods can be used to monitor treatment effects during the course of the disease. Moreover, there are additional MR-sequences like dynamic contrast-enhanced perfusion MRI, which have been demonstrated to correlate with relative cerebral blood volume and VEGF expression in brain tumors and may therefore also be used for tumor-associated brain edema monitoring.[18]

Whereas MRI is now the standard of care for brain tumor evaluation before and after surgery as well as for regular treatment monitoring,[20,21] CT can be used for easy and fast delineation of tumor edema and subsequent brain shift if clinical symptoms occur.

Anatomo-Functional Background

The brain is encased within the rigid bony skull and in contrast to newborn and small children the skull does not expand in adults any more. Therefore, each newly developed or growing mass within the skull no matter if it is benign or malignant will at some point lead to an increase of intracranial pressure. Additionally, the intracranial space consists of two partially separated areas, the supratentorial space containing the cerebrum and an infratentorial space where the cerebellum and brainstem are located. This separation is of high clinical relevance, since the mechanical barrier between supra- or infra-tentorial space may further aggravate a mass effect within one of these compartments, which will cause different clinical symptoms. Due to the functional anatomy of the brain, lesions within the cerebrum, cerebellum or brainstem will lead to a functional impairment dependent on the exact location within the brain. Noteworthy, especially the infratentorial area in the posterior fossa of the skull is difficult to monitor clinically but at the same time may result in rapid neurological deterioration and ultimately to life-threatening brain herniation. Another important factor is the daily production of approximately 500–700 mL of CSF from the choroid plexus within the brain ventricles which is almost four times the volume of CSF normally present in the skull (120–200 mL). A mismatch between CSF production and resorption due to occlusion of the draining passages will obviously result in massive backwater in the brain and skull. Finally, of the three compartments which fill the intracranial space brain, blood and cerebrospinal fluid (CSF), only brain tissue can be compressed easily. This situation can be accelerated if CSF will not be reabsorbed or transported out of the skull. In this case, hydrocephalus will evolve within a few hours especially due to obstruction of the CSF passages in the posterior fossa leading to fatal aggravation of brain herniation at the tentorial rim. Therefore, either by the tumor itself and/or tumor-associated brain edema, each mass effect within the cranial vault will lead to a rise of intracranial pressure (ICP), which has to be compensated initially by the brain.

MEDICAL TREATMENT OF BRAIN EDEMA IN NEUROONCOLOGY

Corticosteroids

Corticosteroids are the most common used agents in the management of tumor-associated brain edema, since edema formation in oncology is mainly vasogenic in nature, as already outlined above. Steroids can be divided into mineralocorticoids, glucocorticoids, and adrenocorticoids all of which are synthesized from cholesterol within different areas of the adrenal gland. Each of these hormones carries out different predominant actions within the body, but due to their similarity, transition is fluent. The use of steroids was established in the 1950s and 1960s due to clinical oberservation.[22] When prednisolone was administered to patients with brain tumors benefits regarding the relief of neurological function were reported. Shortly afterwards, a clinical trial on the use of dexamethasone for tumor-associated brain edema was performed and confirmed with positive effects on neurological deficits.[23] Even if dexamethasone was widely used thereafter, it was not before the 1990s that the one and only randomized controlled trial on the therapeutic efficacy of dexamethasone in the management of cerebral edema of brain tumor patients was published.[24] Patients with metastatic brain tumors on CT were randomized between 4, 8, and 16 mg/day of dexamethasone in two series; 1–8 mg vs. 16 mg and 2–4 mg vs. 16 mg/day. It turned out that patients receiving 4 mg/day had <50% less Cushingoid faces as compared to patients who received 16 mg/day ($P = 0.03$). Moreover there was no significant difference in the dosing regimens regarding improvement of the clinical status measured by the Karnofsky performance score (KPS), whereas toxic effects were more pronounced if 16 mg/day were administered for prolonged time periods. Therefore, the authors recommendations for dosing regimen were 10 mg dexamethasone intravenously followed to be 4 × 4 mg/day orally, if clinical deterioration or imminent brain herniation and elevated ICP were present. Otherwise, if no clinical neurological deficits were present and no signs of elevated ICP on CT visible, 4 mg/day orally were deemed enough. Of note, more recent data suggest that due to the biological half-life of approximately 48 hours a single daily dose may be sufficient. In accordance with the circadian rhythm of the internal steroid secretion dexamethasone should be administered mainly in the early morning with a second optional dose at noon but not in the evening. Thereby, side effects like sleep disorders may be avoided.[25]

Treatment of patients with tumor-associated brain edema is often long-lasting. Whereas newly formatted tumor-associated brain edema may encounter rapid solution by surgical removal of the tumor and short-term steroid application, brain edema which may also be persistent during the course of the disease or may follow radiation therapy and accompany radiation necrosis. Therefore, besides their positive effects on brain edema resolution and in prevention and treatment of chemotherapy-associated nausea and vomiting, there are a lot side effects which may lead to serious complications during and after the treatment. Some of these side effects will develop only after a prolonged duration and comprise arterial hypertension, hyperglycemia, negative impact on the immune system which results in a higher susceptibility to fungal infection, osteoporosis, myopathy and psychiatric disorders or avascular bone necrosis.

Boswellia

Frankincense or Boswellia serrata has been used traditionally as a medication against inflammatory diseases for a long time.[26,27] Besides α- and β-boswellic acid, it contains other pentacyclic triterpenic acids, which inhibit pro-inflammatory processes via 5-lipoxygenase and cyclo-oxygenase.[26,28] Moreover, acetyl-11-keto-β-boswellic acid (AKBA) was shown to potentiate apoptosis and inhibit invasion in different human cancer cell lines[29] and was also demonstrated to be cytotoxic to glioma cells in vitro.[30] However, a recent systemic review of the literature identified only seven randomized clinical trials of boswellia therapy for any human disease which demonstrated some evidence for clinical effectiveness.[27] Clinical data for treatment of tumor-associated brain edema was limited to some observational reports[31] until a recently published prospective, randomized, placebo-controlled double-blind study demonstrated a significant reduction of cerebral radiotherapy-related brain edema of >75% in 60% of patients receiving boswellia as compared to 26% of patients receiving placebo only ($P = .023$).[32] However, boswellia did not have a significant impact on dexamethasone dose-reduction, quality of life or cognitive function during chemoradiation. Even if the exact mode of action of boswellia in tumor-associated brain edema treatment remains to be elucidated, more recent data suggest that boswellia may inhibit VEGF-induced tumor growth and angiogenesis.[33] Despite some encouraging results, it is obvious that larger and more clinical trials will be necessary in order to identify the best time-point for boswellia administration, since data for initial or long-term treatment independent of radiation therapy are still missing. Additionally, the often hypothesized steroid-sparing effect could not be demonstrated so far and needs to be analyzed in more detail in future trials.

Antiangiogenic Therapy

Brain tumor-associated edema is largely promoted by expression of VEGF.[34] As already outlined above, VEGF leads to elevated vascular permeability and increases disruption of the BBB leading to a vasogenic brain edema by an increase in extravascular fluid in the brain thereby exacerbating the mass effect of the tumor. This mode of action depicts an excellent target for antiedema treatment since various VEGF-inhibitors either via direct interaction of VEGF or its different receptor subtypes have been synthesized in the past.[35] Even if they had been intended for direct antiangiogenic and antitumor treatment in various cancer types, their potential for antiedema treatment in the brain has been recognized early.[36]

Bevacizumab is a recombinant human monoclonal antibody directed against VEGF. It was first introduced in oncology for metastatic colorectal carcinoma due to its antiangiogenic properties.[37] In neurooncology, bevacizumab was approved in the United States for treatment of recurrent glioblastomas by the FDA in 2009. However, two large recently published trials failed to reach significant overall survival benefit for treatment of patients with newly diagnosed glioblastomas.[5,6] On the other hand, both studies were able to demonstrate prolonged progression-free survival but critiques argued that this might be due to its effect on BBB stabilization by reduced angiogenesis and vascular permeability leading to reduced vasogenic brain edema and less extravasation of contrast agents without direct antitumorigenic activity.[38] These and other findings have made bevacizumab an interesting agent also for treatment of vasogenic brain edema. There are indeed recent single-institution reports, which demonstrated that administration of BEV in patients with primary malignant brain tumors and acute deterioration due to massive tumor-associated edema formation may be life-saving and also leading to improvement of neurological symptoms in most of these patients.[39–42] Moreover, one randomized controlled clinical trial on refractory radiation necrosis demonstrated a long-term benefit after 10 months for patients who received intravenous bevacizumab injections (7.5 mg/kg) every three weeks for four cycles (Fig. 24.3). However, side effects were often seen, which points out the critical consideration of such a treatment in these patients.[43] Another potent agent against VEGF activity is the VEGFR-tyrosine kinase inhibitor cediranib. It has been used for antiangiogenic therapy in different tumors and also for treatment of tumor-associated brain edema. However, similar to

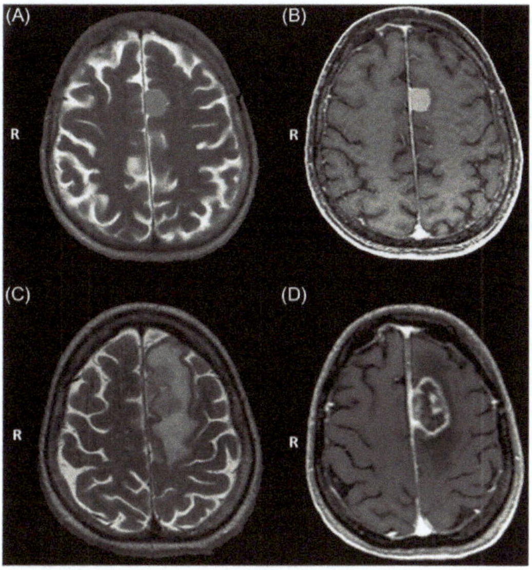

FIGURE 24.3 **Brain edema due to radiation necrosis.** Approximately six months after stereotactic radiosurgery for left-sided parafalcine meningioma (A, B), a 65-year-old patient developed right-sided hemiparesis and headache. MRI revealed hyperintense signal alteration surrounding the irradiated meningioma representing an extensive brain edema of the whole left fronto-parietal cerebrum (C). T1w-MRI demonstrated substantial CE due to massive breakdown of the BBB representing radiation necrosis (D), which turned out to be therapy-refractory to steroids for 6 months and was subsequently treated successfully by bevacizumab.

bevacizumab, there are no randomized controlled trials available which clearly demonstrate efficacy of such a treatment. Therefore there is also no approval for cediranib in the therapy for peritumor brain edema and treatment has to be considered experimental regarding this indication until sufficient clinical trials are available.

Aquaporin

Aquaporins (AQPs) are a family of transmembrane water channels, which have been demonstrated to be effective mediators of different types of brain edema in the past years.[44] In brain tumors, especially the aquaporins 1, 4, and 9 seem to play an important role in brain edema formation[45,46] but also for tumor cell invasion and migration, as has been shown only recently.[47,48] Even if it is still unknown whether VEGF-inhibition via bevacizumab, cediranib or other agents may reduce AQP-expression and with that reduction of brain water motility and edema formation, steroids have been shown to lead to a decrease of cerebral AQP-4 protein expression, which may explain at least in part the effect of steroids on edema reduction.[44,47]

In malignant gliomas and brain metastasis it has been demonstrated that expression of AQP-4 was upregulated and correlated with BBB breakdown and brain edema formation.[49] Moreover, AQP-4 levels seem to correlate with VEGF expression in primary brain tumors, whereas AQP-5 is highly expressed in meningiomas.[50,51] Even if the exact mechanism of AQPs regarding brain edema formation has not been elucidated in detail yet, they seem to be a potential target for development of antiedema drugs. Additionally, other effects on AQPs may also lead to reduced tumor cell migration and invasion and this antitumor effect may also indirectly be a causal therapy against brain tumors and their accompanying brain edema, although at present anti-AQPs therapy not of clinical relevance yet but may be in the future.

Osmodiuretic

Osmodiuretic agents have been used for treatment of brain edema for a long time, especially in traumatic brain injury or stroke and elevated ICP due to cytotoxic brain edema. The most widely used osmodiuretic agent in traumatic brain injury is mannitol[52–54] but glycerol has also been used for treatment of ischemia-associated brain edema after stroke for a long time.[55] Tumor-associated brain edema is also always in part cytotoxic since compression of the surrounding brain parenchyma leads to local hypoxia resulting in the formation of cytotoxic edema. However, since peritumoral edema due to dirupture of the BBB is mainly vasogenic in nature, the application of osmotic agents can be detrimental, since osmotically active substances will easily pass the open BBB and accumulate within the interstitial space thereby further aggravating an already existing mass effect. Therefore, osmotic agents are not recommended in this situation and have to be considered with great caution, if administration is indicated in a life-threatening emergency situation of a brain tumor patient. Moreover, osmodiuretic agents like mannitol also have a very serious side effect. A "rebound phenomenon" will occur after termination of treatment which will not only lead to a loss of function of the medication but also aggravate an already existing edema by interstitial water capture.[56] Since the broken down BBB will not be a barrier for these large osmotically active substances, they will equally distribute between the intra- and extra-vascular space, making it even more difficult to eliminate these proteins from the site of the edema which will then cause severe secondary brain injury by exaggeration of a often fatal vicious circle.

SURGICAL INTERVENTION

Brain tumors lead to peritumoral brain edema by uncontrolled production of cytokines and other relevant factors like VEGF, which are responsible for increased vascular permeability and vasogenic edema as already outlined in the previous chapters. Since therapeutic options for pharmacological reduction of tumor-associated brain edema are limited to transient symptomatic relief as demonstrated above, one of the most important methods to reduce further aggravation or the return of peritumoral edema is to reduce further production of edema-supporting agents by removing the origin of cytokine and growth factor production (Fig. 24.4). Therefore, surgical tumor resection is one of the most important and most effective ways treat tumor-associated brain edema and to terminate the cause of the edema formation if it is directly caused by primary brain or CNS

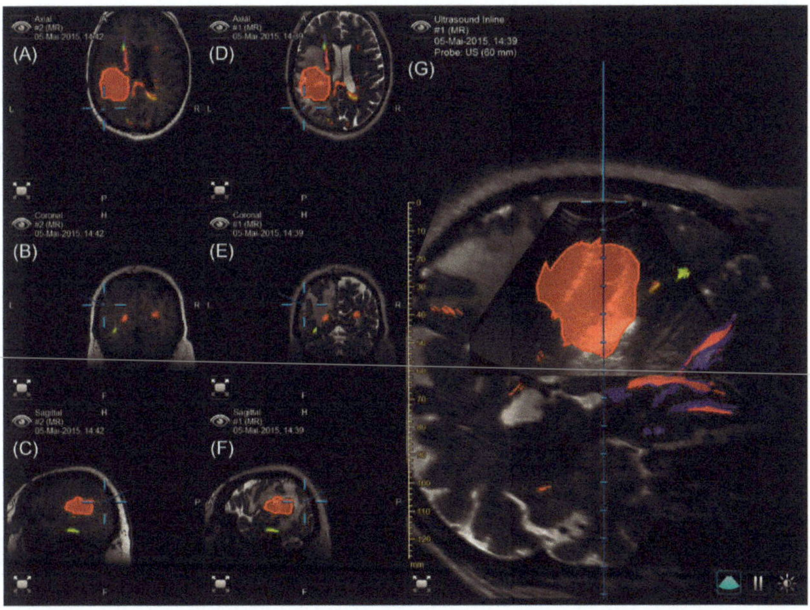

FIGURE 24.4 **Intraoperative screenshot of neuronavigation with MRI and integrated ultrasound.** Treatment-induced peritumoral left fronto-parietal brain edema due to radiation necrosis after stereotactic radiosurgery for brain metastasis. Preoperative delineation of tumor volumes according to T1 + CE and T2-weighted MRI (*red*, A–G) with integrated intraoperative ultrasound (G). Preoperative imaging modalities with T1 + CE MRI (A-C), T2 MRI (D-F) are fused and displayed for neuronavigation in axial (A, D), coronal (B, E) and sagittal (C, F) planes. Additionally, intraoperative ultrasound can be integrated into the neuronavigation system in order to depict the delineated tumor volumes in accordance to the ultrasound image (coronal iUS image overlay in G).

tumors. However, due to several aspects and the anatomo-functional background tumor resection of primary brain or other CNS tumors are not always feasible due to eloquent location of the tumors within the brain and skull. Decision for open tumor resection should be considered with great caution, since malignant primary brain tumors or brain metastasis are prominent in older patients often suffering from serious comorbidities with an elevated surgery-related morbidity rate. Moreover, from an oncological point of view only complete brain tumor resection seems to be associated with better prognosis in patients with malignant primary brain tumors. On the other hand, for better control of tumor-associated brain edema and its aftermath it may also be necessary to perform subtotal resection, if large mass effect exists and adjuvant oncological treatment cannot be initiated otherwise. This may especially be relevant in patients where neurological deficits at initial presentation may be overcome by medical treatment with steroids. If functional deficits are caused by the edema and not infiltration of eloquent areas by the tumor itself, this may enable tumor resection.[57]

Surgical resection is considered the treatment of choice if complete resection at least of all CE on MRI can be achieved safely. Several reports support this approach but none of these studies provided level I evidence due to their retrospective and nonrandomized study design.[58] Whereas some studies reported about positive prognostic relevance of surgical resection with as much as 78% of tumor volume,[59] there are others showing no prognostic benefit from surgical tumor resection in malignant gliomas if complete resection cannot be achieved.[60] Moreover, there is only one randomized trial comparing tumor biopsy with surgical resection; however, it was underpowered and not able to reach a clear recommendation.[61] Moreover, tumors can sometimes not be removed at all, which may have only minor impact on the resolution of peritumoral brain edema and has to be balanced against surgery-induced brain irritation and edema formation risk. If eloquent areas are infiltrated by an intracerebral or at least intracranial mass, tumor resection may be carried out in selected cases via awake surgery in order to have a permanent control of functional integrity of brain functions especially in speech-relevant areas. However, mass effect and large tumor-associated brain edema is considered a knock-out criterion for awake surgery for many neurosurgeons since the risk for further brain swelling and herniation with life-threatening complications may be suspected.

Treatment-associated brain edema and especially therapy-refractory radionecrosis is another indication for surgical intervention. However, as already outlined earlier in this chapter, medical treatment with bevacizumab or other VEGF-inhibitors may become a less invasive and probably more effective therapeutical alternative.[62]

MANAGEMENT OF INTRACRANIAL PRESSURE

Space occupying lesions within the skull, no matter if the underlying pathology is a malignant or benign tumor of the brain, meninges or other stromal origin, as well as infection, abscess, hemorrhage will at least locally lead to elevated ICP and displacement of un- or less-effected brain areas. Whereas tumor infiltration or destruction of eloquent areas of the cerebrum results in permanent deficits independent of treatment, involvement of eloquent areas by brain edema formation that will induce functional deterioration and promote seizures due to the elevated excitability of the edematous brain but may be reversible if properly treated. Moreover, due to high velocity of brain edema formation (acquisition of extracellular fluid equals 90mL/day), each mass effect in the cerebrum can promptly result in brain shift across the midline or lead to brain herniation comprising indispensable brainstem function. Since treatment of such a life-threatening situation has to be started immediately and must be effective from the beginning, often national guidelines exist to ensure best possible care.[63]

In some cases, either the underlying primary brain or CNS tumor itself, of the peritumoral brain edema may lead to hydrocephalus due to reduced drainage of CSF from the supratentorial brain ventricles to the extracerebral CSF space. If hydrocephalus of the supratentorial ventricles occurs due to a supratentorial compression of the foramen of Monroi within a short time period, this may be treated in an emergency case by external ventricular drainage. However, this is not a permanent solution of the underlying cause and draining volume that has to be limited in order to avoid immediate release of pressure followed by secondary herniation or the risk of consecutive intracerebral hemorrhage. This may be particularly fatal, if an already existing herniation of the cerebellum in the posterior fossa and in the direction of the tentorial opening may be pronounced. Therefore, mass effect and brain edema within the cerebellum and brainstem have to be considered with great care since there is a high risk of life-threatening damage of key pacemaker areas in the brainstem. Additionally, brain edema in the posterior fossa is eminent to manage early since dislocation of cerebellar tonsils may cause inferior brain herniation in the foramen magnum which in the worst case results in internal decapitation and decerebration. If mass effect and elevated ICP will not be terminated at this point, ICP may further increase until it reaches a certain level where reduction of blood flow and brain perfusion will result in hypoxia of the brain. Since hypoxia or ischemia trigger additional cytotoxic brain edema, a vicious circle is initiated which speeds up further mass effect and increased edema formation which may further aggravate into a deadly situation.

If either directly tumor-associated brain edema or therapy-induced edema cannot be controlled by conservative treatment, decompressive surgery may be the ultimate option to avoid further aggravation of hypoxia and brain herniation and finally death in selected patients. However, this intervention is discussed controversially and may only be recommended if certain requirements are fulfilled. In case of an emergency in an otherwise healthy patient with a most probably benign CNS tumor like meningioma or acoustic schwannoma in the posterior fossa with tumor-associated brain edema and mass effect leading to hydrocephalus and herniation, decompressive surgery may be a life-saving intervention until further treatment regarding the tumor itself or the peritumoral edema is started. The same applies to assumed benign tumors within the cerebrum, if immediate resolution of ICP will be necessary in order to repel a life-threatening situation and to be able to deliver further tumor-specific therapy to the patient afterwards. On the other hand, if the same situation occurs in patients with already diagnosed malignant brain tumors and uncontrollable tumor-associated brain edema, decompressive surgery should not be considered, since this will only prolong the patients suffering and the underlying disease will not be changed in any way.

FUTURE PERSPECTIVES

The discovery of AQPs and their functional impact in the tumor cell migration and invasion but more importantly also the formation of tumor-associated brain edema may be an important step in order treat brain tumors and the accompanying peritumoral edema more effectively. Nevertheless, more translational research and clinical trials are necessary to learn more about the clinical relevance of these findings and possible interactions with other important factors in the pathophysiology of tumor-associated brain edema like, e.g., VEGF and its receptors.

Furthermore, corticorelin acetate (CrA) may be a promising future treatment option for patients suffering from severe side as a consequence of long-term treatment of tumor-associated brain edema by cortocosteroids.[62,64] CrA is a synthetic analog of human corticotropin-releasing factor (hCRF), which is a naturally occurring neuropeptide produced in the hypothalamus. hCRF leads to release of hydrocortisone from the adrenal gland via adrenocorticotropin hormone (ACTH) release from the pituitary gland. In an experimental setting, hCRF was able to reduce vascular permeability and tumor-associated vasogenic brain edema independent of corticosteroid release.[65,66] In a prospective, randomized, controlled, double-blind trial of 200 patients, CrA was able to reduce steroid requirement of patients with peritumoral brain edema as well as incidence and severity of common steroid-associated side effects including myopathy.[64]

Another very recent finding, which will have a major impact on neuroimmunology may also be an important aspect regarding the formation of tumor-associated brain edema. The discovery of a CNS lymphatic system has to be considered a major breakthrough for future research and may have clinical relevance for a lot of neurodegenerative and neuroinflammatory diseases but also other CNS pathologies associated with neuroimmunological processes like benign or malignant brain tumors and the formation of peritumoral edema. Interestingly, VEGF seems to play a role in the regulation of lymphatic endothelial cells (LEC) in the brain, since experimental administration of recombinant VEGF-c resulted in an increase of the diameter of lymphatic meningeal vessels via VEGF receptor 3.[67] Moreover, together with other findings this suggests that interstitial fluid may be drained from the brain parenchyma through a recently discovered glymphatic system into the CSF[68,69] and from there via the meningeal lymphatic vessels into the periphery. Malfunction of this system may lead to lymphedema of the brain in various CNS diseases, but on the other hand, this may also be an important target for novel treatment strategies.

CONCLUSION

Patients with CNS tumors may suffer from symptoms which are not only caused by the tumors themselves but also by the formation of a peritumoral vasogenic brain edema. Moreover, cytotoxic edema may add on if a rise in ICP leads to cerebral ischemia. The clinical management of tumor-associated brain edema formation and its solution remains a difficult challenge. Therapeutic options are limited and until present, steroids are the only medication which has demonstrated efficacy in clinical trials. However, the long-term administration of steroids results in serious side effects which have to avoided.

Brain edema in patients with CNS tumors may result from various sources, e.g., directly from the tumors, but also due to surgical, radiooncological or chemotherapeutical interventions. Therefore, patients are often dependent on antiedema treatment during the whole course of their disease. There is an urgent need for new drug development in order to have a broader armamentarium to treat tumor-associated brain edema and reduce side effects despite suggested long-term dependency. Therefore, targeting VEGF or its receptors and especially AQPs may represent promising new strategies in order to overcome these issues in the future.

References

1. Ostrom QT, Gittleman H, Liao P, Rouse C, Chen Y, Dowling J, et al. CBTRUS statistical report: primary brain and central nervous system tumors diagnosed in the United States in 2007–2011. *Neuro Oncol* 2014;**16**(Suppl 4):iv1–63.
2. Stupp R, Mason WP, van den Bent MJ, Weller M, Fisher B, Taphoorn MJ, et al. Radiotherapy plus concomitant and adjuvant temozolomide for glioblastoma. *N Engl J Med* 2005;**352**:987–96.
3. Gilbert MR, Wang M, Aldape KD, Stupp R, Hegi ME, Jaeckle KA, et al. Dose-dense temozolomide for newly diagnosed glioblastoma: a randomized phase III clinical trial. *J Clin Oncol* 2013;**31**:4085–91.
4. Stupp R, Hegi ME, Gorlia T, Erridge SC, Perry J, Hong YK, et al. Cilengitide combined with standard treatment for patients with newly diagnosed glioblastoma with methylated MGMT promoter (CENTRIC EORTC 26071-22072 study): a multicentre, randomised, open-label, phase 3 trial. *Lancet Oncol* 2014;**15**:1100–8.
5. Gilbert MR, Sulman EP, Mehta MP. Bevacizumab for newly diagnosed glioblastoma. *N Engl J Med* 2014;**370**:2048–9.
6. Chinot OL, Wick W, Mason W, Henriksson R, Saran F, Nishikawa R, et al. Bevacizumab plus radiotherapy-temozolomide for newly diagnosed glioblastoma. *N Engl J Med* 2014;**370**:709–22.
7. Louis DN, Ohgaki H, Wiestler OD, Cavenee WK, Burger PC, Jouvet A, et al. The 2007 WHO classification of tumours of the central nervous system. *Acta Neuropathol* 2007;**114**:97–109.
8. Gaspar L, Scott C, Rotman M, Asbell S, Phillips T, Wasserman T, et al. Recursive partitioning analysis (RPA) of prognostic factors in three Radiation Therapy Oncology Group (RTOG) brain metastases trials. *Int J Radiat Oncol Biol Phys* 1997;**37**:745–51.
9. Sperduto PW, Kased N, Roberge D, Xu Z, Shanley R, Luo X, et al. Effect of tumor subtype on survival and the graded prognostic assessment for patients with breast cancer and brain metastases. *Int J Radiat Oncol Biol Phys* 2012;**82**:2111–7.
10. Shaffrey ME, Mut M, Asher AL, Burri SH, Chahlavi A, Chang SM, et al. Brain metastases. *Curr Probl Surg* 2004;**41**:665–741.
11. Hirano A, Kawanami T, Llena JF. Electron microscopy of the blood-brain barrier in disease. *Microsc Res Tech* 1994;**27**:543–56.
12. Osawa T, Tosaka M, Nagaishi M, Yoshimoto Y. Factors affecting peritumoral brain edema in meningioma: special histological subtypes with prominently extensive edema. *J Neurooncol* 2013;**111**:49–57.
13. Strugar JG, Criscuolo GR, Rothbart D, Harrington WN. Vascular endothelial growth/permeability factor expression in human glioma specimens: correlation with vasogenic brain edema and tumor-associated cysts. *J Neurosurg* 1995;**83**:682–9.
14. Pietsch T, Valter MM, Wolf HK, von Deimling A, Huang HJ, Cavenee WK, et al. Expression and distribution of vascular endothelial growth factor protein in human brain tumors. *Acta Neuropathol* 1997;**93**:109–17.
15. Iwado E, Ichikawa T, Kosaka H, Otsuka S, Kambara H, Tamiya T, et al. Role of VEGF and matrix metalloproteinase-9 in peritumoral brain edema associated with supratentorial benign meningiomas. *Neuropathology* 2012;**32**:638–46.
16. Dobrogowska DH, Lossinsky AS, Tarnawski M, Vorbrodt AW. Increased blood-brain barrier permeability and endothelial abnormalities induced by vascular endothelial growth factor. *J Neurocytol* 1998;**27**:163–73.
17. Jain RK, di Tomaso E, Duda DG, Loeffler JS, Sorensen AG, Batchelor TT. Angiogenesis in brain tumours. *Nat Rev Neurosci* 2007;**8**:610–22.
18. Ginat DT, Mangla R, Yeaney G, Schaefer PW, Wang H. Correlation between dynamic contrast-enhanced perfusion MRI relative cerebral blood volume and vascular endothelial growth factor expression in meningiomas. *Acad Radiol* 2012;**19**:986–90.

19. Marmarou A. A review of progress in understanding the pathophysiology and treatment of brain edema. *Neurosurg Focus* 2007;**22**:E1.
20. Macdonald DR, Cascino TL, Schold Jr. SC, Cairncross JG. Response criteria for phase II studies of supratentorial malignant glioma. *J Clin Oncol* 1990;**8**:1277–80.
21. Wen PY, Macdonald DR, Reardon DA, Cloughesy TF, Sorensen AG, Galanis E, et al. Updated response assessment criteria for high-grade gliomas: response assessment in neuro-oncology working group. *J Clin Oncol* 2010;**28**:1963–72.
22. Kofman S, Garvin JS, Nagamani D, Taylor 3rd SG. Treatment of cerebral metastases from breast carcinoma with prednisolone. *J Am Med Assoc* 1957;**163**:1473–6.
23. Galicich JH, French LA, Melby JC. Use of dexamethasone in treatment of cerebral edema associated with brain tumors. *J Lancet* 1961;**81**:46–53.
24. Vecht CJ, Hovestadt A, Verbiest HB, van Vliet JJ, van Putten WL. Dose-effect relationship of dexamethasone on Karnofsky performance in metastatic brain tumors: a randomized study of doses of 4, 8, and 16mg per day. *Neurology* 1994;**44**:675–80.
25. Elder GJ, Wetherell MA, Barclay NL, Ellis JG. The cortisol awakening response--applications and implications for sleep medicine. *Sleep Med Rev* 2014;**18**:215–24.
26. Ammon HP. Boswellic acids in chronic inflammatory diseases. *Planta Med* 2006;**72**:1100–16.
27. Ernst E. Frankincense: systematic review. *BMJ* 2008;**337**:a2813.
28. Knaus U, Wagner H. Effects of boswellic acid of Boswellia serrata and other triterpenic acids on the complement system. *Phytomedicine* 1996;**3**:77–80.
29. Takada Y, Ichikawa H, Badmaev V, Aggarwal BB. Acetyl-11-keto-beta-boswellic acid potentiates apoptosis, inhibits invasion, and abolishes osteoclastogenesis by suppressing NF-kappa B and NF-kappa B-regulated gene expression. *J Immunol* 2006;**176**:3127–40.
30. Glaser T, Winter S, Groscurth P, Safayhi H, Sailer ER, Ammon HP, et al. Boswellic acids and malignant glioma: induction of apoptosis but no modulation of drug sensitivity. *Br J Cancer* 1999;**80**:756–65.
31. Streffer JR, Bitzer M, Schabet M, Dichgans J, Weller M. Response of radiochemotherapy-associated cerebral edema to a phytotherapeutic agent, H15. *Neurology* 2001;**56**:1219–21.
32. Kirste S, Treier M, Wehrle SJ, Becker G, Abdel-Tawab M, Gerbeth K, et al. Boswellia serrata acts on cerebral edema in patients irradiated for brain tumors: a prospective, randomized, placebo-controlled, double-blind pilot trial. *Cancer* 2011;**117**:3788–95.
33. Pang X, Yi Z, Zhang X, Sung B, Qu W, Lian X, et al. Acetyl-11-keto-beta-boswellic acid inhibits prostate tumor growth by suppressing vascular endothelial growth factor receptor 2-mediated angiogenesis. *Cancer Res* 2009;**69**:5893–900.
34. Berkman RA, Merrill MJ, Reinhold WC, Monacci WT, Saxena A, Clark WC, et al. Expression of the vascular permeability factor/vascular endothelial growth factor gene in central nervous system neoplasms. *J Clin Invest* 1993;**91**:153–9.
35. Schiff D, Lee EQ, Nayak L, Norden AD, Reardon DA, Wen PY. Medical management of brain tumors and the sequelae of treatment. *Neuro Oncol* 2015;**17**:488–504.
36. Klagsbrun M, D'Amore PA. Vascular endothelial growth factor and its receptors. *Cytokine Growth Factor Rev* 1996;**7**:259–70.
37. Hurwitz H. Integrating the anti-VEGF-A humanized monoclonal antibody bevacizumab with chemotherapy in advanced colorectal cancer. *Clin Colorectal Cancer* 2004;**4**(Suppl 2):S62–8.
38. Verhoeff JJ, van Tellingen O, Claes A, Stalpers LJ, van Linde ME, Richel DJ, et al. Concerns about anti-angiogenic treatment in patients with glioblastoma multiforme. *BMC Cancer* 2009;**9**:444.
39. Ellingson BM, Cloughesy TF, Lai A, Nghiemphu PL, Lalezari S, Zaw T, et al. Quantification of edema reduction using differential quantitative T2 (DQT2) relaxometry mapping in recurrent glioblastoma treated with bevacizumab. *J Neurooncol* 2012;**106**:111–19.
40. Kaley T, Nolan C, Carver A, Omuro A. Bevacizumab for acute neurologic deterioration in patients with glioblastoma. *CNS Oncol* 2013;**2**:413–18.

41. Berghoff AS, Sax C, Klein M, Furtner J, Dieckmann K, Gatterbauer B, et al. Alleviation of brain edema and restoration of functional independence by bevacizumab in brain-metastatic breast cancer: a case report. *Breast Care (Basel)* 2014;**9**:134–6.
42. Dashti SR, Spalding A, Kadner RJ, Yao T, Kumar A, Sun DA, et al. Targeted intraarterial anti-VEGF therapy for medically refractory radiation necrosis in the brain. *J Neurosurg Pediatr* 2015;**15**:20–5.
43. Levin VA, Bidaut L, Hou P, Kumar AJ, Wefel JS, Bekele BN, et al. Randomized double-blind placebo-controlled trial of bevacizumab therapy for radiation necrosis of the central nervous system. *Int J Radiat Oncol Biol Phys* 2011;**79**:1487–95.
44. Badaut J, Fukuda AM, Jullienne A, Petry KG. Aquaporin and brain diseases. *Biochim Biophys Acta* 2014;**1840**:1554–65.
45. Saadoun S, Papadopoulos MC, Davies DC, Bell BA, Krishna S. Increased aquaporin 1 water channel expression in human brain tumours. *Br J Cancer* 2002;**87**:621–3.
46. Saadoun S, Papadopoulos MC, Davies DC, Krishna S, Bell BA. Aquaporin-4 expression is increased in oedematous human brain tumours. *J Neurol Neurosurg Psychiatry* 2002;**72**:262–5.
47. Nico B, Ribatti D. Role of aquaporins in cell migration and edema formation in human brain tumors. *Exp Cell Res* 2011;**317**:2391–6.
48. Ding T, Gu F, Fu L, Ma YJ. Aquaporin-4 in glioma invasion and an analysis of molecular mechanisms. *J Clin Neurosci* 2010;**17**:1359–61.
49. Warth A, Simon P, Capper D, Goeppert B, Tabatabai G, Herzog H, et al. Expression pattern of the water channel aquaporin-4 in human gliomas is associated with blood–brain barrier disturbance but not with patient survival. *J Neurosci Res* 2007;**85**:1336–46.
50. Mou K, Chen M, Mao Q, Wang P, Ni R, Xia X, et al. AQP-4 in peritumoral edematous tissue is correlated with the degree of glioma and with expression of VEGF and HIF-alpha. *J Neurooncol* 2010;**100**:375–83.
51. Lambertz N, Hindy NE, Adler C, Rump K, Adamzik M, Keyvani K, et al. Expression of aquaporin 5 and the AQP5 polymorphism A(-1364)C in association with peritumoral brain edema in meningioma patients. *J Neurooncol* 2013;**112**:297–305.
52. Ghajar J, Hariri RJ, Narayan RK, Iacono LA, Firlik K, Patterson RH. Survey of critical care management of comatose, head-injured patients in the United States. *Crit Care Med* 1995;**23**:560–7.
53. Jeevaratnam DR, Menon DK. Survey of intensive care of severely head injured patients in the United Kingdom. *BMJ* 1996;**312**:944–7.
54. Sorani MD, Manley GT. Dose-response relationship of mannitol and intracranial pressure: a metaanalysis. *J Neurosurg* 2008;**108**:80–7.
55. Cantore G, Guidetti B, Virno M. Oral glycerol for the reduction of intracranial pressure. *J Neurosurg* 1964;**21**:278–83.
56. Palma L, Bruni G, Fiaschi AI, Mariottini A. Passage of mannitol into the brain around gliomas: a potential cause of rebound phenomenon. A study on 21 patients. *J Neurosurg Sci* 2006;**50**:63–6.
57. Stummer W, Reulen HJ, Meinel T, Pichlmeier U, Schumacher W, Tonn JC, et al. Extent of resection and survival in glioblastoma multiforme: identification of and adjustment for bias. *Neurosurgery* 2008;**62**:564–76. Discussion 76.
58. Lacroix M, Abi-Said D, Fourney DR, Gokaslan ZL, Shi W, DeMonte F, et al. A multivariate analysis of 416 patients with glioblastoma multiforme: prognosis, extent of resection, and survival. *J Neurosurg* 2001;**95**:190–8.
59. Sanai N, Polley MY, McDermott MW, Parsa AT, Berger MS. An extent of resection threshold for newly diagnosed glioblastomas. *J Neurosurg* 2011;**115**:3–8.
60. Kreth FW, Thon N, Simon M, Westphal M, Schackert G, Nikkhah G, et al. Gross total but not incomplete resection of glioblastoma prolongs survival in the era of radiochemotherapy. *Ann Oncol* 2013;**24**:3117–23.

61. Vuorinen V, Hinkka S, Farkkila M, Jaaskelainen J. Debulking or biopsy of malignant glioma in elderly people—a randomised study. *Acta Neurochir (Wien)* 2003;**145**:5–10.
62. Roth P, Regli L, Tonder M, Weller M. Tumor-associated edema in brain cancer patients: pathogenesis and management. *Expert Rev Anticancer Ther* 2013;**13**:1319–25.
63. JüttlerE.Intrakranieller Druck. In: Diener H-C, Weimar, C. (Hersg.); Neurologie KLdDGf. *Leitlinien für Diagnostik und Therapie in der Neurologieed.*: Thieme Verlag, Stuttgart, 2012.
64. Recht L, Mechtler LL, Wong ET, O'Connor PC, Rodda BE. Steroid-sparing effect of corticorelin acetate in peritumoral cerebral edema is associated with improvement in steroid-induced myopathy. *J Clin Oncol* 2013;**31**:1182–7.
65. Tjuvajev J, Uehara H, Desai R, Beattie B, Matei C, Zhou Y, et al. Corticotropin-releasing factor decreases vasogenic brain edema. *Cancer Res* 1996;**56**:1352–60.
66. Moliterno JA, Henry E, Pannullo SC. Corticorelin acetate injections for the treatment of peritumoral brain edema. *Expert Opin Investig Drugs* 2009;**18**:1413–19.
67. Louveau A, Smirnov I, Keyes TJ, Eccles JD, Rouhani SJ, Peske JD, et al. Structural and functional features of central nervous system lymphatic vessels. *Nature* 2015;**523**:337–41.
68. Yang L, Kress BT, Weber HJ, Thiyagarajan M, Wang B, Deane R, et al. Evaluating glymphatic pathway function utilizing clinically relevant intrathecal infusion of CSF tracer. *J Transl Med* 2013;**11**:107.
69. Xie L, Kang H, Xu Q, Chen MJ, Liao Y, Thiyagarajan M, et al. Sleep drives metabolite clearance from the adult brain. *Science* 2013;**342**:373–7.

CHAPTER

25

Perspectives on Future Translational Research on Brain Edema

J. Marc Simard[1], *Jérôme Badaut*[2] *and Nikolaus Plesnila*[3,4]

[1]University of Maryland School of Medicine, Baltimore, MD, United States [2]CNRS-University of Bordeaux, Bordeaux, France [3]University of Munich Medical Center, Munich, Germany [4]Munich Cluster of Systems Neurology, Munich, Germany

OUTLINE

Brain Edema.
DOI: http://dx.doi.org/10.1016/B978-0-12-803196-4.00025-4

INTRODUCTION

For many decades the only available medical therapies for the treatment of brain edema have been mannitol and hypertonic/hyperoncotic saline, with the notable exception of steroids for brain tumor-induced brain swelling. The application of hypertonic solutions is based on the pathophysiological concept that water enters the brain mainly due to an intravascular/parenchymal osmotic gradient, and therefore intravascular hyperosmolarity should reverse the process. Empirically, this therapeutic concept proved to be correct, and generations of physicians used these substances to successfully treat patients suffering from intracranial hypertension and its often deadly consequences. As we all know, however, the treatment of brain edema with hypertonic solutions has severe limitations–intracranial pressure-induced reduction or even cessation of the cerebral circulation still represents one of the major causes of death and severe disability in patients suffering from such frequent diseases as cardiac arrest, ischemic stroke, cerebral hemorrhage, or traumatic brain injury.[1,2]

This woeful therapeutic situation regarding the treatment of brain edema has been mainly attributed to the lack of experimental and clinical methods to investigate the mechanisms of brain edema formation on the cellular and molecular level during the past decades. With the broader availability of MR-based imaging of water movements in the living brain of patients and experimental animals in the late 1990s,[3–5] with the growing possibilities for genetic manipulations in vitro and in vivo, with the availability of a variety of specific antibodies for immunohistochemistry, with the development of better in vitro models, and with the recent possibility to image cerebral vessels deep within the brain parenchyma in vivo in experimental animals,[6] the field of brain edema research has taken huge strides forward. Today we know that vasogenic brain edema is not only a purely mechanical event, but is also driven by molecular mechanisms, that cytotoxic brain edema has a compensatory and most likely neuroprotective component mediated by astrocytes, that water flux across cell membranes can be facilitated by aquaporin water channels, and that the classical forms of brain edema, i.e., vasogenic and cytotoxic, are accompanied by ionic edema, a process that occurs as a consequence of the relative hypoosmolarity of the extracellular space in the brain caused by intracellular uptake of ions during the process of cytotoxic cell swelling[1,2,7–16] (Section one and two).

Despite the significant conceptual and methodological advances made in the field of brain edema research during the past 30 years, fundamental questions and enormous challenges remain. Prominent among these is the question of the relative contribution of various molecular mechanisms, some of which were discussed in foregoing chapters, and others of which, undoubtedly, are yet to be discovered. To what extent do these various molecular mechanisms contribute to edema formation in different disease contexts? Do different edema-associated pathologies produce edema in the same manner, regardless of the underlying disease or the CNS region (gray vs. white matter), or are molecular mechanisms of edema formation "disease specific"? Do the mechanisms of edema formation differ with age? Will treatments developed for a given condition be applicable to other conditions? This chapter will try to dissect some of the most fundamental issues and will try to suggest future direction of research.

SOURCE OF BRAIN EDEMA FLUID

There remains active debate regarding the precise source(s) of edema fluid that contributes to brain swelling, the answer to which will likely profoundly influence future research and future therapies. Until recently, for example, it was not even clear that trauma-induced vasogenic edema derives not only from intraparenchymal arterioles and capillaries, but to a large extent also from intraparenchymal veins.[6] The classic explanation, the microvascular endothelium, has been at the center of extensive study for decades, spurred on by the enormous contributions made over a century ago by the British physiologist, Ernest Starling, who studied edema formation not in the brain but in the hind limb.[17] With methodological advances during the past decades, only recently have we started to understand the molecular mechanisms responsible for swelling of endothelial cells, disruption of cerebrovascular tight junctions, and the entry of edema fluid into the brain parenchyma (see Part II: Techniques to Investigate Cerebral Blood of this book). Equally little is known about the spatial and temporal heterogeneity of transporters and ion channels expressed during the formation of brain edema. For most clinically relevant conditions, such as ischemic and hemorrhagic stroke and traumatic brain injury, our current specific knowledge on edema formation is quite limited and mainly based on pharmacological studies, i.e., merely on edema reducing effects of pharmacological compounds with often questionable selectivity and specificity profiles on the molecular level in vivo. We still do not know the exact time course of opening and closing of the blood–brain barrier following traumatic or ischemic brain injury and in most cases, we do not know if the applied pharmacological compounds do indeed what they are expected to do.

This partial knowledge is mainly caused by the particularly complex and expensive methodology necessary to address these issues, e.g., small animal MRI and two-photon microscopy. However, due to the increasing broader availability of these technologies and due to the more collaborative nature of current and future research efforts, more researchers interested in the pathogenesis of brain edema will have access to these techniques. Therefore, we anticipate a significantly improved understanding of the function of the BBB during pathological conditions during the years to come. A good example of how technology, in this case two-photon microscopy, is able to broaden our understanding of fluid dynamics in the brain, is the recent discovery that the glymphatic system drains CSF through the brain parenchyma along paravascular arteriolar and venular routs.[18–20] The role of the glymphatic system in the formation of brain edema is unknown, but already, this new concept challenges the unique stature of the classic theory on edema formation. Will one of these two schemas predominate, or will both be found to contribute to cerebral edema formation?

RESOLUTION OF EDEMA: WHERE AND HOW IS THE WATER LEAVING THE CNS?

Brain edema in many cases resolves with time without any tissue damage. This process piqued the interest of researchers to understand the mechanisms that contribute to the drainage of water from the brain. Historically, it has been proposed that clearance of brain edema occurs into the subarachnoid space.[21] More recently, AQP4 was suggested to be involved in clearance of brain edema,[22,23] since AQP4 seems to be essential for glymphatic water flux through the brain parenchyma.[24] In addition to edema resolution, the clearance of CSF is still a subject of discussion and it is not clear if it occurs only via the arachnoid villi or also along cervical lymph nodes, the space surrounding cranial nerves, or the olfactory submucosa.[25] With the recent methodological advances that have occurred in the field of brain edema, a better understanding of the mechanisms of edema resolution will be an important issue to be tackled in the future.

THE LINK BETWEEN CYTOTOXIC CELL SWELLING AND EDEMA

It has been known for a long time that cerebral ischemia causes swelling of cells and subsequent shrinkage of the extracellular space in the brain.[1,2] Based on the concept that lack of energy causes cell swelling, this process was termed "cytotoxic." Later on, however, it was recognized that it was not simply lack of energy that caused cell swelling, but the attempt of astrocytes to remove toxic substances from the extracellular space. Since the uptake of such substances, e.g., glutamate or protons, occurs together with ions, this process resulted in cell swelling.[26,27] Consequently, ischemia-induced cell swelling is not necessarily toxic, but according to our current understanding, is in large part neuroprotective.

Despite our deeper understanding of the basic principles of ischemic cell swelling, some fundamental questions are still unresolved. One important issue is whether, in view of its neuroprotective nature, ischemic cell swelling should be inhibited at all. In vitro data suggest that only acidosis-induced cell swelling may be inhibited without compromising the regulation of intracellular pH,[28] while inhibition of glutamate- and K^+ ion-induced cell swelling are purely protective and should not be inhibited. However, no in vivo data are available to address this issue. Another important question is the link between ischemic cell swelling and brain edema, i.e., how the quantitatively neutral movement of water between the intra- and the extracellular space of the brain causes a net increase in the brain water content, i.e., brain edema. So far this process has been explained by the concept of ionic edema, which suggests that ischemic cell swelling causes an increase in extracellular osmolarity which then pulls water from the intravascular space into the brain.[29,30] Despite the beauty of this concept, further data obtained under clinically relevant experimental conditions, and possibly in patients, are needed to ultimately prove that ionic edema is indeed the consequence of ischemic cell swelling. Initial attempts in patients using brain microdialysis are on their way,[31] but more clinical and experimental studies need to follow in order to obtain a clear answer.

EDEMA FORMATION IN DIFFERENT REGIONS OF THE CNS

The mechanisms of edema formation differ significantly between the various regions of the CNS. For example, white matter is highly sensitive to swelling after injuries[32] due to its higher transport capacity for fluids as compared to gray matter.[33] This property of white matter may be caused by a more parallel organization of fibers in white matter as compared to cortex, or by the presence of "fibrous" astrocytes in white matter, while gray matter harbors more "protoplasmic" astrocytes,[34] or by different expression of water channels and ion transporters in these different brain regions.[35] The latter point seems to be of particular relevance under pathological conditions, since the subcellular location and expression of AQP4 differs significantly between gray and white matter after cerebral ischemia, and may serve as a molecular explanation for the different degree of edema formation in different areas of the brain.[36] Hence, the better understanding of brain region differences may also be critical for future treatment development.

TUMOR AND IMMUNE CELL-MEDIATED BRAIN EDEMA

Dramatic examples exist of the enormous impact that cerebral edema can have on the neurological function of patients with a metastatic tumor to the brain (*see Figure*). Although remarkable, such examples are still within the realm of common clinical experience, with important progress being made in the search for molecular targets.[37] However, there are malignancies in which cerebral edema plays a newly discovered, completely unexpected role, and in which edema can be deadly. An emerging therapy for cancer involves utilizing a patient's own T-lymphocytes, molecularly engineered to fight the patient's own cancer. This innovative therapy is termed chimeric antigen receptor T (CAR-T) cell therapy. T-lymphocytes are removed from a patient, modified so that they express receptors specific to the patient's particular cancer, then are reintroduced into the patient. In the case of B-cell acute lymphoblastic leukemia, a form of leukemia that is routinely fatal, complete remission of disease in up to 90% of patients has been observed following CAR-T cell therapy.[38] The dark side in this otherwise brilliant picture is that some patients administered CAR-T therapy have died, due not to their cancer, but to their treatment.[39] For unknown reasons, a subset of patients administered CAR-T therapy experience massive cerebral edema with fatal consequences. Solving the enigmatic problem of potentially fatal cerebral edema associated with CAR-T therapy could open the door for aggressive, highly promising advanced immunotherapy for a variety of cancers.

MOLECULAR TARGETS FOR THE TREATMENT OF BRAIN EDEMA

As noted, we are in a new era wherein cerebral edema is and will increasingly be targeted by molecularly specific therapies. Bradykinin receptor, MMP, VEGF receptor, ET_B receptor, and vasopressin V_{1a} receptor inhibitors, on the one hand, and $Na^+/K^+/2Cl^-$ cotransporter, Na/H antiporter, and K^+ channel antagonists, on the other hand, have been suggested as potential compounds for the treatment of vasogenic brain edema and ischemic cell swelling, respectively.[7] The most translationally promising results derive from the recently published GAMES-RP trial.[40] Using an intravenously injectable form of glibenclamide, which inhibits the sulfonylurea binding subunit of the SUR1-TRPM4 channel, Sheth and colleagues show that a molecularly directed therapy can be remarkably effective at reducing cerebral edema, indeed far more effective than conventional nonspecific therapies such as mannitol, hypertonic saline, or dexamethasone. The

GAMES-RP trial studied patients with a large hemispheric infarction, the classic antecedent of massive brain swelling due to edema, also known as "malignant" cerebral edema due to its associated high mortality.[41] Swelling was assessed by measuring "midline shift," a standard clinical correlate of swelling and an excellent predictor of patient outcome. In GAMES-RP, treatment with RP-1127 reduced midline shift by nearly half. Not surprisingly, this degree of reduction in swelling was associated with a commensurate reduction in mortality. However, despite this encouraging result, the mechanism for the residual swelling *not* affected by RP-1127 remains unknown. Preclinical studies suggest that residual swelling is unlikely to be accounted for simply by under-dosing with RP-1127.[42] Other molecular mechanisms of edema formation, yet to be identified, likely are at play to account for "RP-1127-insensitive" swelling. The challenge in large hemispheric infarction will be to identify that mechanism.

In summary, the field of brain edema research is entering an exciting and novel era, leaving osmotherapy behind and moving toward specific treatments based on clear, molecularly based pathophysiological concepts. Significantly more work needs to be done in order to fully understand the mechanism(s) of vasogenic and cytotoxic/ionic edema formation in detail. However, the necessary experimental tools are available and now need to be combined with clinically relevant in vivo animal models in order to achieve results with translational potential. These efforts undoubtedly will be justified by the high likelihood that resulting insights will lead to innovative, novel treatment strategies for a large variety of acute, subacute and chronic CNS conditions associated with edema formation.

References

1. Klatzo I. Presidental address. Neuropathological aspects of brain edema. *J Neuropathol Exp Neurol* 1967;**26**:1–14.
2. Fishman RA. Brain edema. *N Engl J Med* 1975;**293**:706–11.
3. Ito J, Marmarou A, Barzo P, Fatouros P, Corwin F. Characterization of edema by diffusion-weighted imaging in experimental traumatic brain injury. *J Neurosurg* 1996;**84**:97–103.
4. Barzo P, Marmarou A, Fatouros P, Corwin F, Dunbar J. Magnetic resonance imaging-monitored acute blood-brain barrier changes in experimental traumatic brain injury. *J Neurosurg* 1996;**85**:1113–21.
5. Marmarou A, Barzo P, Fatouros P, Yamamoto T, Bullock R, Young H. Traumatic brain swelling in head injured patients: brain edema or vascular engorgement? *Acta Neurochir Suppl* 1997;**70**:68–70.
6. Schwarzmaier SM, Gallozzi M, Plesnila N. Identification of the vascular source of vasogenic brain edema following traumatic brain injury using in vivo 2-photon microscopy in mice. *J Neurotrauma* 2015;**32**:990–1000.
7. Michinaga S, Koyama Y. Pathogenesis of brain edema and investigation into anti-edema drugs. *Int J Mol Sci* 2015;**16**:9949–75.
8. Klatzo I. Evolution of brain edema concepts. *Acta Neurochir Suppl (Wien)* 1994;**60**:3–6.
9. Kimelberg HK. Current concepts of brain edema. Review of laboratory investigations. *J Neurosurg* 1995;**83**:1051–9.

10. Amiry-Moghaddam M, Ottersen OP. The molecular basis of water transport in the brain. *Nat Rev Neurosci* 2003;**4**:991–1001.
11. Marmarou A. Pathophysiology of traumatic brain edema: current concepts. *Acta Neurochir Suppl* 2003;**86**:7–10.
12. Unterberg AW, Stover J, Kress B, Kiening KL. Edema and brain trauma. *Neuroscience* 2004;**129**:1021–9.
13. Bardutzky J, Schwab S. Antiedema therapy in ischemic stroke. *Stroke* 2007;**38**:3084–94.
14. Kahle KT, Simard JM, Staley KJ, Nahed BV, Jones PS, Sun D. Molecular mechanisms of ischemic cerebral edema: role of electroneutral ion transport. *Physiology (Bethesda)* 2009;**24**:257–65.
15. Verkman AS. Aquaporins at a glance. *J Cell Sci* 2011;**124**:2107–12.
16. Walcott BP, Kahle KT, Simard JM. Novel treatment targets for cerebral edema. *Neurotherapeutics* 2012;**9**:65–72.
17. Starling EH. On the absorption of fluids from the connective tissue spaces. *J Physiol* 1896;**19**:312–26.
18. Iliff JJ, Lee H, Yu M, et al. Brain-wide pathway for waste clearance captured by contrast-enhanced MRI. *J Clin Invest* 2013;**123**:1299–309.
19. Thrane AS, Thrane VR, Nedergaard M. Drowning stars: reassessing the role of astrocytes in brain edema. *Trends Neurosci* 2014;**37**:620–8.
20. Morris AW, Sharp MM, Albargothy NJ, et al. Vascular basement membranes as pathways for the passage of fluid into and out of the brain. *Acta Neuropathol* 2016;**131**:725–36.
21. Ohata K, Marmarou A. Clearance of brain edema and macromolecules through the cortical extracellular space. *J Neurosurg* 1992;**77**:387–96.
22. Fukuda AM, Pop V, Spagnoli D, Ashwal S, Obenaus A, Badaut J. Delayed increase of astrocytic aquaporin 4 after juvenile traumatic brain injury: possible role in edema resolution? *Neuroscience* 2012;**222**:366–78.
23. Tourdias T, Mori N, Dragonu I, et al. Differential aquaporin 4 expression during edema build-up and resolution phases of brain inflammation. *J Neuroinflammation* 2011;**8**:143–8.
24. Iliff JJ, Wang M, Liao Y, et al. A paravascular pathway facilitates CSF flow through the brain parenchyma and the clearance of interstitial solutes, including amyloid beta. *Sci Transl Med* 2012;**4**:147ra111.
25. Bradbury MW, Cserr HF, Westrop RJ. Drainage of cerebral interstitial fluid into deep cervical lymph of the rabbit. *Am J Physiol* 1981;**240**:F329–36.
26. Kempski O, Zimmer M, Neu A, von RF, Jansen M, Baethmann A. Control of glial cell volume in anoxia. In vitro studies on ischemic cell swelling. *Stroke* 1987;**18**:623–8.
27. Kempski O, von RF, Weigth H, et al. Glial swelling during exposure to putative mediators of ischemic brain damage. *Acta Physiol Scand Suppl* 1989;**582**:27–35.
28. Plesnila N, Haberstok J, Peters J, Kolbl I, Baethmann A, Staub F. Effect of lactacidosis on cell volume and intracellular pH of astrocytes. *J Neurotrauma* 1999;**16**:831–41.
29. Simard JM, Kent TA, Chen M, Tarasov KV, Gerzanich V. Brain oedema in focal ischaemia: molecular pathophysiology and theoretical implications. *Lancet Neurol* 2007;**6**:258–68.
30. Stokum JA, Kurland DB, Gerzanich V, Simard JM. Mechanisms of astrocyte-mediated cerebral edema. *Neurochem Res* 2015;**40**:317–28.
31. Martinez-Valverde T, Vidal-Jorge M, Montoya N, et al. Brain microdialysis as a tool to explore the ionic profile of the brain extracellular space in neurocritical patients: a methodological approach and feasibility study. *J Neurotrauma* 2015;**32**:7–16.
32. Fenske A, Samii M, Reulen HJ, Hey O. Extracellular space and electrolyte distribution in cortex and white matter of dog brain in cold induced oedema. *Acta Neurochir (Wien)* 1973;**28**:81–94.
33. Cserr HF, Cooper DN, Suri PK, Patlak CS. Efflux of radiolabeled polyethylene glycols and albumin from rat brain. *Am J Physiol* 1981;**240**:F319–28.

34. Sensenbrenner M, Lucas M, Deloulme JC. Expression of two neuronal markers, growth-associated protein 43 and neuron-specific enolase, in rat glial cells. *J Mol Med (Berl)* 1997;**75**:653–63.
35. Badaut J, Fukuda AM, Jullienne A, Petry KG. Aquaporin and brain diseases. *Biochim Biophys Acta* 2014;**1840**:1554–65.
36. Stokum JA, Mehta RI, Ivanova S, Yu E, Gerzanich V, Simard JM. Heterogeneity of aquaporin-4 localization and expression after focal cerebral ischemia underlies differences in white versus grey matter swelling. *Acta Neuropathol Commun* 2015;**3**:61.
37. Thompson EM, Pishko GL, Muldoon LL, Neuwelt EA. Inhibition of SUR1 decreases the vascular permeability of cerebral metastases. *Neoplasia* 2013;**15**:535–43.
38. Davila ML, Brentjens RJ. CD19-Targeted CAR T cells as novel cancer immunotherapy for relapsed or refractory B-cell acute lymphoblastic leukemia. *Clin Adv Hematol Oncol* 2016;**14**:802–8.
39. Hu Y, Sun J, Wu Z, et al. Predominant cerebral cytokine release syndrome in CD19-directed chimeric antigen receptor-modified T cell therapy. *J Hematol Oncol* 2016;**9**:70–0299.
40. Sheth KN, Elm JJ, Molyneaux BJ, et al. Safety and efficacy of intravenous glyburide on brain swelling after large hemispheric infarction (GAMES-RP): a randomised, double-blind, placebo-controlled phase 2 trial. *Lancet Neurol* 2016;**15**:1160–9.
41. Hacke W, Schwab S, Horn M, Spranger M, De GM, von KR. 'Malignant' middle cerebral artery territory infarction: clinical course and prognostic signs. *Arch Neurol* 1996;**53**:309–15.
42. Simard JM, Tsymbalyuk N, Tsymbalyuk O, Ivanova S, Yurovsky V, Gerzanich V. Glibenclamide is superior to decompressive craniectomy in a rat model of malignant stroke. *Stroke* 2010;**41**:531–7.

Index

Note: Page numbers followed by "*f*" and "*t*" refer to figures and tables, respectively.

C

J

K

N

R

S

FSC
www.fsc.org
MIX
Papier | Fördert
gute Waldnutzung
FSC® C083411

Kolibri 360 GmbH
Ferdinand-Jühlke-Straße 7
99095 Erfurt, Deutschland
produktsicherheit@kolibri360.de